Quantitative Sensory Analysis

To Michael and Patrick

Quantitative Sensory Analysis

Psychophysics, Models and Intelligent Design

Harry T. Lawless, Ph.D.
Professor Emeritus, Cornell University

WILEY Blackwell

This edition first published 2013 © 2013 by John Wiley & Sons, Ltd

Wiley-Blackwell is an imprint of John Wiley & Sons, formed by the merger of Wiley's global Scientific, Technical and Medical business with Blackwell Publishing.

Registered Office
John Wiley & Sons, Ltd, The Atrium, Southern Gate, Chichester, West Sussex, PO19 8SQ, UK

Editorial Offices
9600 Garsington Road, Oxford, OX4 2DQ, UK
The Atrium, Southern Gate, Chichester, West Sussex, PO19 8SQ, UK
111 River Street, Hoboken, NJ 07030-5774, USA

For details of our global editorial offices, for customer services and for information about how to apply for permission to reuse the copyright material in this book please see our website at www.wiley.com/wiley-blackwell.

Library of Congress Cataloging-in-Publication Data

Lawless, Harry T.
Quantitative sensory analysis / Harry T. Lawless.
pages cm.
Includes bibliographical references and index.
ISBN 978-0-470-67346-1 (cloth)
1. Food–Sensory evaluation. 2. Chemistry, Analytic–Quantitative. I. Title.
TX546.L378 2014
664′.07–dc23
2013008677

A catalogue record for this book is available from the British Library.

Wiley also publishes its books in a variety of electronic formats. Some content that appears in print may not be available in electronic books.

Cover design and illustration by Sophie Ford www.hisandhersdesign.co.uk

Set in 10/12pt Times by SPi Publisher Services, Pondicherry, India
Printed and bound in Malaysia by Vivar Printing Sdn Bhd

1 2013

Contents

Preface

It was my intention that the book might serve as a text or companion for an upper level course in sensory evaluation, perhaps as a second semester offering following the introductory course in sensory evaluation. As such, it is aimed at beginning graduate students, advanced undergraduates, and practitioners with a solid background in theory. That being said, it is nearly impossible to write a book that is truly state of the art because the field is so rapidly developing and new models and theories abound each year. This is especially true in sensometrics. Rather than dwell on the latest popular theory or development, I have focused on fundamentals, mathematical principles, and models that I suspect will stand the test of time.

This book makes some assumptions about the experience and background of the reader. It is assumed that the reader has some familiarity with applied sensory evaluation methods as they are used in the foods and consumer products industries. Therefore, the book spends a minimal amount of space defining the methods, and it avoids elaborate descriptions of all the variations, pros and cons, pitfalls, and controversies of sensory testing procedures. For further information about the practical aspects of sensory testing, many books are available, notably *Sensory Evaluation of Foods, Principles and Practices* by Lawless and Heymann (2010), *Sensory Evaluation Practices* by Stone et al., (2006), and *Sensory Evaluation Techniques* by Meilgaard et al. (2006). These three are lengthy textbooks. Shorter guides for the novice reader are include the ASTM document *Sensory Testing Methods* by Chambers and Wolf (1996) and the practical guide *Sensory Evaluation, A Practical Handbook* by Kemp et al. (2009). A brief overview of sensory techniques is provided in Appendix A.

I have also assumed that the reader has some passing familiarity with the foods or consumer products industries, and possibly the flavor and fragrance industry. This is not meant to be a text on psychophysics per se, so the areas of visual and auditory sensation and perception are rarely mentioned. A fundamental background in basic chemistry and biology is also assumed. Mathematically, I have tried to avoid calculus and matrix algebra as much as possible, so that a reader with basic algebra skills can understand the models and equations, their variations and transformations. It is difficult to know the level of detail that different readers will see as mathematically appropriate or challenging. So the information and models herein are those I deem necessary to understand for the well-trained sensory scientist operating in the food, beverage, or consumer products industries or in an academic setting. To that extent, the book is intended to form a basis or starting point for further study of individual issues and models. Worked examples are provided in most sections to illustrate the application of the equations that have been presented.

The obvious technical area of sensory evaluation that is neglected in this book is the realm of descriptive analysis. I chose not to deal with issues in panel leadership and terminology development and take only a quick look at panelist monitoring. The first topic is primarily qualitative and deals with techniques and problems in human interaction. It is best taught by going to a workshop, participating in panels with a good leader, and then leading a panel yourself. The old rule for learning an operation or procedure among surgical residents was "watch one, do one, teach one." Perhaps that rule applies in being a descriptive panel leader as well. Terminology development is primarily conceptual, qualitative, and language driven, and as such is not truly a quantitative technique. Panelist monitoring is primarily a statistical issue. However, descriptive techniques are touched upon in the sections on scaling and panelist calibration.

Another controversial topic was multivariate analyses and perceptual mapping. There has been an explosion in techniques used for multivariate analysis of sensory data, and most of these result in the production of a biplot, a two-dimensional representation of sensory space, with products and assessors as points and with attributes as vectors. Probably more than half the methodological papers in journals like *Food Quality and Preference* pertain to these techniques. Certainly, it is a central theme of sensometrics. There is nothing wrong with this data reduction and mapping approach, and it can be most valuable in looking at competitive monitoring (where are my products relative to the competition?) in situations like product category appraisals. Yet, I remain somewhat skeptical as to how much these techniques can help a sensory practitioner on a day-to-day basis. Most of the normal and often tedious sensory testing that goes on in a large food company concerns issues like cost reduction, ingredient substitution, improvements in the nutritional profile of a product, producing successful variations on a current product, quality control, and shelf-life testing. Very few of these endeavors require any multivariate techniques. So, they are dealt with in the topic of product optimization, and one later chapter gives a general overview of perceptual mapping, which necessarily entails some multivariate techniques.

It should also be clear to the reader that I did not intend to write a sensory statistics guide. Good resources are available, such as the appendices to Lawless and Heyman (2010), the excellent short treatise on analysis of variance by Lea et al. (1998), the sensory statistics text by Naes et al. (2010), and the book on discrimination testing by Jian Bi (2006). An older resource is the sensory statistics book by O'Mahony (1986), which contains much pithy advice. Some less well known techniques are presented, however, where I felt the penetration of these tests into the mainstream was insufficient. An example is the use of the Stuart–Maxwell test for analysis of categorical or nominal data in a simple two-product test and a complete block design (analogous to the paired or dependent t-test for scaled data).

The structure of the book proceeds as follows. The first four chapters are foundational, dealing with psychophysics. The next four deal with basic topics from a sensory evaluation perspective, namely discrimination, sensory equivalence, scaling, and hedonic testing. I have tried to incorporate current theoretical and practical developments, so their titles include the word "progress." Chapter 9 is an overview of some principles in experimental design that I refer to as "intelligent." The next three chapters cover some applied topics, including check-all-that-apply and other frequency count data, time–intensity modeling, and shelf-life testing. More advanced topics come next, including product optimization methods, perceptual mapping, and consumer segmentation. Note that some of these use multivariate techniques; however, I chose not to include a specific chapter on multivariates (there are several useful books) but rather to focus on how they are applied. At the end, I have provided a forward-looking chapter on the topic of Bayesian analysis, because I felt a graduate student in sensory evaluation should at least understand the principles involved. Appendices are provided, first as a primer in sensory evaluation, second to show some common experimental designs not mentioned elsewhere, and third to provide a glossary.

Many mentors and collaborators have helped, directly or indirectly with the construction of this work. The inspiration for my career in quantitative sensory work starts with a course in sensory psychology that was team-taught in the 1970s by the scientists at the John B.

Pierce Foundation, notably (in alphabetical order) Ellie Adair, Linda Bartoshuk, Bill Cain, Larry Marks, and Joe Stevens. I owe a special debt to Linda Bartoshuk for teaching me that psychophysics could provide a window into sensory physiology and to Bill Cain for providing so many good examples of the parametric quantification of sensory function.

Other mentors in psychophysics have been major influences during my career, notably Trygg Engen, Don McBurney, David Stevens, Michael O'Mahony, Armand Cardello, Howard Schutz, Herb Meiselman, and Howard Moskowitz. This book is aimed at workers in food science, and so a debt must be acknowledged to the late Rose Marie Pangborn, to David Peryam, and to Hildegarde Heymann. The writings of Daniel Ennis and colleagues, and also of Jian Bi, were highly influential in many sections of this book, as seen in the citations and reference lists. I thank them for bringing the field to a higher level. The Institute for Perception is thanked for providing the valuable compendium "Short Stories in Sensory and Consumer Science."

Specific assistance was received from many sources. George Gescheider was a helpful correspondent during the writing phase, and his book on psychophysics was a major resource. I would like to give a special note of appreciation to Dr. Guillermo Hough for producing his excellent little book on sensory shelf life, and conducting workshops. I was fortunate to take his workshop on shelf-life methods in Florence in 2009. Much of Chapter 12 is derived from his book and workshop course notes. My former student Michael Nestrud was instrumental in providing the outline for the section on graph theory. Nort Holschuh of General Mills suggested the scoring scheme for the same–different test with a sureness rating scale. Kevin Blot and Rui Xiong from Unilever assisted in analysis of simulation data. Benoit Rousseau provided valuable discussions concerning Thurstonian analysis. Thierry Worch and Pieter Punter provided useful literature for the section on ideal profiling.

Humans are natural comparators. We are fundamentally change-detectors. The status quo is rarely of interest. The smell of the other apes in the cave is not important. The sudden smell of the saber-toothed cat at the mouth of the cave is. Sensory test methods that take into account this human ability and use it appropriately are destined to be useful tools.

Harry T. Lawless
Ithaca, New York, 2013

References

Chambers, E.C., IV, and Wolf, M.B. 1996. Sensory Testing Methods. Second edition. ASTM Manual Series MNL 26. ASTM, West Conshohocken, PA.

Kemp, S.E., Hollowood, T., and Hort, J. 2009. Sensory Evaluation. A Practical Handbook. John Wiley & Sons, Ltd, Chichester, UK.

Lawless, H.T. and Heymann, H. 2010. Sensory Evaluation of Foods. Principles and Practices. Second edition. Springer Science+Business, New York, NY.

Lea, P., Naes, T., and Rødbotton, M. 1998. Analysis of Variance for Sensory Data. John Wiley & Sons, Ltd, Chichester, UK.

Meilgaard, M., Civille, G.V., and Carr, B.T. 2006. Sensory Evaluation Techniques. Third edition. CRC Press, Boca Raton, FL.

Naes, T., Brockoff, P.B., and Tomic, O. 2010. Statistics for Sensory and Consumer Science. John Wiley & Sons, Ltd, Chichester, UK.

O'Mahony, M. 1986. Sensory Evaluation of Food, Statistical Methods and Procedures. Marcel Dekker, New York, NY.

1 Psychophysics I: Introduction and Thresholds

PORTIA: *That light we see is burning in my hall*
How far that little candle throws its beams.
So shines a good deed in a naughty world.
NERISSA: *When the moon shone we did not see the candle.*
PORTIA: *So doth the greater glory dim the less.*
A substitute shines brightly as a king
Unto the king be by and then his state
Empties itself as doth an inland brook
Into the main of waters.
The Merchant of Venice, Act V, Scene 1.

1.1 Introduction and Terminology

Psychophysics is the study of the relationship between energy in the environment and the response of the senses to that energy. This idea is exactly parallel to the concerns of sensory evaluation – how we can measure peoples' responses to foods or other consumer products. So in many ways, sensory evaluation methods draw heavily from their historical precedents

Quantitative Sensory Analysis: Psychophysics, Models and Intelligent Design, First Edition. Harry T. Lawless.

in psychophysics. In this chapter we will begin to look at various psychophysical methods and theories. The methods have close resemblance to many of the procedures now used in sensory testing of products.

Psychophysics was a term coined by the scientist and philosopher Gustav Theodor Fechner. The critical event in the birth of this branch of psychology was the publication by Fechner in 1860 of a little book, *Elemente der Psychophysik*, that described all the psychophysical methods that had been used in studying the physiology and limits of sensory function (Stevens, 1971). Psychophysical methods can be roughly classified into four categories having to do with:

- absolute thresholds,
- difference thresholds,
- scaling sensation above threshold, and
- tradeoff relationships.

A variety of methods have been used to assess absolute thresholds. An absolute threshold is the minimum energy that is detectable by the observer. These methods are a major focus of this chapter. Difference thresholds are the minimum amount of change in energy that are necessary for an observer to detect that something has changed. Scaling methods encompass a variety of techniques used to directly quantify the input–output functions (of energy into sensations/responses), usually for the dynamic properties of a sensory system above the absolute threshold. Methods of adjustment give control of the stimulus to the observer, rather than to the experimenter. They are most often used to measure tradeoff functions. An example would be the tradeoff between the duration of a brief flash of light and its photometric intensity (its light energy). The observer adjusts one or the other of the two variables to produce a constant sensation intensity. Thus, the tradeoff function tells us about the ability of the eyes to integrate the energy of very brief stimuli over time. Similar tradeoff functions can be studied for the ability of the auditory system to integrate the duration and sound pressure of a very brief tone in order to product a sensation of constant loudness.

Parallels to sensory evaluation are obvious. Flavor chemists measure absolute thresholds to determine the biological potency of a particular sweetener or the potential impact of an aromatic flavor compound. Note that the threshold in this application becomes an inverse measure of the biological activity of the stimulus – the lower the threshold, the more potent the substance. In everyday sensory evaluation, difference testing is extremely common. Small changes may be made to a product, for example due to an ingredient reduction, cost savings, nutritional improvement, a packaging change, and so forth. The producer usually wants to know whether such a change is detectable or not. Scaling is the application of numbers to reflect the perceived intensity of a sensation, and is then related to the stimulus or product causing that sensation. Scaling is an integral part of descriptive analysis methods. Descriptive analysis scales are based on a psychophysical model and the assumption that panelists can track changes in the product and respond in a quantitative fashion accordingly.

The differences between psychophysics and sensory evaluation are primarily in the focus. Psychophysics focuses on the response of the observer to carefully controlled and systematically varied stimuli. Sensory evaluation also generates responses, but the goal is to learn something about the product under study. Psychophysical stimuli tend to be simple (lights or tones or salt solutions) and usually the stimulus is varied in only one physical attribute at a time (such as molar concentration of salt in a taste perception study). Often, the resulting change is also unidimensional (such as salty taste). Products, of course, are multidimensional, and changing

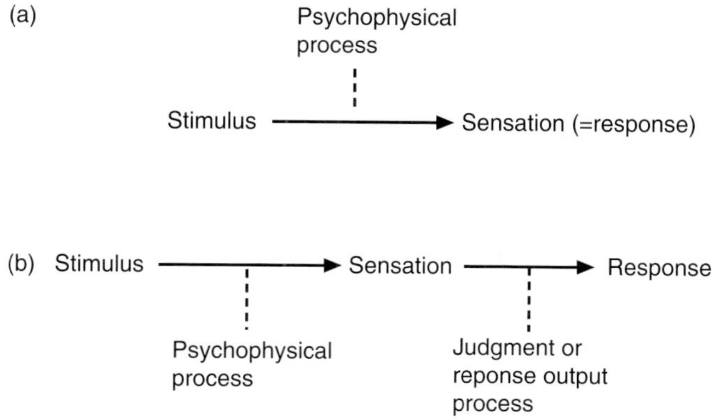

Figure 1.1 Stimulus response models in psychophysics. A classical model has it that the stimulus causes a sensation which is directly translated into an accurate response. A more modern model recognizes that there can be decision processes and human judgment involved in translating the sensation in a response (a data point) and, thus, there are at least two stages and two processes to study.

one ingredient or aspect is bound to have multiple sensory consequences, some of which are hard to predict. Thus, the responses of a descriptive analysis panel, for example, often include multiple attributes. However, the stimulus–response event is necessarily an interaction of a human's sensory systems with the physical environment (i.e., the product or stimulus), and so psychophysics and sensory evaluation are essentially studying the same phenomena.

The reader should be careful not to confuse the stimulus with the sensation or response. This dichotomy is critical to understanding sensory function: an odor does not cause a smell sensation, but an odorant does. Thus, an odor is an experience and an odorant is a stimulus. Human observers or panelists do not measure sugar concentrations. They report their sweetness experiences. We often get into trouble when we confuse the response with the stimulus or vice versa. For example, people often speak of a sweetener being "200 times as sweet as sucrose" when this was never actually measured. It is probably impossible for anything to be 200 times as intense as something else in the sense of taste – the perceptual range is just not that large. What was actually measured was that, at an iso-sweet level (such as the threshold), it took a concentration 200 times higher to get the same impact from sucrose (or 1/200th for the intensive sweetener in question). But this is awkward to convey, so the industry as adopted the "*X* times sweeter than *Y*" convention.

The reader should also be careful to distinguish between the subjective experience, or sensation itself, and the response of the observer or panelist. One does not directly translate into the other. Often, the response is modified by the observer even though the same sensation may be generated. An example is how a stimulus is judged in different contexts. So there really are at least two processes at work here: the psychophysical event that translates energy into sensation (the subjective and private experience) and then the judgment or decision process, which translates that experience into a response. This second process was ignored by some psychophysical researchers, who assumed that the response was always an accurate translation of the experience. Figure 1.1 shows the two schema.

The rest of this chapter will look at various classical psychophysical methods, and how they are used to study absolute and difference thresholds. Some early psychophysical theory, and its modern variations, will also be discussed. A complete review of classical psychophysics can be found in *Psychophysics, The Fundamentals* (Gescheider, 1997).

1.2 Absolute Sensitivity

1.2.1 The Threshold Concept

In the early 1800s, at about the time when scientists were becoming better able to control the energy in a stimulus (light, sound, heat), a popular notion took hold that there was an energy level below which a human sensory system would not be able to detect anything, and above which a sensation would be noticed. This idea is often attributed to the philosopher Herbart, who wrote in 1824 that mental events had to be stronger than some critical amount in order to be consciously experienced (Gescheider, 1997). Thus, at any given moment, for a single observer, a stimulus with energy below the threshold is not detected, and a stimulus above the threshold is. This is an all-or-nothing concept of threshold, as shown in Figure 1.2.

As appealing as this concept seemed, it was almost impossible to demonstrate. Attempts to measure the threshold soon encountered a major problem: the sensitivity of the observer seemed to change from moment to moment. Although one might still like the idea that at any moment there was a fixed threshold, and that crossing it caused a sensation, attempts to bring the concept into the laboratory rendered the idea questionable, and of limited utility. Let us assume you are doing a study with changing sound pressure levels and that you ask your observer to respond "yes" when they hear something and "no" when they do not. We present all the stimulus tones of different levels many times in random order. When we plot the percentage of "yes" responses against the sound pressure levels, we do not see the all-or-nothing function of Figure 1.2, but rather a curve resembling an ogive or sigmoid shape, as shown in Figure 1.3a.

This was the first **psychometric function**, a term used to describe the probability of response as a function of the stimulus, such as its energy or sound pressure level. From this time forward, most people conceived of a threshold in practical terms, as a statistical entity rather than a fixed point. As a practical matter then, it became useful to call this empirical or experimental threshold the level at which detection occurs 50% of the time. This fits well with what we now know about the human nervous system: first, that it has a spontaneous

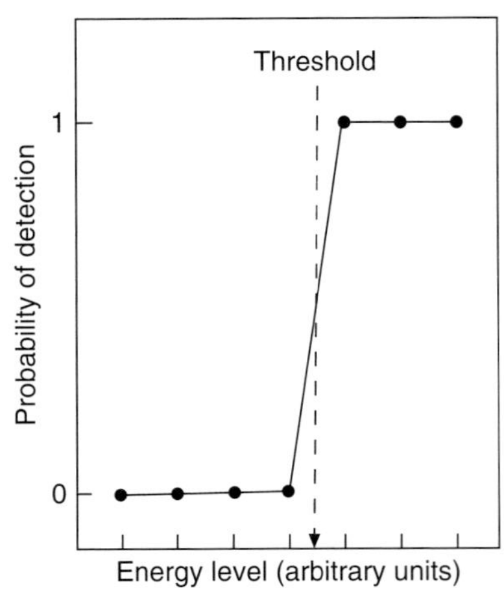

Figure 1.2 The all-or-none concept of threshold. Detection never occurs below threshold. Above threshold, it always reaches consciousness.

background level of activity (your nerves are not quiet, even in a completely dark soundless room) and that this activity level appears to vary randomly.

As the normal distribution is well understood, has known properties, and describes a host of natural phenomena, it is perhaps not surprising that this curve is often used to describe the common psychometric function. As you might recall from statistics, the function has a location parameter (the mean μ) and a spread parameter (the standard deviation σ or variance σ^2). The exact form that generates our familiar bell shaped curve is given by

$$f(x) = \frac{1}{\sqrt{2\pi\sigma^2}} e^{-(x-\mu)^2/2\sigma^2} \tag{1.1}$$

And if we standardize the function to $\sigma = 1$ and $\mu = 0$ (we will use ϕ instead of f) we get

$$\varphi(x) = \frac{1}{\sqrt{2\pi}} e^{-\frac{1}{2}z^2} \tag{1.2}$$

This is the equation, then, that gives us the relationship for *Z*-scores, or the conversion of any observation x into its distance from the mean in standard deviation units as $Z = (x - \mu)/\sigma$. If we integrate this function over a given interval, we can find the cumulative proportion of the area under the curve from negative infinity to the upper bound of the interval. This gives us a useful relationship of proportions to *Z*-scores, as shown in Appendix 1.A. Taking our data from the hypothetical psychometric function, we can convert the proportions to *Z*-scores, as shown in Table 1.1. Plotting *Z*-scores instead of proportions will give us a linear relationship, amenable to least-squares fitting or other simple curve-fitting methods. Using our *Z*-score conversion, we see the linearization of the data from our curve in Figure 1.3b. The reader should feel comfortable with the conversion of proportions to *Z*-scores and vice versa, as this relationship will form the basis for several psychophysical calculations later in the book. Appendix 1.A illustrates these relationships.

Because the data are actually describing a probabilistic set of events that are bounded by zero and one, another attractive option is to use the logit function, which is similar to the

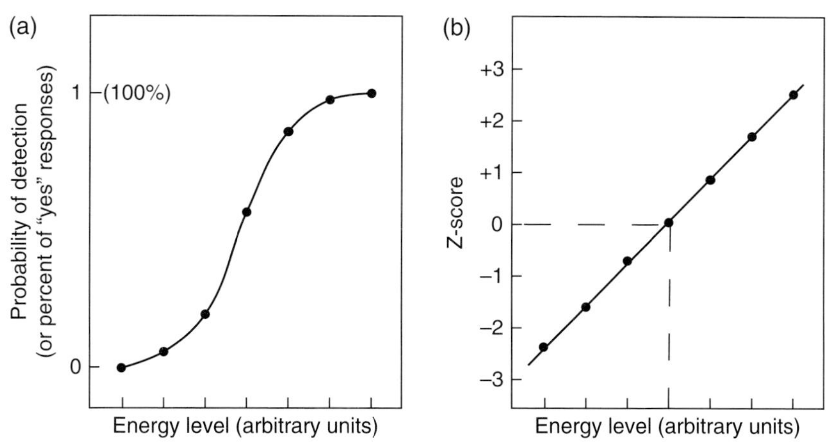

Figure 1.3 A psychometric function. The probability of response is plotted against the stimulus energy level, and often forms an S-shaped curve similar to the cumulative normal distribution.

Table 1.1 Conversion of threshold curve proportions to Z-scores

Energy level	Proportion detecting	Z-score
1	0.01	–2.33
2	0.05	–1.64
3	0.20	–0.84
4	0.50	0.0
5	0.70	+0.84
6	0.95	+1.64
7	0.99	+2.33

cumulative standard normal curve. It is often used to model probabilistic events in medicine and actuarial science, such as the rate of population growth and life expectancy. It takes the following form:

$$f(x) = \frac{e^x}{e^x + 1} = \frac{1}{1 + e^{-x}} \tag{1.3}$$

where x can take on any value from negative to positive infinity and $f(x)$ varies from zero to one. If we convert our probability of detection $f(x)$ to an odds ratio $(p/1-p)$ and take the logarithm, the model becomes roughly linear and amenable to least-squares or other simple fitting methods:

$$\ln\left(\frac{p}{1-p}\right) = b_0 + b_1 x \tag{1.4}$$

where b_0 and b_1 are slope and intercept parameters. Additional terms (e.g., quadratic) may be added for greater accuracy. Examples of fitting logistic functions to threshold data can be found in Appendix 1.B, as well as in Walker et al. (2003) and Lawless (2010). The function is generally useful for modeling binomial events where there are two outcomes, such as correct versus incorrect responses in a choice task, or response versus nonresponse in a detection experiment.

Whichever bell-shaped curve we choose to use, and its cumulative S-form one chooses to adopt, it is clear that the mean and variance (location and spread parameters) of the distributions will affect the position of the threshold and the steepness of the S-curve. Figure 1.4 shows how the distributions of variability will affect the resulting psychometric function. Observers A and B have the same variability, but A has greater sensitivity and thus a lower threshold. Observers B and C have the same sensitivity, but observer C has less neural noise or other factors contributing to variance, and the smaller variability leads to a steeper psychometric function.

1.2.2 Threshold Theories

There were several theories that arose to explain this behavioral phenomenon of the apparent probabilistic nature of the threshold. One hypothesized that the actual momentary threshold of the observer was a fixed quantity, but that this varied according to a normal distribution. If so, over many trials, the data would look like the cumulative version of the normal distribution. Classical psychophysics calls this the phi–gamma

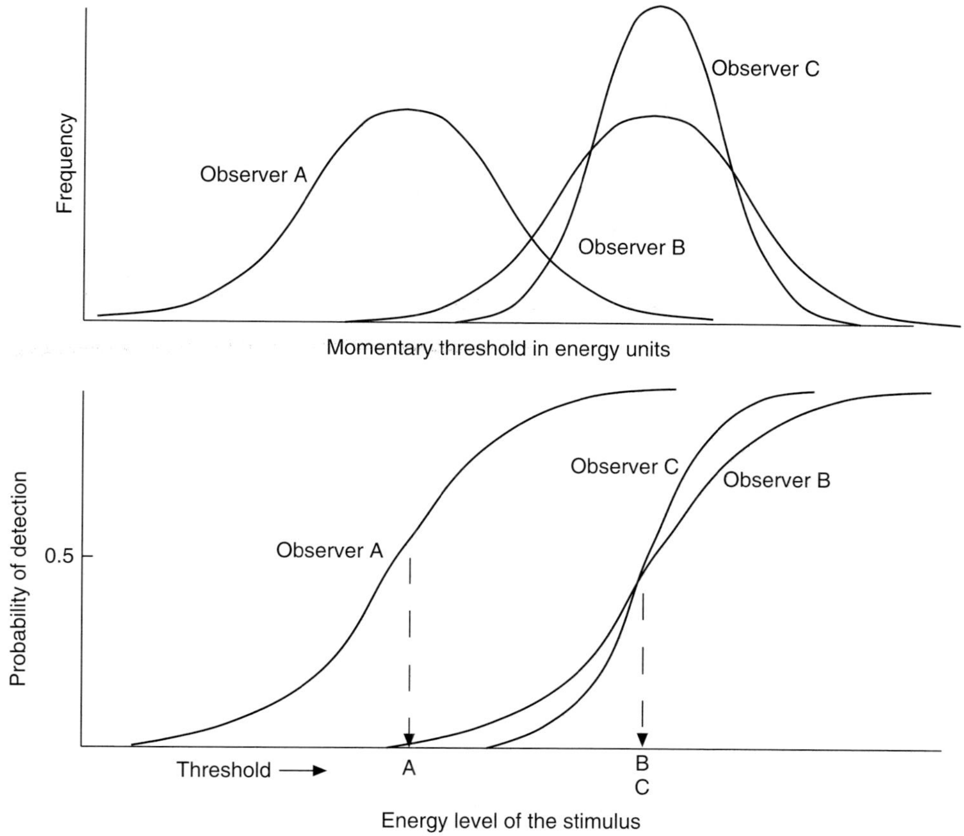

Figure 1.4 Momentary and obtained thresholds from three hypothetical observers. Observer A has higher sensitivity than B or C because this person responds at a lower energy level. Observer C has lower inherent variability than Observer B, giving a steeper psychometric function.

hypothesis (Gescheider, 1997). Some workers held that if Fechner's log function held true (as discussed below) then the psychometric function obtained would take the form of the normal ogive when plotted against the log of stimulus intensity (thus the phi–log-gamma hypothesis). An example is shown in Figure 1.3 and one from actual data for odor thresholds in Figure 1.8. A second theory held that the observer's threshold was actually fixed, but the stimulus itself had some random variation that was normally distributed. If so, you could obtain the familiar psychometric function as a result of stimulus variation even from a fixed, static observer. In reality, probably both sources of variability are in play in most studies. Even the output of carefully controlled and well-engineered olfactometers can produce a variable stimulus to the nose, as Cain (1977) found when he actually measured the output of his olfactometer with a gas chromatograph.

1.2.3 Other Types of Thresholds

The absolute or detection threshold was only one type of threshold considered in psychophysics. The **recognition threshold** describes the level of energy needed for the observer to correctly name or identify the stimulus. For example, in the sense of taste, the detection threshold is often lower than the recognition threshold. A common response is "I know I can taste something, but I can't name it." At a slightly higher concentration, say of sodium chloride, the observer correctly discerns a salty taste. The **difference threshold**, discussed at length below, is the level of energy change needed for the observer to perceive that the stimulus has become stronger or weaker. Because some sensory systems are known to reach a maximum or to saturate, the **terminal threshold** is used to describe the energy level at which subsequent increases in physical intensity fail to create any concomitant increase in sensation strength. This is rarely studied because it is difficult to present high-intensity stimuli without invoking a pain response or other changes in the stimulus quality. Recently, Prescott et al. (2005) have come up with a useful notion of the **rejection threshold**. This is the level at which a taste or odor would be found objectionable in a food or beverage product. They recognized that taints or off-flavors are not always objectionable at low levels. The rejection threshold is measured by conducting preference tests at increasing concentrations, comparing a spiked sample containing the off-flavor with a control without the offending substance.

1.3 Methods for Measuring Absolute Thresholds

1.3.1 The Method of Limits

The most widely used method for measuring thresholds in classical psychophysics was probably the **method of limits**. In this method, the stimulus energy is raised or lowered until the response of the observer changes. In an ascending series, the stimulus energy is increased until the observer responds that they detect the stimulus (i.e., get a perceived sensation). Descending trials are alternately conducted and these trials continue until the stimulus is no longer sensed. Because the individual's sensitivity is variable from moment to moment, several ascending and descending series would be presented, as shown in Figure 1.5. After the response appears to stabilize, the empirical threshold is obtained from an average of the change points in the last few runs.

Many experimenters realized that the method of limits, although appealing in its systematic presentations, was not very efficient. A lot of trials could be wasted when the observer sensed the higher levels of the stimulus and perception was fairly obvious. Why not use the previous reversal point as a starting level for the next series? This notion led to various adaptive methods for measuring threshold. They are adaptive in the sense that the next series is adapted to take into account the information we get from the current series of stimuli. An example of this is the *staircase procedure*, so named because when the stimulus sequence is connected by horizontal lines the record resembles a rising and falling series of steps (Cornsweet, 1962; see also Linschoten et al. (1996) for a modern variant). In its simplest form, the stimulus is raised on the next trial if no sensation is detected (i.e., the observer responds negatively) and the stimulus is lowered on the next trial if the observer responds affirmatively, as shown in Figure 1.6. Many variations on this procedure were developed, such as a random double staircase in which trials would randomly be drawn from two intertwined staircases, one starting from a high level of stimulus intensity and the other from a low level. Another popular option was to descend only after two positive responses, but to ascend after one negative, the so-called up–down transformed-response rule (UDTR). This method titrated around a 71% detection level. These methods are discussed at greater length in Lawless and Heymann (2010: chapter 6), and the reader is referred there for further methodological details.

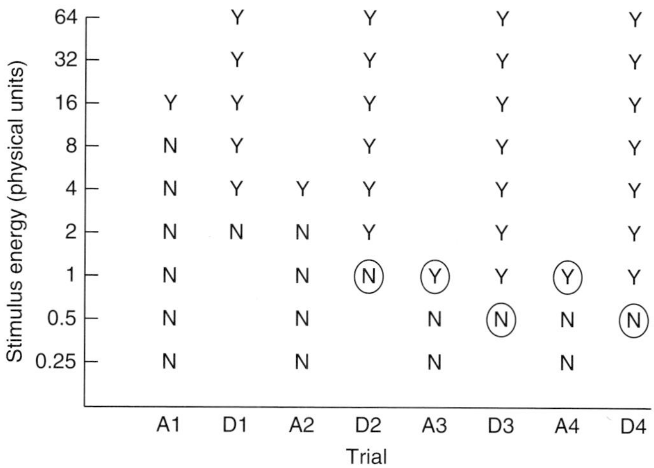

Figure 1.5 The method of limits. Stimuli were presented in alternating ascending (A1, A2, etc.) and descending (D1, D2, etc.) series and the observer responds either "no" if they do not sense anything (N) or "yes" on each trial. After responses appear to stabilize, the last few points of change (circled) are averaged to obtain the threshold. In this case, we average the change points 1, 1, 0.5, 1, and 0.5, giving a mean of 0.8 as the individual's threshold estimate.

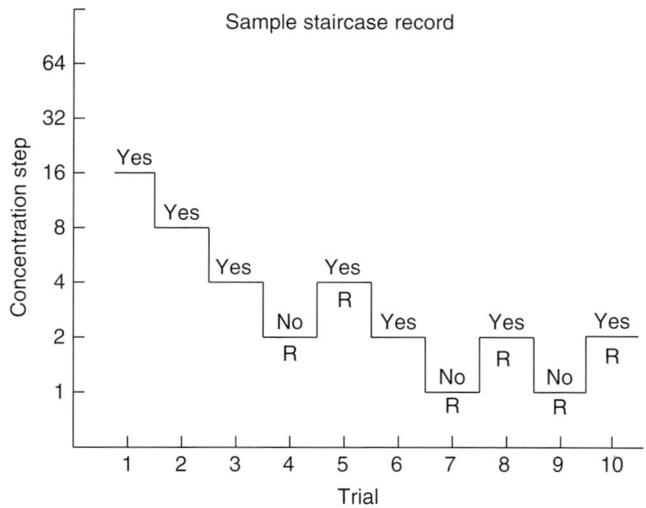

Figure 1.6 An example (hypothetical) of a record from a staircase procedure.

1.3.2 Forced-Choice Method of Limits

Several problems remained with the classical method of limits. One was that the observer could become adapted or fatigued to the high levels of the stimulus in the descending series. For example, with a bitter substance, the taste is difficult to remove, and thus lower level trials became increasingly difficult to discern due to a buildup of residual bitter taste in the mouth. So, in the chemical senses, the method of limits is often performed with only ascending trials. The second problem is that the actual response is both a function of the sensitivity of the observer (what you are really trying to measure) and their criterion for how much of a sensation is needed for a positive response. Some observers could be quite conservative and only respond when they are absolutely sure they heard or saw or tasted something, while another person might respond if they had even the slightest inkling that something was sensed. Thus, the individual's proclivity to respond or not respond with various levels of evidence is a confounding influence in the threshold we obtain. This issue of separating response bias or one's individual criterion from the actual sensitivity is addressed by signal detection theory, as explained in detail in Chapter 3.

A simple solution to this problem is to force the observer to prove to us that they can detect the stimulus by correctly choosing it from amongst a group of other stimuli that have only the background noise. Let us call the stimulus with the higher energy present the target, and the stimulus at the background level the blank. An example of a blank in the sense of taste would be a solution of distilled water that contained none of the tastant whose threshold you are trying to measure. Typically, this is done with two, three, or four total stimuli at each level, although a variety of numerical combinations of targets and blanks has been used in different sensory studies (e.g., Lawless et al., 1995). A good example of such a procedure is the **ascending forced-choice method of limits** described by ASTM procedure E-679 (ASTM, 2008). In this case, each observer in a group of from 10 to 25 individuals is given an ascending series with one target and two blank stimuli at each level, as is asked to choose the target item. The individual threshold estimate becomes the first correct response in the series, given that all subsequent responses are also correct (a correct choice followed by an incorrect choice is discounted as probably a lucky guess). Then the final reversal points are tabulated across the group of individuals to obtain a group estimate threshold for a particular taste or odor substance (see Stocking et al. (2001) for an example). Because the stimulus series is often in a geometric progression, such as concentration steps in factors of two or three, the geometric rather than the arithmetic mean is often taken. The geometric mean is the *N*th root of the product of *N* items. A convenient calculation for the geometric mean is to convert the data via logarithmic transformation, take the mean of the log data, and then the antilog of that mean, as shown in eqns 1.5–1.7.

$$\text{Mean}_{\text{geometric}} = \sqrt[n]{x_1 \cdot x_2 \cdot x_3 \cdot \cdots \cdot x_n} \tag{1.5}$$

where x_1, x_2, and so on are the individual best-estimate thresholds from a panel of n individuals.

We can also take the arithmetic mean of the logs (let us call it "ML") and then exponentiate to get the antilog:

$$\text{ML} = \frac{\sum_{i=1}^{n} \log x_i}{n} \tag{1.6}$$

Assuming our logarithms were to the base 10:

$$\text{Mean}_{\text{geometric}} = 10^{\text{ML}} \tag{1.7}$$

As a measure of central tendency, the geometric mean has the property (like the median) that it is less influenced by high outliers in the data than the arithmetic mean is. Also, because the data are in a geometric progression due to the doubling or tripling of concentration steps, they are in equal steps when converted to logarithms. This makes the data handling easy if you have to calculate geometric means by hand.

The *N*-alternative ascending forced choice method is a useful procedure. However, the data handling outlined above (i.e., by getting individual threshold estimates and then averaging them) still does not deal with the fact that the participants can guess correctly (and do so one-third of the time for the ASTM E-679 procedure). A solution to this problem is to use the common correction for guessing, sometimes called **Abbott's formula**, as shown in eqns 1.8 and 1.9:

$$P_D = \frac{P_{Obs} - P_{chance}}{1 - P_{chance}} \tag{1.8}$$

where P_{D} is the true proportion detecting, P_{obs} is the proportion choosing the target sample correctly, and P_{chance} is the chance level or 1/3 for the ASTM method. This can also be recast as

$$P_{\mathrm{obs}} = P_{\mathrm{chance}} + P_{\mathrm{D}}\left(1 - P_{\mathrm{chance}}\right) = P_{\mathrm{D}} + P_{\mathrm{chance}}\left(1 - P_{\mathrm{D}}\right) \tag{1.9}$$

One can think of the P_{obs} in these equations as *the level you need to get to* in order to obtain your threshold after the correction; in other words, the adjusted percentage correct. Note that this assumes there are only two classes of individuals at any given level, a group that detects the stimulus and those that do not, some of whom guess correctly. This two-state model is sometimes attributed to Morrison (1978).

Now all we need to do to obtain the group threshold is to count the proportion correct at each concentration level and then interpolate, graphically or from a fitted equation. The interpolation point for any forced-choice method is that which satisfies Abbott's formula at 50% actual discriminators (see Antinone et al. (1994) for an example). We do not need to worry about estimating individual thresholds or whether a person made a lucky guess or not at each trial. Every person's data contributes equally. To obtain the historical criterion of 50% detection for threshold, we need to interpolate at 2/3 or 66.6% correct in the ASTM choose-one-out-of-three type of procedure. This interpolation is shown in Figure 1.7. A second advantage of this method of data analysis is that we can also estimate other levels of detection after correction for guessing. For example, we might want to know the level at which 25% of the population can detect a substance or even 10% (requiring 50% observed correct and 40% for the ASTM method, respectively). For a food scientist, trying to track down the source of an off-flavor, finding that your food has an offending chemical present at the consumer threshold at 50% is not too useful. One might want to limit the amount to insure that far fewer people can detect it, thus leading to fewer complaints and less loss of the business franchise due to consumers' dissatisfaction.

Figure 1.8 shows the method applied to the data of Stocking et al. (2001) on odor thresholds for methyl tertiary butyl ether (MTBE, a water contaminant). Using a simple linear fit of the equation shown, we get a threshold value of about 14 ppb (parts per billion). Fitting a logistic equation (plotting $\ln(p/1-p)$ versus log concentration), the interpolated value is about 12 ppb. These values agree reasonably well with the level estimated by the ASTM procedure of taking individual best estimates and then finding the geometric mean, yielding a value of 14.3 ppb in the Stocking et al. data. A method for fitting a logistic equation such as eqn 1.4 to the data is shown in Appendix 1.B. Another alternative for forced-choice data is given in the paper by Harwood et al. (2012) using a four-parameter logistic equation as follows:

$$P_{\mathrm{observed}} = \mathrm{Min} + \left[\frac{\mathrm{Max} - \mathrm{Min}}{1 + 10^{k(\log(T) - X)}}\right] \tag{1.10}$$

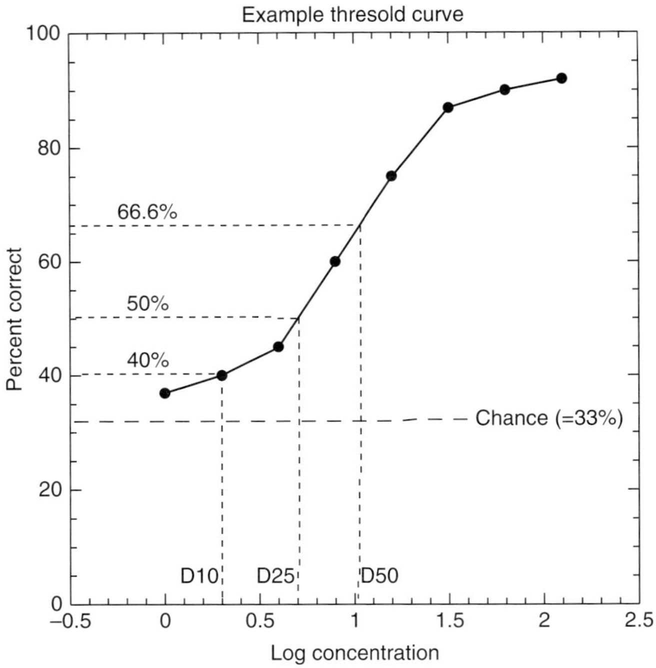

Figure 1.7 Using the 3-AFC method of ASTM E-679, we can interpolate on a plot of proportion correct versus concentration (or log concentration) to find the chance-corrected threshold level (66.6% correct for 50% detection) and also other levels; for example, for 25% detection (in this case 50% correct) and 10% detection (40% correct).

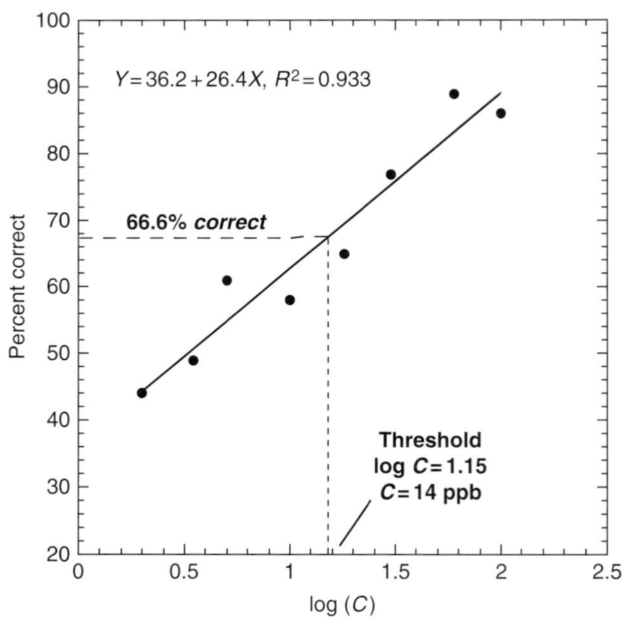

Figure 1.8 An example of interpolation on the percentage correct from a three-alternative forced-choice test to obtain the chance-corrected 50% discrimination level, in this case using the data of Stocking et al. (2001). Using the equation shown, we get a threshold of about 14 ppb; and using a logistic regression, we get about 12 ppb.

where Min and Max are the minimum and maximum levels of response, k is a slope parameter, $\log(T)$ is the chance corrected threshold level to be found, and X is the log concentration of the stimulus. The values for Max and Min are generally known from the data, reducing the solution to a search for values of $\log(T)$ and k. Another fitting method for these kinds of data, using maximum likelihood methods, is found in Peng et al. (2012: appendix A).

Yet one problem still remains. We have changed the definition of threshold when we get to a group-averaged or population measure, when we speak of 50% *of the group* detecting. Remember that when we began our discussion of classical thresholds we talked about a single observer detecting 50% *of the time*. While reanalyzing the data of Stocking et al. (2001), a statistician from the US Environmental Protection Agency, Andrew E. Schulman, recognized this embedded issue that was being overlooked in the world of practical threshold estimation (USEPA, 2001). He stated (USEPA, 2001: 49):

> Odor detection thresholds should be defined as the concentrations at which a certain percent of people can detect the contaminant a certain percent of the time. Both the time and subject fractions must be specified in order for a threshold to be interpretable.

After extensive statistical modeling of the Stocking et al. (2001) data, he concluded that the ASTM method of analysis appeared to find the point at which 50% of the group would detect 50% of the time. Unfortunately, if Schulman's analysis is correct, the overall probability of any detection event would thus be 0.25 rather than our classical definition of 0.50. Subsequent researchers will probably need to sort out this issue in future methodological studies.

Many modifications of the ascending forced-choice methods can be found in the literature. One important decision is whether to invoke a **stopping rule**. A stopping rule allows the researcher to terminate the ascending series after a sequence of correct answers, typically three in a row. This prevents the panelist or subject from experiencing very high levels of the stimulus that could be unpleasant, painful, or cause fatigue or adaptation as the experimental series progresses. The exact choice for the stopping rule, as well as the definition of threshold, will influence the values obtained (Peng et al., 2012).

1.4 Differential Sensitivity

1.4.1 The Difference Threshold

The second most important psychophysical phenomenon in classical psychophysics was the different threshold. The difference threshold is the minimum amount of stimulus change (in energy or physical units) that is necessary for the observer to note that the stimulus has become stronger or weaker. Other terms for this concept include the difference limen (DL) and the just-noticeable difference (JND). In some ways the DL was of greater practical utility than the absolute threshold, because it dealt with sensations across the full stimulus range that the observer could respond to. The entire dynamic range of stimulus and response could now be measured and quantified, and not just the very weak sensations around threshold. In comparing stimuli to measure a difference threshold, we shall call the baseline or starting stimulus the "standard" stimulus and the second stimulus, whose level will be varied, the "comparison" stimulus.

Table 1.2 Examples of Weber fractions for different sensory modalities*

Sensory system or modality	Weber fraction
Electric shock	0.013
Saturation, red	0.019
Heaviness	0.020
Finger span	0.022
Length	0.029
Vibration, 60 Hz	0.036
Loudness	0.048
Brightness	0.079
Taste, NaCl	0.083

*From Teghtsoonian (1971).

In the 1830s, the German physiologist E.H. Weber studied the size of the difference threshold, mostly working with lifted weights. He found that, as the stimulus became heavier, it took a larger increase in weight to be just-noticeably heavier. In other words, heavier weights became harder to discriminate or to tell apart (Stevens, 1971; Gescheider, 1997). But the truly valuable insight came when he looked at the size of the difference threshold relative to the weight of the standard stimulus. The ratio of the stimulus change to the level of the standard stimulus was virtually constant. If we let ΔE be the size of the change necessary to produce a JND and E be the energy (or weight or concentration) of the standard, then the following relationship held:

$$\frac{\Delta E}{E} = c \quad \text{and thus} \quad \Delta E = cE \tag{1.11}$$

In other words, the weights had to change by a constant proportion or percentage c in order to cross the perceivable boundary. The DL or JND was found to be a constant fraction or proportion of the starting level. This was the first quantified psychophysical relationship and has become known as **Weber's law** (Stevens, 1971; Gescheider, 1997). The Weber fraction c could be used as a measure of the resolving power of any sensory system or modality. Researchers could apply the techniques of psychophysics to compare the functioning of different sensory modalities. It became apparent, for example, that the visual and auditory senses were able to discriminate much smaller percentage changes in a stimulus energy level, than touch, taste, or olfaction senses could. Some examples are shown in Table 1.2. Measuring the DL and determining the Weber fraction for different senses and different conditions would keep graduate students busy for many decades to come. Fechner was once touted as having made no less than 24,576 judgments in testing Weber's law for lifted weights (James, 1892).

But how general was Weber's "law"? A common observation was that at very low stimulus levels, those approaching the absolute threshold, that the Weber fraction increased (Stevens, 1971). That is, it took a larger percentage change at very low levels to be detectably different than the change that was needed over the middle of the functional stimulus range. Sometimes this departure (increase) was also noted at high levels of the stimulus as well. So when $\Delta E/E$ was plotted as a function of E, it did not always form a flat horizontal

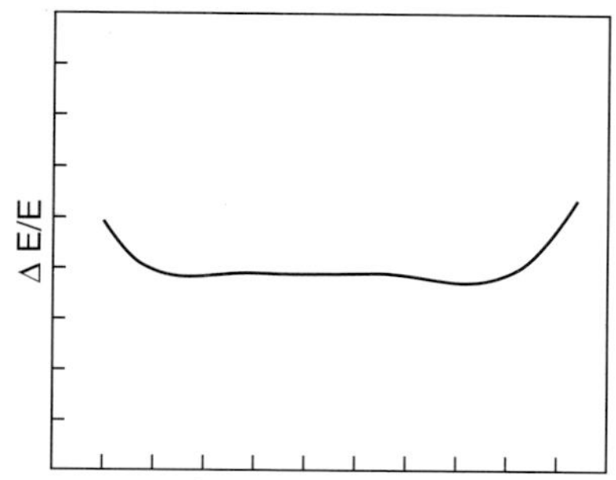

Figure 1.9 The theoretical relationship of Weber's law and the common finding of departures at low and high levels of energy.

line, but might curve up at the ends as in Figure 1.9. In order to reconcile this discrepancy, sometimes an additive constant was included in Weber's law:

$$\frac{\Delta E}{E+k} = c \quad \text{and thus} \quad \Delta E = c\left(E+k\right) \tag{1.12}$$

This constant *k* adjusted for the curvature of the Weber fraction plot at the lower end of the range. The exact meaning of this convenient constant is not clear, but it could represent some background noise in the sensory system that becomes more important at low levels around threshold measurement, and less of a factor as the stimulus level becomes stronger (Gescheider, 1997). Note that the additive constant does not help with the breakdown of Weber's law at very high levels. However, working at high stimulus levels is not easy, and other factors come into play. For example, a very strong stimulus may become painful, and thus the stimulus is now multidimensional in terms of the sensations it is causing.

Note that the absolute threshold can also be conceived of as a special case of the difference threshold. When we experimentally determine an absolute threshold, we are usually making a comparison with some blank, background, or "pure noise" stimulus, such as distilled water in the case of taste studies. Thus, the detection or absolute threshold is empirically the difference threshold against this neutral background.

1.4.2 Methods for Measuring Difference Thresholds

The common method for measuring difference thresholds was to present a long sequence of paired comparisons in which one member of the pair was held constant (the standard stimulus) and the second (the comparison stimulus) was varied, usually both above and below the standard. This was the "bread and butter" of classical psychophysics when studying Weber fractions, and was known as the **method of constant stimuli**. The term is perhaps a bit unfortunate, because only one item is really constant. The task for the observer is to say whether the comparison stimulus is heavier (in the case of lifted weights) or lighter than the standard. Note that you can also conduct this experiment as a blind comparison, and just ask the observer to choose which item is heavier. A choice is forced, although some variations of the method allow for an "equal" judgment as well. A random order of comparison stimuli in the pairings is presented; that is, there is no fixed order, increasing or decreasing, of the comparison items.

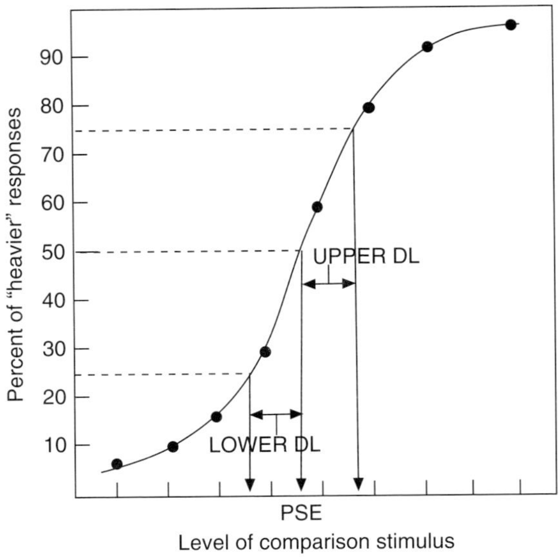

Figure 1.10 The method of constant stimuli. The standard stimulus is a fixed value and the comparison stimulus is varied around it. In an experiment on lifted weights, the observer may be forced to respond either "heavier" or "lighter" when comparing the comparison item with the standard. A plot of the proportion "heavier" will give an upper and a lower difference threshold, found by interpolating at the 75% and 25% points, respectively. A sigmoid or ogive curve is a common observation.

Figure 1.10 shows the critical plot of the data. We consider only one of the responses, in this case "heavier." When the comparison is lower in weight than the standard, the percentage of "heavier" responses will usually be below 50% and drop off as the comparison weight decreases. Conversely, the percentage rises above 50% as the comparison is physically greater in weight. Upper and lower difference thresholds (DLs) can be determined by the physical change necessary to produce a 75% and 25% response, and taking their difference from the 50% point. Note that the DL is stated in physical units.

Often, the 50% point does not exactly correspond to the standard stimulus. This was known as the **point of subjective equality** (PSE), and the difference from the physical value of the standard was know as the **constant error** (CE).

The method of limits could also be adapted to the measurement of difference threshold. The sequence of pairs would now be increasing or decreasing, rather than random. In modern times this would be considered less than optimal, because it would very likely induce some expectations on the part of the observer that things are going to change in a certain direction. It was also common to allow the observer to respond "equal." This introduces a further problem, in that the observer has to decide what level of change is necessary to leave the comfortable "equal" option. A conservative observer might want to be sure that the comparison was different, while a more lax observer might be inclined to respond with very little perceived difference. So there is a criterion problem here again, which will be dealt with in Chapter 3.

Figure 1.11 shows how the data might look from a hypothetical experiment on lifted weights. The comparison stimulus is changed to move it toward the standard in ascending and descending runs. Note that the run continues past the level of the standard. At some

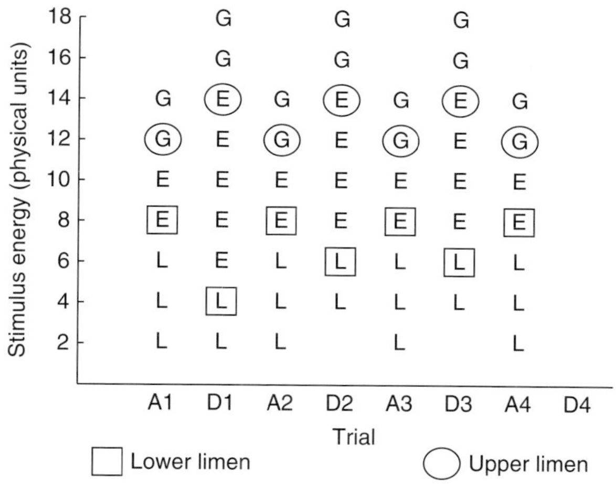

Figure 1.11 The method of limits with an "equal" response used to obtain the difference threshold. The IU is found from the average of the lower limens subtracted from the average of the upper limens, the points at which the response changes. The DL is then one-half the IU.

point, while increasing from a low level, the observer's response changes from "lighter" to "equal" and, as the series progresses, from "equal" to "heavier." Thus, there are two change points in each series, and these define the **interval of uncertainty** (IU). The upper change point was deemed the upper limen or L_u and the lower change point was the lower limen or L_l. After several series, the mean lower limen would be subtracted from the mean upper limen to get the IU and then the difference threshold would be one-half of the IU. The PSE could also be defined as the midpoint of the two means.

1.5 A Look Ahead: Fechner's Contribution

G.T. Fechner is considered the father of psychophysics, largely due to the publication of his book on the topic in 1860. Fechner had studied medicine, mathematics, and physics and later in his life turned to philosophy. He was troubled by the mind–body dichotomy of René Descartes, and felt that mind and matter were equal in the sense they were two manifestations of the same reality. On 22 October in 1848, or so the story goes, Fechner had the insight that Weber's law would connect the two. He was also interested in finding a way to represent subjective experience; that is, the loudness of a sound or the brightness of a light. Until this time, we only knew how much to change the light or the sound to create a sensation or a sensation difference.

The problem of providing a numerical representation of subjective magnitude could be solved, according to Fechner, if one used the JND as the unit of subjective increase. Assuming that all JNDs were subjectively equal, one merely had to add them up (starting with absolute threshold) to provide a measure of how strong a sensation appeared to the observer. A plot of the JNDs would provide the ruler needed to define a given subjective magnitude for some stimulus continuum, as shown in Figure 1.12. Note that this would only hold over the range in which Weber's law was accurate. Fechner was also familiar with the branch of mathematics we have come to call calculus, and understood that, as one accumulated a larger and larger number of smaller and smaller intervals, one could integrate

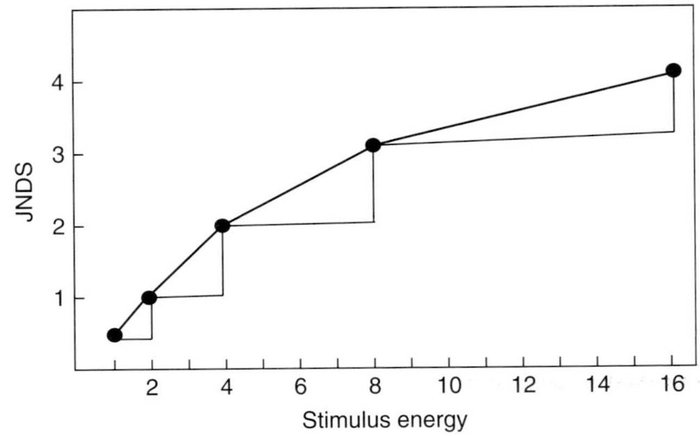

Figure 1.12 A plot of JNDs is accumulated to obtain the psychophysical function relating subjective intensity against physical stimulus intensity, as according to Fechner.

to get this summed relationship. Thus, since the integral of dx/x is the natural logarithm of x (plus a constant of integration), Fechner's law was given as a logarithmic relationship between subjective intensity S and stimulus energy E:

$$S = k \log E \tag{1.13}$$

This logarithmic relationship was considered a psychophysical "law" and is sometimes combined with Weber's law to be called the Weber–Fechner relationship. As a rule of thumb, it is not a bad idea to keep in mind. It tells us that, in order to make equal subjective changes in sensation strength, we need to vary the stimulus strength in a geometric progression, and not equal arithmetic steps. If I want to increase the sweetness of a beverage by equal steps, I had better not use 8%, 10%, and 12% sucrose. The second jump will probably be smaller than the first. A proportional change across the steps, for example, by a multiplicative factor of 1.2, would be more likely to produce equal steps of sweetness increases. As our quote from Shakespeare at the outset of the chapter shows, the candle is overshadowed in comparison with the light of the moon. The rule is, as the stimulus level increases, it generally takes larger steps to notice a difference. Against the background of a dark night, the candle is observed, but not against the light of the shining moon.

Appendix 1.A Relationship of Proportions, Areas Under the Normal Distribution, and Z-Scores

Because the exact shape of the normal distribution is known, the area under the curve from $-\infty$ to $+\infty$ to any value (usually expressed in Z-scores) can be estimated. Thus, there is a simple relationship between area or probability p and any value's distance from the mean, expressed in standard deviation units (Z-scores). The relationship between p and Z is shown in Table 1.A.1 and in Figure 1.A.1.

Table 1.A.1 Proportions and *Z*-scores*

Proportion	*Z*-score	Proportion	*Z*-score	Proportion	*Z*-score	Proportion	*Z*-score
0.01	−2.33	0.26	−0.64	0.51	0.03	0.76	0.71
0.02	−2.05	0.27	−0.61	0.52	0.05	0.77	0.74
0.03	−1.88	0.28	−0.58	0.53	0.08	0.78	0.77
0.04	−1.75	0.29	−0.55	0.54	0.10	0.79	0.81
0.05	−1.64	0.30	−0.52	0.55	0.13	0.80	0.84
0.06	−1.55	0.31	−0.50	0.56	0.15	0.81	0.88
0.07	−1.48	0.32	−0.47	0.57	0.18	0.82	0.92
0.08	−1.41	0.33	−0.44	0.58	0.20	0.83	0.95
0.09	−1.34	0.34	−0.41	0.59	0.23	0.84	0.99
0.10	−1.28	0.35	−0.39	0.60	0.25	0.85	1.04
0.11	−1.23	0.36	−0.36	0.61	0.28	0.86	1.08
0.12	−1.18	0.37	−0.33	0.62	0.31	0.87	1.13
0.13	−1.13	0.38	−0.31	0.63	0.33	0.88	1.18
0.14	−1.08	0.39	−0.28	0.64	0.36	0.89	1.23
0.15	−1.04	0.40	−0.25	0.65	0.39	0.90	1.28
0.16	−0.99	0.41	−0.23	0.66	0.41	0.91	1.34
0.17	−0.95	0.42	−0.20	0.67	0.44	0.92	1.41
0.18	−0.92	0.43	−0.18	0.68	0.47	0.03	1.48
0.19	−0.88	0.44	−0.15	0.69	0.50	0.94	1.55
0.20	−0.84	0.45	−0.13	0.70	0.52	0.95	1.64
0.21	−0.81	0.46	−0.10	0.71	0.55	0.96	1.75
0.22	−0.77	0.47	−0.08	0.72	0.58	0.97	1.88
0.23	−0.74	0.48	−0.05	0.73	0.61	0.98	2.05
0.24	−0.71	0.49	−0.03	0.74	0.64	0.99	2.33
0.25	−0.67	0.50	0.00	0.75	0.67	0.995	2.58

*Calculated in Excel®.

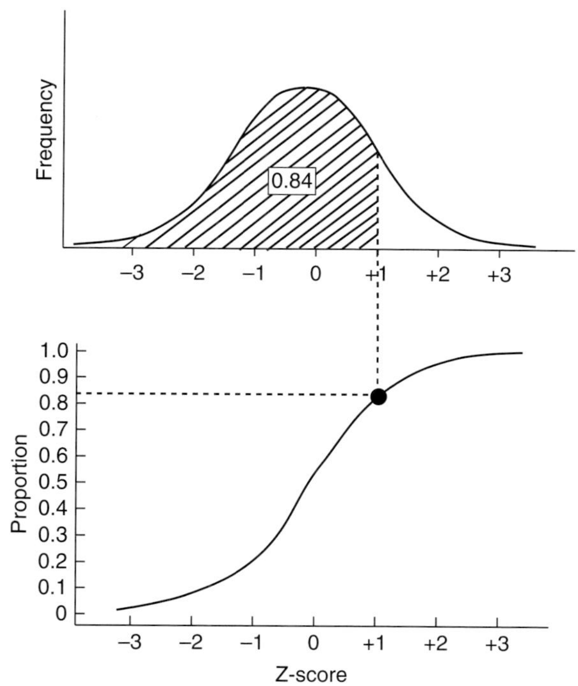

Figure 1A.1 Each point on the ogive curve corresponds to an area under the normal curve, or proportion of the total area that is below and to the left of that point.

Appendix 1.B Worked Example: Fitting a Logistic Function to Threshold Data

In this example, we will take the data from Stocking et al. (2001), fit a logistic function using ordinary least squares, and then interpolate to find the chance-corrected 50% detection level. Recall that Stocking et al. used the ASTM E-679 procedure, which is a hybrid of a 3-AFC and triangle test. It resembles the 3-AFC in that only one of the three samples contains the odorant, in this case MTBE. It resembles a triangle test in that participants are asked to choose the odd sample, rather than the strongest smelling. The actual strategy adopted by testers is usually unknown. Stocking et al. used 57 testers and eight concentration steps, differing roughly by a factor of 2. The data that we will use consists of the percentage of correct choices at each concentration level. The actual ASTM procedure requires finding individual thresholds based on some heuristic rules, and then averaging, but this discounts some correct judgments, so we will use the entire data set and percentages correct. The data set can be found in Stocking et al. (2001) and also in the appendix to Chapter 6 in Lawless and Heymann (2010).

Recall that for a three-alternative procedure, we need to apply the correction for guessing, also known as Abbott's formula, in order to find the true proportion detecting using the following expression:

$$P_{\mathrm{D}} = \frac{P_{\mathrm{obs}} - P_{\mathrm{chance}}}{1 - P_{\mathrm{chance}}} \tag{1.B.1}$$

where P_{D} is our true percentage detecting and P_{obs} is the proportion observed in the data that would yield the desired P_{D}. With a chance probability of 1/3, we require 2/3 correct to get a true detection rate of 0.5, 50% correct to get a detection rate of 0.25, and 40% correct to get a true proportion of 0.10. After we fit our equation to the data, we will interpolate at these three levels. In this example, we will fit the function to the raw proportion correct, fit the function, and then interpolate at the corrected levels. We could have corrected the proportion before fitting as well, but this does not seem to affect the results very much (USEPA, 2001).

The logistic function takes the form of

$$f(x) = \frac{\mathrm{e}^x}{\mathrm{e}^x + 1} = \frac{1}{1 + \mathrm{e}^{-x}} \tag{1.B.2}$$

We can "linearize" this relationship by converting our proportion correct to an odds ratio, $p/(1-p)$:

$$\ln\left(\frac{p}{1-p}\right) = b_0 + b_1 x \tag{1.B.3}$$

Table 1.B.1 shows the raw proportion correct, the odds ratio (our Y variable), the concentration steps in parts per billion, and the log of the concentrations (our X variable). The logarithm of concentration is used because there is a roughly geometric progression in the concentration steps.

The method of least squares, or ordinary least squares, fits a straight line to the data by minimizing the squared residual deviations in the Y-direction of the actual data points to the

Table 1.B.1 Proportions correct versus log concentration from the Stocking et al. (2001) data

Conc.	log(*c*) "*X*"	Prop. *p*	*p*/(1 − *p*)	ln[*p*/(1 − *p*)] "*Y*"	X^2	Y^2	*XY*
2	0.301	0.44	0.786	−0.241	0.091	0.058	−0.072
3.5	0.544	0.49	0.961	−0.039	0.296	0.002	−0.022
6	0.778	0.61	1.564	0.447	0.606	0.199	0.348
10	1.000	0.58	1.381	0.323	1.000	0.104	0.322
18	1.255	0.65	1.857	0.618	1.576	0.383	0.777
30	1.477	0.77	3.348	1.208	2.182	1.460	1.784
60	1.778	0.89	8.091	2.091	3.162	4.371	3.717
100	2.000	0.86	6.143	1.815	4.000	3.295	3.631
Sum	9.133		24.131	6.223	12.91	9.874	10.49
Mean	1.142			0.778			

intercepted value on the fit line. Although there are many ways to fit a function to the data, this is probably the most commonly used method.

The sum of the squared residual deviations $\sum R^2$ will be minimized when the partial derivatives of R with respect to the slope b and intercept a are set to zero as follows:

$$y = a + bx \tag{1.B.4}$$

$$\sum R^2(a, b) = \sum_{i=1}^{N} [y_i - (a + bx_i)]^2 \tag{1.B.5}$$

And setting the partial derivatives to zero:

$$\frac{\partial R^2}{\partial b} = -2\sum_{i=1}^{N} [y_i - (a + bx_i)]x_i = 0 \tag{1.B.6}$$

and

$$\frac{\partial R^2}{\partial a} = -2\sum_{i=1}^{N} [y_i - (a + bx_i)] = 0 \tag{1.B.7}$$

Solving the two equations in two unknowns yields the following expressions for a and b (dropping the summation index as we are always summing from 1 to N data points:

$$b = \frac{N\sum xy - \left(\sum x\right)\left(\sum y\right)}{N\sum \left(x^2\right) - \left(\sum x\right)^2} \tag{1.B.8}$$

$$a = \frac{\left(\sum x^2\right)\sum y - \sum x \sum xy}{N\sum x - \left(\sum x\right)^2} \tag{1.B.9}$$

These become a little simpler if we work in the means of x and y (using the bar symbol for the mean) as follows:

$$b = \frac{\sum xy - N\overline{xy}}{\sum \left(x^2\right) - N\left(\overline{x}\right)^2} \tag{1.B.10}$$

Table 1.B.2 Summary of the procedure

Percentage detecting	Percentage correct required after correction	Converting to $\ln[p/(1-p)]$	Solving for log(*c*) via $x=(y-a)/b$	10^x (ppb)
50	66.7	0.693	1.079	12.0
25	50.0	0.0	0.571	3.7
10	40.0	–0.405	0.274	1.9

and

$$a = \overline{y} - b\overline{x} \tag{1.B.11}$$

Returning to the Stocking et al. data, we can now make our calculations using eqns 1.B.10 and 1.B.11:

$$b = \frac{10.49 - 8(1.142)(0.778)}{12.91 - 8(1.142)^2} = \frac{3.382}{2.478} = 1.364$$

and

$$a = 0.778 - 1.364(1.142) = -0.779$$

Now we can solve for log(c) for our three points of interest. Solving the equation for x gives us $x=(y–a)/b$, so $\log(c)=(y–0.778)/1.364$. Our proportions of detectors again are 0.667, 0.5, and 0.4. Converting to the natural log of the odds ratio, $\ln[p/(1-p)]$, gives us y values of 0.693, 0.0, and –0.405 for y-values. Our interpolations then are as follows:

- $[0.693-(-0.779)]/1.364=1.079$, and so $10^{1.079}=12.0$ ppb for our threshold estimate;
- for our 25% detection rate, we get $[0-(-0.779)]/1.364=0.571$ and$10^{0.571}=3.7$ ppb;
- and for our 10% detection level we get $[-0.405-(0.779)]/1.364=0.274$ and $10^{0.274}=1.9$ ppb.

The procedure is summarized in Table 1.B.2.

These estimates are actually quite close to those you can obtain via the ASTM method, which gives a threshold in the range of 14–15 ppb, reasonably close to our estimate of 12 ppb via curve fitting and interpolation. Simple inspection of the raw data also provides some validation. For example, there were 10 people who chose the correct sample all the way through the data set and who might have been detecting the MTBE sample at the lowest level of 2 ppb. If we take that proportion of 10/57 we get about 17.5%. Correcting this for guessing gives us about 10% as a likely estimate for true detectors at the 2 ppb level, which is quite close to our interpolated estimate of 1.9 ppb!

References

Antinone, M.A., Lawless, H.T., Ledford, R.A., and Johnston, M. 1994. The importance of diacetyl as a flavor component in full fat cottage cheese. Journal of Food Science, 59, 38–42.

ASTM. 2008. Standard practice for determining odor and taste thresholds by a forced-choice ascending concentration series method of limits, E-679-04, Annual Book of Standards, Vol. 15.08. ASTM International, Conshocken, PA, pp. 36–42.

Cain, W.S. 1977. Differential sensitivity for smell: noise at the nose. Science, 195, 795–798.
Cornsweet, T.M. 1962. The staircase method in psychophysics. American Journal of Psychology, 75, 485–491.
Gescheider, G.A. 1997. Psychophysics. The Fundamentals. Third edition. Lawrence Erlbaum, Mahwah, NJ.
Harwood, M.L., Ziegler, G.R., and Hayes, J.E. 2012. Rejection threshold in chocolate milk: evidence for segmentation. Food Quality and Preference, 26, 128–133.
James, W. 1892. Psychology, Briefer Course. Holt, Rinehart and Winston, New York, NY.
Lawless, H.T. 2010. A simple alternative analysis for threshold data determined by ascending forced-choice method of limits. Journal of Sensory Studies, 25, 332–346.
Lawless, H.T. and Heymann, H. 2010. Sensory Evaluation of Food. Springer Publishing, New York, NY.
Lawless, H.T., Thomas, C.J.C., and Johnston, M. 1995. Variation in odor thresholds for L-carvone and cineole and correlations with suprathreshold intensity ratings. Chemical Senses, 20, 9–17.
Linschoten, M.R., Harvey, L.O., Eller, P.A., and Jafek, B.W. 1996. Rapid and accurate measurement of taste and smell thresholds using an adaptive maximum-likelihood staircase procedure. Chemical Senses, 21, 633–634.
Morrison, D.G. 1978. A probability model for forced binary choices. American Statistician, 32, 23–25.
Peng, M., Jaeger, S.R., and Hautus, M.J. 2012. Determining odour detection thresholds: Incorporating a method-independent definition into the implementation of ASTM E679. Food Quality and Preference, 25, 95–104.
Prescott, J., Norris, L., Kunst, M., and Kim, S. 2005. Estimating a "consumer rejection threshold" for cork taint in white wine. Food Quality and Preference, 18, 345–349.
Stevens, J.C. 1971. Psychophysics. In: Stimulus and Sensation, Readings in Sensory Psychology. W.S. Cain and L.E. Marks (Eds). Little, Brown and Co., Boston, MA, pp. 5–18.
Stocking, A.J., Suffet, I.H., McGuire, M.J., and Kavanaugh, M.C. 2001. Implications of an MTBE odor study for setting drinking water standards. Journal of the American Water Works Association, 93, 95–105.
Teghtsoonian, R. 1971. On the exponents in Steven's law and the constant in Ekman's law. Psychological Review, 78, 71–80.
USEPA. 2001. Statistical analysis of MTBE odor detection thresholds in drinking water. National Service Center for Environmental Publications (NSCEP) No. 815R01024, available from http://nepis.epa.gov.
Walker, J.C., Hall, S.B., Walker, D.B., Kendall-Reed, M.S., Hood, A.F., and Nio, X.-F. 2003. Human odor detectability: new methodology used to determine threshold and variation. Chemical Senses, 28, 817–826.

2 Psychophysics II: Scaling and Psychophysical Functions

> *For nearly 150 years, the goal of measuring sensation magnitude has been fundamental to psychophysics, yet many problems remain today. For example, different scaling techniques yield different results. Either all or some of the methods produce invalid results, or the different methods are measuring different aspects of perception.*
>
> George Gescheider (1997: 369).

2.1 Introduction

This chapter will continue our discussion of classical psychophysics as a measurement science. Leaving behind the topics of thresholds and just-noticeable differences (JNDs), we enter the realm of direct scaling of sensory intensity or magnitude. Historically, this was viewed as an intractable problem – how to quantify the strength of subjective experiences,

Quantitative Sensory Analysis: Psychophysics, Models and Intelligent Design, First Edition. Harry T. Lawless.

which after all are a private matter and not subject to external observation. Yet humans are constantly observing and adjusting their external world to provide sensory intensities that they find desirable. We add sugar to our coffee to sweeten it. We decide the room is too cool and adjust the thermostat. We judge the size of today's waves at the beach when we decide to go surfing. Furthermore, we discuss these impressions with others and expect them to understand what we are talking about. The statement, "there is a large mouse running up the trunk of a small elephant" is understood by all, and (perhaps remarkably) it does not imply that the rodent is larger than the pachyderm. So our experiences are common among people and able to be communicated. These facts lead us to the possibility that subjective events can be meaningfully quantified.

Inextricably linked with the attempts to provide numerical quantification of sensory intensity was a search for the general mathematical relationship that would connect energy in the world to the strength of our sensory experiences. We call these **psychophysical functions**. Various functions have been proposed over the years, mostly based on some kind of empirical observation and then mathematical fitting. These attempts were not based in any underlying neurological or physical or chemical theory, although they were sometimes discussed as if they did have a theoretical basis. Toward the end of the chapter we will contrast this curve-fitting approach with a chemically based theory used to describe psychophysical functions for taste and smell stimuli.

Simply because we apply numbers to stimuli to represent something about our sensations does not mean we have measured anything. We can put numbers on the uniforms or jerseys of players on a sports team. This merely differentiates them. Similarly, we could codify the rooms of a house in which we consume a snack food with numerical labels for counting purposes after a consumer survey. But this implies nothing other than commonality or difference. Such application of numbers is sometimes referred to as a **nominal scale**, because we have used numbers to name something. The numbers on our football players do not imply anything about their size, speed, or talent. Somewhat more useful is the application of numbers to represent rankings. We do not know anything about the magnitude or strength of things when we rank them, only that one item has more or less of a certain property than another item. An analogy is describing the finishers in a horse race as first, second, and third. This is called **ordinal scaling**. When we use numbers to represent a constant unit of difference, we have an even more useful application. For example, in a horse race we commonly say that one horse finished two lengths in front of another and three lengths from a third. We do not yet know anything about how fast they ran or the time it took to traverse the racecourse, but we have some comparable idea of the distances at the end. This is an example of an **interval scale**. Even more useful is an application of numbers to represent ratios or proportions. If we measure the time taken to run a race, we could say that, on the average, one experienced marathoner ran twice as fast as a specific novice runner. Much of our physical measurements, such as length, time, and mass, are made with these **ratio scale** properties. Whether psychophysical or sensory evaluation data ever achieve ratio properties is an object of long and ongoing debate, but this would be desirable in any measurement system. Table 2.1 shows these four kinds of measurements and some of the allowable statements and transformations. Numerification is not limited to only these four categories. For example, pH measurement uses equal numerical changes to represent logarithmic steps in hydrogen ion concentration in an aqueous solution. It is thus an example of a logarithmic interval scale. Stellar magnitude is another example.

Psychophysical researchers were also quick to distinguish between different kinds of sensory continua. Not all changes in the sensory world were simple increases or decreases in intensity (strength). Some were qualitative rather than intensive, such as pitch or hue.

Table 2.1 Classifications of numerical measurement

Scale type	Example	Allowable statements	Transformations
Nominal	Numbers on football jerseys	$A=B$ $A \neq B$	Any that preserve identity
Ordinal	Ranking; finishers in a race	$A>B$ $A<B$	Any that preserve rank order
Interval	Centigrade, Fahrenheit (arbitrary zero)	A "differs from B by X amount" $A-B=X$	$A=XB+C$ (linear)
Ratio	Mass, length, time (fixed zero)	A is X times B	$A=XB$ (multiplicative)

Stevens (1986) distinguished between **prothetic continua** for those that were intensive and **metathetic continua** that were more of kind, place, or type of sensation. Metathetic continua usually have JNDs that seemed subjectively equal, whereas prothetic continua might not. This terminology is used rarely now, but a sensory scientist would be well advised to keep in mind that not all sensory changes fit the simple psychophysical model. In setting up a descriptive analysis ballot, for example, we often use scaling methods as if they were all representing intensive changes, but this is not always true. An orange color may vary from more or less red to more or less yellow, for example.

Further discussion of measurement theory and the history of scaling can be found in Stevens (1986), Gescheider (1997) and Baird and Noma (1978). A divergent view can be found in the functional measurement literature (Anderson, 1974).

2.2 History: Cramer, Bernoulli, Weber, and Fechner

Perhaps the first attempt to quantify a subjective experience concerned the value of money. Sometimes referred to as the St. Petersburg paradox, after a wagering game, people wondered why someone would risk a small gain against a potentially large loss. Stated another way, the (subjective) value of money appears to be relative to the amount you start with. So, \$5 may be a lot to someone who is poor man but it is insignificant to someone who is rich. Suppose you leave a building and find a \$20 bill lying on the ground. No one is around and thus you cannot return it. It is rightfully yours. Think for a minute: How happy did that make you? Now suppose you were to find more money. How much would it take to make you twice that happy? When I do this as a classroom exercise, the median amount lies somewhere between \$50 and \$100; let us say \$75 for the sake of argument. The point is that the relationship is not linear. It takes more and more dollars for you to make equal jumps in happiness or in perceived wealth.

Economists refer to this as a **utility function**. As a graph, it would plot subjective value against the numerical amount of money. The function is clearly compressive – subjective utility grows more slowly than actual wealth. There is decreasing "marginal utility." In consideration of the St. Petersburg paradox, the mathematician Gabriel Cramer, in 1728, suggested that a square-root relationship would work (Stevens, 1986). For our example of \$20, a doubling in value would require four times the amount or about \$80, which is not far from my historical classroom estimate of \$75. Note that a square-root relationship is also a power function with an exponent of 0.5. About a decade later, Daniel Bernoulli (1738/1954) wondered about the same issue. As a mathematician, he was patronized by European nobility in order to help them with their gambling strategies. He often wondered how it was possible for these wealthy people to wager such large sums on the single turn of a card. Clearly, the value seemed inversely proportional to the amount you owned. Bernoulli proposed a logarithmic function. This is note-worthy because of the following relationship:

$$\frac{\Delta M}{M} = k \quad \text{and} \quad \Delta M = kM \tag{2.1}$$

where ΔM is the amount of change needed to create a constant change in your subjective wealth (utility) and M is the amount of money you start with. If we integrate this function to get a relationship between utility U and actual wealth M, we get

$$U = \int \frac{\Delta M}{M} = \ln(M) + C \tag{2.2}$$

In other words, utility should rise as the natural logarithm of M. The base of the logarithm is not so important at this point, but the compressive relationship is key.

As discussed in Chapter 1, this led to the finding of Weber's law for JNDs and Fechner's logarithmic relationship for subjective intensity S as a function of physical stimulus intensity E:

$$\frac{\Delta E}{E} = k_1 \quad \text{and} \quad S = k_2 \log(E) + C \tag{2.3}$$

where ΔE is the energy increase needed to create a JND; that is, to cross the difference threshold. Note that the constants k have different subscripts. One (k_1) is the important Weber fraction and the other (k_2) is merely a proportionality constant that depends upon the units of measurement and the base of the logarithm. Assuming that the JND is a constant unit of subjective magnitude (a big assumption), we obtain the general psychophysical law that subjective intensity increases in proportion to the logarithm of physical stimulus energy. This logarithmic relationship produced a useful rule of thumb and generally went unchallenged until the mid-20th century. Once again, the compressive relationship was the important finding, as it permits the sensory system to respond to a wide range of stimulus energy. It makes sense for a sensory system to be engineered to respond to a wide range of stimulus energy. From the auditory threshold to the loudest sound we can tolerate, there are about 15 log units of sound pressure change (150 dB). A similar range is found in vision, from the dimmest light we can perceive when fully dark-adapted to the approximate luminance of the sun, the brightest thing we probably every look at (Stevens, 1986: 33).

2.3 Partition Scales and Categories

The first category scale for a sensory phenomenon is attributed to Hipparchus (Stevens, 1986). About 150 BC he invented a category scale to classify the brightness of stars: stellar magnitude. Hipparchus used six "buckets" to categorize visible stars, starting with the brightest at category 1 and the faintest at category 6. Now that we can measure the actual photometric luminance of the stars, we know that this upside-down category scale forms a roughly logarithmic function of photometric intensity. Each step is a 4 dB change in light intensity, or four tenths of a logarithmic unit (see Appendix 2.A).

Nowadays, we often assume that when faced with a category scale (i.e., a fixed number of response options representing increasing sensory intensity) people will figure out the range of stimuli and sensations they are working with and then attempt to distribute them across the response options to represent about equal subjective intervals. This might be called a "differencing strategy" because equal intervals represent equal sensory differences, rather than ratios. Remember that, in the psychophysical laboratory, lights or sounds would be presented many, many times, often with some additional warm-up stimuli or discarded data points for practice. Thus, the subject in the study would have a chance to learn the stimulus range and use the available response scale effectively. Often, verbal labels would be applied to reinforce this equal interval property, such as "weak–moderate–strong." To most English speakers, these three adjectives seem to be about equally spaced, perceptually.

Other tasks were invented to try to measure, at least indirectly, the psychophysical relationship. The blind physicist, Plateau, asked various artists to produce a gray color that would lie halfway between black and white. Even though they worked in their own studios under presumably different conditions, they produced remarkably similar levels of gray. Note that the halfway point represents a ratio instruction, and Plateau thought the psychophysical relationship might follow a power function. In 1873, Delboeuf continued this work to construct other contrast steps of approximately equal intervals or partitions, and the results favored a logarithmic relationship (Stevens, 1986). So, based upon partitioning, Fechner's insight seemed to be correct.

If people can bisect the sensory distance between two stimuli, why not try other ratios? This was attempted by the physicist Merkel, in 1888, for doubling of brightness, but he did not construct a psychophysical function with the results (Stevens, 1986). The notion of a ratio instruction reappears in the 1930s, with a series of studies using apparatus derived from the recently invented electrical telephone. It was possible at last to control sound stimuli very accurately, and a number of people began to study loudness perception and phenomena like binaural summation. Richardson and Ross (1933) are usually credited with the first study where people were presented with two tones, and they estimated the ratio of the loudness of the tones.

A number of related studies using ratio instructions were conducted, so by the late 1930s, S.S. Stevens and his students were able to try to construct a loudness scale based upon ratios. They noted two important observations. First, the loudness did not exactly conform to sound pressure in decibels, as Fechner's log function would predict (see Appendix 2.A for a discussion of decibels). Second, when asked to partition or bisect the difference of two tones by adjusting a third, the chosen level did not correspond to what was predicted (one-half loudness) by the ratio-derived loudness scale. When asked to partition things, people behaved differently than when they were asked to judge ratios.

2.4 Magnitude Estimation and the Power Law

2.4.1 The Method Evolves

By 1953, Stevens was ready to make the jump from adjusting stimuli that formed fixed ratios to letting the subject or observer give their own impressions. This was the birth of **magnitude estimation**. Stevens assumed that people could report numbers that would reflect the perceived ratios of the strengths of two perceptions, and in the laboratory they appeared to do this. There was both internal agreement with one's previous judgments and a fair amount of consistency amongst subjects as well. He also noted that the usual arithmetic mean was a poor choice for pooling data across subjects. The data were positively skewed; that is, there were high outliers. For this reason, he began to take the medians or geometric means as measures of central tendency, both being less influenced by high outliers in the data.

Typical instructions went something like this:

> You will be presented with a series of stimuli in irregular order. Your task is to tell how intense they seem by assigning numbers to them. The first stimulus will be called a standard, and will be assigned the number 10. If the next stimulus seems five times as strong, give it a number five times as big. If half as strong, give it a number half as big. You can use any numbers you wish including fractions and decimals. Try to match each number to the intensity as you perceive it.

The standard was called a **modulus**, and had the effect of bringing all the subjects' number usage into the same general range. Later, Stevens allowed the subject to choose any number they wished, as long as they preserved ratios in their subsequent judgments. Some subjects remarked that this was easier, and it became a commonplace procedure. The problem arose, however, of what to do with the data to bring them into the same range and still preserve the ratio properties of the judgments. Suppose I chose to use numbers from 1 to 100 and you chose to use numbers from 1 to 1000? Your higher range would give you more weight if simply pooled without further adjustment. One solution (shown in detail in Appendix 2.B) is to construct an individual multiplicative factor for each individual subject and then multiply their data by that factor, such that the overall or geometric mean for all subjects was equated before averaging (Lane et al., 1961). Frequently, the data were converted to logarithms before statistical treatment, in which case an additive constant would do instead of a multiplier. Converting to logarithms had the additional benefit of taking a positively skewed distribution and making it more symmetric and a better approximation to the normal distribution, thus better satisfying some assumptions needed for statistical analysis.

2.4.2 The Power Law and the Method Become Linked

After some tweaking, Stevens and colleagues seemed satisfied that the new method was simple, direct, and that data were reproducible. He then took a step that would influence decades of graduate students and psychophysical researchers. Plotting the data on logarithmic coordinates produced a straight line. Many functions will appear straight on log–log paper, but remembering Cramer and Plateau, Stevens chose a **power function** as follows:

$$S = kE^n \tag{2.4}$$

where S is the sensation intensity (i.e., response) and E is the physical measurement of stimulus energy. The constant k was not very interesting, just a proportionality constant that would change if the units of measurement changed. The key was the exponent n. Taking logarithms of both sides, we get

$$\log(S) = k\log(E) + \log(k) \tag{2.5}$$

And so the critical exponent, now the slope of a straight line, could easily be found by least squares or other fitting methods. Note that in the Fechnerian logarithmic relationship the sensation intensity goes up arithmetically, in equal steps, with ratios of the stimulus energy. In this case, there needs to be a ratio progression on both sides. This need not be the same ratio, of course, but there is a geometric rather than an arithmetic progression on both sides that will determine equal steps.

The log function and power function are not interchangeable. In a semi-log plot, the log function is a straight line but the power function will be concave upward. In the log–log plot, the log function will be concave downward while the power function is linear.

There were occasional deviations, notably near threshold, so an adjustment was sometimes introduced to bring the physical stimulus into a proper range as follows:

$$S = k\left(E - E_0\right)^n \tag{2.6}$$

where E_0 represents the stimulus energy at absolute threshold.

One attractive feature of the power function was that it could accommodate sensory continua that were expansive as well as compressive. As shown in Figure 2.1, exponents greater

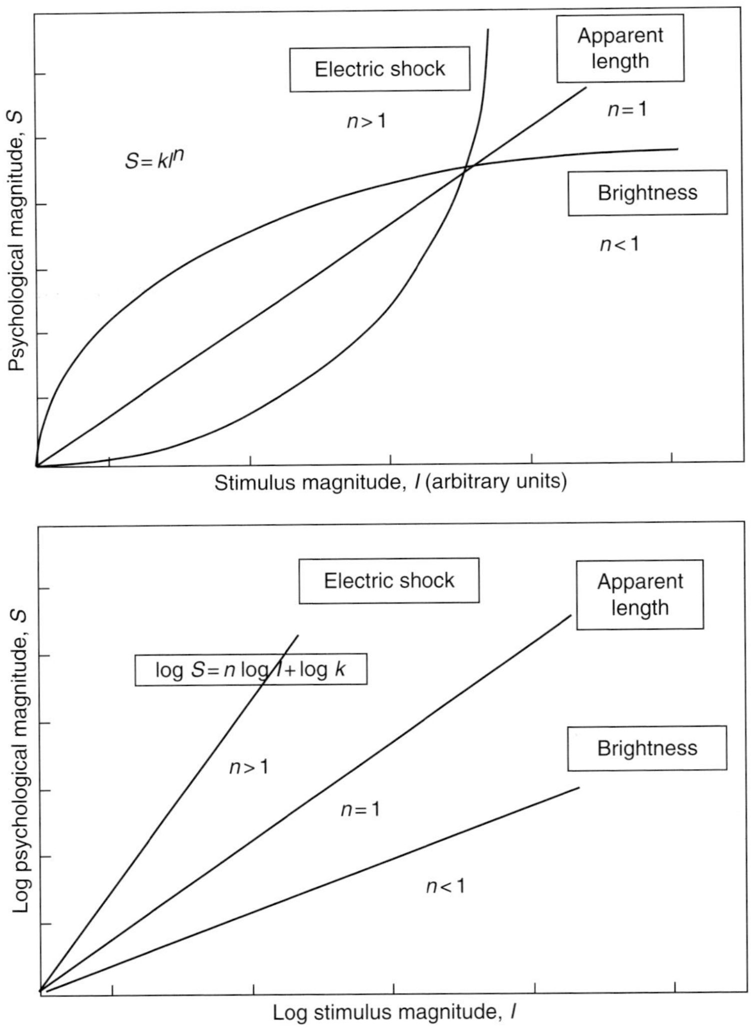

Figure 2.1 Power functions and log–log plots for three continua. Magnitude estimation data usually conform to a power function and form a straight line in a log–log plot. The power function will accommodate expansive sensory continua with exponents greater than one, linear continua with exponents of one, and compressive continua (probably the most common) with exponents less than one.

Table 2.2 Power function exponents and Weber fractions

Modality	Weber fraction	Exponent (power function)
Electric shock	0.013	2.5
Saturation, red	0.019	1.7
Heaviness	0.020	1.4
Finger span	0.022	1.3
Length	0.029	1.0
Vibration, 60 Hz	0.036	0.95
Loudness	0.048	0.60
Brightness	0.079	0.33
Taste, NaCl	0.083	0.41

From Teghtsoonian (1971).

than one showed positive acceleration, exponents of one were linear relationships, and exponents less than one (perhaps the majority) were compressive or negatively accelerating. This versatility was seen as a big asset.

The method caught on, and soon scores of graduate students were doing thesis work measuring exponents of various sensory continua. As shown in Table 2.2, sometimes there appeared to be an inverse relationship with the Weber fraction. As one might expect, continua that had high Weber fractions were slow to differentiate as stimulus energy increased, and this corresponded to a lower power function exponent. Continua that were expansive would likely have small Weber fractions – small changes are required to feel a difference. One should not make too much of this comparison, as there are plenty of exceptions, and remember that they are really measuring different things.

One informative area of research was to examine the conditions under which the power function would change. For example, changing a person's adaptation level will affect the exponent of the brightness function. Masking sounds with white noise has a systematic effect on the loudness exponent, and so on. Thus, important functional properties like adaptation, inhibition, and masking could be better understood and quantified using this simple method (see Stevens (1986: chapter 3) for many examples).

Enthusiasm ran high. However, there were critics, as we will see below. The philosophical stance became more or less circular. How do we know that magnitude estimation is valid? Because it produces the power function. How do we know that the power function is accurate? Because it is the result we get when we ask people to judge ratios. The best method was assumed to be the one that produces the purest power function, and even Stevens' choice of reference sounds was driven by this consideration! Furthermore, there were some simplistic assumptions. The first was that people were actually giving you numbers that reflected the ratios of their experienced sensations. Many of Stevens' subjects did not talk like that; rather, they were just matching numbers, as in the following quote (Stevens, 1986: 28):

> I felt freer to use numbers over a wide range. I liked the idea that I could just relax and contemplate the tones. When there was a fixed standard I felt more constrained to try to multiply and divide loudnesses, which is hard to do; but with no standard I could just place the tone where it seemed to belong.

Just because you have given ratio instructions, does not mean that your data have generated a ratio scale. Nonetheless, it was assumed to do so. The logic was so enticing because the result was so potentially useful. It would be of great practical value to be able to say "this solution is twice as sweet as that one" and have confidence in the ratio statement.

2.4.3 Ratio Scales for Hedonics?

In addition to measuring sensory intensity, some researchers were tempted to expand the method to include hedonics; that is, scaling of likes and dislikes (Moskowitz, 1971). If sensations could stand in ratio relationships to one another, why not one's pleasure or displeasure? One issue concerned the problem that hedonics are bipolar – that is, they seem to have a neutral point or indifference, with positive and negative sides. Some attempts were made at unipolar magnitude estimation of hedonics (the more you like it, the bigger the number) (Pearce et al., 1986). But most people who tried this settled on using a bipolar scale. This made the choice of the modulus or reference standard somewhat problematic. If one used a positive standard, you could certainly try to judge how much more or less you liked something else, but what if you got a stimulus you disliked? The way around this seemed to be to get some idea of the strength of your reaction to the standard, and use the negative side of the scale relative to that strength, just in the other direction, a kind of "distance from zero" concept. In one attempt to use a line scale with ratio properties, Lawless (1977) gave subjects ratio instructions (if you like it twice as much, make a mark twice as far from zero) and offered to tape together a longer line if they needed more room. For all practical purposes, ratio scaling of hedonics did not catch on in the food or consumer products industries, although it did have its proponents. Most people were still using simple category scales, which seemed to work quite well (Peryam & Girardot, 1952). We will revisit the notion of a hedonic ratio scale when we discuss labeled affective magnitude scales in Chapter 8.

2.5 Cross-Modality Matching; Attempts at Validation

2.5.1 Magnitude Production

A variation on magnitude estimation is to have the experimenter give the numbers and the subject adjust the stimulus to reflect the intensity suggested by the number. This is called magnitude production. If the power function is valid, and the ratio-scaling methods are accurate, you would expect the same exponent to be generated by either estimation or production. For the most part, they are, with one consistent bias or exception. Stevens was troubled by a persistent phenomenon. When asked to match stimuli to numbers, as in magnitude production, and when the power function was compared with magnitude estimation results, the product experiment produced a slightly higher exponent (steeper slope in a log–log plot). This change also appeared in cross-modality matching. When given control over one of the two continua, people used a smaller range for the stimulus (or the numbers in the case of estimation) that they had control over. Stevens called this the "**regression effect**" because it resembled a regression toward the mean in the range of values produced (Stevens, 1986). In his writing, you get a sense of his frustration with this tendency; he called it "ever present." Various theories were proposed to explain it and various methods to minimize it, with limited success. For our purposes, it simply illustrates that the notion of a true fixed exponent is not realistic. Human response tendencies and biases need to be considered in any scaling study.

2.5.2 Cross-Modality Matching

If subjects can both estimate and produce ratio judgments, why not have them match the sensation intensities from two continua? This would seem to eliminate the need for numbers altogether. For example, one could express any sensation (taste, smell, light, touch) in terms of a loudness match to some sound pressure level. Note that both variables are now expressed in physical units. There is no longer a perceptual variable like ratings or numerical estimates. This idea had some appeal to those who would rather stick to physical measures that can be objectively validated. One popular method was to use perceived force with a hand-held dynamometer that provided a simple and easy way to match intensities (Stevens et al., 1960). The method was called **cross-modality matching**.

Perhaps more importantly, this procedure provided a way to validate the ratio-scaling methods, or so Stevens and his colleagues believed. One should be able to predict the exponent (slope in a log–log plot) of the matching function from a ratio of the exponents from two individual ratio scalings of the continua judged alone. Thus, if we have two power functions for the two continua:

$$S_1 = k_1 E_1^{n_1} \quad \text{and} \quad S_2 = k_2 E_2^{n_2} \tag{2.7}$$

taking logs we get

$$\log S_1 = n_1 \log E_1 + \log k_1 \quad \text{and} \quad \log S_2 = n_2 \log E_2 + \log k_2 \tag{2.8}$$

Setting $S_1 = S_2$ (so $\log S_1 = \log S_2$) and solving for log E_1, we see the ratio of the exponents (slopes) should appear in the new matching plot:

$$\log E_1 = \frac{n_2}{n_1} \log E_2 + C \tag{2.9}$$

where C represents some combination of the not-so-interesting proportionality constants. In the handgrip study of J.C. Stevens et al. (1960) (no relation to S.S.) the handgrip exponent was found to be 1.7. This could then be used as a divisor to predict what the exponents should be for the other matching continua. When compared with the observed exponents, the average error was only about 4%. This level of agreement turned out to be so reliable that it was used as an undergraduate laboratory exercise in sensory psychology classes.[1]

2.5.3 Magnitude Matching

One important conclusion we can draw from the cross-modal scaling studies is that people seem to have a generalized notion of sensory intensity, one that can be used to compare sensations of very different qualities. Even if they do not have a verbal concept of universal sensation strength, they are able to match sensations. If so, we can also instruct them to use the same numbers for two continua. That is, if a taste is equally as strong as a sound, you should use the same number to represent that strength. This notion of putting two continua on the same scale led to the technique of magnitude matching (Stevens & Marks, 1980). We can cross-reference any sensory continuum to any other via numbers if people are told to put them on the same response scale.

[1] Unfortunately this was not the validation of the power function that Stevens hoped it would be. A log function will also produce this result. Suppose that we have two log functions $S_1 = k_1 \log E_1$ and $S_2 = k_2 \log E_2$; once again setting $S_1 = S_2$, we can also produce a ratio slope for the matching function, where $\log E_1 = (k_2/k_1)\log E_2 + C$. This result can also be obtained in other psychophysical models, such as the two-stage models discussed below.

A clever application of this technique appeared in an experiment comparing the bitter taste intensity of PROP (propylthiouracil) tasters and nontasters. This is the classic genetic marker for one type of bitter taste, with about 30% of the population falling into the less-sensitive nontasting group, generally believed to be homozygous recessive for this trait (see Bufe et al. (2005) for a discussion of the actual mutations and genotypes). In this study (Marks et al., 1988), subjects scaled the intensity of sounds, the taste of NaCl, and the taste of PROP. Having been instructed to use the same number scale for sounds and tastes, everyone's data were adjusted by an individual multiplicative constant to place the average response to the sounds at 10. After doing so, they observed the classical difference between tasters and nontasters in response to PROP, but no difference in response to NaCl. This proved to be a useful way to compare groups, as long as one can make the assumption that the reference continuum, in this case loudness of sounds, was experienced as the same strength for all people in the study. A similar method of using a reference continuum to compare different groups was employed by Berglund (1991) for environmental noise estimation.

2.5.4 Trouble and Worry. The Variable Exponent

A fundamental assumption in most ratio-scaling work was that the power function exponent was a fact of nature, as long as one conducted the experiment correctly. The exponent must represent some fundamental process in the sensory system, such as the way that receptors transduce energy into a neural signal. It was a constant that one could rely upon. The exponent, of course, could change as a function of the physical stimulus properties, such as the size of the area on the skin that was stimulated in a warmth experiment. Other stimuli and sensations present were a factor, as in the case of visual and auditory masking that changed in the exponent in predictable and systematic ways. But what if the exponent changed as a function of some arbitrary choice of the experimenter in the procedure? What if this choice or alteration did nothing to change the sensations, but the responses changed anyway?

A number of such seemingly insignificant methodological choices have been shown to affect the power function exponent (see Baird and Noma (1978: chapter 6) for a complete discussion). Thus, they call into question the notion that the exponent is a fixed constant. Furthermore, the exponent is not only a function of the physiological properties of the sensory systems, such as receptor sensitivity or density, but is likely to be affected by decision processes and other cognitive factors. Probably the most thoroughly documented is the shift with stimulus range. Using a wide range of stimuli yields a lower exponent than a short range of stimuli (Poulton, 1968). The position of the standard within the stimulus series also shifts the exponent (Engen & Levy, 1955). When the standard is chosen from the middle of the series, the exponent is higher (Engen & Ross, 1966). The number assigned to the standard can also change the exponent (Poulton, 1968). Whether the stimuli are judged as fractions or multiples of the standard also appears to have an effect (multiples give a higher exponent). All of these sorts of incidental or arbitrary decisions can change the scaling results, a finding not consistent with the idea that magnitude estimates and other ratio-type judgments are in fact a veridical translation of the actual sensory experience.

Number usage can also differ among people, and perception of numbers themselves suggests that they are not used in strictly mathematically correct ways. Poulton (1989) describes the relationship between numbers and their perceived magnitude as "nearly logarithmic." By this we mean that it takes greater increases in numbers themselves to suggest an equal increase in magnitude, as the magnitude or number itself rises. The difference (perceived change) between 1 and 2 seems larger than the difference between 101 and 102, although mathematically we know from childhood that they are equivalent. One way this manifests itself is that, as the stimuli in a scaling study increase in intensity, number usage begins to make larger jumps for most people. That is, we might start with changes from single digits (4, 5, 6, 7), but then when we get to 10 we start to make bigger jumps (10, 15, 20, 30). There is also a kind of perceptual discontinuity when we add another digit; that is, cross a numerical boundary with a step change, such as from 99 to 100. A third kind of number bias appears as a kind of favoritism toward multiples of 2, 5, and 10 (see Poulton (1989: 2, figure 1.1)). It is much more likely that a subject in a magnitude estimation study will choose numbers such as 12, 25, and 50, than 37.5 or 117 (Giovanni & Pangborn, 1983).

2.6 Two-Stage Models and Judgment Processes

2.6.1 The Personal Exponent

All of the effects mentioned in Section 2.5.4 should give a person reason to doubt that simple numerical judgments with ratio-scaling instructions are a simple veridical reflection of subjective experience. So what is going on? A classic paper in scaling found that individual subjects tended to have some consistency in how they deviated from the group averages, when they participated in more than one scaling study (Jones & Marcus, 1961). For example, if I happen to have a rather expansive usage of numbers when scaling loudness, I will probably show a steeper function for brightness as well. If my use of numbers is conservative and compressive, I will exhibit this behavior and have a shallower function in both continua. Jones and Marcus went on to model this as an additional exponent in the power function, and thus the exponent we measure is actually a combination of the psychophysical exponent and a personal numerification exponent peculiar to that subject as follows:

$$R = kE^{bc_i} \tag{2.10}$$

where R is the response, E is the physical stimulus energy, b is the psychophysical exponent, and c_i is the personal exponent for that subject that governs their number usage. The actual measured exponent, which we have called n, is a function of the product of the two exponents.

An interesting solution to this problem was suggested by Gesheider and colleagues (Collins & Gescheider, 1989). Based on the observation that the perceived length of lines is proportional to line length (Zwislocki, 1983) and thus having and exponent of 1.0, it should be possible to estimate a person's idiosyncratic number usage and correct for it. In Figure 2.2 we see the data from a loudness study where corrections were made for the individual number usage based on a separate scaling of line length. Clearly, the corrected functions are much less variable than the uncorrected or raw magnitude estimates.

But what about cross-modality matching? Is not that still evidence of the validity of magnitude estimation? Well, given the combined exponents such as the multiplicative personal exponent of Jones and Marcus, one can still obtain the correspondence between scaling of individual continua, and the predicted ratio exponent of the cross-modality match (see eqn 2.9). Because the number or response function exponent is in both the numerator and denominator of the ratio, it cancels, thus producing the required prediction.[2]

[2] If we take our two power functions $\log S_1 = n_1 \log E_1 + \log k_1$ and $\log S_2 = n_2 \log E_2 + \log k_2$ and now substitute $b_1 c$ for n_1 and $b_2 c$ for n_2, we have the new equations $\log S_1 = b_1 c \log E_1 + \log k_1$ and $\log S_2 = b_2 c \log E_2 + \log k_2$, and so our cross-modality match is predicted to be $\log E_1 = (b_2 c / b_1 c) \log E_2 + C$ with a slope equal to $b_2 / b_1 \times 1$ (c/c). The important conclusion is that a two-stage model with a number output exponent will also accommodate the cross-modality match prediction.

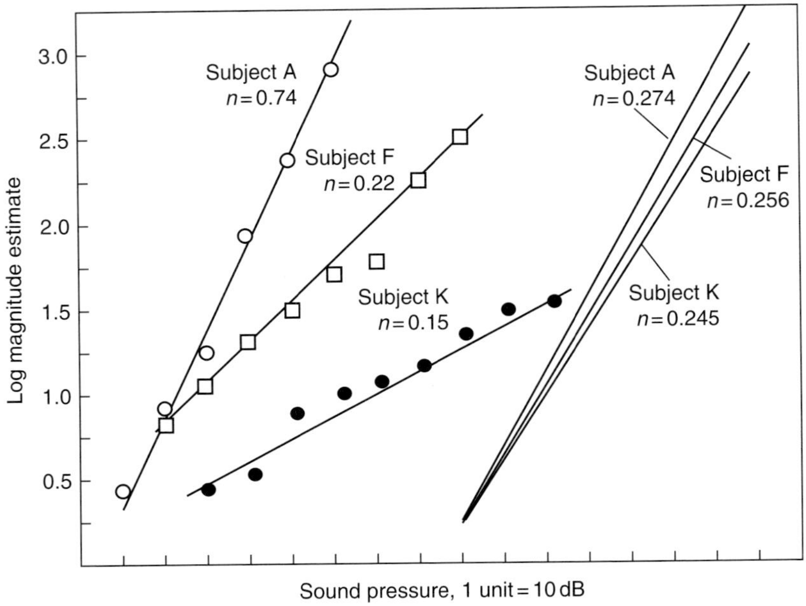

Figure 2.2 Magnitude estimates of loudness from three individual subjects; data from Collins and Gescheider (1989). The three functions on the left show the subjects with lowest, highest, and one intermediate exponent for loudness determined by magnitude estimation. The three functions on the right show the slopes after correction for their number usage habits via a line-length scaling task.

The theory was advanced one step farther by the formal concept of a response output function to describe the process of number assignment by a subject. This is sometimes called a judgment function or response production. A number is applied to the sensation. Curtis et al. (1968) formalized this idea and attempted to estimate the mathematical form of the response output (this had not yet been measured by Jones and Marcus, only suggested). They called it a two-stage model of magnitude judgment, to allow for a psychophysical sensation function and a response output function. The experimental procedure that allowed this was to have people make ratio judgments of the differences between pairs of stimuli, in this case lifted weights. Now we have two relationships: a sensation–stimulus relationship given by

$$S = k_1 E^m \tag{2.11}$$

and a **response output function**, for which they chose another power relationship:

$$R = k_2 S^p \tag{2.12}$$

with the exponent p for the number production function. Thus, the observed exponent n would once again be a product of the two contributing exponents. In the judgment of the magnitude of differences, we have

$$R_{ij} = k_3 \left(S_j^m - S_i^m\right)^p \tag{2.13}$$

where S_j and S_i are the two sensations from the lifted weights.

Now the value of the differencing task is that you can get a least-squares estimate of the values of m and p. You can also use them to compare with a separate estimate of the power function exponent n as a test of validity and fit. With lifted weights, the differencing exponent was estimated at 0.736 (the product of the estimate of 0.645 for m and 1.141 for p) and the simple exponent from the magnitude estimates of the same lifted weights was 0.746. Not bad. Another idea was that we can use the number output function estimate to "correct" the magnitude estimates to see the true sensation function by raising them to the power of $1/p$. The only problem with this study was that the number exponent was greater than one, which would not be consistent with a logarithmic number bias that should be a compressive function! Attneave (1962: 626) himself had earlier argued that the number exponent was surprisingly low (estimated at 0.4), so there was a discrepancy. Most likely the reason had to do with the range of the comparisons that were made, and so we suspect that the stimulus range effect was once again in play.

2.6.2 Anderson's Approach: Functional Measurement

Another criticism of the simple stimulus–response model for scaling came from a theoretical foundation called **functional measurement**, whose major proponent was Norman Anderson (1970, 1974, 1977). The functional measurement experiment was essentially a combination task. One would be presented with two weights, for example, and asked to give a rating for their average weight. The combination judgment could also be one of difference or summing. The experimental design is factorial; that is, every level of one member of the pair is at some time paired with every other. The overall judgment (such as the average or sum) is plotted as a function of one member of the pair, with a family of lines connecting constant values of the other member. The typical result is shown in Figure 2.3, with a set of lines that might or might not be parallel.

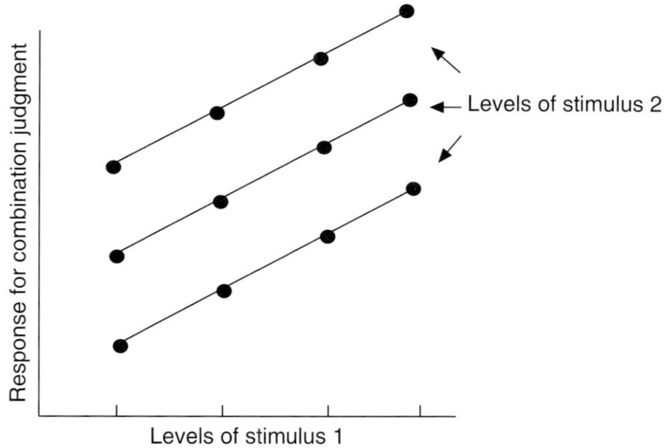

Figure 2.3 Data from a hypothetical factorial experiment showing the pattern of parallel functions. This pattern indicates an additive integration rule and a linear response output function for whatever scaling method was employed.

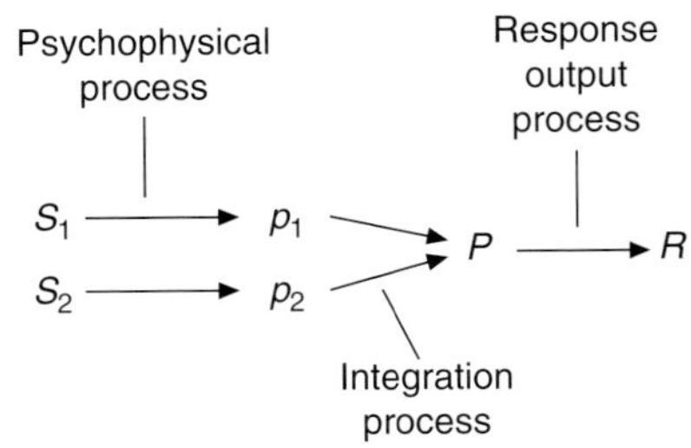

Figure 2.4 The stimulus–response integration scheme of functional measurement (Anderson, 1974).

The theoretical basis for Anderson's next logical step is shown in Figure 2.4. As in the two-stage model, there is a psychophysical process which translates the physical stimulus energy into a perceived sensation and a judgment process by which the integrated impression is translated into an overt judgment (i.e., a response, a data point). A third process is, of course, the integration or combination process. Now here is the critical reasoning: Anderson postulates that if the integration process is additive (rather than some other rule such as multiplicative) *and* the response output function is a linear translation of the internal experience, then and only then would one see a set of parallel lines in the data plot. If either the combination rule or the judgment function were nonlinear, one would not obtain the set of parallel functions. One simple mathematical test, of course, is the lack of an interaction term in a two-way analysis of variance. The telling result was that, in most of his work, the parallel functions were achieved with category scales or line scales, whereas magnitude estimates tended to give fan-shaped functions indicative of a nonlinear response output function.

So Anderson routinely found that category (and line) scales gave data consistent with his requirements of an additive integration mechanism and a linear response output function. How this was achieved, however, entails a critical set of experimental details. Often, he would use a 20-step rating scale. Furthermore, he would show the subject the lowest stimulus in the series and ask them to call that "4" on the scale. Similarly, the highest example (shown to the subject but unknown to the subject that it was the highest) would be assigned the value "16" or the approximate equivalent on a line scale. This procedure has several important effects. Obviously, it gives subjects room to move should they feel the need to use higher or lower numbers; that is, counteracts the end-of-scale avoidance and ceiling and floor effects that are common with category ratings (Anderson, 1974). Such practice stimuli also instruct the subject as to the frame of reference for their subsequent judgments (Anderson, 1974: 231–32). The use of end-anchor examples and further practice with the stimulus set has a strong stabilizing effect and promotes the equal-interval use of the scale.

Two opposing camps evolved in this discourse: one favoring the ratio-instructed scaling methods and the other favoring categories or lines (see Giovanni and Pangborn (1983) for an interesting blurring of the two methods – line scales are referred to as "categories" in that paper). The line and category scales already had a long history in applied sensory evaluation (Baten, 1946; Caul, 1957). The problem is that when a given sensory continuum was scaled using both methods, they were curvilinearly related (Stevens & Galanter, 1957). So according to the logic of Anderson, one or the other could be a valid linear translation of sensation intensity, but not both. This is about what one would expect if categories usually conformed to a log function and magnitude estimates to a power function. The category = log function was roughly true as a generalization, but not always, and the point was argued further (see Stevens (1986: 143)). Category scales and scales derived from adding JNDs sometimes matched (McBride, 1983), but once again not always, with JND scales somewhat more concave than category data in a log–log plot (see Stevens (1986: 229)).

2.7 Empirical Versus Theory-Based Functions

The fitting of log or power functions to a set of data is really nothing more than an exercise in curve fitting. The functions do not have the status of "laws" except for the fact that they may be generally a good fit across many sensory continua. But they are not derived from any underlying theory of a physical, neural, physiological, or chemical nature, although Stevens (1986: chapter 7) did try to connect the power function to some neural and receptor phenomena. For example, the concentration functions from recordings from human taste nerves during a middle-ear operation showed a striking correspondence to magnitude estimation functions collected earlier from the same individuals (Borg et al., 1967). Once again, simple curve fitting.

However, some taste physiology moved in a more mechanistic direction. Notably, the work of Lloyd Beidler used chemical and kinetic theory to derive a general equation for taste responses. The functions were initially applied to taste nerve recordings to describe the relationship between concentration and response (Beidler, 1961). However, they are a potentially useful description of taste psychophysical responses as well. This is the simple semi-hyperbolic relationship also used to describe the velocity of enzyme–substrate reactions in Michaelis–Menten enzyme kinetics (Lehninger, 1975). This makes sense if one thinks about the reaction of tastant molecules to a taste receptor protein as an event much like the binding of a substrate to an enzyme (usually a protein). Beidler's published function was a transformation of the Michaelis–Menten function, but we will use the latter here because its interpretation is more direct. In its simplest form, the equation relates response R to concentration C using two parameters: R_{max}, the maximal response at the level of receptor saturation, and an association constant (or dissociation constant if one uses the reciprocal) K, indicative of the strength of the binding of the tastant to the receptor. K is also the concentration at which the function reaches half-maximal, a useful index of biological potency. So we have

$$R = \frac{R_{max} C}{K + C} \tag{2.14}$$

There are various ways to linearize this relationship, notably the double-reciprocal or Lineweaver–Burke plot. Figure 2.5 shows the so-called Beidler function. Note that when it is plotted as a function of log concentration, as we often do, we get the familiar S or

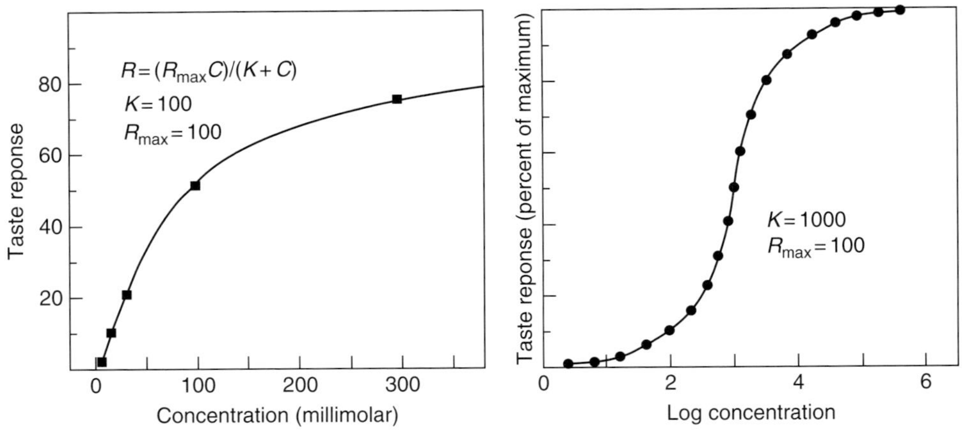

Figure 2.5 The so-called Beidler function, semi-hyperbolic relationship between taste response R and concentration C, shown in arithmetic and semi-log coordinates. The function shown is hypothetical in terms of the chosen parameters (chosen for simplicity).

ogive shape typical of so many psychometric functions, a curious coincidence. The S-shaped function also makes sense, as there is a zone of threshold and a zone of saturation. Such functions are often seen in studies of taste response and neural recordings (e.g., Bufe et al., 2005). Ennis (1991) provided an extension of the Beidler functions for taste mixtures, and further elaborated the model to include a ligand, a receptor, and a transducer.

2.8 Hybrid Scales and Indirect Scales: A Look Ahead

The dispute about scale validity might lead one to think that one is faced with an either/or choice when looking for a scaling method to use. However, various attempts have been made to construct hybrid scales, with the same properties as magnitude estimation (i.e., alleged ratio properties) but with convenient category labels such as weak, moderate, and strong. Based on earlier work by Borg (1982, 1990), Green and co-workers made the clever jump of having people scale the category words using magnitude estimation (see Gracely et al. (1978a, b) for a similar approach), and then spaced these words along a line scale to construct a **labeled magnitude scale**. Since the spacing of the words reflected ratio judgments, one would expect data from such a scale to correspond to traditional magnitude estimates, and they do (Green et al., 1993). These scales will be discussed further in Chapter 7. Various attempts have been made to extend the approach to hedonic scaling (likes and dislikes) as well (Schutz & Cardello, 2001; Cardello & Schutz, 2004). The labeled magnitude scale is shown in Figure 2.6. It has been proposed that this scale facilitates between-person and between-group comparisons due to people's similar perception of the verbal high end anchor, "greatest imaginable," especially when cross-referenced to *sensations of any kind*; that is, a very general frame of reference (Bartoshuk et al., 2004).

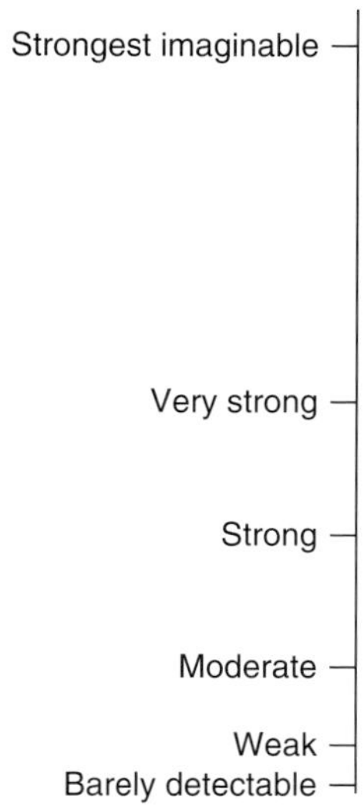

Figure 2.6 The labeled magnitude scale from Green et al. (1993).

2.9 Summary and Conclusions

Throughout this literature, there is an underlying doctrine that posits the following: "Sure the scaling results may change with conditions, but if you just choose the correct conditions and do the right experiment, you will get the true scale values." In the absence of any compelling evidence for independent validation, such a position seems largely a matter of opinion. Just because you have given ratio instructions to a subject does not mean they are generating data that have ratio properties; that is, the numbers do not necessarily reflect ratios of the actual experienced sensations.

In summary, we are left with the following facts and conclusions:

1. Magnitude estimation results change with arbitrary changes in conditions such as the range of stimuli or the specific numerical examples used for instructions or for the modulus.
2. Magnitude scales and category scales (as well as line scale data) are curvilinearly related to one another. As the number of categories on the category scale gets larger, the discrepancy becomes smaller.
3. People use numbers in odd ways that probably do not reflect a linear use, but rather something closer to a logarithmic translation of actual numerical magnitude.
4. Use of numbers in a scaling task is also idiosyncratic, as shown by the correlation within individuals in different studies and their "personal exponent."
5. A two-stage model with a response output process seems to be a more reasonable model for cognitive processing during a scaling task than taking ratings at face value.

Many sensory scientists would still like to find a scaling method that would allow ratio statements and conclusions such as "this product is twice as sweet as that product." I am not sure food and consumer products manufacturers are in dire need of such an ability. It seems they have functioned perfectly fine without ratio statements. Rather than keep searching for the "perfect scale," it is perhaps better simply to study and understand scaling behavior and all its biases and foibles (Poulton, 1989). This puts scaling in a behavioral context and views the human measuring instrument as one that operates under certain rules, and that it has operating characteristics that depend upon the conditions of measurement, the stimulus context, the scaling method, training and calibration, and so on.

One should also question the either/or dichotomy of the validity of magnitude versus category scales. It is equally reasonable to think that people work in different modes with different scaling tasks. Judging ratios is a different thought process than judging intervals or differences. So there may be two or more modes of scaling. Other differences seem obvious, but have received scant attention. To my knowledge, few if any researchers have looked at the simple fact that category and line scales are necessarily bounded (although see Lawless (1977) for an interesting methodological twist using a theoretically unbounded line scale). Magnitude estimates are not. Why should it be a surprise that a bounded scale gives a different psychophysical function than a method with no upper limit to the possible responses?

What about leaving direct scaling altogether and just constructing a JND-type of scale from discrimination studies? One could certainly do this, although it would be laborious to say the least. For example, Yamaguchi (1967) was able to document the synergistic taste summation of monosodium glutamate and inosinate by a series of overlapping paired comparisons analyzed by Thurstonian methods (Chapter 4). But the student should keep in mind that discriminability is in part a function of the degree of experimental control in the study (Cain, 1977). You are measuring all sources of error and noise in the experiment (not just those intrinsic to the human observer) and basing your scale values on error and variability. As the experimental conditions get "cleaner" one would find greater sense distance between any two items. So the question to the end user is whether the validity potentially gained by doing an indirect scaling task is worth the effort, when all one is really measuring is variability anyway.

Finally, we would probably be well advised to decouple the questions of scaling method validity and the form of the psychophysical function. These are two separate issues and need not be logically linked (Poulton, 1989).

Appendix 2.A Decibels and Sones

Both sound and light energy are sometimes specified in decibels, where 1 dB is one-tenth of a log unit, sometimes called a decilog. So a 10 dB change in energy or sound pressure is one log unit. All such logarithmic scales must be specified relative to the energy or pressure level that forms the bottom of the range or 0 dB. In the case of sound, this is usually a pressure level of 20 $\mu N/m^2$ (micro-newtons per square meter) or 20 μPa. This is roughly the human threshold for a sine wave stimulus at 1000 Hz.

The equivalent sound energy level is 10^{-6} W/m^2. Pressure is proportional to the square root of the energy intensity, so we have the following relationship:

$$I\ (\mathrm{dB}) = 10 \log_{10}\left(\frac{I}{I_0}\right) \text{ where } I \text{ and } I_0 \text{ on the right-hand side are in W/m}^2 \quad (2.A.1)$$

A common measure is the sound pressure level (SPL; expressed in decibels) as

$$\text{SPL (dB)} = 10\log_{10}\left(\frac{p^2}{p_0^2}\right) = 20\log_{10}\left(\frac{p}{p_0}\right) \tag{2.A.2}$$

Figure 2.A.1 shows the decibel value of some common sounds.

Stevens attempted to construct a convenient sound intensity unit based on the idea that loudness L formed a power function of sound pressure with an exponent of 0.3. This was called a *sone*, and had a reference value of 40 dB SPL. Since the antilog of 0.3 is about 2 ($\log_{10}2 = 0.301$), the value in sones doubles for every 10 dB change:

$$L\ (\text{sone}) = \left[10^{(\text{dB}-40)/10}\right]^{0.3} = 2^{(\text{dB}-40)/10} \quad \text{for} > 40\ \text{dB SPL}$$
$$50\,\text{dB} = 2\,\text{sones},\ 60\,\text{dB} = 4\,\text{sones},\ 70\,\text{dB} = 8\,\text{sones},\ \ldots \tag{2.A.3}$$

Such units of equal sensory change were popular for a time, including a bril scale for brightness and mel scale for pitch. Bear in mind that the sone scale is also referenced to a 1000 Hz tone, and thus its utility is limited when one starts to use other pitches, other real sounds or bands of noise, and so on. Also, sones are not SI units.

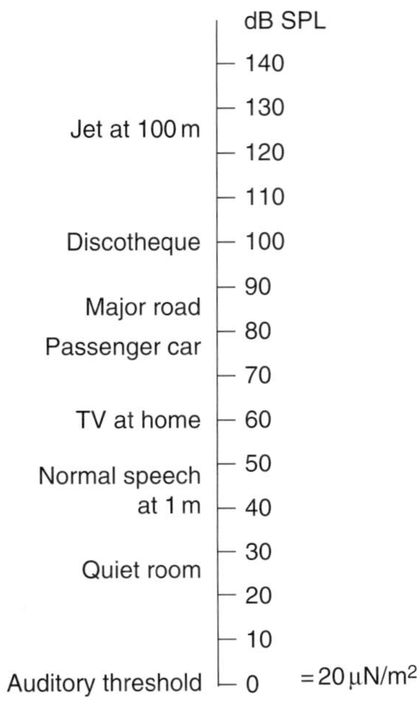

Figure 2.A.1 Decibel values for some common sounds.

Appendix 2.B Worked Example: Transformations Applied to Non-Modulus Magnitude Estimation Data

Table 2.B.1 shows the data from three hypothetical subjects, similar to the three shown in Figure 2.2. Note that subject 1 has an expansive range of number choices and subject 3 spans

Table 2.B.1 Data from three hypothetical subjects

	Subject 1		Subject 2		Subject 3		
	Data	**log data**	**Data**	**log data**	**Data**	**log data**	**Grand mean**
Stimulus							
10	3	0.477	5	0.699	3	0.477	
20	8	0.903	12	1.079	8	0.903	
30	20	1.301	20	1.301	10	1.000	
40	50	1.699	30	1.477	12	1.079	
50	200	2.301	40	1.602	15	1.176	
60	600	2.778	60	1.778	20	1.301	
70	1000	3.000	150	2.176	30	1.477	
80	2000	3.301	300	2.477	40	1.602	
Mean		1.970		1.574		1.127	1.557
Antilog		93.32		37.49		13.39	36.06
Multiplicative factor		0.3864		0.9619		2.6931	
Additive factor		−0.4131		−0.0168		0.4299	
Transformed	1.16	0.064	4.81	0.682	8.070	0.907	
	3.09	0.490	11.54	1.062	21.520	1.333	
	7.73	0.888	19.24	1.284	26.900	1.430	
	19.32	1.286	28.86	1.460	32.280	1.509	
	77.26	1.888	38.48	1.585	40.350	1.606	
	231.78	2.365	57.71	1.761	53.800	1.731	
	386.30	2.587	144.29	2.159	80.700	1.907	
	772.60	2.888	288.57	2.460	107.600	2.032	
Midpoint (new)	48.5	38.6	33.5	33.3	36.3	35.9	

Notes:
1.The actual number used for the multiplicative factor is arbitrary. Using a value derived from the geometric mean is representative of the original data, but any constant value would do.
2.The procedure sets all the subjects' data to approximately the same midpoint. This is shown in the final row. It does not fully adjust for different ranges of numbers chosen by subjects. The adjustment creates an intersection of all subjects at the grand geometric mean but still allows their slopes to vary. This is the equivalent of taking out the main effect of subjects in analysis of variance, but allowing the subject-by-stimulus interaction to remain. If an analysis of variance is performed, you should probably reduce your error degrees of freedom by the number if subjects, as that variance has now been eliminated.
3.A data value of zero is problematic for this method because the log of zero is undefined and the geometric mean with a zero in the data set is meaningless. Various solutions have been proposed, including choosing an arbitrary small positive value to replace the zero such as one-half the next smallest positive value in the data. Another solution, as long as you do not have to work in logs, is to use the median for each individual, rather than the geometric mean in the calculation for the transformation constant. You can still derive a multiplicative factor for each person based on the average (or geometric mean) of the medians. This has the benefit of preserving the zero in the data as, after all, it may be a real value to that person – they may have perceived no sensation at all from that stimulus.

a much smaller range. The method is based on the procedure of Lane et al. (1961). The steps are as follows:

1. Convert the data to logarithms.
2. Take the mean of the logarithms and then the antilog (10^x) of that to obtain the geometric mean for each subject.
3. Find the mean of the geometric means ("grand mean" above).
4. Take the antilog of the grand mean (in this case 30.06).
5. Derive an individual multiplicative factor for each subject as follows: divide the grand mean by the individual mean. For subject 1, this is 30.06/93.32 = 0.3864. The expansive range chosen by this subject will be reduced by this factor.
6. As an alternative you can use an additive factor to work with the log data. For subject 1, this is –0.4131. This is convenient if you wish to do your statistical analysis on the log data.

References

Anderson, N.H. 1970. Functional measurement and psychophysical judgment. Psychological Review, 77, 153–70.

Anderson, N.H. 1974. Algebraic models in perception. In: Handbook of Perception. II. Psychophysical Judgment and Measurement. E.C. Carterette and M.P. Friedman (Eds). Academic Press, New York, NY.

Anderson, N.H. 1977. Note on functional measurement and data analysis. Perception & Psychophysics, 21, 201–15.

Attneave, F. 1962. Perception and related areas. In Psychology: A Study of a Science. Volume 4. Biologically Oriented Fields: Their Place in Psychology and in Biological Science. S. Koch (Ed.). McGraw-Hill, New York, NY.

Baird, J.C. and Noma. E. 1978. Fundamentals of Scaling and Psychophysics. John Wiley & Sons, Inc., New York, NY

Bartoshuk, L.M., Duffy, V.B., Chapo, A.K., Fast, K., Yiee, J.H., Hoffman, H.J., Ko, C.-W., and Snyder, D.J. 2004. From psychophysics to the clinic: missteps and advances. Food Quality and Preference, 15, 617–32.

Baten, W.D. 1946. Organoleptic tests pertaining to apples and pears. Food Research, 11, 84–94.

Beidler, L.M. 1961. Biophysical approaches to taste. American Scientist, 49, 421–31.

Berglund, M.B. 1991. Quality assurance in environmental psychophysics. In Ratio Scaling of Psychological Magnitude: In Honor of the Memory of S. S. Stevens. S.J. Bolanowski and G.A. Gescheider (Eds). Lawrence Erlbaum, Hillsdale, NJ, pp. 140–62.

Bernoulli, D. 1738/1954. Exposition of a new theory on the measurement of risk. Econometrica, 22, 23–35 (translation, originally published in Latin, 1738).

Borg, G. 1982. A category scale with ratio properties for intermodal and interindividual comparisons. In: Psychophysical Judgment and the Process of Perception. H.-G. Geissler and P. Petzold (Eds). VEB Deutscher Verlag der Wissenschaften, Berlin, pp. 25–34.

Borg, G. 1990. Psychophysical scaling with applications in physical work and the perception of exertion. Scandinavian Journal of Work and Environmental Health, 16, 55–8.

Borg, G., Diamant, H., Ström, L., and Zotterman, Y. 1967. The relation between neural and perceptual intensity: a comparative study on the neural and psychophysical response to taste stimuli. Journal of Physiology (London), 192, 13–20.

Bufe, B., Breslin, P.A.S., Kuhn, C., Reed, D.R., Tharp, C.D., Slack, J.P., Kim, U.-K., Drayna, D., and Meyerhof, W. 2005. The molecular basis of individual differences in phenylthiocarbamide and propylthiouracil bitterness perception. Current Biology, 15, 322–7.

Cain, W.S. 1977. Differential sensitivity for smell: noise at the nose. Science, 195, 795–8.

Cardello, A.V. and Schutz, H.G. 2004. Research Note. Numerical scale-point locations for constructing the LAM (labeled affective magnitude) scale. Journal of Sensory Studies, 19, 341–6.

Caul, J.F. 1957. The profile method of flavor analysis. Advances in Food Research, 7, 1–40.

Collins, A.A. and Gescheider, G.A. 1989. The measurement of loudness in children and adults by absolute magnitude estimation and cross-modality matching. Journal of the Acoustical Society of America, 85, 2012–21.

Curtis, D.W., Attneave, F., and Harrington, T.L. 1968. A test of a two-stage model of magnitude judgment. Perception & Psychophysics, 3, 25–31.
Engen, T. and Levy, N. 1955. The influence of standards on psychophysical judgments. Perceptual and Motor Skills, 5, 193–7.
Engen, T. and Ross, B.M. 1966. Effect of reference number on magnitude estimation. Perception & Psychophysics, 1, 74–6.
Ennis, D.M. 1991. Molecular mixture models based on competitive and non-compeitive agonism. Chemical Senses, 16, 1–17.
Gescheider, G.A. 1997. Psychophysics. The Fundamentals. Third edition. Lawrence Erlbaum, Mahwah, NJ.
Giovanni, M.E. and Pangborn, R.M. 1983. Measurement of taste intensity and degree of liking of beverages by graphic scaling and magnitude estimation. Journal of Food Science, 48, 1175–82.
Gracely, R.H., McGrath, P., and Dubner, R. 1978a. Ratio scales of sensory and affective verbal-pain descriptors. Pain, 5, 5–18.
Gracely, R.H., McGrath, P., and Dubner, R. 1978b. Validity and sensitivity of ratio scales of sensory and affective verbal-pain descriptors: manipulation of affect by Diazepam. Pain, 5, 19–29.
Green, B.G., Shaffer, G.S., and Gilmore, M.M. 1993. Derivation and evaluation of a semantic scale of oral sensation magnitude with apparent ratio properties. Chemical Senses, 18, 683–702.
Jones, F.N. and Marcus, M.J. 1961. The subject effect in judgments of subjective magnitude. Journal of Experimental Psychology, 61, 40–4.
Lane, H.L., Catania, A.C., and Stevens, S.S. 1961. Voice level: autophonic scale, perceived loudness and the effects of sidetone. Journal of the Acoustical Society of America, 33, 160–7.
Lawless, H.T. 1977. The pleasantness of mixtures in taste and olfaction. Sensory Processes, 1, 227–37.
Lehninger, A.L. 1975. Biochemistry. Second edition. Worth Publishers, New York, NY.
Marks, L.E. , Stevens, J.C., Bartoshuk, L.M., Gent, J.F., Rifkin, B., and Stone, V.K. 1988. Magnitude matching: the measurement of taste and smell. Chemical Senses, 13, 63–87.
McBride, R.L. 1983. A JND-scale/category scale convergence in taste. Perception and Psychophysics, 34, 77–83.
Moskowitz, H.R. 1971. The sweetness and pleasantness of sugars. American Journal of Psychology, 84, 387–405.
Pearce, J.H., Korth, B., and Warren, C.B. 1986. Evaluation of three scaling methods for hedonics. Journal of Sensory Studies, 1, 27–46.
Peryam, D.R. and Girardot, N.F. 1952. Advanced taste-test method. Food Engineering, 24, 58–61, 194.
Poulton, E.C. 1968. The new psychophysics: six models for magnitude estimation. Psychological Bulletin, 69, 1–19.
Poulton, E.C. 1989. Bias in Quantifying Judgments. Lawrence Erlbaum Associates, Hove, Sussex, UK.
Richardson, L.F. and Ross, J.S. 1930. Loudness and telephone current. Journal of General Psychology, 3, 288–306.
Schutz, H.G. and Cardello, A.V. 2001. A labeled affective magnitude (LAM) scale for assessing food liking/disliking. Journal of Sensory Studies, 16, 117–59.
Stevens, J.C. and Marks. L.E. 1980. Cross-modality matching functions generated by magnitude estimation. Perception & Psychophysics, 27, 379–89.
Stevens, J.C., Mack, J.D., and Stevens, S.S. 1960. Growth of sensation on seven continua measured by force of handgrip. Journal of Experimental Psychology, 59, 60–7.
Stevens, S.S. 1936. A scale for the measurement of psychological magnitude: loudness. Psychological Review, 43, 405–16.
Stevens, S.S. 1957. On the psychophysical law. Psychological Review, 64, 153–81.
Stevens, S.S. 1986. Psychophysics. Introduction to its Perceptual, Neural and Social Prospects. Transaction Books, Oxford, UK.
Stevens, S.S. and Galanter, E.H. 1957. Ratio scales and category scales for a dozen perceptual continua. Journal of Experimental Psychology, 54, 377–411.
Teghtsoonian, R. 1971. On the exponent in Steven's law and the constant in Ekman's law. Psychological Review, 78, 71–80.
Teghtsoonian, R. and Teghtsoonian, M. 1978. Range and regression effects in magnitude scaling. Perception & Psychophysics, 24, 305–14.
Yamaguchi, S. 1967. The synergistic taste effect of monosodium glutamate and disodium 5′-inosinate. Journal of Food Science, 32, 473–5.
Zwislocki, J.J. 1983. Group and individual relations between sensation magnitudes and their numerical estimates. Perception and Psychophysics, 33, 460–8.

3 Basics of Signal Detection Theory

The demonstrated constancy of the sensitivity indices d' *and D(Δm,s) obtained by any of the several procedures we have discussed, represents a substantial advance in psychophysical technique. Prior to this development, it had become commonplace that the prominent psychophysical procedures yield different values for the sensory threshold, and there was a tendency to regard the results of a psychophysical experiment as meaningless except in relation to the specific procedure used. It is now the case that half a dozen laboratories, which have been using the procedures and the measures associated with the theory of signal detectability in auditory experiments, generally obtain results that are within one decibel of one another. … Such consistency of results obtained by different techniques is not easy to obtain in the measurement of complex physical phenomena, and it has been very rare, perhaps nonexistent, in the measurement of human behavior.*

Green and Swets (1966/1988: 113–15)

Quantitative Sensory Analysis: Psychophysics, Models and Intelligent Design, First Edition. Harry T. Lawless.

3.1 Introduction

Signal detection theory (SDT) represented a great leap forward in our thinking about sensory phenomena and how to measure them. Its primary contribution was a model and a methodology to separate a person's true sensitivity to a stimulus from their response bias or judgmental tendencies. The theory was actually anticipated by the formulations of Thurstone in the 1920s in converting choice data, usually paired judgments, into estimates of sensory difference or distance. However, the considerations of engineering and decision theory in the 1950s and 1960s were more concerned with responses to a single event (i.e., one stimulus at a time), so the experimental basis for the theory was somewhat different. Only later did the SDT founders recognize the connection between their models and those of Thurstone (Baird & Noma, 1978).

As discussed in Chapter 2 on scaling, we also have a two-stage model for psychophysical measurement. The observed behavior is considered a result of both a sensory experience and a judgment process. Thus, one half of the scheme is a cognitive process. The goal is to obtain a measure of a person's discrimination or resolving power or detectability, uncontaminated by that subject's response bias (Baird & Noma, 1978). In the most common experimental design, a subject is presented with one of two possible stimuli over many trials, and they must give one of two responses. In sensory evaluation, this would be an equivalent of the "A–not-A" method (Lawless & Heymann, 2010), but with many, many trials for the same person. Because the methods usually dealt with the detection of very weak stimuli, they were somewhat like threshold tests. However, there was a big conceptual difference. Thresholds are seen by most people as some kind of boundary to cross, going from nondetection to detection. In SDT, this idea is replaced by the notion that the signal emerges from the background in a continuous manner. Thus, there is no need for a threshold concept at all!

The connection to Thurstonian theory will be made in Chapter 4, although it historically precedes SDT. In that chapter we will connect the Thurstonian theory to the SDT measures of discriminability, and show how they are related to sensory evaluation tests of discrimination. This provides a common yardstick for sensory differences, one that adjusts for the inherent differences in the difficulty or variability of sensory tests. Importantly, it provides a framework for looking at the relative power of different test methods. Power is the ability to detect differences and not miss them when they are in fact present. It behooves a sensory practitioner to choose the most powerful test that can be used in any given situation, in order to provide the highest test sensitivity, avoid missing a difference, and/or design a test that allows for the use of fewer panelists for the same output.

Several important texts have been written on signal detection, and the reader is directed to them for further information. The classical text is *Signal Detection Theory and Psychophysics* by Green and Swets (1966/1988). A good general text is by Macmillan and Creelman (1991). Shorter, but very well -written summaries can be found in Baird and Noma (1978) and Gescheider (1997).

There are three central ideas to SDT. The first is that any signal is embedded in a background of **noise**. This noise is variable, and sometimes is strong enough to be mistaken for a signal. We will model the noise as a normally distributed event; that is, a bell-shaped frequency distribution with greater or lesser strength on each trial. The signal also is variable, and includes the noise or background stimulation. So the model represents the signal stimulus effects as a signal-plus-noise normal distribution. The observers' responses are binary. On any given trial, they have to decide whether the sensations came from a presentation of the signal or were just from the noise. The second important idea is that a subject in such a situation will try to maximize the amount of times they choose correctly, all other things being equal. But sometimes the situation is not so equal, and there can be different payoffs and penalties for correct answers versus mistakes. For example, there might be a slight penalty for a **false alarm** (calling a noise trial a signal) but a big penalty for missing the signal (calling a signal trial only noise). An example might be a radar observer in a military field maneuver. Missing the signal for an enemy aircraft could have devastating consequences.

Consider the following practical example. Suppose you have just gone for a long jog and you are back home enjoying a hot shower. You think you hear the phone ring, but you are not sure because the noise from the water sometimes sounds like the phone might be ringing. By the way, your answering machine is broken. In one case you are expecting a call from a potential employer about a second interview for a really good job. In another case, you are expecting a call from a cousin, who gets a stipend once a month but always seems to run out of money near the end of the month and routinely asks to borrow funds from you. What are the relative payoffs and penalties for responding or not responding given those two different expectations? What would make you jump out of the shower and grab the phone?

3.2 The Yes/No Experiment

3.2.1 Experimental Design; Responses and Payoff Matrices

In the simplest version of a signal detection experiment, we have two versions of the stimulus. For a very weak stimulus, as might be in a threshold experiment, we can consider the weaker of the two as a background or blank stimulus (e.g., pure air in an odor study or deionized water in a taste study). Presentations of this background give rise to a distribution of sensations called the noise distribution. Since the target we are trying to discern is embedded in that background stimulus, its distribution of sensations is the signal strength plus the strength of the noise at that moment. Your job is to decide whether on any given trial the sensation you are getting arises from the signal presentation or the noise. All you know on any trial is how strong the sensation seemed at that moment. You do not know from which distribution we have sampled. Given two stimuli and two responses, There are two ways to be correct and two ways to be wrong, as shown in Table 3.1.

This design is sometimes called a yes/no study and is analogous to the "A–not-A" test in sensory evaluation in which one of two products is given singly and there are two possible responses. In a typical signal detection study, there would be many presentations of signal and noise, sometimes several hundred (obviously, this is not a practical method with food products). Because we have set up the experiment ourselves, we know the number of signal trials and noise trials. Usually, they are equally frequent. The behavior of the subject is completely specified in either column – you do not really need both. In signal detection studies we look at the first column, which shows proportion of hits (i.e., saying signal when in fact one was presented) and the proportion of false alarms (i.e., saying noise when the signal was presented). Because we have specified the totals, the other two boxes are then defined; that is, there is only one degree of freedom in any row.

Table 3.1 Response matrix for a signal detection study

Stimulus	Response: "signal" (or "yes, I sense something")	Reponse: "noise" (or "no, I do not sense anything")
Signal presented	Hit	Miss
Noise presented	False alarm	Correct rejection

Table 3.2 Response matrix inducing a lax criterion

Stimulus	Response: "signal" (or "yes, I sense something")	Reponse: "noise" (or "no, I do not sense anything")
Signal presented	Hit +$5	Miss −$5
Noise presented	False alarm −$1	Correct rejection +$1

Table 3.3 Response matrix inducing a conservative criterion

Stimulus	Response: "signal" (or "yes, I sense something")	Reponse: "noise" (or "no, I do not sense anything")
Signal presented	Hit +$1	Miss −$1
Noise presented	False alarm −$5	Correct rejection +$5

Most of the time people want to maximize their correct responses. However, we could easily manipulate the rewards or punishments for correct and incorrect answers, respectively. Table 3.2 and Table 3.3 show two different payoff matrices.

In Table 3.2, there is a high payoff for saying "yes" and being correct and a low penalty for saying "yes" and being wrong. Conversely, you would win relatively little and potentially be penalized heavily for saying "no." This kind of payoff matrix tends to make people say "yes" a lot. That is, they need relatively weak evidence for their being a signal presentation to call it such. In Table 3.3 we see the opposite situation. There is a small penalty for missing a signal but a big penalty for a false alarm. Thus, it behooves the wise gambler to choose to be absolutely certain that a signal was presented before responding positively. In other words, the **criterion** for saying "yes" becomes quite conservative.

3.2.2 The Dice Game Metaphor

Swets et al. (1961) offered the following example to illustrate a signal detection study and human response tendencies. Suppose we were to play a dice game with three dice. Two were normal dice, with from one to six spots. The third die was unusual: it had no spots on three sides and three spots on the other three sides. In this game, I am going to tell you only the total showing after each throw of the three dice. Your job is to guess whether the unusual die was showing no spots or three spots (analogous to a noise versus a signal trial, embedded in the background "noise" of the total from the other two dice). A little bit of math and probability calculations would yield the distributions shown in Figure 3.1.

The sensible way to maximize your correct guesses would be to consider the height of each distribution for each outcome. Given a total of six, it is more likely that the odd die had zero spots than three. Given a total of 11, the three-spot distribution is higher. Drawing the boundary for changing your response between eight and nine will produce an optimal criterion for the most correct responses. The likelihood of each outcome, then, is given by the height of the distribution at that point, and the position of the criterion can be specified by

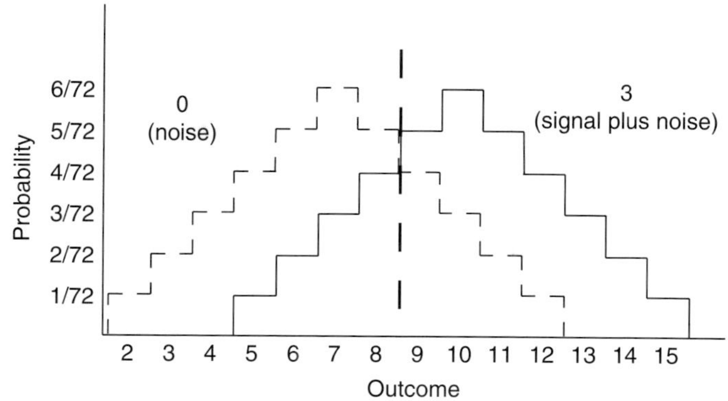

Figure 3.1 The hypothetical dice game with two normal dice and one showing three spots on one side and no spots on the other three sides. The task is to guess whether the third die showed no spots or three spots, given that you are only told the total number of spots on each trial (after Swets et al., 1961). The dashed line shows the probability distribution when no spots are showing and the solid lines shows the distribution when three spots are showing. Note that the probabilities are shown as probabilities of the total number of trials. Conditional probabilities (of any outcome given a noise trial, for example) should multiply these values by two.

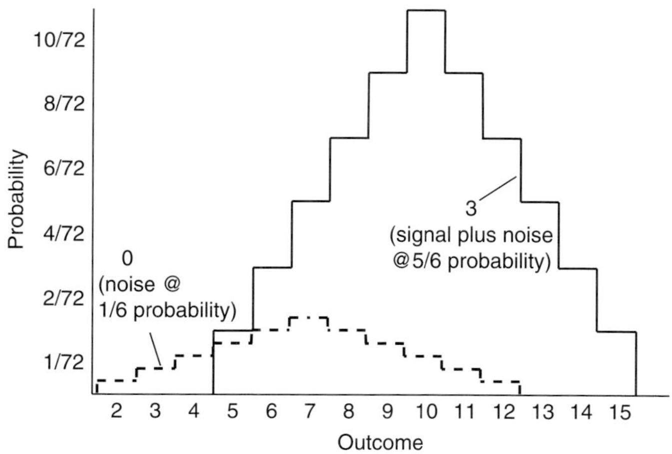

Figure 3.2 The dice game with the probability of a noise trial reduced to 1/6 by leaving only one side of the third die blank.

the relative levels of the two curves, sometimes given as a ratio or likelihood ratio. If we made the payoffs and penalties all equal in weight, it would make sense to place the criterion at the point where the curves cross and the likelihood ratio is one.

But in real life, there are several ways to perturb this nice symmetrical situation that could change your behavior. Suppose we used a third die with five sides having three spots and only one side blank. Now the probability of getting a signal trial is 83% (5/6). Figure 3.2 shows this situation. Now it would make sense to call anything higher than a four a signal trial, because the signal distribution is higher than the noise distribution from that point on. Suppose we put five spots on the third die instead of three. Now the signal is "stronger." What happens to the totals? Figure 3.3 shows the distributions. Now anything higher than

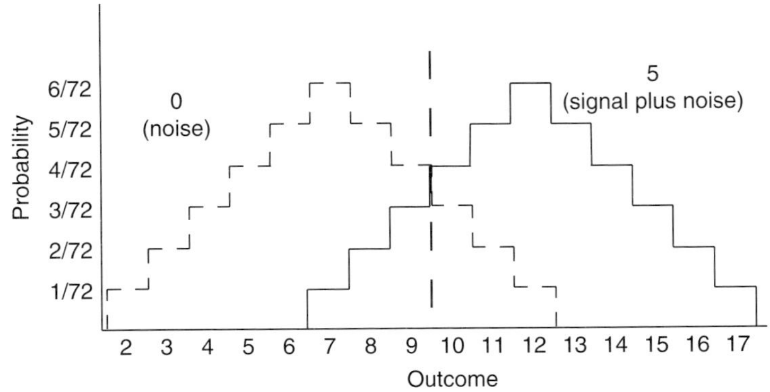

Figure 3.3 The dice game with three sides painted with five spots instead of three spots. This has the effect of making the signal "stronger" as it has a larger contribution to the total. Note that the errors will be fewer than in Figure 3.1.

a total of 10 is likely to be a signal trial. Note that you will make fewer errors than the situation in Figure 3.1 – given this optimal criterion you will be wrong only 1/6 (17%) of the time, as opposed to 5/18 (28%) for our optimal criterion with the three-spot situation.

Now here is the key: what we really want to know in any signal detection situation is how much these two distributions overlap. If they sit right on top of one another, you have no way of knowing which kind of trial was presented. The more they pull apart, as in Figure 3.3, the more discriminable is the signal from the noise. Furthermore, this degree of overlap has nothing to do with where you set your criterion, be it conservative, lax, or balanced. In Section 3.3 we will show how the theory can give us a measure of this overlap, or the discriminability of the two stimuli. Furthermore, the yardstick we use will be independent of where any particular person sets their criterion. That is, we will have separated the measure of discrimination from the contaminating influence of response bias, yielding a true objective measure of discriminability (or conversely, the sensitivity of the subject to the sensory difference).

3.3 Connecting the Design to Theory

3.3.1 The Model and Measuring Sensory Distance

Let us assume that we have a continuous distribution of sensory events from signal and from noise, rather than the discrete outcomes in our dice game. For now, let us also assume they have equal variance or spread. This situation is shown in Figure 3.4. If we think of the horizontal axis as the strength of the sensation, then the noise distribution is sometimes stronger and sometimes weaker, and sometimes it is mistaken for a signal. Likewise, the signal experiences will vary from moment to moment and sometimes be mistaken for a noise event. This variability can arise from many sources. It can be part of the variation in the stimulus or part of the variation in the observer from moment to moment. Our nervous system has a background of spontaneous activity, and this varies over time. So any neural signal must be embedded in that background of spontaneous and randomly varying neural events.

Remember that our subject or observer is going to set some criterion for responding, as shown in Figure 3.5. You can think of it as a kind of boundary or cutoff level, above which the subject responds "signal" or "yes" and below which the subject responds "no" or "noise."

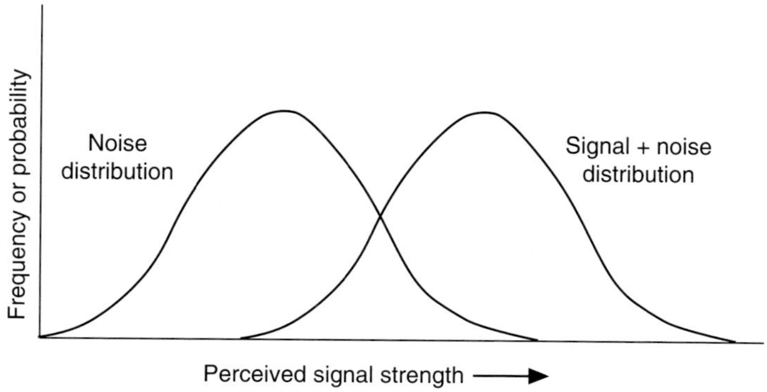

Figure 3.4 Signal and noise as continuous distributions.

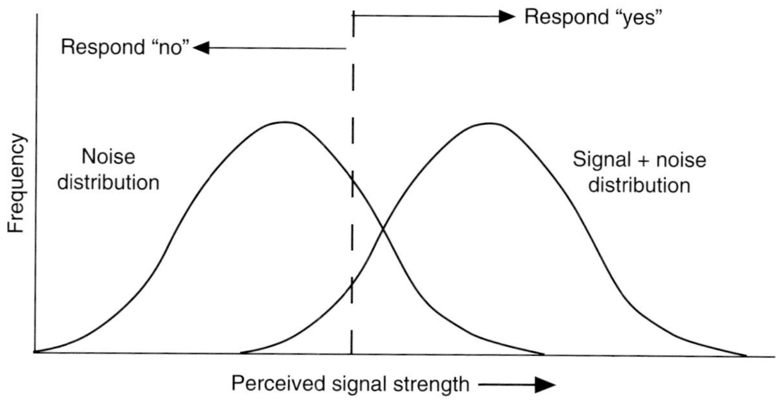

Figure 3.5 Signal and noise distributions with a response criterion imposed.

This criterion is sometimes called beta, and can be specified by the relative height of the signal and noise distributions at the criterion point:

$$\beta = \frac{f(\mathrm{SN})}{f(\mathrm{N})} \tag{3.1}$$

That is, the likelihood ratio given by the height (probability) of the signal plus noise distribution divided by the height of the noise distribution.

So the subject sets some criterion or cutoff. Once again, sensory experiences stronger than that cutoff are called signal and sensory experiences that are weaker than that level are responded to as noise. Remember that the subject does not know from which distribution any given trial has been sampled. They only know the sensation that is experienced. Relating this scheme to the data set is now straightforward. We know the proportion of hits as a function of the total signal trials and the proportion of false alarms from the noise trials (note: not the total number of all trials). We have this in our data matrix at the end of the experiment, as tabulated in Table 3.1. Now all we need to do is convert the areas to some measure of the overlap or separation of the two distributions.

Figure 3.6 shows how the criterion is related to the proportions of hits and false alarms. Because we know the exact shape of the normal distributions, we can relate the area underneath the curve at any given cutoff or segment to a *Z*-score, which is the distance from the mean of the distribution to the cutoff. You can use a common *Z*-score table that

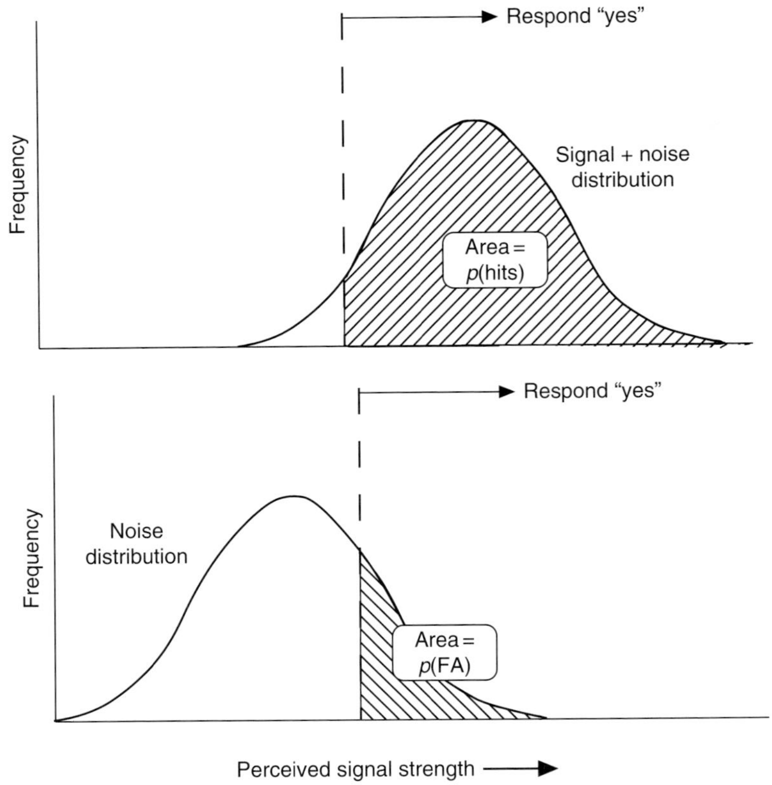

Figure 3.6 Proportions of hits and proportions of false alarms related to the areas under the signal and noise curves, respectively, to the right of the criterion.

converts proportions to distances. A convenient shortcut table is presented in Table 3.A.1 in Appendix 3.A. Thus, we can specify distance in standard deviation units, which conveniently accounts for the degree of spread in our distributions, and not simply the separation on some arbitrary scale. This distance is called d' **(d-prime)**. Because of the way that Z-scores are usually tabulated, we have to subtract the Z-score for false alarms from the Z-score for hits. In our example from Figure 3.7, the false-alarm rate was 20% of the total area of the noise distribution, or 0.20, which gives us a Z-score of −0.84. The hit rate was 85% of the signal distribution, corresponding to a Z-score of 1.04. So the overall d' value is $1.04-(-0.84)=1.88$.

3.3.2 Changing Criteria

As shown in Tables 3.2 and 3.3, we could manipulate the payoffs and penalties and this would likely induce some change in the behavior of our subject. That is, they would shift the criterion to maximize their "winnings." This can also be achieved by changing the probability of signal and noise, as in Figure 3.2. Even if they are not told what the payoffs are, or what the baseline probabilities are, they will figure it out if you start to impose penalties or give rewards, and then they shift accordingly. This shifting behavior is such a reliable effect that you can use it as an undergraduate laboratory demonstration.

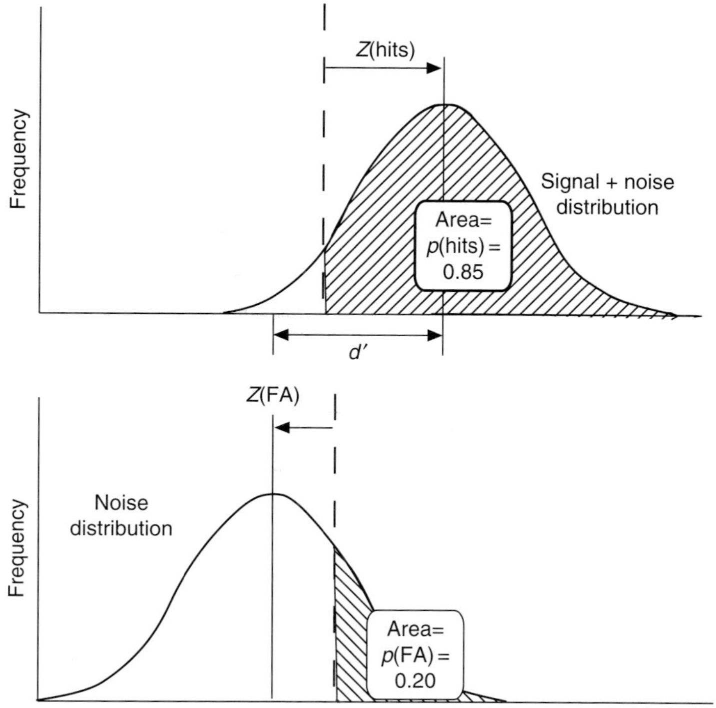

Figure 3.7 Conversion of proportions to Z-scores provides a distance measure in standard deviation units, and calculation of the total separation by the quantity d' ("d-prime").

Figure 3.8 shows our situation from Figure 3.7, but now with a more conservative criterion. The false-alarm rate drops to 5%, but at the expense of a hit rate that falls to a little less than 60%. The Z-scores also adjust, to give us the same d' as Z(FA) = −1.64 and Z(hits) = +0.24, so d' comes out to be +0.24–(−1.64) = 1.88, as before. Figure 3.9 shows a more lax criterion, in which the subjects responds "yes" almost all the time. Now the proportion of hits is quite high at 95% (Z = 1.64), but at the expense of a higher false-alarm rate of 39.5% (Z = −0.24), and so we have 1.64–(−0.24) = 1.88.

It is possible for a subject to set a criterion that will maximize the payoffs and minimize the penalties. The mathematical solution for this optimal response strategy is given by

$$\beta = \frac{p(n)}{p(s)}\left(\frac{O_{\text{CR}} - O_{\text{FA}}}{O_{\text{hit}} - O_{\text{miss}}}\right) \tag{3.2}$$

where O is the outcome (i.e., payoff or penalty) associated with each of the four events: hits, misses, false alarms, and correct rejections. Remember that this is a likelihood ratio that specifies the height of the signal distribution relative to the height of the noise distribution at the cutoff or criterion. In terms of the rather extreme payoff matrix in Table 3.2, we have (0.5/0.5)[+1 − (−1)]/[+5 − (−5)] = 2/10; so, we should choose a criterion point where the noise distribution is five times higher than the signal distribution, or quite far to the left. The opposite is true for the example in Table 3.3, since we have (0.5/0.5)[+5 − (−5)]/[+1 − (−1)] = 10/2, or a point rather far to the right where the signal distribution is five times higher than the noise distribution; or in other words, responding "yes" infrequently.

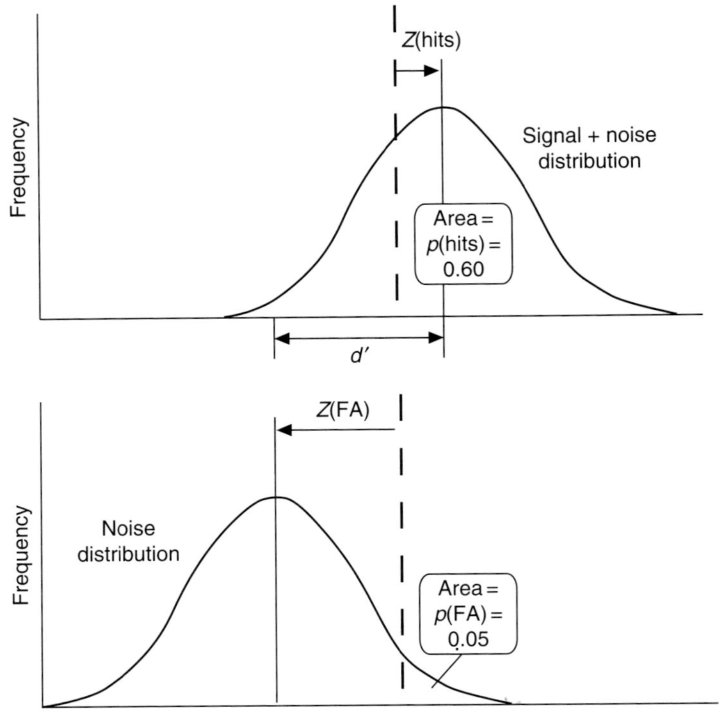

Figure 3.8 Changing the criterion to a more conservative level (responding "yes" only when very sure, for example) decreases the false-alarm rate, but at the expense of a lower hit rate. However, *d′* stays the same.

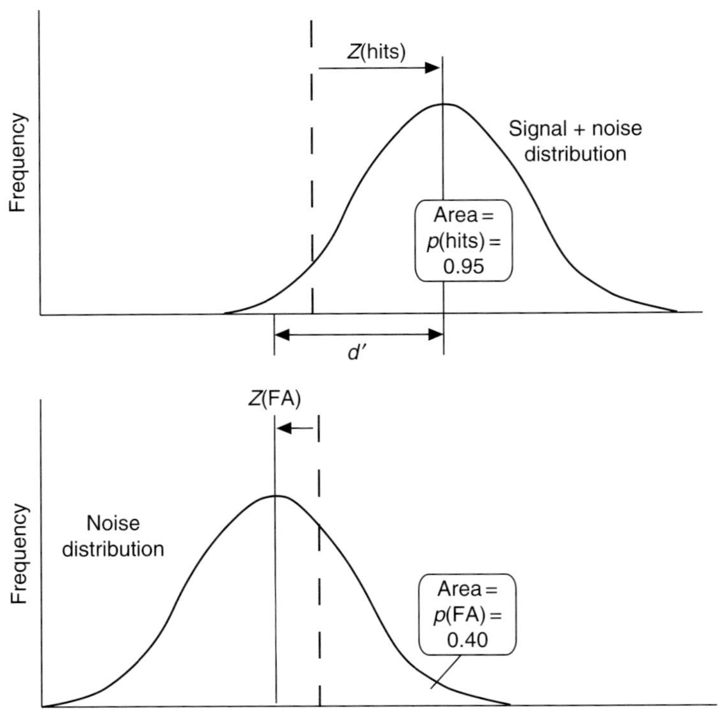

Figure 3.9 Changing the criterion to a less conservative (lax) level (responding "yes" when there is any evidence at all of a signal) increases the hit rate, but at the expense of a higher false-alarm rate. However, *d′* stays the same.

An alternative specification of the criterion (as C) can be found from the following expression (Gescheider, 1997):

$$C = 0.5\left(Z_{\mathrm{SN}} + Z_{\mathrm{N}}\right) \qquad (3.3)$$

where Z_{SN} is the Z-score for misses (i.e., the Z-score for $1-p(\text{hits})$) and Z_{N} is the Z-score for correct rejections (i.e., the Z-score for $1-p(\text{FA})$). This specifies the position of the criterion in Z-score units, rather than by the heights of the curves, as in the case of β. At the point where signal and noise distributions cross, $C=0$. If C is negative, the response bias favors more frequent "yes" or "signal" responses; if positive, it favors more "no" or "noise" responses. Another way to specify the criterion point is in terms of the d' separation, giving C' as follows:

$$C' = \frac{C}{d'} = 0.5\left(\frac{Z_{\mathrm{SN}} + Z_{\mathrm{N}}}{Z_{\mathrm{N}} - Z_{\mathrm{SN}}}\right) \qquad (3.4)$$

3.4 The ROC Curve

3.4.1 The ROC Curve: A Plot of Changing Criteria

Setting up an experiment in which criteria could be shifted allowed further insights into both behaviors and the SDT models. If we do an experiment several times using the same subject and the same stimuli, and just get the subject to change criteria, we should get a constant estimate of d', but at changing hit and false-alarm proportions. This behavior can be described in a simple graph called a receiver operating characteristic (ROC) curve. The ROC curve plots the proportion of hits (saying yes to signal) versus the proportion of false alarms (saying yes to noise) over several experimental conditions.

Figure 3.10 shows an ROC curve for a subject with a d' of 1.0 for some pair of stimuli. As the criterion changes from more conservative (A) to more lax (B to C) the proportions of hits and false alarms both increase. This traces a characteristic curve. If d' was zero and the signal and noise distributions were completely overlapped, the hit rate would always equal the false-alarm rate, and would follow the diagonal dashed line in Figure 3.10. As d' increases, and the signal and noise distributions pull apart, the ROC curve traces a line that increasingly bows toward the upper left. That is, there are more hits per false alarm (or fewer false alarms for a given hit rate). It should be apparent that the larger the area is below the ROC curve and to the right, the greater the discrimination is. So an area measure is an alternative way of looking at the discrimination performance, and one that does not depend in any way upon the assumptions of the SDT model (such as normal distributions and equal variance). This area measure will be explored further in discussion below.

A common graphical representation of the ROC curve is to convert the probabilities of hits and false alarms to Z-scores. Figure 3.11 shows such a plot for our hypothetical subject shown in Figure 3.10. The function is now linear, or approximately so in most data sets (Green & Swets, 1988/1966). Sometimes the convention is to reverse the Z-score plot numerically, so they plot from positive on the left (or bottom) to negative on the right (or top), but in this case we will keep it in the same direction as the ROC curve itself. We can

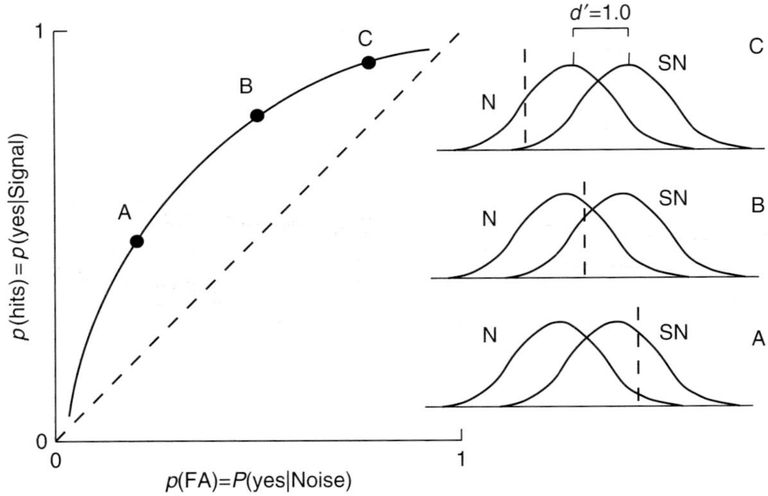

Figure 3.10 The ROC curve plots the probability of hits against the probability of noise. It is obtained by varying the criterion level and conducting the signal detection session several times on the same subject with the same stimuli. As the curve bows further to the upper left, the discrimination is better and the *d′* level is higher. At the diagonal dashed line, the probability of hits and false alarms is equal, which describes the situation in which signal and noise distributions have the same mean, there is no discrimination, and *d′* is zero.

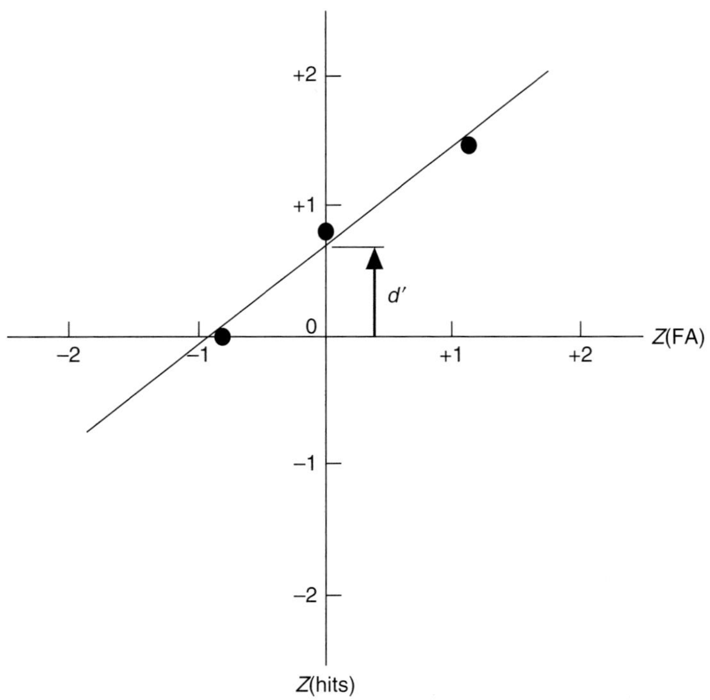

Figure 3.11 The ROC curve plotted as *Z*-scores. Note that the intersection of the line at *Z*(FA) = 0 gives the estimated value of *d′*.

now find d' graphically by finding the intersection of the straight line with the ordinate; that is, the value where $Z(\text{FA})=0$. The slope of the line also indicates the relative size of the variance of the two distributions. Specifically, it is the ratio of the standard deviation of the noise distribution to the standard deviation of the signal distribution. This value, s ($\sigma_N/\sigma_{SN}=s$), will become useful when we calculate alternative measures of discriminability, discussed in Section 3.4.2.

3.4.2 Unequal Variance of Signal and Noise and Alternative Measures of Discriminability

The ROC curve can also give us information about the nature of the variance of signal and noise distributions. Recall that, in the simplest form of the SDT model, we have assumed equal variance of signal and noise. But Weber's law (see Chapter 1) teaches that the variance around any stimulus will increase as stimulus energy increases. Thus, we might expect the signal distribution to have a higher variance than the noise distribution. This situation is shown in Figure 3.12. The ROC curve now becomes asymmetric, is steeper initially, and flattens out at higher levels of more lax criteria.

Unequal variance poses some problems for the common SDT model, because the d' measure is expressed in standard deviation units. The ROC curve of Figure 3.12 is replotted as a Z-score line in Figure 3.13. Several alternative measures have been proposed for an index of discriminability, and these are nicely summarized by Gescheider (1997). One measure is Δm, defined as the distance between signal and noise in standard deviation units of the noise distribution. This can be found at the point where the Z-score line intersects the value for $Z(\text{hits})=0$ and we take the absolute value. This expresses the discriminability index in terms of the standard deviation units of the noise distribution. For our hypothetical observer in Figure 3.13, we see this is now a rather large value near -1.5. So Δm is $|-1.5|$. But this does not tell the whole story. The slope of the line is such that the standard deviation of the signal distribution is almost 1.5 times that of the noise distribution (or a slope of 0.66). Green and Swets (1966/1988) suggest that when using Δm we also specify the slope value

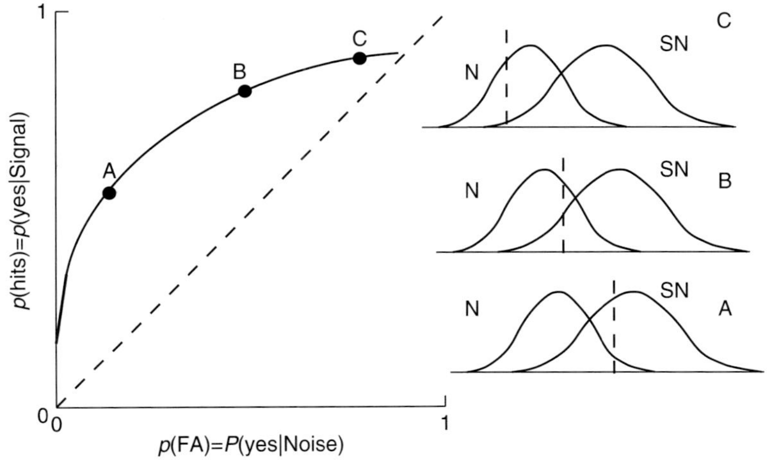

Figure 3.12 The ROC curve becomes asymmetric when the signal distribution has a higher variance than the noise distribution.

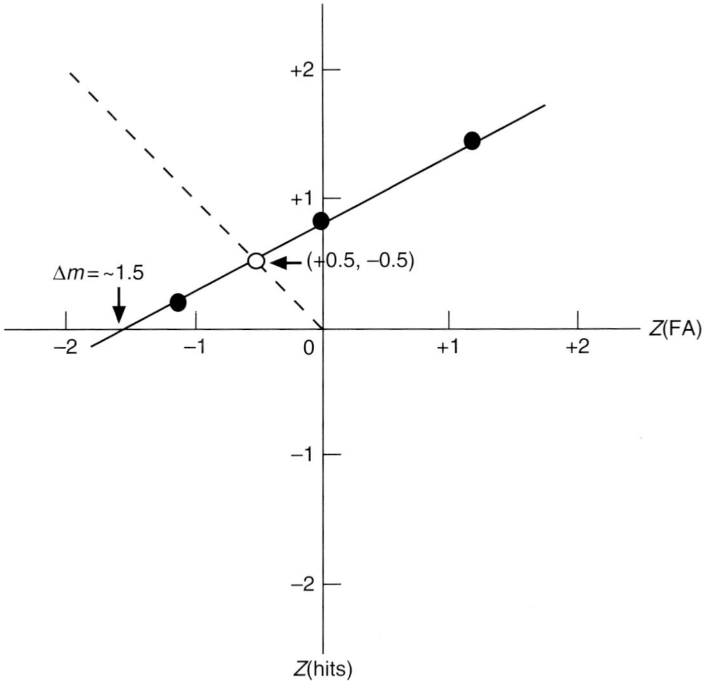

Figure 3.13 The *Z*-score plot with an unequal variance situation. The slope of the line gives the ratio of the standard deviation of the noise distribution over the standard deviation of the signal distribution. The reciprocal is a quantity *s* useful in calculations of the alternative measures of discriminability.

and use the uppercase letter *D* for this index. So the *D*-value for our subject ($\Delta m,s$) would be specified as (1.5,0.66).

Another alternative is to give equal weight to the standard deviations of signal and noise. This value, called d'_e, can be found graphically from the intersection of the negative diagonal in Figure 3.13 with the *Z*-score line. We just take the value of d' (*Z*(hits) − *Z*(FA)) at that point. There is also a general mathematical relationship between Δm, *s*, and d'_e, given as follows:

$$d'_e = 2\Delta m\left(\frac{s}{1+s}\right) \tag{3.5}$$

and

$$\Delta m = \frac{d'_e}{2}\left(\frac{1}{s}\right) + \frac{d'_e}{2} \tag{3.6}$$

So, for our subject, we now get d'_e of 2(1.5)(0.66/1.66), or about 1.1. Looking at our intersection point with the negative diagonal, we see it is near the point (+0.5, −0.5) so applying our formula for d', we get +0.5 − (−0.5) = 1.0. Note that this is a different estimate for d' than the +1.5 value we got for Δm, due to the large discrepancy in variance for the (larger) signal and (smaller) noise distributions. In practice, this would rarely be found, because the two stimuli are generally not that different (or we would not be doing a discrimination experiment).

Gescheider (1997) gives an alternative way to find Δm. We can work in *Z*-scores based on the location of the means of the signal and noise distributions, rather than the *Z*-scores based on *p*(hits) and *p*(FA). To obtain these we take the *Z*-score equivalents of the following expressions:

$$Z_{\text{N}} = \Phi^{-1}\left(1 - p(\text{FA})\right) \quad \textit{and} \quad Z_{\text{SN}} = \Phi^{-1}\left(1 - p(\text{hits})\right) \tag{3.7}$$

where Φ^{-1} means "convert from *p* to *Z*" or find the *Z*-score that corresponds to that proportion on the cumulative normal distribution. Note that $1-p(\text{FA})$ is simply the proportion of correct rejections and $1-p(\text{hits})$ is the proportion of misses. Now in order to find Δm, we merely convert our points on the ROC curve to values of Z_{N} and Z_{SN}. As shown above, plotting these *Z*-scores usually forms a straight line. Now find the point at which $Z_{\text{SN}}=0$. The corresponding value of Z_{N} will be Δm.

In addition to Δm and d'_e, several other measures have been suggested for discriminability under the unequal variance situation. In Gescheider's plot of Z_{N} and Z_{SN}, d'_e is the absolute difference between Z_{N} and Z_{SN} at a point where our plot of Z_{SN} versus Z_{N} crosses a diagonal with slope equal to one. Again, d'_e represents the difference in terms of the average of the standard deviations of the two distributions. A third estimate is d_{a}, which is the difference expressed in the root-mean-square of the two standard deviations (the square root of the average of the squared standard deviations – in other words, square them, divide by two and take the square root). Experimentally, this can also be found from our plot of Z_{SN} versus Z_{N} as an alternative way of looking at the ROC curve. First, we find the slope *s* of the line, which tells us that the standard deviation of the N distribution is *s*-times the standard deviation of the SN distribution. This is equivalent to setting the standard deviation of the SN distribution to one and the standard deviation of the noise distribution to *s*. We also find the absolute value of the intercept *b* of the line. The value of d_{a} then becomes

$$d_{\text{a}} = \frac{b}{\sqrt{\dfrac{1+s^2}{2}}} \tag{3.8}$$

Two alternative measures of discriminability make no assumptions about the form of the distributions (normal or otherwise) or about equal variances. These are related to the area underneath the ROC curve. This area is sometime symbolized by *p*(*A*). It can be found by graphical estimation (e.g., drawing the curve and counting blocks on graph paper). Another alternative is to find the area under the *Z*-score plot, called A_Z. Both of these measures arise from the plot of *p*(hits) versus *p*(FA). A_Z can be found from the following relationship (Gescheider, 1997):

$$A_z = \Phi\left(Z(A)\right), \quad \text{where} \quad Z(A) = \frac{s\Delta m}{\sqrt{1+s^2}} \tag{3.9}$$

And where *s* is the slope of the ROC curve when expressed as *Z*-scores, Δm was just defined, and Φ means "take the value of the cumulative normal distribution corresponding to that *Z*-score" (convert from *Z* to *p*). These alternative measures are summarized in Table 3.4.

Table 3.4 Alternative measures of discriminability

Measure	Definition
Δm	Distance (difference) of signal and noise distributions in standard deviations of the noise distribution
d'_e	Distance (difference) of signal and noise distributions in units of the average of the standard deviations of the signal and noise distributions
d_a	Distance (difference) of signal and noise distributions in units of the root-mean-square of the standard deviations of the signal and noise distributions
$p(A)$	Area under the ROC curve, plotting probability of hits against probability of false alarms
A_Z	Area under the ROC curve converted to Z-score units

3.5 ROC Curves from Rating Scales; the *R*-Index

3.5.1 Confidence Ratings

A convenient way to obtain an ROC curve is to use confidence ratings, rather than doing several, usually tedious, experiments and getting the subject to change their criterion several times. The rating scale consists of different levels of confidence as to whether the subject believes the current stimulus to arise from signal or noise. In other words, we have single stimulus presentation, but instead of a yes/no reponse we have allowed the subject to response with some graded levels of sureness or certainty. For example, the scale might have five points, as shown in Table 3.5.

We can use the data as if the subject was working with five different criterion levels. To make the math a little easier, let us set up a study with 100 trials of signal and 100 trials of noise. The example in Table 3.6 shows how the data might look.

The calculations work as if we are dividing the responses into four 2×2 matrices. Starting at scale point 1, we have 15/100 hits and 5/100 false alarms. Assuming a somewhat looser criterion, at scale point 2 we have 37 hits and 18 false alarms. But wait! The data from category 1 would also fit this criterion since the subject was even more sure those were signal trials. So the next hit and false-alarm rate is (15+37)/100, or 0.52, and the false-alarm rate is (5+18)/100, or 0.23. We can do this operation two more times. For another more lax criterion we can collapse the first three categories to give (15+37+24) hits, or 0.76, and (5+18+26) false alarms, or 0.49. The final boundary gives 96 hits and 84 false alarms. At this point we run out of options, because the final row sums to 100, where the Z-score becomes infinite. But we have already used that category as the miss and correct rejection rate for the fourth boundary, so the data are in fact used. Table 3.7 summarizes these operations and shows the conversion to Z-scores. Figure 3.14 shows the boundaries as four criteria, C1 through C4.

We could of course compute a d' value for each one of our boundary conditions (giving 0.60, 0.79, 0.74, and 0.67) and average them, but it is somewhat more satisfying to look at this as an ROC curve as if we had done four experiments. Figure 3.15 shows the Z-score plot of the ROC curve for these data. So the intercept of the Z-score line at Z(FA)=0 is about +0.74 as our graphical estimate of d'.

3.5.2 The *R*-index

One of the primary difficulties in adapting a signal detection experiment to foods and consumer products is that the number of trials for a given subject is far too many to be practical

Table 3.5 A confidence rating scale

Scale category	Label or interpretation
1	Very certain it was a signal trial
2	Somewhat certain it was a signal trial
3	Not sure it was signal or noise
4	Somewhat certain it was a noise trial
5	Very certain it was a noise trial

Table 3.6 Hypothetical data in a confidence rating study

	Number of responses	
Scale category	Signal trials	Noise trials
1	15	5
2	37	18
3	24	26
4	15	35
5	9	16

Table 3.7 Summary of calculations for rating scale data

Category	Signal trials	Noise trials	"Boundary"	Cumulative hit rate	Cumulative FA rate	Z(hits)	Z(FA)
1	15	5	1	0.15	0.05	−1.04	−1.64
2	37	18	2	0.52	0.23	+0.05	−0.74
3	24	26	3	0.76	0.49	+0.71	−0.03
4	20	35	4	0.96	0.84	+1.75	+1.08
5	4	16					

for most food testing. The solutions to this problem are partly given in Chapter 4, where we will show how the SDT measures of discriminability can be derived from common forced-choice methods such as the triangle, duo–trio, and n-AFC procedures. Another way to shorten the testing burden on subjects is to use a rating scale procedure, as shown above. One popular method for this approach is called the ***R*-index** procedure (O'Mahony, 1979, 1992). This is simply a straightforward rating scale, usually with four categories or three boundary criteria. It is often set up as shown in Table 3.8.

The letters A through H indicate the counts of responses in each category and N is the total number of responses, with $N/2$ signal trials and $N/2$ noise trials. The derivation of the R-index arises from making hypothetical paired comparisons of every signal with every noise trial, even though they were single presentations and not paired. Thus, for 10 signal and 10 noise trials, we can forge $10 \times 10 = 100$ hypothetical paired comparisons, as if we had paired all the experiences and asked the subject which one seemed more like a signal.

The calculation works as follows. Take the leftmost box (count A) for the signal trials in Table 3.8, and imagine that each one of those experiences was paired with all noise boxes to its right (i.e., F, G, and H). In all cases, the experiences for box A would have seemed more like a signal than any experience in boxes F, G, and H; so, in a paired comparison test,

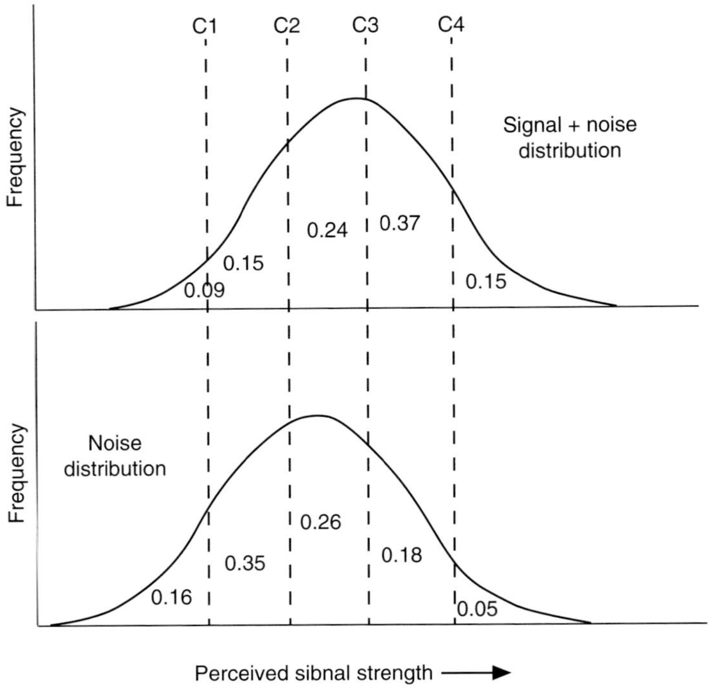

Figure 3.14 The behavior shown in Table 3.6 represented graphically. The subject has set four boundary criteria for the five possible response categories.

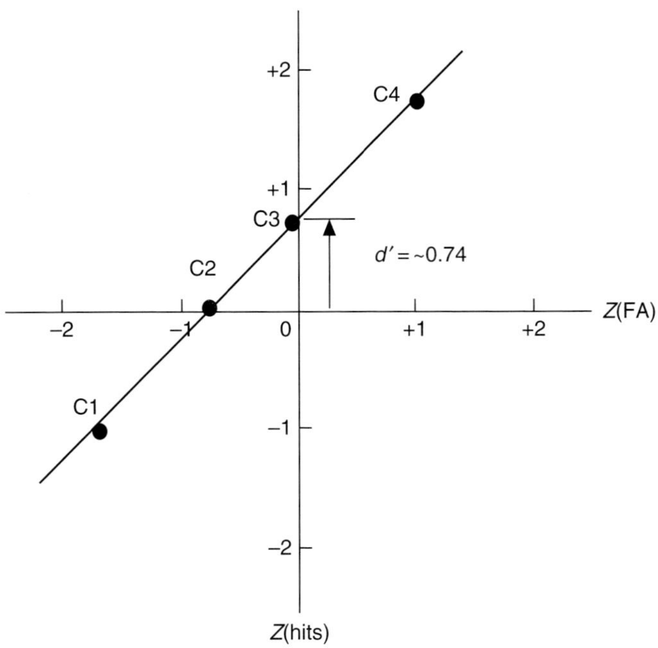

Figure 3.15 The *Z*-score plot of the rating scale data from Table 3.7. Once again, finding the intercept at $Z(\text{FA}) = 0$ provides an estimate of d'.

Table 3.8 The *R*-index rating scale

Trial	Response			
	Signal, sure	Signal, unsure	Noise, unsure	Noise, sure
Signal	*A*	*B*	*C*	*D*
Noise	*E*	*F*	*G*	*H*

Table 3.9 An *R*-index data set

Trial	Response			
	Signal, sure	Signal, unsure	Noise, unsure	Noise, sure
Signal	4	3	2	1
Noise	0	2	3	5

a correct response would have been given for $A\times(F+G+H)$ total paired trials. Similarly, the experiences in box B would have seemed more like a signal than the experiences in boxes G and H, leading to $B\times(G+H)$ "correct" paired comparisons. The same is true for boxes C and H, with $C\times H$ correct answers. For the items that are tied (i.e., in the same column), we can make no such inferences, so we will simply split them in half. This splitting gives us $(A\times E)/2$, $(B\times F)/2$, $(C\times G)/2$, plus $(D\times H)/2$ for a total from the "ties." So the total of these "correct" hypothetical responses, divided by the total number of possible trials, and then multiplied by 100, gives us R. Here is the equation, and a sample calculation is shown in Table 3.9:

$$R=\frac{A(F+F+H)+B(G+H)+C(H)+\text{ties}}{(A+B+C+D)(E+F+G+H)}\times 100 \qquad (3.10)$$

where the number of ties is equal to

$$\text{ties}=\frac{AD+BE+CG+DH}{2} \qquad (3.11)$$

A quick worked example. Suppose we have 10 trials of signal and 10 trials of noise distributed as in Table 3.9.

Our subject seems to be discriminating rather well, and the R-value comes out to be

$$R=\frac{4(2+3+5)+3(3+5)+2(5)+8.5}{10\times 10}\times 100=82.5$$

where there were $[(4\times 0)+(3\times 2)+(2\times 3)+(1\times 5)]/2$ contributed by ties, or 8.5.

The R-index has several convenient properties. First, it is an estimate of the area under the ROC curve, and is thus analogous to our measure $p(A)$ in Table 3.4. It is thus a measure of discriminability that does not depend upon the assumptions of the standard signal detection model, such as normal distributions of equal variance. R varies, at least in theory, from 50 to 100, and thus is also related to what we would expect in a paired

Table 3.10 A subject with an unusual response bias

	Response			
Trial	**Signal, sure**	**Signal, unsure**	**Noise, unsure**	**Noise, sure**
Signal	10	0	0	0
Noise	0	10	0	0

comparison test for the percentage correct. R is easily converted to a d' value by the following relationship (Bi, 2006b):

$$R = \Phi\left(\frac{d'}{\sqrt{2}}\right) \tag{3.12}$$

A simple statistical test for the significance of R is given by

$$Z = \frac{R - 0.5}{\sqrt{\frac{R(1-R)}{N-1}}} \tag{3.13}$$

where N is the number of signal or noise trials (or the smaller of the two if different) (Bi & O'Mahony, 1995; Bi, 2006b). A full discussion of various error estimates for the standard deviation of R can be found in Bi (2006b).

Another attractive feature of this index is that it is impervious to any extreme or unusual response biases from a given subject. For example, we might have a subject who thinks that everything seems like a signal, but calls all the noise trials "signal, unsure," as shown in Table 3.10.

This person obviously has a strange criterion placement, but they are discriminating perfectly, and $R = 100$.

We can also get a similar measure from ranking data (Brown, 1974). Suppose we have six signal and six noise stimuli, and a subject gets to rank them as to how strong they seem. Let us work with the hypothetical data set in Table 3.11, ranked from strongest on the left to weakest on the right.

We can do a similar hypothetical paired comparison, looking at all the noise trials that are ranked as weaker than the signal trials. So for S1 and S2, there are six pairings each that are judged stronger than a noise trial. For S3 we have five pairings, for S4 and S5 there are three pairs each, and two for S6. This gives us a total of $(6+6+5+3+3+2)=25$. We have to divide this by the total number of possible pairs and multiply by 100, giving us $(25/36)\times 100$ or $R=69.4$.

O'Mahony (1992) showed that the ranking data can also be used to fill in a response matrix, similar to the R-index by ratings. Given two products that are presented N times and ranked by X judges (or X times by one judge), one simply fills in the $2\times N$ matrix of frequencies. In the case of the data set in Table 3.11, you would obtain a set of 12 categories for each product, and perform a simple R-index calculation on the frequencies as if they had been placed on a rating scale. This produces the same R-index as the ranking method if there are only two stimuli. However, if there are more than two signal stimuli, the indices will differ (Ishii et al., 1992) and will be somewhat higher than the values of R obtained from the true rating task.

Table 3.11 Data from a hypothetical ranking experiment

S1	S2	N1	S3	N2	N3	S4	S5	N4	S6	N5	N6

3.6 Conclusions and Implications for Sensory Testing

3.6.1 Lessons for Sensory Evaluation

The classical SDT experiment, with dozens or hundreds of trials on a single individual, is a far cry from the kinds of tests done in applied sensory evaluation of foods and consumer products. In those arenas, we typically test just a few products, but on many individuals. But this gap can be bridged, as we will see in Chapter 4. We can link the common discrimination tests such as the triangle, duo–trio, and n-AFC tests to signal detection measures of discrimination, as long as we understand the strategies used by subjects for each test and we construct an appropriate model. The advantage of a signal detection analysis is that it gives us a measure of sensory difference such as d', the R-index, or $p(A)$, which can be applied to any test method, even if some are inherently more difficult or have inherent higher variability. In other words, we have found a universal yardstick, and one that is free from the influences of response bias.

The critical aspect of an SDT study is the use of the inherent control condition, the noise trial, and its resulting data in terms of a false-alarm level. This teaches us that if we are to separate response bias and criterion choices from actual discrimination, we must have that baseline or control condition. This is especially critical when we have a yes/no or same/different kind of response. An example in sensory evaluation is the "A–not-A" test. Samples are presented one at a time, and the subject's task is to say whether it is an example of a previously viewed sample or target ("A") or something else ("not-A"). This method makes no sense whatsoever without the appropriate control condition, of actually presenting examples of items which are, in fact, "not-A." Now we have a 2×2 matrix of stimulus and response, just like the common SDT method. The same is true of a same/different study in which pairs are given. It is necessary to have control pairs of identical samples as a baseline. Also, extensions of the same/different test, such as rated degree of difference from a control sample, must have a control pair of identical samples. Otherwise, the rating of the test pair on its own is impossible to interpret and, in fact, is meaningless.

Another situation where this problem arises is in quality control or shelf-life studies where the panelist makes pass/fail judgments about a product. It is incumbent upon the sensory professional in that kind of situation to give some kind of catch trials, in which confirmed in-spec products are presented to assess the false-alarm rate. It is also useful in that situation to give purposefully spiked samples with known defects, to make sure that the panel is detecting the kinds of problems that it ought to detect. Think of this as an estimate of the miss rate.

Note that such experiments usually demand that both types of stimuli be given to the same individuals. That is, the optimal design is a within-subjects or complete block. If we give the test and control samples to different groups, we do not know whether the difference is due to actual sensory differences between the samples or to different response criteria in the different groups. The advantages of the complete block experiment are discussed further in Chapter 9.

3.6.2 Case Study: A Signal Detection Tragedy

The following story is based on an actual court case. The details have been changed slightly to protect the identities of the individuals involved. A family went on vacation from their home in the northern USA to a popular resort in Florida by car. After a week or so, they began to drive home, and stopped for a period of time at an amusement park several hundred miles from their starting point and several hundred miles from home. After a rather full day, they resumed their trip. The parents were smokers and were smoking inside the closed van for most of the trip. Upon arriving home, they parked the van in the garage and left the motor running for a little while for it to cool itself. When the mother got out of the van, she remarked to the father that she thought she might be smelling a gas leak. The father dismissed it as just a smell from the catalytic converter. Upon entering the house the family found that the basement had flooded, and the sump pump had stopped working due to lack of electricity. The mother went upstairs to attend to the children and make the beds while the father went to his garage to get an extension cord to try to get the sump pump working again. At this point, he was very tired and also quite frustrated with the basement situation. Upstairs the mother was shaking out some sheets, and once again noticed that there seemed to be the smell of leaking gas. She instructed one of the children to go downstairs and warn the father. Unfortunately, at this point he connected the extension cords, causing a spark and ignited a propane explosion that blew up the house. Several children were badly burned.

The family later sued a number of parties for damages, including the propane company, the parent oil company, the manufacturers of the tanks, valves and other equipment, and the company that sold the odorization agent, mostly on the grounds that these were defective products that caused danger and harm. The case went to a jury trial, which is allowed at the discretion of the plaintiff in that state. The lawyers for the plaintiff argued that the gas odorization substance failed, because no one who smelled it would purposely blow up their house, so obviously it was not effective. The defense argued that, according to the SDT, one cannot argue that since the person did not respond, they did not actually smell the signal. He might simply have set his criterion too high.

Appendix 3.A Table of *p* and *Z*

Table 3.A.1 Proportions and *Z*-scores*

Proportion	*Z*-score	Proportion	*Z*-score	Proportion	*Z*-score	Proportion	*Z*-score
0.01	−2.33	0.26	−0.64	0.51	0.03	0.76	0.71
0.02	−2.05	0.27	−0.61	0.52	0.05	0.77	0.74
0.03	−1.88	0.28	−0.58	0.53	0.08	0.78	0.77
0.04	−1.75	0.29	−0.55	0.54	0.10	0.79	0.81
0.05	−1.64	0.30	−0.52	0.55	0.13	0.80	0.84
0.06	−1.55	0.31	−0.50	0.56	0.15	0.81	0.88
0.07	−1.48	0.32	−0.47	0.57	0.18	0.82	0.92
0.08	−1.41	0.33	−0.44	0.58	0.20	0.83	0.95
0.09	−1.34	0.34	−0.41	0.59	0.23	0.84	0.99
0.10	−1.28	0.35	−0.39	0.60	0.25	0.85	1.04
0.11	−1.23	0.36	−0.36	0.61	0.28	0.86	1.08
0.12	−1.18	0.37	−0.33	0.62	0.31	0.87	1.13
0.13	−1.13	0.38	−0.31	0.63	0.33	0.88	1.18

Table 3.A.1 *(Continued)*

Proportion	*Z*-score	Proportion	*Z*-score	Proportion	*Z*-score	Proportion	*Z*-score
0.14	−1.08	0.39	−0.28	0.64	0.36	0.89	1.23
0.15	−1.04	0.40	−0.25	0.65	0.39	0.90	1.28
0.16	−0.99	0.41	−0.23	0.66	0.41	0.91	1.34
0.17	−0.95	0.42	−0.20	0.67	0.44	0.92	1.41
0.18	−0.92	0.43	−0.18	0.68	0.47	0.03	1.48
0.19	−0.88	0.44	−0.15	0.69	0.50	0.94	1.55
0.20	−0.84	0.45	−0.13	0.70	0.52	0.95	1.64
0.21	−0.81	0.46	−0.10	0.71	0.55	0.96	1.75
0.22	−0.77	0.47	−0.08	0.72	0.58	0.97	1.88
0.23	−0.74	0.48	−0.05	0.73	0.61	0.98	2.05
0.24	−0.71	0.49	−0.03	0.74	0.64	0.99	2.33
0.25	−0.67	0.50	0.00	0.75	0.67	0.995	2.58

*Calculated in Excel[R].

Appendix 3.B Test for the Significance of Differences Between *d'* Values

A simple test for the difference of two experimentally obtained d' values is discussed in Bi (2006a: 248–54). A simple and convenient test is based on a *Z*-score, where

$$Z = \frac{d_1' - d_2'}{\sqrt{V(d_1') + V(d_2')}} \tag{3.B.1}$$

where the variance estimates $V(d')$ are obtained from *B*-factors divided by the sample size N for that d'. The *B*-factors have been tabulated for different forced-choice sensory tests. For the simple yes/no task, the appropriate table is given for the "A–not-A" method in Bi (2006a: 193; table 9.6).

If the absolute value of Z obtained is greater than the tabulated Z for $1 - \alpha/2$, the conclusion is that the two d' values are significantly different at the α significance level.

References

Baird, J.C. and Noma , E. 1978. Fundamentals of Scaling and Psychophysics. John Wiley & Sons, Inc., New York, NY.

Bi, J. 2006a. Sensory Discrimination Tests and Measurements. Blackwell Publishing, Ames, IA.

Bi, J. 2006b. Statistical analyses for *R*-index. Journal of Sensory Studies, 21, 584–600.

Bi, J. and O'Mahony, M. 1995. Tables for testing the significance of the *R*-index. Journal of Sensory Studies, 10, 341–7.

Brown, J. 1974. Recognition assessed by rating and ranking. British Journal of Psychology, 65, 13–22.

Gescheider, G.A. 1997. Psychophysics. The Fundamentals. Third edition. Lawrence Erlbaum, Mahwah, NJ.

Green, D.M. and Swets, J.A. 1966/1988. Signal Detection Theory and Psychophysics. John Wiley & Sons, Inc., New York, NY.

Ishii, R., Vie, A., and O'Mahony, M. 1992. Sensory difference testing: ranking *R*-indices are greater than rating *R*-indices. Journal of Sensory Studies, 7, 57–61.

Lawless, H.T. and Heymann, H. 2010. Sensory Evaluation of Foods, Principles and Practices. Second edition. Springer, New York, NY.

Macmillan, N.A. and Creelman, C.D. 1991. Detection Theory: A User's Guide. Cambridge University Press, Cambridge.

O'Mahony, M.A. 1979. Short-cut signal detection measures for sensory analysis. Journal of Food Science, 44(1), 302–3.

O'Mahony, M. 1992. Understanding discrimination tests: a user-friendly treatment of response bias, rating and ranking *R*-index tests and their relationship to signal detection. Journal of Sensory Studies, 7, 1–47.

Swets, J.A., Tanner, W.P., Jr., and Birdsall, T.G. 1961. Decision processes in perception. Psychological Review, 68, 301–40.

4 Thurstonian Models for Discrimination and Preference

The sensory professional must be able to understand the theories and strategies of behavioral measurement. This is essential if sensory evaluation is to advance and be used to its full potential. It will involve the introduction of new methods and approaches. It will mean that some of the traditional ways of testing will need to change.

O'Mahony (1992: 38)

4.1 The Simple Paired-Choice Model

4.1.1 Introduction

Chapter 3 introduced a new way of looking at discrimination, based on variability in sensory experience. This variability affects the perception of each stimulus or product that is tested (Green & Swets, 1966/1988). The variability arises both within and between individuals. In Chapter 3 on signal detection theory (SDT), the assumption has been strictly that this variability arises within an individual over many observations of each of two stimuli. However, that is not the usual sensory evaluation testing scenario, in which products are

Quantitative Sensory Analysis: Psychophysics, Models and Intelligent Design, First Edition. Harry T. Lawless.

sampled only once (or a very few times) by each person on a panel of testers. In this chapter we will attempt to connect the SDT model to those common kinds of sensory tests as performed in the foods and consumer products industries. The chapter will focus primarily on discrimination tests, although extension of these models to preference will also be discussed.

In the first section, no consideration will be given to the issue of whether the percentage correct has been derived from many trials with one subject, from one trial with many subjects, or some intermediate combination. That is, we will not worry for now about whether the variability that contributes to our data arises from variable processes within an individual or those between an individual. Thurstone (1927) assumed that the size of the sensory difference would be the same if one observer made many judgments or many observers made one judgment – the variability would be comparable. This remains an unresolved issue in **Thurstonian modeling**, and we have no applicable model such as the beta-binomial or Dirichlet multinomial as they have been applied to replicated discrimination data. This should be a fertile area for future research. For now, we will treat all variance as the same and derive our measures of discriminability without regard to their origins. The student and practitioner should bear in mind that any process that reduces either intra-individual or inter-individual variance is likely to improve the performance and, thus, the numerical estimate of discrimination. Such processes could involve training panelists, screening them to make the panel more homogeneous, practice, warm-up or replicated trials, re-tasting, and so on.

An extensive discussion of Thurstonian models applied to discrimination and preference can be found in *Sensory Discrimination Tests and Measurement* by Bi (2006a), and further details about Thurstonian history and derivations in Baird and Noma (1978) and Gescheider (1997). The original papers by Thurstone (1927) ("A law of comparative judgment") and by Frijters (1979) (The paradox of discriminatory nondiscriminators resolved), are quite readable, clear, and are recommended to the student. The paper by Ennis (1993) remains a classic in the field.

An underlying philosophy asserts that there is an advantage that people have in a comparison task such as the 2-AFC (paired comparison or two alternative forced-choice) versus the yes/no task discussed in Chapter 3. That philosophy states that humans are naturally comparative measuring devices. We are much better at looking at two things and comparing them than making an absolute judgment of stimuli one at a time (Macmillan & Creelman, 1991). In fact, our performance often improves when we have the two stimuli to compare. In this chapter we shall see how this intuitive idea of humans as comparators is supported (or not) by the variance estimation in Thurstonian models.

4.1.2 The Law of Comparative Judgment

Let us begin with a visit to the old psychophysics. Recall that the bread-and-butter method for determining difference thresholds was the method of constant stimuli, which is basically a random sequence of paired comparisons against a fixed standard stimulus. It would probably occur to someone that, when looking at the data, the comparison stimulus that was judged stronger 95% of the time was more discriminable from the standard than one judged stronger only 60% of the time. But how can we put a ruler or measurement scheme onto this situation? The simple percentage difference does not seem like a good idea – the data often formed a curved function similar to the ogive of the cumulative normal distribution. Furthermore, we would like to get into the subject's head and see what kinds of sensations probably gave rise to these overt, observable behaviors. It is the sensation difference that we are really

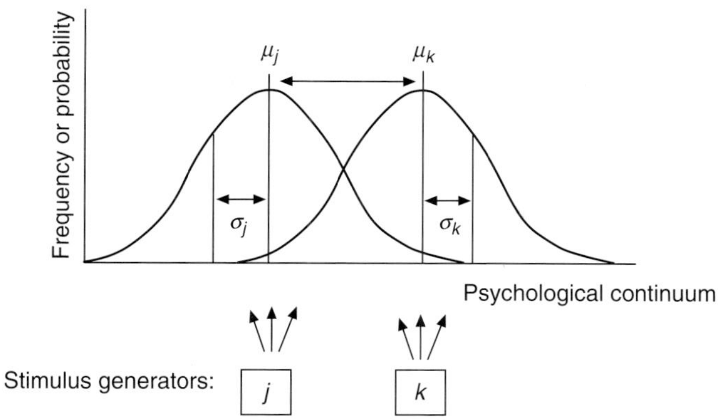

Figure 4.1 The two stimuli give rise to discriminal dispersions, the distributions of which have means and standard deviations. These dispersions are sampled in the paired comparison task, with one experience arising from each source. The goal is to determine the psychological difference, defined by the overlap of these two distributions. This difference or distance is operationalized as the difference between the mean values expressed in standard deviation units.

interested in. Consider one of the pairs from the method of constant stimuli. What would be an appropriate model?

A few basic assumptions can get us there. First, let us assume that each version of the two stimuli give rise to a variable experience, sometimes stronger and sometimes weaker. Thurstone (1927) called these **discriminal dispersions**, and they are exactly analogous to the signal and noise distributions we considered in Chapter 3. The diagram in Figure 4.1 shows how the two stimuli produce two distributions of experiences, with two mean values and two standard deviations. That is, there are four variables that define the system (Baird & Noma, 1978).

So if the subject has these two experiences, that person must choose one of them as stronger. This choice, in fact, looks like a process of subtraction (after all, we are looking for a difference). If we model this differencing as a subtractive process, sometimes we will get positive values when the physically more intense item is actually sensed as stronger, and sometimes we will get negative values when the physically more intense item is mistakenly viewed as weaker. Sometimes, then, there are reversals or mistakes. The "percentage correct" will reflect the proportion of times the former situation is true, rather than the latter. So once again we have data that look like a proportion or percentage. But how can we connect this to the difference of our discriminal dispersion distributions from the original stimuli? That difference is what we would really like to know.

Figure 4.2 shows the difference sampling distribution we will get when we draw random experiences in pairs from the discriminal dispersion distributions. For convenience, we can set the left boundary to zero; after all, we have an interval scale (Baird & Noma, 1978). Thus, the *Z*-score distance from the mean of the difference distribution to the zero point will correspond to the percentage of choice of one stimulus over the other (i.e., stimulus *k* greater than stimulus *j*). We also know from statistics that the mean of the difference distribution, expressed this way, is related to the difference of the means of the discriminal dispersions. Only the variance of the difference distribution and the discriminal dispersions are different, which we will solve below. So far, we have the following expression:

$$Z_{jk} = \frac{\mu_k - \mu_j}{\sigma_{jk}} \tag{4.1}$$

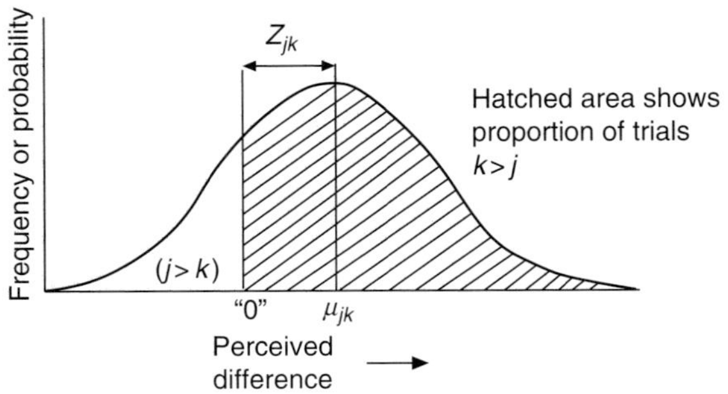

Figure 4.2 The difference sampling distribution arising from the trial-by-trial sampling from the original discriminal dispersions.

Thurstone proposed that the discriminal dispersions were normally distributed (sounds familiar). The difference distribution would arise from the process of our comparison and doing our mental subtraction. But it will have a different standard deviation, because we are looking at combining the variance from the two contributing distributions that underlie our results. Fortunately, the laws of statistics tell us what the difference distribution should look like, if we know the variance of the contributing ones. That is given by this relationship:

$$\sigma_{jk} = \sqrt{\sigma_j^2 + \sigma_k^2 - 2r\sigma_j\sigma_k} \tag{4.2}$$

where σ_{jk} is the standard deviation of the difference distribution and σ_j and σ_k are the standard deviations of the underlying discriminal dispersions. The correlation coefficient r has to do with whether the experience of one item affects the experience of the other. For example, they could contrast, in which case r would be negative. Or one stimulus might assimilate the other, in which case the r value could be positive.

If we let μ_j and μ_k be the means of the two discriminal dispersions (i.e., our best estimator for the average experience from each of the stimuli), we can obtain the complete law of comparative judgment as follows:

$$Z_{jk} = \frac{\mu_k - \mu_j}{\sqrt{\sigma_j^2 + \sigma_k^2 - 2r\sigma_j\sigma_k}} \tag{4.3}$$

And rearranging we get

$$Z_{jk}\sqrt{\sigma_j^2 + \sigma_k^2 - 2r\sigma_j\sigma_k} = \mu_k - \mu_j \tag{4.4}$$

where Z is the Z-score that corresponds to the proportion of times item k is called stronger than item j, as shown in Figure 4.2.

4.1.3 Simplifying Assumptions

Thurstone went on to review a set of "cases" that involved variations on this model. They are usually labeled as upper case Roman numerals. Case I assumed that one observer made many judgments, and this provided the paired comparison data from which we could derive the scale value for the sensory difference. Case II assumed that many observers made one judgment each, and that the level of variability one would see from that experiment would be about the same as in Case I. However, he did leave room for individual differences, especially when scaling something affective, like the goodness of handwriting samples. Case III would occur when the correlation coefficient r was effectively zero (i.e., there was no successive contrast or assimilation). Case IV explored the situation when the two standard deviations differed by a small amount. **Case V**, which is what most people use in applying Thurstone's model, also assumed that the standard deviations were equal. Furthermore, since we are using them as a unit of measurement, we can conveniently set them to equal one. This reduces eqn. 4.3 to the following relationship:

$$\mu_k - \mu_j = Z_{jk}\sqrt{\sigma_j^2 + \sigma_k^2} = Z_{jk}\sqrt{2} \tag{4.5}$$

In some cases, we can forget about the multiplicative constant, and work directly with the Z-scores. This is merely setting our scale to 1.414 times the discriminal standard deviations, but otherwise preserves the relative size of differences among multiple stimuli that were compared.

Looking back at the signal detection model, there is a correspondence of the performance in the 2-AFC task or paired comparison, and the area we expect to find under the receiver operating characteristic (ROC) curve, called $p(A)$ or A_Z. Given all the simplifying assumptions, the simple relationship of percentage correct $p(c)$ to A_Z is that they are equal. Furthermore, we can relate our performance in the 2-AFC to d' by two simple relationships:

$$A_Z = p(c) = \Phi\left(d'/\sqrt{2}\right) \tag{4.6}$$

and

$$d' = \sqrt{2}\Phi^{-1}\left(p(c)\right) \tag{4.7}$$

where $\Phi(x)$ means find the cumulative normal distribution proportion and $\Phi^{-1}(x)$ means find the cumulative normal distribution Z-score. Another way to think about this is that the 2-AFC task is somehow easier (i.e., provides a clearer picture of the sensory difference) than the yes/no task, and that the factor for this advantage is related to the square root of two (Macmillan & Creelman, 1991). If the variances are unequal, we can still connect the percentage correct and A_Z to our estimate of d_a, the d' estimate given the root-mean-square average of the two variances, by substituting d_a for d' in eqn. 4.5 (Macmillan & Creelman, 1991).

4.1.4 Case Studies

4.1.4.1 *Case Study 1: Engen's Odor Preference Data*

In the early 1970s, the father of olfactory psychophysics, Trygg Engen, did a study of odor preferences among young children. There were surprising reports that children did not react as strongly as adults to the hedonic or pleasurable aspects of odors, nor to the unpleasant character of things the adults found very distasteful. Engen (1974) decided to pursue this

Table 4.1 Proportions and *Z*-scores from the odor preference study for 6-year-olds compared with adults

Odorant	Butyric acid	Rapeseed oil	Diethyl phthalate	Neroli oil	Safrole
Proportion preferred					
Butyric acid	(0)	0.786	0.821	0.786	0.929
Rapeseed oil	0.214	(0)	0.679	0.750	0.679
Diethyl phthalate	0.179	0.321	(0)	0.607	0.786
Neroli oil	0.214	0.250	0.393	(0)	0.536
Safrole	0.071	0.321	0.214	0.464	(0)
Conversion to Z-scores					
Butyric acid	0	0.81	0.92	0.81	1.48
Rapeseed oil	−0.81	0	0.47	0.67	0.47
Diethyl phthalate	−0.92	−0.47	0	0.28	0.81
Neroli oil	−0.81	−0.67	−0.28	0	0.10
Safrole	−1.48	−0.47	−0.81	−0.10	0
Mean	−0.80	−0.16	+0.06	+0.33	0.57
Scale value	0.0	0.64	0.86	1.11	1.37
Adult mean	−1.60	−0.43	+0.34	+0.65	+1.04
Scale value	0.0	1.17	1.94	2.25	2.64

question, but he needed a method that would work equally well for children and adults. So he chose a method of paired comparisons, knowing that the choice proportions could be converted to *Z*-scores and thus a signal detection-related measure of hedonic distance.

The proportions and *Z*-scores are shown in Table 4.1. Note that we can order them from least to most preferred and also set the bottom item to a scale value of zero (Baird & Noma, 1978).

In this case we see that the hedonic range for children seems to be less than that of adults. The unadjusted scale values range over 1.37 units for the 6-year-olds and 2.64 units for the adults. There is one very important consideration, however, in the interpretation of such data. Remember that the proportions are really a measurement of across-subject agreement, rather than true hedonic reactions. So, one equally valid interpretation of these results is just that children are less consistent than adults in what they like or dislike as a group. That is, the adults were more homogeneous.

4.1.4.2 *Case Study 2: Synergistic Effects of Monosodium Glutamate and 5′ Ribosides*

Another early study using Thurstonian comparisons was done by Yamaguchi (1967) on mixtures of monosodium glutamate (MSG) and disodium 5′-inosinate (IMP), one of the 5′ ribonucleotides or “ribosides” that are alleged to have synergistic taste properties when mixed with glutamate salts. Yamaguchi realized that the taste intensities were quite different as the mixtures neared a 50/50 ratio, in terms of their relative intensity contributions. That is, the intermediate mixtures were quite intense and very easily distinguishable from the single compounds, or from mixtures with only a minor component from the second tastant. The problem arises that if there is 100% correct discrimination, the *Z*-score is undefined. So she had to devise an experimental procedure in which similar mixtures would be compared; that is, in cases in which discrimination would be less than perfect. The solution was to come up with an incomplete block design in which only adjacent mixtures along the

Table 4.2 Study design for MSG/IMP mixture comparisons

		Comparisons				
Sample no.	**IMP (%)**	**1st block**	**2nd block**	**3rd block**	**...**	**Final block**
1	0	+				
2	0.15	+	+			
3	0.45	+	+	+		
4	1.15		+	+		
5	1.65			+	...	
⋮					⋮	
45	99.4				...	+
46	99.9					+
47	100.0					+

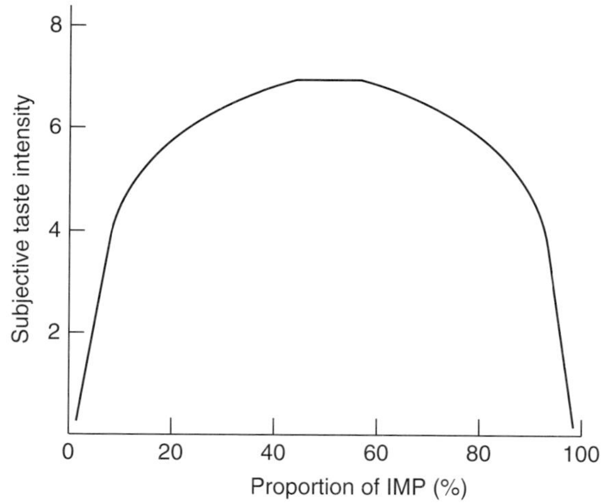

Figure 4.3 The synergistic effect in mixtures of MSG and IMP from Yamaguchi (1967). Mixtures were constructed so that the total concentration of MSG plus IMP would equal 0.5 g/l. Thus, mixtures from the center of the series had approximately equal contributions on a weight basis from the two compounds. Proportions were converted to sensory differences using Thurstone's Case V assumptions and least-squares fits. The curve shown is an approximation of her data as shown in figure 4 of the original paper.

concentration axis would be compared, or those only one step away. From this design, a set of overlapping Thurstonian values could be constructed, as shown in Table 4.2. The results confirmed that the mixtures in the intermediate range were far above any predictions based on the scale values for the individual components, or the estimated sensory intensities when there were mixtures with only minor component contributions. Figure 4.3 shows the results for the mixtures in her study.

This study raises an important point about scales that are based on discriminability or variance. As we shall see below, there are limitations to such scales when the differences become very large or obvious. That is, when we leave the range of discrimination errors, it is appropriate to use more direct scaling measures of sensory intensity and difference, as discussed in Chapter 2. This issue comes up in the measurement of odor units. Odor units are a multiple of threshold for a particular odorous chemical in a complex food, beverage, natural product, or flavor combination. One determines the threshold for the substance, and then the concentration of that substance in the product and the ratio of the two gives the odor units as a multiple of threshold. At the low end, this can be an important piece of information – substances with an odor unit value less than one are unlikely to contribute to the perceived aroma or flavor of a product, because they are below threshold. On the high end, however, they become a rather poor indicator of odor intensity or contribution. It is generally said that around six or so odor units the measure becomes not very meaningful. Once again, at that level it is time to do some direct scaling.

4.2 Extension into *n*-AFC: The Byer and Abrams "Paradox"

4.2.1 The Paradox of the Discriminating Nondiscriminators

Thurstonian modeling can explain or resolve apparent inconsistencies in discrimination testing results. This was first demonstrated in the paradox of the discriminating nondiscriminators (Frijters, 1979). The experimental result that gave rise to the paradox was the following. In 1953, Byer and Abrams asked subjects to pick out the odd sample from a triad containing 0.005% quinine sulfate ("A") and 0.006% ("B"). Twenty-one of 45 answered correctly in this triangle procedure. There were two surprising results. First, 32 out of 45 subjects correctly chose stimulus A as the weakest from the ABB triads or correctly chose stimulus B as the strongest from the AAB triads. Furthermore, 17 of the 24 subjects who incorrectly chose the wrong odd stimulus were able to select the strongest or weakest when asked to do so. They considered this a contradictory set of findings, and later Gridgeman (1970) dubbed it the paradox.

Frijters, however, was able to show that the findings were not inconsistent and that the two procedures, now known as the triangle or oddity test and the 3-alternative forced-choice test (3-AFC), gave consistent results. It was necessary to apply a Thurstonian model in order to show that the results were predictable. There were two key aspects that had to be dealt with. First, the correct psychometric model had to be devised for each task. Second, the psychometric model had to be linked to the statistical probability of a correct or incorrect choice.

Frijters reasoned as follows. Assume stimulus A produces a set of sensations. We will designate each one as S_A and further assume that they are normally distributed around some mean value μ_A. Likewise, stimulus B produces a set of sensations, each designated S_B, and they are normally distributed around some mean μ_B. Correct performance in the triangle procedure will only occur when two conditions are met, when there are two stimuli A and A′ and when B is the odd sample, as follows:

$$\left|S_A + S_{A'}\right| < \left|S_A - S_B\right| \text{ and } \left|S_A - S_{A'}\right| < \left|S_{A'} - S_B\right| \tag{4.8}$$

That is, the sensory difference of the duplicate samples must be smaller than the sensory difference from both of the duplicates when they are compared with the odd item. Another way to put this is that, on a single underlying sensory dimension, the odd sample is the outlier. For the task of picking out the strongest sample (from an AAB triad), the following must be true:

$$S_B > S_A \text{ and } S_B > S_{A'} \tag{4.9}$$

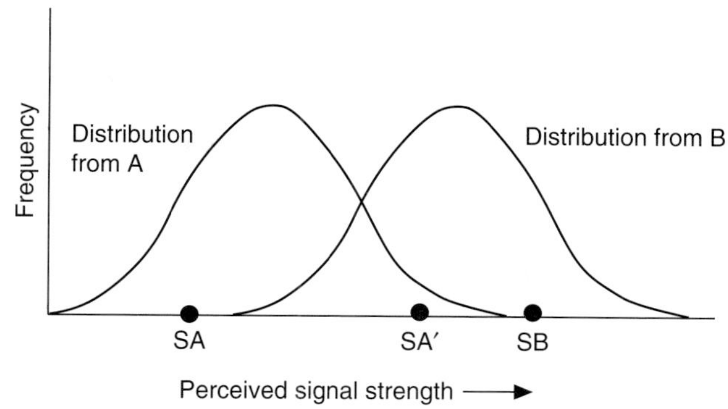

Figure 4.4 A graphical example of how the paradox of discriminating nondiscriminators can produce an apparently contradictory result. In this case, stimulus B is the strongest, but not the outlier. Stimulus A is incorrectly chosen as the odd sample in a triangle, when the momentary sensations S_A, S_A', and S_B are in the positions shown.

In other words, the sensation from item B must be stronger than both the duplicate weaker samples. These two decision processes are sometimes referred to as a **differencing strategy** and a **skimming strategy**, respectively.

These strategies apply to other tests as well. A common division between tests is whether they are specified or unspecified (Van Hout et al., 2011). The triangle and duo–trio procedures are unspecified because they do not direct the subject's attention to any particular attribute, nor to overall intensity. In a specified test, the subject is directed to pick out the strongest or weakest, or to choose the sample highest or lowest in a specific attribute such as saltiness. Because the subject uses a different strategy for the two different types of tests, a different psychometric model is appropriate; and as we shall see below, this leads to vast differences in the overall sensitivity of the test to a given size of sensory difference. Conversely, the unspecified tests require far greater numbers of subjects in order to detect a given size of a difference.

As shown in Figure 4.4, a combination of results could easily be obtained in a situation where the triangle test requirements would not be satisfied but the 3-AFC requirements are. As long as stimulus B is the strongest, eqn. 4.9 will be satisfied, even if it is not the outlier that satisfies eqn. 4.8.

The only remaining task in solving the paradox is to provide a statistical model that predicts the proportion correct, based on the sensory difference δ (where $\delta = (\mu_A - \mu_B)/\sigma$). So we need a model in which two random events are drawn from the distribution of stimulus A and one from B when B is the odd item, and two from B and one from A when A is the odd item. The solution to this in terms of a relation between the proportion correct P_c and δ is shown in eqn. 4.10:

$$P_c = 2\int_0^{\infty}\left[\Phi\left(\frac{-Z}{\sqrt{3}} + \delta\sqrt{\frac{2}{3}}\right) + \Phi\left(\frac{-Z}{\sqrt{3}} - \delta\sqrt{\frac{2}{3}}\right)\right]\frac{e^{-Z^2/2}}{\sqrt{2\pi}}\,dZ \qquad (4.10)$$

where Φ is the cumulative normal distribution function. This allows us to create a table of P_c versus δ (Frijters et al., 1980), as shown in Table 4.B.1. In a similar fashion, we can construct a statistical model that satisfies the relationships in eqn. 4.9 for 3-AFC, which turns out as follows:

$$P_c = 2\int_0^{\infty}\left[\Phi^2(Z+\delta)+\Phi^2(Z-\delta)\right]\frac{e^{-Z^2/2}}{\sqrt{2\pi}}\,dZ \qquad (4.11)$$

This allows us to construct a similar table for the 3-AFC test, as also shown in Table 4.B.1. Once we have these values, we can see that it takes a bigger sensory difference in the triangle test to get the same percentage correct in the 3-AFC. Conversely, the same sensory difference will create a lower percentage correct in the triangle than the 3-AFC due to the inherent variability of making three paired comparisons in the differencing strategy. Now Gridgeman's "paradox" can be resolved. We see from Table 4.B.1 that the proportion correct of 0.47 for the triangle gives us a δ value of about 1.31, and the proportion correct for the 3-AFC of 0.71 gives us a δ value of 1.28. So there is really no discrepancy (these values are well within their variability estimates), and there is no paradox after all.

4.2.2 Extension to Other Discrimination Tests

Once the relationship between the triangle and 3-AFC was determined, the door was open to extend the Thurstonian analysis to other tasks. There are tables of d' or δ values for a number of tasks including A–not-A, duo–trio, and same–different discrimination methods. An extensive discussion of this can be found in Bi (2006a), as well as in many tables for different discrimination test methods. A variety of software programs are commercially available or as freeware that will also give d' value and their variance estimates including the sensR program suite in the R-platform. Tables for the common tests (2-AFC, 3-AFC, duo–trio, and triangle) are given in Appendix 4.B.

The value of this for the sensory professional is that any discrimination result may now be viewed from an additional perspective, and not just from the question of statistical significance. That is, we can specify the size of the sensory difference on a universal yardstick. If management can set certain cutpoints or requirements based on d' values, they have extra leverage in making informed decisions about product differences and similarities (Rousseau, 2010). Furthermore, we can determine the power of different tests to detect true differences. Also, we have a new criterion for making sample size determinations. Both of these factors will be considered in Section 4.3. After that, the method will be extended to same–different tasks and to preference testing.

4.3 A Breakthrough: Power Analysis and Sample Size Determination

4.3.1 Test Power

Given that we now know how to estimate the size of a sensory difference from a particular test, it is also possible to specify the conditions under which a statistically significant difference will be obtained. This is most useful if the conditions are specified in terms of the size of the difference (a d' or δ value), the known or estimated standard deviations, and the chosen α level. Once these items have been specified, the power of the statistical test can be determined. Power is the ability to avoid missing a difference that is real, or in quantitative terms, $1-\beta$, where β is the probability of Type II error. The advantage of using a δ value to specify the size of the difference is that it is not dependent upon the test method, but applicable to all test methods in which the cognitive strategy is known and the correct psychometric model has been established.

The breakthrough in this area came from the work of Daniel Ennis and colleagues in the 1990s. A critical paper appeared in the *Journal of Sensory Studies* in 1993, comparing the relative power of different common discrimination tests. The paper used a normal approximation to the binomial for analysis, and a more recent paper by Ennis and Jesionka (2011) uses an exact binomial calculation and provides some updated tables for power and sample size requirements. For purposes of showing the relative merits of a Thurstonian analysis, we will use the original approximations of Ennis (1993). However, the reader is referred to the more recent paper for the updated tables. Rather than the smooth curves shown in the 1993 paper, the updated analysis shows that the power curves, although generally increasing with sample size, produced a kind of ragged saw-tooth function with small reversals – see also Bi (2011).

Ennis (1993) constructed a set of general relationships for power and sample size based on the conversion of proportion correct P_c to the sensory distance or difference δ. First, we need to find the critical point u that corresponds to $1-\alpha$ under the standard normal distribution. This is, of course, a function of the number of alternatives m and the sample size N. So we can find u using the following relationship:

$$u=\frac{1}{m}+\frac{\Phi^{-1}(1-\alpha)\sqrt{m-1}}{m\sqrt{N}}=P_{\text{chance}}+\frac{Z_{(1-\alpha)}\sqrt{P_{\text{chance}}(1-P_{\text{chance}})}}{\sqrt{N}} \quad (4.12)$$

The rightmost expression merely substitutes P_{chance} for $1/m$ for those readers who prefer to think in terms of the normal approximation to the binomial.

Then, for any given sensory difference δ, we can find the power by the following function:

$$\text{Power}=1-\Phi\left[\frac{u-p}{\sqrt{p(1-p)/N}}\right] \quad (4.13)$$

where we find the value of the proportion correct p from our tables converting δ to P_c.

4.3.1.1 *Worked Examples*

Here, we examine two worked examples, for the triangle test and 3-AFC, using $N=100$, $a=0.05$, and $d=0.74$ (from Ennis (1993)).

Triangle. First, we find the proportion correct corresponding to our δ value of 0.74, which is 0.38. Thus, for a δ value of this size we will get (only!) 38% correct choices. Finding u from eqn. 4.12, we obtain

$$u=\frac{1}{3}+\frac{\Phi^{-1}(1-0.05)\sqrt{2}}{3\sqrt{100}}=0.41$$

and the power is

$$1-\Phi\left[\frac{0.41-0.38}{\sqrt{0.38(0.62)/100}}\right]=0.262$$

What is the interpretation of this result? First, in order to be sure at the $1-\alpha$ level that we have equaled or exceeded our required value of 0.38, we need to obtain an observed proportion in the data of 0.41 or greater. Second, the chance of missing this difference, with $N=100$ is about 74% (=1−0.262). This is not very encouraging, but we have looked for a very small difference with a test that is not all that sensitive.

3-AFC. For the same values of N, a, and δ we get the same value of u, but now our proportion correct obtained will be 0.56 instead of 0.38. This is the same relationship that we saw in the "paradox" in the section above. Solving for power we obtain

$$1-\Phi\left[\frac{0.41-0.56}{\sqrt{0.56(0.44)/100}}\right]=0.998$$

The result of switching to the specified test is that we are almost certain to detect a difference of $d=0.74$ with 100 judges in a 3-AFC, and stand only a one-in-four chance of doing so with a triangle test. Figure 4.5 shows the power curves for $N=50$ and $N=100$ for the four common sensory tests (triangle, duo–trio, 2-AFC, and 3-AFC) for various δ values using the usual $\alpha=0.05$. Once the critical point u is calculated, it is relatively straightforward to get the power from eqn. 4.13 as long as one has the tables or a way of calculating the expected percentage correct for each method from a given δ value. For example, with $N=100$, the u values are 0.41087 for a test with $P_{\text{chance}}=1/3$ and 0.58225 for a test with $P_{\text{chance}}=1/2$.

4.3.2 Sample Size *N*

Turning this around, the sample size necessary for a certain power and a given size of sensory difference can be determined. As one might expect, it takes a smaller sample size with the 3-AFC to obtain the same power as it does with the triangle test. A general formula for sample size in terms of the specified δ value and its corresponding proportion correct can be formulated as follows. Remember that β is given by

$$\beta=\Phi\left[\frac{u-p}{\sqrt{p(1-p)/N}}\right] \tag{4.14}$$

And the Z-score equivalent then becomes

$$Z_{\beta}=\Phi^{-1}\beta=\frac{u-p}{\sqrt{p(1-p)/N}} \tag{4.15}$$

A general equation for sample size is the following:

$$N=\left(\frac{Z_{\alpha}\sigma_{\alpha}+Z_{\beta}\sigma_{\beta}}{p_{\text{c}}-p_{\text{a}}}\right)^{2} \tag{4.16}$$

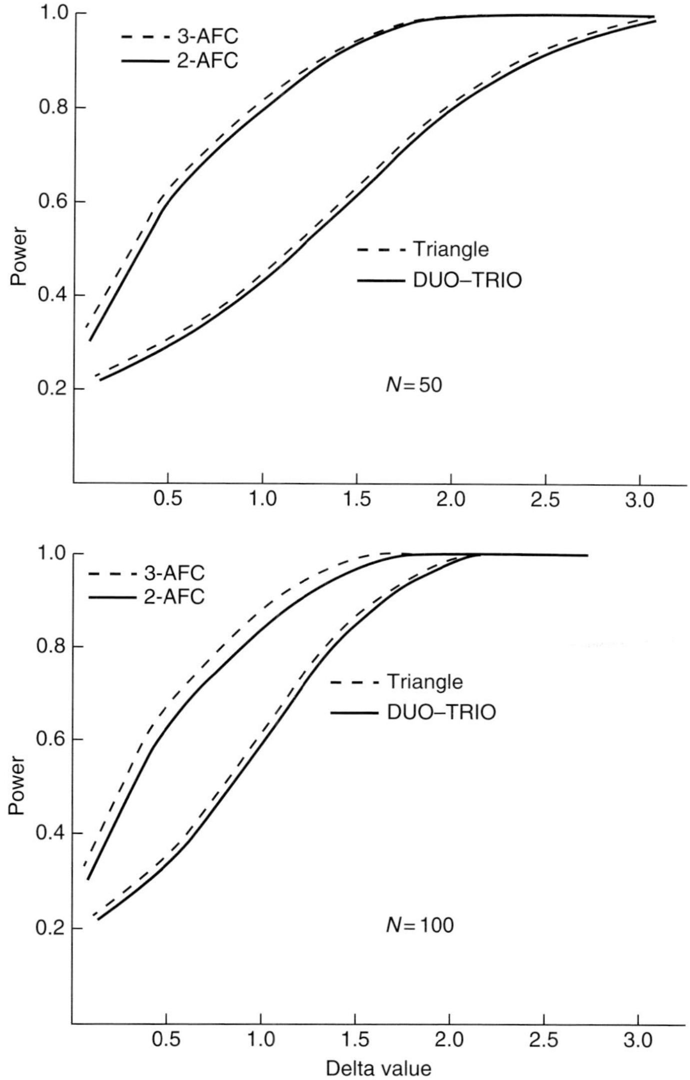

Figure 4.5 Power curves for different levels of sensory difference δ for four common sensory discrimination test methods (the duo–trio, triangle, 3-AFC, and 2-AFC) for $N=50$ and $N=100$, $\alpha=0.05$.

where p_c is the chance probability or $1/m$, p_a is the probability associated with the alternative hypothesis, or our Thurstonian estimate of the proportion correct that would be obtained based on the method we use and its corresponding δ value, and the σ values are the standard deviations of the proportions; so

$$\sigma_\alpha = \sqrt{p_c(1-p_c)} \text{ and } \sigma_\beta = \sqrt{p_a(1-p_a)} \tag{4.17}$$

For our example using the Thurstonian analysis of any discrimination test, the sample size calculation becomes a little easier if we first compute the quantity y instead of $Z_\alpha \sigma_\alpha$ (Ennis, 1993):

$$y = \frac{\Phi^{-1}(1-\alpha)\sqrt{m-1}}{m} \tag{4.18}$$

And then the required sample size N becomes

$$N = \left[\frac{\Phi^{-1}(\beta)\sqrt{p(1-p)} - y}{(1/m) - p} \right] \tag{4.19}$$

The subtraction in the numerator is due to the sign of the Z-score for β. Trust me, it works.

Returning to our triangle and 3-AFC tests (so $m=3$), and letting $\alpha=0.05$, a δ value of 1.0 and 90% power ($\Phi^{-1}(\beta)=-1.28$), we get the following relationships:

$$y = \frac{\Phi^{-1}(0.95)\sqrt{2}}{3} = 1.645\frac{\sqrt{2}}{3} = 0.775$$

For a $\delta=1.0$ for the triangle we get $p=0.418$. This gives us an N estimate of

$$N = \left[\frac{-1.28\sqrt{0.418(0.582)} - 0.775}{0.333 - 0.418} \right]^2 \cong 276$$

So for 90% power ($\Phi^{-1}(\beta)=-1.28$), we would need 276 panelists.

For the 3-AFC, we get a p value of 0.634. The y value stays the same, but now the N calculation yields

$$N = \left[\frac{-1.28\sqrt{0.634(0.366)} - 0.775}{0.333 - 0.634} \right]^2 \cong 21.4$$

So we would need only about 22 people; once again, a much better bargain if we can do a 3-AFC instead of a triangle. Bi (2006b) suggested a continuity correction to be added to the sample size estimate equal to $2/(p-p_{chance})$. For the triangle test, this comes to $2/(0.418-0.333)=24$, so we should revise up to 300 to be on the safe side for the triangle and 46 for the 3-AFC.

4.4 Tau Versus Beta Criteria: The Same–Different Test

4.4.1 The Same–Different Paired Test

In Section 4.2, the difference between the triangle test and 3-AFC was explored and a key factor was the difference in cognitive strategy. In the 3-AFC task, like the yes/no procedure of signal detection, the observer is simply looking at the overall intensity experienced from each item. In the triangle procedure, one is obligated to look at the differences amongst pairs of items. In the yes/no task, one simply decides whether the current experience is above or below one's criterion, which we called β (**beta criterion**) In this section we will examine some models for another kind of paired test, the same–different test. This test can involve two different strategies. One of them involves a consideration of the degree of difference experienced in each pair. If it is greater than a certain amount, the subject will respond "different"; if less than that amount, the subject will respond "same." This amount involves a new kind of criterion, one for the size of the experienced difference. We will designate this as the letter tau (τ, or **tau criterion**). Sometimes a comparison of differences is referred to as a τ strategy, to differentiate it from a comparison to simple intensity, called a β strategy. In the β strategy for the same–different test, we would consider each item individually at first and then decide if they are on the same side of our boundary or cutoff or on a different side.

The same–different task has enjoyed some popularity as a discrimination test method for several reasons. First, it is relatively simple for people to understand and to perform (Stillman & Irwin, 1995). Second, it can be used when the underlying dimension cannot be specified and when the perceived attribute or attributes that will change cannot be predicted. Thus, it provides another alternative to the unspecified overall difference methods such as the triangle and duo–trio. Third, it is amenable to a detection theory analysis (Irwin et al., 1993; Hautus & Irwin, 1995; Irwin & Hautus, 1997). Also, the method can be extended to include a rating scale for sureness or certainty judgments (Delwiche & O'Mahony, 1997). This is similar to a degree of different rating scale, although one can argue that degree of difference and certainty are two different matters. Finally, the same–different method may be more sensitive to differences than the triangle test, as shown by higher d' values (Rousseau et al., 1998). The only potential drawback to the method occurs when a balanced or complete design is used. In that case, each subject or panelist would see both an identical pair and a different pair, so the total number of products to be evaluated is four, rather than three in the triangle procedure, or three in the duo–trio (without warm up). So there would be a greater opportunity for fatigue or adaptation.

A statistical analysis of the same–different test is also straightforward. A common design gives both the identical (control) pair and the test (different) pair to each panelist. This permits a cross-classification in a 2×2 table as to whether they responded same or different to each pair. An appropriate statistic, then, is the McNemar test for changes (Bi, 2006a; Lawless & Heymann, 2010). This essentially compares the size of the cell (number of panelists) in which they responded "different" to the test pair and "same" to the control pair with the size of the cell in which both responses were reversed; that is, "same" to the different pair and "different" to the control pair. Panelists who responded with both "different" or both "same" to the two pairs are not counted. Although this deletion lowers the effective N and power of the test, it makes sense because they are not transmitting any information if they respond to the two pairs in the same way.

4.4.2 A Thurstonian Model for Same–Different

There are a couple of strategies for subjects to use when they are given the same–different task. First, we will examine a simple strategy in which the subject compares the two stimuli. This is the same idea as in our 2-AFC analysis, in that a subtraction is required. If the difference obtained is greater than some value τ, the subject will respond "different." If less than τ, they will respond "same." In this section, whenever an adjective is placed in quotation marks, it will be that the word is a response. If not, it refers to the stimulus pair.

For simplicity, let us assume we have discriminal dispersions with variance equal to one. Then with a differencing procedure, the difference distributions will have variance equal to two, and standard deviations equal to the square root of two. Figure 4.6 shows that the identical pairs will generate a difference distribution with mean zero. The different pairs will generate distributions with means $+d'$ and $-d'$. Given a cutoff value for "different" of $\pm k$ $(=\tau/2)$, the hit and false-alarm rates become as follows (Macmillan & Creelman, 1991):

$$p(\text{hit}) = p(\text{“different”} \mid \text{different}) = \Phi\left(\frac{-k+d'}{\sqrt{2}}\right) + \Phi\left(\frac{-k-d'}{\sqrt{2}}\right) \tag{4.20}$$

and

$$p(\text{FA}) = p(\text{“different”} \mid \text{same}) = 2\Phi\left(\frac{-k}{\sqrt{2}}\right) \tag{4.21}$$

A further complication to this analysis is that the ROC curves do not have unit slope when plotted as Z-scores due to an asymmetry in the curve. So we cannot simply find d' from subtracting Z(FA) from Z(hits). However, Kaplan et al. (1978) made calculations based upon different hit and false-alarm values, and these tables were reproduced in Macmillan and Creelman (1991).

Another common analysis of same–different data is to convert the performance to a single estimate of the proportion correct P_c. This is given by

$$P_c = \frac{H + (1 - \text{FA})}{2} \tag{4.22}$$

where H is the proportion of hits ("different" | different) and FA is the proportion of false alarms ("same" | different). The variance for P_c is given by

$$\sigma^2_{P_c} = \frac{H(1-H) + \text{FA}(1-\text{FA})}{4N} \tag{4.23}$$

according to Hautus and Irwin (1995). They caution, however, that P_c is highly dependent upon the method from which it is generated, unlike d'. So the interpretation needs to take into account the details of the procedure, or restrict one's interpretation to only compare those methods with exactly the same task.

4.4.3 An Alternative Strategy: Independent Observations and/or Classification

The interpretation of same–different results is further complicated by different strategies that can be adopted by the panelists. An alternative strategy for panelists in the same–different task is to judge each item separately. If they both fall on the same side of some criterion level β, then they are reported as "same." If they fall on opposite sides, the subject would respond "different." In some cases this strategy shows an advantage over the differencing strategy. Macmillan and Creelman (1991) give the following example.

Suppose we are told the heights of two children sampled from the fourth and fifth grades. Our task is to judge whether pairs of random children come from the same grade or different grades. We know that the mean heights are 52 and 54 inches, and so we decide in the

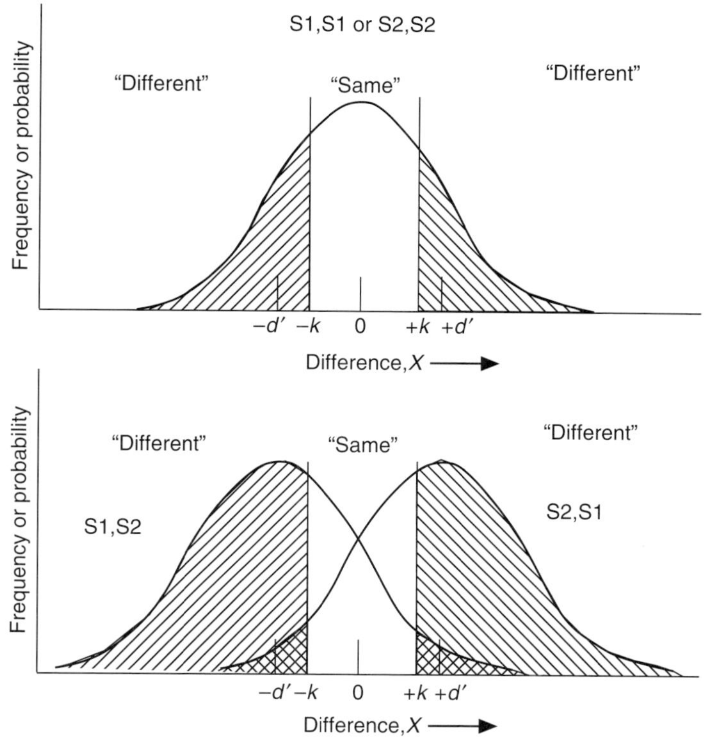

Figure 4.6 A Thurstonian model for the same–different task based on a differencing strategy: (a) false-alarm rates in the cross-hatched areas bounded by the criterion boundaries $\pm k$ (where $k = \tau/2$); (b) the hit-rate areas.

differencing strategy to call any difference of 2 inches or greater "different" and any difference less than 2 inches "same." Suppose we have a couple of really tall girls in the fifth grade, who have started their pre-teen growth spurt and are 62 and 65 inches tall. The differencing strategy would incorrectly classify them as "different," creating a false alarm. However, if we looked at them independently and decided that they fell on the same side of a single cutoff, we would correctly classify them as fifth graders, and then respond "same."

An optimal strategy for the independent observations model, then, is to make some cutoff at a criterion level where the two sampling functions cross, or in other words where the likelihood ratio is about 1.0. If our two observations come from zones where the likelihood ratios are both greater than or both less than one, we will maximize our performance. A likelihood ratio model for this task was developed by Irwin and Hautus (1997).

Separate tables for d' have been generated for the differencing and independent observations models (Macmillan & Creelman, 1997). How can we decide which strategy people are using? One approach is to look at ROC curves, which can be generated from rating scale data similar to the R-index method or by using sureness judgments. The differencing model predicts an asymmetric ROC curve (flatter at the high end), and in at least one set of experimental observations this seems to be the case (Hautus & Irwin, 1995). However, with a group of individuals who are each contributing only two or very few judgments, it is not possible to determine their strategies with any degree of certainty. So the sensory professional should consider both d' values, and then make a decision based upon whether Type I or

Type II errors are more serious. Am I more likely to suffer consequences if I declare a difference when none exists, or when I miss a difference that is really present in the larger population? Of course, if both models produce d' values that exceed some critical value for our decision making (or both do not), then the decision is straightforward. A further advantage of generating an ROC curve is that the area under the curve, even with a differencing strategy, corresponds to the maximal proportion correct expected with the more optimal independent-observations strategy (Irwin et al., 1999).

4.4.4 A Further Complication: ROC Curves and *R*-Index Revisited

For a given d' value (i.e., equal sensory differences), does the ROC curve for the yes/no task and the same–different task look identical? Unfortunately, the answer may be no (see Macmillan and Creelman (1991: 152, figure 6.5) for an example). The same–different task is less efficient, and produces a much flatter ROC curve, with less area below it. In Macmillan and Creelman's example, it takes a d' of about 2.0 to get the same ROC curve from same–different as it does for a d' of 1.0 in the yes/no task.

The same issue arises with an *R*-index generated from the same–different task. In one version of this method, the panelists use a rating scale with four responses: "same, sure," "same, not sure," "different, not sure," and "different, sure" (Irwin et al., 1993). Does this generate the same *R*-index as the yes/no task? Once again, perhaps not. Ennis et al. (2011) show curves relating d' to the *R*-index for the two tasks (reprinted from Rousseau (2006)). The *R*-index from the same–different rating scale method can trail the *R*-index from the yes/no task by almost 20% for a given d' value. Thus, for a given d', the *R*-index generated from this task is much less efficient. This should caution anyone using the simple binomial test for the significance of *R*,[1] and to not use tables developed for the *R*-index as if it came from a yes/no procedure (Bi & O'Mahony, 1995) when in fact it came from the same–different test when it contains sureness ratings. In other words, your result will be method dependent. The exact relationship of *R* to d' will also vary with the number of categories used on the sureness scale (Rousseau, 2006).[2]

Given the correspondence of the yes/no *R*-index to the proportion correct in 2-AFC, the same discrepancy holds when we try to compare the same–different *R*-index with 2-AFC or the area under the ROC curve $p(A)$. Irwin et al. (1993) gave the following example: the mean value of $p(A)$ from their data was 0.78 from the same–different task in that paper, which corresponded to a d' value of 2.36. However, in a 2-AFC task, the same d' value would yield a proportion correct of 0.95! Converting from rating scale data in the same–different tasks was discussed using maximum-likelihood techniques to estimate d' in Irwin et al. (1993), and a theorem for calculating areas under the ROC curve was presented in Irwin et al.

[1] The simple binomial formula for the significance of R, as appropriate to the yes/no task is

$$z = (R - 0.5)/\sqrt{R(1-R)/(n-1)}.$$

[2] A further cautionary note, however. The previous argument (about discrepant d' from the same *R*-value and vice versa) appears to hinge on models that use the correspondence of the 4IAX task to the same–different task. In a 4IAX procedure, the subject is presented with two intervals, one containing the same pair and one a different pair, and must just choose between the two. Such a strategy may or may not be adopted in the same–different task, in which each pair may in fact be judged independently, and not as part of a larger quartet. Further models may be required. See Irwin et al. (1999) for a model for these tasks and the proposed relationship to 4IAX.

(1999). An original discussion of signal detection measures from same–different and other tasks can be found in Macmillan et al. (1977).

4.5 Extension to Preference and Nonforced Preference

As noted above in the case study from Engen's data with children, proportions of preference can also be converted to Z-scores and a kind of hedonic d'-value associated with the degree of preference (Bi, 2006a). In the case of data from a single individual, this might be interpreted as a degree of internal consistency or, conversely, the amount of hedonic variation experienced by that person. Such variation, of course, might arise as a function of the discriminal dispersions of the sensory intensities from the stimuli. However, they could also just reflect a change in attitude due to boredom, need for change, or any other kinds of reasons for a hedonic shift. In the case of group data, the most reasonable interpretation is simply the homogeneity versus heterogeneity of the likes and dislikes of the group, but once again this could arise from differing sensory experiences as well as from different preferences for the same sensory experience. We just do not know without further information. So the sensory professional should be careful in interpreting the meaning of a hedonic d'.

In the case of nonforced preference, the Thurstonian model can be very helpful in interpretation, as well as providing an alternative test for a significant preference. Nonforced preference occurs when we allow a "no-preference" option as one of the responses, rather than forcing a choice. This is analogous to allowing an "uncertain" response in a 2-AFC (making it a 2-AC test instead). Assuming each product in a paired preference gives rise to a covert response or experience along a unidimensional hedonic continuum, we can apply a difference distribution model and a τ criterion to fit a d' value (Ennis & Ennis, 2010; see Braun et al. (2004) for an application of this analysis to the 2-AC task).

Figure 4.7 shows how a Thurstonian analysis is applied to the no-preference data. First, we can assign a Z-value to each proportion. These proportions will correspond to areas under the difference distribution, as shown in Figure 4.7(a). Next, we transform this distribution by subtracting d' from everything and by dividing by the square root of two, getting back to our unit variance. This produces the transformed distribution shown in Figure 4.7(b). Assuming a result where 20% prefer B, 20% have no preference, and 60% prefer A, we can assign a Z-value to each of our transformed proportion boundaries. Specifically, the leftmost boundary corresponds to the Z-score for 0.2 or –0.84. The next boundary corresponds to the Z-score the sum of 0.2 and 0.2, or 0.4, which equals –0.25. We now have two equations in two unknowns, so we can solve for τ and d' given the following:

$$\frac{-\tau - d'}{\sqrt{2}} = -0.84 \quad \text{and} \quad \frac{+\tau - d'}{\sqrt{2}} = -0.20$$

Solving for τ and d', we get $\tau = 0.42$ and $d' = 0.77$. We can now use this information to test for a nonzero d', in order to see whether there was a significant preference.

Power analysis has been performed to compare this analysis and test (for nonzero d') to other strategies for dealing with no preference votes (Ennis & Ennis, 2010). This analysis shows that the Thurstonian model is less powerful than splitting the votes in proportion to the remaining preferences (i.e., fudging the data), but more powerful than dropping them from the analysis (which lowers N and essentially discounts the information) or splitting

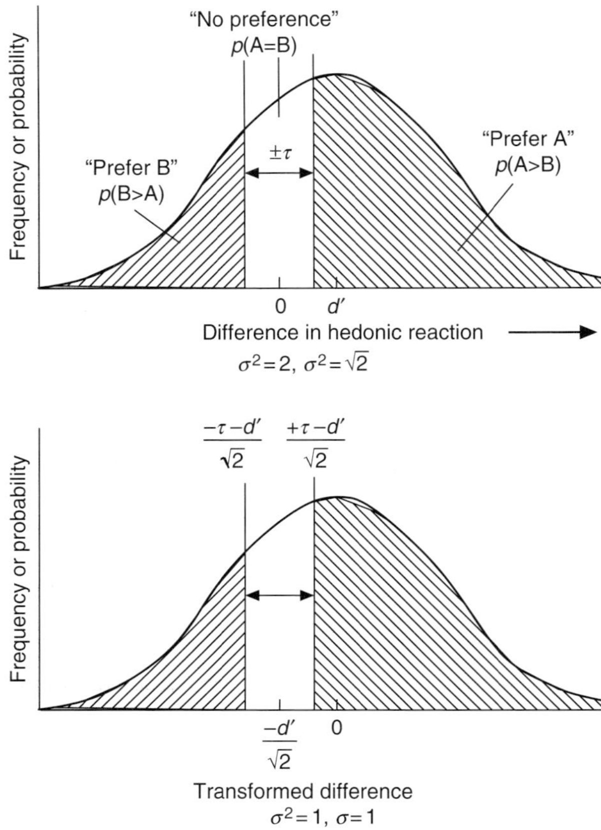

Figure 4.7 Thurstonian model for preference testing with a no-preference option. The transformed difference distribution (b) has had d' subtracted from each value to set the mean equal to zero and been divided by $\sqrt{2}$ to create a unit standard deviation.

them equally (which dilutes the proportions obtained). Of course, whether one takes a more liberal or conservative approach depends upon the relative liabilities of committing Type I or Type II error. All things being otherwise equal, α and β are inversely related, so a decision for high power usually entails greater risk of a false alarm.

4.6 Limitations and Issues in Thurstonian Modeling

4.6.1 From Breakthrough to Breakdown

There are significant advantages of a Thurstonian model for sensory difference measurement. Most importantly, they provide a method-free measure of sensory differences so that the results from different discrimination test methods and different panelist strategies can be directly and appropriately compared. So one might question whether there are any problems at all with these methods, or issues and complications worth noting. This section will mention a few.

We have seen that a d' or δ value can be generated from discrimination data. But what properties do these measures have? Do they correspond to the pattern of responses we would see with direct scaling methods? One problem is that they are based on variability, so that any experimental modification that changes the variability of the situation is likely to give a different value, such as improved stimulus control, panel training, and re-tasting. However, models can be developed to accommodate such parameters. A more fundamental question is whether you wish to base your estimate of sensory changes on what is essentially error in the system?

As noted above, one of the drawbacks of using odor units (multiples of threshold) as an odor intensity measure is that they tend to break down somewhere in the range of five or six units. That is, they become of questionable utility and direct scaling seems more appropriate. δ values are like odor units in a way, just based on a kind of difference threshold determination rather than a detection threshold. Does a similar limitation apply do δ values? Look again at the design in the taste synergy study of Yamaguchi (1967). She could not compare mixtures that were too different, but only those that were confusable. We can also ask what happens when the discrimination becomes too easy and discrimination performance is very good.

A relevant comparison was conducted by Lawless and Schlegel (1984), who looked at discrimination of taste and odor mixtures (sucrose and citral) at two levels (low and high) of each. These mixtures were, for the most part, somewhat discriminable and somewhat confusable, but just at the level that direct scaling (using intensity ratings) also was possible. We asked the simple question of whether the sensory differences, as determined in triangle tests, would yield a pattern where the δ values could be added up, like physical measures of length or mass. If not fully additive, could we construct a simple geometric or vector model that would give a reasonable prediction or sensible model?

Figure 4.8 shows the pattern of results. The d' values show that the results from triangle testing could be fit into a two-dimensional geometric model that generally recaptures the high and low levels of the two mixture components. But there are two important discrepancies. First, when the mixture with the two low levels is compared with the mixture with the two high levels, performance shot way up, generating a large d' value that did not fit the rest of the configuration (comparing A with D). In this case, the changes were not additive. A similar result was found comparing d' data from triangles with data from same–different sureness ratings (Kamerud & Larson, 2010). When the triangle test is very easy, the d' values diverge, with the triangle d' values becoming unexpectedly high. Note that this was not a ceiling effect due to perfect performance in the case of Lawless and Schlegel (1984). Second, there was an asymmetry – changing both levels, but in opposite directions, did not produce the same gain, and in fact this discrimination turned out to be quite difficult (comparing B with C). Now, one might possibly explain the first result by a change in cognitive strategy (from differencing to skimming or from τ to β strategies), but it serves as warning that the simple triangle test model is not necessarily fixed, perhaps not even for the same subjects in the same study, nor does it necessarily produce a simple additive measure of sensory distance. A third discrepancy noted in that study was that, when direct scaling of sweetness and lemon odor were performed, there were some highly significant differences in the direct scaling task that were not apparent in the discrimination test results. The direct and indirect scalings were not comparable. Thus, the sensory professional should be careful of substituting Thurstonian measures for direct scaling results, even when it can be done.

Another interesting discrepancy was noted by Warnock et al. (2006) in a comparison of Thurstonian analysis of ratings data with triangle data on the same stimuli and with a more common or standard analysis of variance (ANOVA) on the ratings data. Consider the case

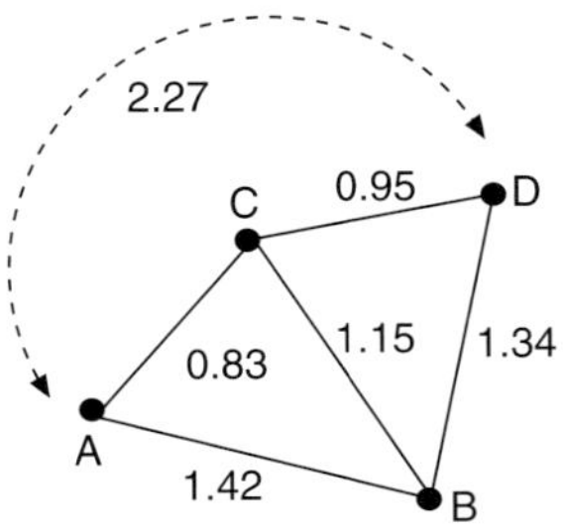

Figure 4.8 The pattern of sensory differences for mixtures of high and low levels of citral and sucrose, as determine from d' values generated from triangle tests. Numbers at each line segment give the approximate d' values. Stimulus A was lowest in both sucrose and citral; stimulus D was highest in both. B and C were high in one and low in the other attribute.

of the preference data with a no-preference option. This situation can be viewed as a kind of ratings data on a three-category scale. So, in theory, any category scale used to generate a set of data can be subjected to a similar analysis, in which the category boundaries, analogous to the τ values in the no-preference analysis, can be found and any two products compared on the basis of d' values, by using the frequency counts for each category. This type of analysis has certain advantages over treating the category values as plain integers, due to panelists' unequal use of categories, end-category avoidance, uneven variance for high and low samples across the scale, and unequal psychological spacing if the categories are labeled with numbers. Software has become available for Thurstonian analysis of ratings data (Ennis et al., 2011) so d' values can be compared as a potentially useful alternative to traditional ANOVA.

In the case of the Warnock et al. (2006) data, however, Thurstonian analysis and ANOVA produced similar results. Somewhat more troubling was the fact that the d' values obtained from ratings data and from triangle tests were discrepant, and triangle tests produced surprisingly higher d' values. This was true even when the triangle data were analyzed as if they were 3-AFC; that is, as if panelists had changed from a differencing to a skimming strategy. One potential reason for this discrepancy was that the number of categories was quite large (18), more than would usually be used for a simple category scale. So some of the categories may have been used very infrequently, adding some uncertainty to their estimated parameters.

4.6.2 Effect of Multidimensionality

Many of the writings on Thurstonian models assume that there is a single underlying perceptual continuum upon which discrimination and judgments are made. For example, we find the following quote from a recent paper by Ennis and Ennis (2011) on preference modeling:

> In this example we assume that there is a single variable driving preference such as sweetness, and that the products differ on this attribute.

However, foods and consumer products are usually multidimensional. Furthermore, consumers or even trained panelists may decide to focus on different aspects of the product among each other, and even within the same person at different times. It is not known whether Thurstone ever worried about this issue; but it seems unlikely, because he extended his modeling to include such multidimensional scenarios as the excellence of handwriting samples, which issue must surely be complex.

There are significant theoretical publications on multidimensional models for discrimination from Ennis and coworkers (Ennis & Mullen, 1985, 1986; Ennis et al., 1988). Macmillan and Creelman (1991) devoted a chapter to multidimensional detection models in their classic text. However, these considerations seem not to have made the transition in applications for everyday sensory testing. One relevant finding is potentially troubling. Irrelevant variation in other dimensions (attributes not intended to vary or having a d' of zero) can create a set of false-positive impressions of differences and lead to incorrect choices degrading the value of δ that would be estimated if they were not present, or not varying (Ennis & Mullen, 1985). Another way to think about this is that the percentage correct for a given d' value will be less, due to the variation in these other dimensions. These facts, of shifting attention and shifting criteria for product preferences, are well known in the marketing research literature, but need further attention and better modeling from quantitative sensory analysis. Ennis and Mullen (1986), for example, showed how the number of dimensions, dimensional weighting, and the covariance of two attributes could affect the relationship between d' and the percentage correct. The covariance factor may have played an important part in the results from Lawless and Schlegel (1985) discussed above.

4.6.3 The Individual Versus Group Data

Another issue that is often glossed over in the use of Thurstonian values in common industry tests was mentioned above. That is, the issue of whether the models developed to describe individual performance, and based on individual variability in sensations, are applicable to group data. In the extreme cases, we have one person tested many times versus many people tested only once. Yet the latter situation is not uncommon in applied sensory evaluation. One fact is relevant. Group data with varying criteria should yield lower estimates of d'. Macmillan and Kaplan (1985) showed that averaging response rates across subjects (who may have different criteria) would underestimate the d' values obtained from an averaged d' across subjects. Thus, it would seem that combining group data may not always give the same estimates. This is a cautionary note insofar as we have assumed that the variability across a group of people tested once is similar to the variability within an individual tested many times, as assumed in Thurstone's Case II. A clever way around this issue is to collect only a limited amount of judgments from a few selected observers, but generate more data from them via a resampling or jackknife type of procedure (see Hautus and Irwin (1995) for an example). The extension of models such as the β-binomial or Dirichlet multinomial (which account for individual variation in repeated testing) into the realm of Thurstonian modeling would seem appropriate at this time.

4.6.4 Tests of Significance Using d′ Values

A further issue arises when using nonzero d' as a test for difference or preference. In the case of the unspecified tests such as the triangle or duo–trio, the variance estimates or

B-values for d' go through a minimum (of about $d'=2$) and then begin to increase again as d' approaches zero. Thus, using the variance estimate to do a simple *Z*-test on whether a d' is larger than zero becomes problematic. For low values of d' it is difficult to establish a nonzero d' value unless the sample size *N* is very large. This reduces the effective power of the test if you are using the common discrimination test panel size of $N<100$. Of course, the result of a test of low power or low sensitivity is an effective increase in the risk of Type II error, which should always be weighed as an important concern in product development. Type II errors can lead to lost opportunities in new product development and can lead to potential franchise losses if poor products are developed (say, in a cost-reduction program) and foisted upon loyal consumers who can detect the difference after all.

One solution to this problem is suggested by Bi (2011). One can still use a d' as a criterion for a critical difference or action standard or cutoff for product similarity or equivalence. However, rather than doing the statistical test directly on d', one coverts back to proportion correct. Using the tables for each particular type of discrimination test, one finds the percentage correct equivalent, or P_c value, for that required d'. Then a critical value x_o can be found that indicates the critical proportion that must be obtained in order to be significantly below our P_c cutoff. This value x_o will satisfy the following simple binomial relationship:

$$\sum_{x=0}^{x_o} \binom{n}{x} P_c^x \left(1-P_c\right)^{n-x} < \alpha \qquad (4.24)$$

Now this test is somewhat conservative, and it creates potential problems when the critical value x_o is less than the number expected by chance. In other words, chance performance presents a kind of floor effect that can limit the utility of this approach, especially with small sample sizes. In Bi's published example (Bi, 2011: 152) with a critical d' of only 0.2 and $N=100$, all four common difference tests require values less than chance (less than 50% or less than 33% of the sample) in order to meet the criterion for similarity!

4.7 Summary and Conclusions

In this chapter, we considered how Thurstonian analysis was related to the signal detection measures of Chapter 3 and showed the basis for the simpler univariate models. Thurstonian models provide a powerful tool for sensory analysis. They give us a common ruler for comparing the size of a sensory difference generated by different methods. They help explain previous paradoxical results in sensory testing. The methods can provide action standards for sensory results in quality control, preference testing, and advertising claims. To date, many of the models seem simplistic when one considers the multidimensional nature of foods and consumer products. Surprisingly, the univariate models do an excellent job of fitting many sensory experiments. Software is readily available for generating d' or δ values from sensory data as well as published tables. Further developments are sure to come as the complexity of different products, different testing situations, and human tendencies such as attentional shifts and other foibles are considered.

Appendix 4.A The Bradley–Terry–Luce Model: An Alternative to Thurstone

Given all the sound and fury that has surrounded Thurstonian models in the past two decades, we should ask if there are alternatives, and there are many. One historical model for choice behavior arises from behavioral psychology. It is known as the Bradley–Terry–Luce model after its originators (Suppes & Zinnes, 1963). The model has several axioms and assumptions. One is that the choice should be independent of any selection procedure. That is, the outcome does not depend upon other factors, such as the order in which multiple alternatives are compared or eliminated. This includes situations in which alternatives are grouped into categories first (think about going to several stores and choosing one store above the others) and situations in which alternatives are completely subordinate to others (zero probability of choice in that comparison).

As in Thurstonian scaling, we can derive scale values from choice proportions. The simplest case, of a paired comparison, is illustrated below, borrowed from Baird and Noma (1978). Let us assume that the probability $p(j,k)$ of choosing item j over item k is related to their scale values v_j and v_k. You can think of this as choosing the louder of two loudspeakers. So we have

$$p(j,k) = \frac{v_j}{v_j + v_k} \tag{4.A.1}$$

And dividing by v_j gives

$$p(j,k) = \frac{1}{1 + (v_k / v_j)} \tag{4.A.2}$$

Now, since any number may be represented by a base (in this case e) raised to some power, we can represent the mean scale values, μ_j and μ_k as follows:

$$p(j,k) = \frac{1}{1 + (e^{\mu_k} / e^{\mu_j})} \tag{4.A.3}$$

and then

$$p(j,k) = \frac{1}{1 + e^{-(\mu_j - \mu_k)}} \tag{4.A.4}$$

This expression may look familiar, as it is a form of the logistic distribution.

Solving for the scale difference $(\mu_j - \mu_k)$ we get

$$\frac{1}{p(j,k)} = 1 + e^{-(\mu_j - \mu_k)} \tag{4.A.5}$$

and then

$$\frac{1}{p(j,k)} - 1 = e^{-(\mu_j - \mu_k)} \tag{4.A.6}$$

Taking logarithms, we get

$$-(\mu_j - \mu_k) = \ln\left(\frac{1 - p(j,k)}{p(j,k)}\right) \tag{4.A.7}$$

and then

$$\mu_j - \mu_k = -\ln\left(\frac{1 - p(j,k)}{p(j,k)}\right) \tag{4.A.8}$$

Remember that in Thurstonian modeling, we can represent $(\mu_j - \mu_k)$ on the basis of converting it to its *Z*-score. In this case we are modeling it as an inverse of the logistic function; that is, the natural log of the inverse of an odds ratio, sometimes called the anti-logit.

If we have several alternatives and several paired comparisons, we can obtain the scale values by an averaging technique (Baird & Noma, 1978). The point here is that we should not be eternally wedded to the normal distribution when other models may be equally appropriate or better fits.

Appendix 4.B Tables for delta Values from Proportion Correct

Tables 4.B.1 and 4.B.2 show approximate values for d' or δ, for common discrimination tests. For further precision, users are encouraged to use appropriate software.

Table 4.B.1 Percentage correct versus sensory difference δ for the triangle and 3-AFC procedure. Values are rounded to the second decimal place. Calculated in SensR

Proportion	Triangle	3-AFC	Proportion	Triangle	3-AFC	Proportion	Triangle	3-AFC
0.33	(0)	(0)	0.54	1.67	0.69	0.75	2.80	1.43
0.34	0.27	0.02	0.55	1.72	0.72	0.76	2.86	1.48
0.35	0.43	0.06	0.56	1.77	0.75	0.77	2.92	1.52
0.36	0.55	0.09	0.57	1.82	0.79	0.78	2.99	1.56
0.37	0.64	0.13	0.58	1.87	0.82	0.79	3.06	1.61
0.38	0.73	0.16	0.59	1.92	0.85	0.80	3.13	1.65
0.39	0.81	0.20	0.60	1.98	0.89	0.81	3.20	1.70
0.40	0.88	0.23	0.61	2.03	0.92	0.82	3.28	1.75
0.41	0.95	0.26	0.62	2.08	0.95	0.83	3.35	1.80
0.42	1.01	0.30	0.63	2.13	0.99	0.84	3.44	1.85
0.43	1.07	0.33	0.64	2.18	1.02	0.85	3.52	1.91
0.44	1.13	0.36	0.65	2.23	1.06	0.86	3.61	1.97
0.45	1.19	0.39	0.66	2.29	1.09	0.87	3.71	2.03
0.46	1.25	0.43	0.67	2.34	1.13	0.88	3.81	2.09
0.47	1.31	0.46	0.68	2.39	1.16	0.89	3.91	2.16
0.48	1.36	0.49	0.69	2.45	1.20	0.90	4.03	2.23
0.49	1.41	0.52	0.70	2.50	1.24	0.92	4.29	2.41
0.50	1.47	0.55	0.71	2.56	1.28	0.94	4.61	1.59
0.51	1.52	0.59	0.72	2.62	1.31	0.96	5.03	2.85
0.52	1.57	0.62	0.73	2.68	1.35	0.98	5.71	3.25
0.53	1.62	0.65	0.74	2.74	1.39			

Table 4.B.2 Percentage correct versus sensory difference δ for the duo–trio (D–T) procedure and 2-AFC. Values are rounded to the second decimal place. Calculated in SensR

Proportion	D–T	2-AFC	Proportion	D–T	2-AFC	Proportion	D–T	2-AFC
0.50	(0)	(0)	0.70	1.72	0.74	0.90	3.26	1.81
0.51	0.33	0.04	0.71	1.77	0.78	0.91	3.39	1.90
0.52	0.47	0.07	0.72	1.83	0.82	0.92	3.53	1.99
0.53	0.58	0.11	0.73	1.90	0.87	0.93	3.69	2.09
0.54	0.68	0.14	0.74	1.96	0.91	0.94	3.87	2.20
0.55	0.76	0.18	0.75	2.02	0.95	0.95	4.07	2.33
0.56	0.84	0.21	0.76	2.08	1.00	0.96	4.32	2.48
0.57	0.91	0.25	0.77	2.15	1.04	0.97	4.63	2.66
0.58	0.98	0.29	0.78	2.22	1.09	0.98	5.04	2.91
0.59	1.05	0.32	0.79	2.28	1.14	0.99	5.70	3.29
0.60	1.12	0.36	0.80	2.35	1.19			
0.61	1.18	0.39	0.81	2.43	1.24			
0.62	1.24	0.43	0.82	2.50	1.29			
0.63	1.30	0.47	0.83	2.58	1.35			
0.64	1.36	0.51	0.84	2.66	1.41			
0.65	1,42	0.55	0.85	2.75	1.47			
0.66	1.48	0.58	0.86	2.84	1.53			
0.67	1.54	0.62	0.87	2.93	1.59			
0.68	1.60	0.66	0.88	3.05	1.66			
0.69	1.66	0.70	0.89	3.15	1.73			

References

Baird, J.C. and Noma, E. 1978. Fundamentals of Scaling and Psychophysics. John Wiley & Sons, Inc., New York, NY.

Bi, J. 2006a. Sensory Discrimination Tests and Measurements. Blackwell Publishing, Ames, IA.

Bi, J. 2006b. Statistical analyses for *R*-index. Journal of Sensory Studies, 21, 584–600.

Bi, J. 2011. Similarity tests using forced-choice methods in terms of Thurstonian discriminal distance, d'. Journal of Sensory Studies, 26, 151–7.

Bi, J. and O'Mahony, M. 1995. Tables for testing the significance of the *R*-index. Journal of Sensory Studies, 10, 341–7.

Braun, V., Rogeaux, M., Schneid, N., O'Mahony, M., and Rousseau, B. 2004. Corroborating the 2-AFC and 2-AC Thurstonian models using both a model system and sparkling water. Food Quality and Preference, 15, 501–7.

Byer, A.J. and Abrams, D. 1953. A comparison of the triangular and two-sample taste test methods. Food Technology, 7, 185–7.

Delwiche, J. and O'Mahony, M. 1997. Changes in secreted salivary sodium are sufficient to alter taste sensitivity: use of signal detection measures with continuous monitoring of the oral environment. Physiology and Behavior, 59, 605–11.

Engen, T. 1974. Method and theory in the study of odor preferences. In Human Responses to Environmental Odors. A. Turk, J.W. Johnston, Jr., and D.G. Moulton (Eds). Academic Press, New York, NY, pp. 122–41.

Ennis, D.M. 1993. The power of sensory discrimination methods. Journal of Sensory Studies, 8, 353–70.

Ennis, D.M. and Ennis, J.M. 2010. How to account for "no difference/preference" counts. IFPress, 13(3), 2–3.

Ennis, D.M. and Ennis, J.M. 2011. How to set identicality norms for no preference data. IFPress, 14(1), 3–4.

Ennis, J.M. and Jesionka, V. 2011. The power of sensory discrimination methods revisited. Journal of Sensory Studies, 26, 371–82.

Ennis, D.M. and Mullen, K. 1985. The effect of dimensionality on results from the triangular method. Chemical Senses, 10, 605–8.

Ennis, D.M. and Mullen, K. 1986. Theoretical aspects of sensory discrimination. Chemical Senses, 11, 513–22.

Ennis, D.M., Rousseau, B., and Ennis, J.M. 2011. Short Stories in Sensory and Consumer Science. IFPress, Richmond, VA.

Ennis, D.M., Palen, J., and Mullen, K. 1988. A multidimensional stochastic theory of similarity. Journal of Mathematical Psychology, 32, 449–65.

Frijters, J.E.R. 1979. The paradox of the discriminatory nondiscriminators resolved. Chemical Senses, 4, 355–8.

Frijters, J.E.R., Kooistra, A., and Vereijken, P.F.G. 1980. Tables of d' for the triangular method and the 3-AFC signal detection procedure. Perception & Psychophysics, 27(2), 176–8.

Gescheider, G.A. 1997. Psychophysics. The Fundamentals. Third edition. Lawrence Erlbaum, Mahwah, NJ.

Green, D.M. and Swets, J.A. 1966/1988. Signal Detection Theory and Psychophysics. John Wiley & Sons, Inc., New York, NY.

Gridgeman, N.T. 1970. A re-examination of the two-stage triangle test for perception of sensory differences. Journal of Food Science, 35, 87–91.

Hautus, M.J. and Irwin, R.J. 1995. Two models for estimating the discriminability of foods and beverages. Journal of Sensory Studies, 10, 203–15.

Irwin, R.J. and Hautus, M.J. 1997. Likelihood-ratio strategy for independent observations in the same–different task: an approximation to detection-theoretic model. Perception & Psychophysics, 59, 313–16.

Irwin, R.J., Stillman, J.A., Hautus, M.J., and Huddleston, L.M. 1993. The measurement of taste discrimination with the same-different task: a detection theory analysis. Journal of Sensory Studies, 8, 229–39.

Irwin, R.J., Hautus, M.J., and Butcher, J.C. 1999. An area theorem for the same–different experiment. Perception & Psychophysics, 61, 766–9.

Kamerud, J. and Larson, G. 2010. Use of same–different test and *R*-index to efficiently compare multiple product differences. Poster presented and the Society of Sensory Professionals Meeting, Napa, CA, October 27–29.

Kaplan, H.L., Macmillan, N.A., and Creelman, C.D. 1978. Tables for d' for variable-standard discrimination paradigms. Behavior Research Methods & Instrumentation, 10, 796–813.

Lawless, H.T. and Heymann, H. 2010. Sensory Evaluation of Food, Principles and Practices. Second edition. Springer, New York, NY.

Lawless, H.T. and Schlegel, M.P. 1984. Direct and indirect scaling of taste–odor mixtures. Journal of Food Science, 49, 44–46.

Macmillan, N.A. and Creelman, C.D. 1991. Detection Theory: A User's Guide. Cambridge University Press, Cambridge, UK.

Macmillan, N.A. and Kaplan, H.L. 1985. Detection theory analysis of group data: estimating sensitivity from average hit and false-alarm rates. Psychological Bulletin, 98, 185–199.

Macmillan, N.A., Kaplan, H.L., and Creelman, C.D. 1977. The psychophysics of categorical perception. Psychological Review, 84, 452–471.

O'Mahony, M. 1992. Understanding discrimination tests: a user-friendly treatment of response bias, rating and ranking *R*-index tests and their relationship to signal detection. Journal of Sensory Studies, 7, 1–47.

Rousseau, B. 2006. Indices of sensory difference: R-index and d′. IFPress, 9(3), 2–3.

Rousseau, B. 2010. Action standards in a successful sensory discrimination program. IFPress, 13(4), 2–3.

Rousseau, B., Meyer, A., and O'Mahony, M. 1998. Power and sensitivity of the same–different test: comparison with triangle and duo–trio procedures. Journal of Sensory Studies, 13, 149–173.

Stillman, J.A. and Irwin, R.J. 1995. Advantages of the same-different method over the triangular method for the measurement of taste discrimination. Journal of Sensory Studies, 10, 261–272.

Suppes, P. and Zinnes, J.L. 1963. Basic measurement theory. In Handbook of Mathematical Psychology, vol. 1. R.D. Luce, R.R. Bush, and E. Galanter (Eds). John Wiley & Sons, Inc., New York, NY.

Thurstone, L.L. 1927. A law of comparative judgment. Psychological Review, 34, 273–286.

Van Hout, D., Hautus, M.J., and Lee, H.-S. 2011. Investigation of test performance over repeated sessions using signal detection theory: comparisons of three nonattribute-specified differences tests, 2-AFCR, A–not A and 2-AFC. Journal of Sensory Studies, 26, 311–321.

Warnock, A.R., Shumaker, A.N., and Delwiche, J.F. 2006. Consideration of Thurstonian scalings of ratings data. Food Quality and Preference, 17, 556–561.

Yamaguchi, S. 1967. The synergistic taste effect of monosodium glutamate and disodium 5′-inosinate. Journal of Food Science, 32, 473–478.

5 Progress in Discrimination Testing

What's the difference between a screeching Macaw and a banjo?
One makes loud obnoxious repeated noises, the other is a bird.
What's the difference between an onion and a banjo?
Nobody cries when you cut up a banjo.
What's the difference between …
Anon. (overheard at old-timey music session)

Difference testing methods constitute a major foundation for sensory evaluation and consumer testing. These methods attempt to answer fundamental questions about stimulus and product similarity before descriptive or hedonic evaluations are even relevant. In many applications involving product or process changes, difference testing is the most appropriate mechanism for answering questions concerning product substitutability.

D.M. Ennis (1993)

5.1 Introduction

5.1.1 Overview

Discrimination testing is designed to discover whether there is a perceivable difference between two or more versions of a product. Although most sensory tests can be thought of

Quantitative Sensory Analysis: Psychophysics, Models and Intelligent Design, First Edition. Harry T. Lawless.

as tests for differences between products, discrimination tests are the simplest form of difference testing. Discrimination data can also be used to help establish product equivalence or sensory similarity. Equivalence testing will be discussed further in Chapter 6. Many different kinds of discrimination tests have been invented over the years. Virtually all of them (except rated degree of difference) are some kind of choice tests, with a clearly stated null probability or expectation about chance performance. The null and alternative hypotheses are exact mathematical statements. For example, in the triangle or oddity test, two samples are equivalent (i.e., physically the same) and one is a potentially different test sample, so the null hypothesis becomes "the population proportion correct is equal to one-third." Because there is only one correct answer, and we are looking for performance above chance (never below), the alternative is the one-tailed inequality statement, "the population proportion correct is greater than one-third." Students and practitioners should be careful about the quantitative nature of this relationship, rather than making the casual statement that the null is that "there is no difference." That may be your conclusion, but it is a non-mathematical judgment, not an equation or inequality.

Most discrimination testing is done internally within a company and generally uses a panel of volunteers from the research facility. Such a panel is usually screened for their sensory acuity, mostly to insure that people with poor discrimination skills are excluded. That is, a certain basic level of sensory function is desirable. Product familiarity is considered a plus. Panelists are generally given orientation to familiarize them with the types of tests that they will be called to perform, but training is not extensive and usually does not involve any vocabulary development, in contrast to a more highly trained descriptive analysis panel. Of course, simple difference testing can also be done with untrained consumers and walk-ons. However, the goals of such a test are usually different. In the first scenario, the preselected and screened panel is thought of as an instrument that provides a "safety net." That is, if they do not detect a difference under controlled conditions and having been preselected for good sensory acuity, then it is unlikely that most consumers in the outside world would detect the difference. Such a panel is being used as an analytical instrument. In the latter case, testing with consumers is done to maximize the ability to generalize the result to the target market population. One can think of this as maximizing test sensitivity versus maximizing project-ability. O'Mahony and colleagues have characterized these two styles as Sensory Evaluation I or analytical and Sensory Evaluation II, or consumer oriented (O'Mahony, 1995; O'Mahony & Rousseau, 2002). These goals are sometimes simultaneous and conflicting, and what version of the test you want to conduct will depend upon your specific goals and concerns with Type I versus Type II error. But the decision should be carefully considered and not chosen on the basis of some textbook recommendation, but rather your specific situation.

The specific execution of each test may vary, even though they go by the same name. A good example is the duo–trio procedure, which in Peryam's original presentation showed that there was a warm-up sample, even before the reference standard was presented (Peryam & Swartz, 1950). The idea that **warm-up samples** could be beneficial was supported by O'Mahony and colleagues (O'Mahony & Goldstein, 1986; Theime & O'Mahony, 1990) for two important reasons. First, it often resulted in enhanced discrimination performance; second, it could be used as a way to let the panelist try to identify the sensory attribute to be discriminated (let us call it "X"), thus allowing a 2-AFC protocol to be used. Instructions could then be "choose the item that has the most of X," a very powerful test (in theory), as shown in Chapter 4. However, McClure and Lawless (2010) did not find any advantage to that approach (see also Dacremont et al. (2000)). Further modifications of the duo–trio and

2-AFC protocols will be discussed in Section 5.4. The reader should be aware that there has been a fair amount of "tinkering" with the various procedures, and what works for your particular products may not be the best procedure for a different type of product. Your mileage may vary, as they say.

A second important insight provided by O'Mahony's group was that some sequences of products or stimuli might be more confusing to participants than other sequences. The strong tradition in sensory testing is to use a sequence of products that is randomized or counterbalanced across trials and participants – see Lawless and Heymann (2010: chapter 3). However, if one wished to maximize the discriminative sensitivity of a test, it might make sense to use those stimulus orders that promoted the maximum degree of differentiation on the part of participants (i.e., the least confusion). The theory that was developed became known as "sequential sensitivity analysis" (O'Mahony & Odbert, 1985; O'Mahony & Goldstein, 1986). It was based upon and grew naturally out of principles of taste adaptation. Adaptation is a process of decreased responsiveness under conditions of constant stimulation, such as how the eyes adapt to the ambient light levels – which is very obvious if you enter or leave a dark cinema on a sunny day. Owing to adaptation effects, O'Mahony reasoned that when following a strong stimulus with a weaker one the difference would be less obvious than when a weaker stimulus went first. Most of these experiments were done with simple salt (NaCl) solutions, and results were not always consistent with the theory (Santosa & O'Mahony, 2008). However, it is a potentially valuable consideration for any sensory practitioner who wanted to maximize the sensitivity of the test. Once again, with products that are more or less prone to adaptation, fatigue, or other sequential effects, your results many vary.

5.1.2 Classification of Tests

Discrimination tests can be categorized in various ways. They can all be thought of as a kind of sorting task, and some are explicitly so, such as the **tetrad** procedure. Others are more strictly choice tasks, such as choosing the item from among a triad that has the strongest flavor. When two of the samples are identical, and expected to be less perceivably intense, this is called the three-alternative forced-choice (3-AFC) task. In many cases in food research, the exact nature of the difference is not known, is unexpected, or involves multiple sensory attributes, and in these cases tests of overall difference are applicable. These include both the sorting procedures and matching tasks, such as the **duo–trio** (one reference, two test items), **ABX test** (two reference items, one test item to match) and the **dual standard** (two reference items, two test items). Recently, Hautus and colleagues have conceptualized the duo–trio and ABX as 2-AFCR tests, meaning two-alternative forced-choice with a reference. Their research has studied where the position of the item to be matched is placed in the series: first, in the duo–trio, last in the ABX; and it can also be given second (i.e., in the middle of the series).

Some tests involve choice of a response, rather than a choice of items in a series. These are the **A–not-A test**, which is essentially a yes/no procedure that attempts to classify individually presented items as an example of a product presented earlier in a training session. A similar response classification test is the **same–different** test, in which a pair of items is given and the participant must choose the response "same" or "different." Because the tests are not bias free, it is critical in both of these tests to have control items or control pairs also presented. For example, the actual examples of both the control and test item in the A–not-A test and the identical control pair as part of the same–different test must be presented in order to measure a baseline from which to assess the discrimination ability. The same–different test can be converted to a choice test by presenting both control and test pairs at the same time or in succession and asking the respondent which of the two pairs is the different one, a procedure known as the dual pair test or in psychology as the 4-AIX test. Note that the A–not-A test and the same–different test are not criterion free. A product choice is not forced, so the response choice depends upon the criterion that the respondent sets. For example, in the same–different test, no two products will ever appear to be exactly the same, so a criterion must be set for how much of a perceived difference there must be before one decides to call the samples "different." This makes the control trials absolutely essential, for without them the effect of the criterion setting cannot be separated from the true sensitivity.

A summary and classification of the discrimination tests is shown in Table 5.1.

Further basic information on discrimination tests and models can be found in Lawless and Heymann (2010: chapters 4 and 5). A good treatment of mathematical models and statistical handling of discrimination data is found in *Sensory Discrimination Tests and Measurements* (Bi, 2006) and in the statistical text by Gacula et al. (2009).

Table 5.1 Classification of common discrimination tests

		Samples			
Class of test	**Test**	**Inspection phase**	**Test phase**	**Task/instructions**	**Chance probability**
Matching	Duo–trio	Ref-A	A, B	Match sample to reference	1/2
	2-AFCR-M	Ref given second	A, B	Match samples to item in middle of series	1/2
	ABX	Ref-A, Ref-B	A (or B)	Match sample to reference	1/2
	Dual standard	Ref-A, Ref-B	A, B	Match both pairs	1/2
	Pick-2	Ref-A	A, A′, B, B′	Match two samples to reference	1/6
Forced-choice	Paired comparison	(none)	A, B	Choose sample with most of specified attribute	1/2
	3-AFC	(none)	A, A′, B	Choose sample with most of specified attribute	1/3
	n-AFC	(none)	A_1–A_{n-1}, B	Choose sample with most of specified attribute	1/*n*
	Dual pair	(none)	A, B and A, A′	Choose A, B (different pair)	1/2
Sorting	Triangle	(none)	A, A′, B (or A, B, B′)	Choose the most different sample	1/3
	2 out of 5	(none)	A, A′, B, B′, B″	Sort into two groups	1/10
	4/8 "Harris–Kalmus"	(none)	A_1–A_4, B_1–B_4	Sort into two groups	1/70
	Tetrad/non directional	(none)	A, A′, B, B′	Sort into two groups	1/3
	Tetrad/directional	(none)	A, A′, B, B′	Sort into two groups (told how B is different)	1/3
Yes/no (response choice)	Same–different	(none)	Pairs: A, A′ or A, B	Choose response: "same" or "different"	N/A*
	A, not-A	Ref-A	A or B	Choose response: "A" or "not-A"	N/A*

*For the yes/no tests, a criterion may be set by each individual, and therefore the chance probability may not be equal to 1/2. The correct baseline is set by the false-alarm rate.

5.1.3 Basic Data Handling

As seen in Table 5.1, all of the choice tests are associated with a specific chance probability that defines the null, except for the A–not-A and same–different procedures. This is equivalent to making the assumption that behavior is random, like the flip of a coin, and/or that people who cannot tell the difference are merely guessing. Because there are two functional outcomes, a correct or incorrect answer, the statistical model is based on the binomial distribution, or the normal approximation to the binomial. Because of the relationship between the normal distribution, Z-scores, and the chi-square distribution, the Z-score formula is equivalent to the chi-square formula (for a proof, see Lawless and Heymann (2010: 503, appendix B.6)). If we let p represent the chance probability, N be the total number of observations, and P_{obs} represent the proportion correct observed in the data, we have the following relationship under the guessing model:

$$Z = \frac{(P_{obs} - p) - (1/2N)}{\sqrt{p(1-p)/N}} \tag{5.1}$$

So this relationship is the difference of two proportions, divided by the standard error of the proportions. The continuity correction, $1/2N$, adjusts for the fact that the normal distribution is continuous but our "counting heads" is discrete, not continuous. Some statistical authorities prefer to use the observed proportion in the denominator (P_{obs} and $1-P_{obs}$), but most sensory texts use the chance probability. The difference is generally minor.

Two functional relationships are important here. First, owing to the one-tailed nature of the test, the critical value for Z is 1.645, and thus the right side of eqn 5.1 must exceed this value in order to reject the null hypothesis. This leads to the inequalities shown in eqns 5.2a and 5.2b for chance probability values of 1/3 and 1/2, respectively. Second, because reciprocal of the sample size, $1/N$, is in the denominator, the critical difference between the observed and chance proportions shrinks as the sample size N increases. Thus it takes a smaller difference from chance to attain statistical significance. So a larger N leads to a more powerful test, one that is less likely to miss a true difference. Calculations of power and sample size are discussed in Chapters 4 and 6.

For tests with a chance probability of 1/3,

$$X \geq \frac{2N+3}{6} + 0.775\sqrt{N} \tag{5.2a}$$

where X is the minimum number correct to achieve significance. For tests where $p = 1/2$,

$$X \geq \frac{2N+1}{2} + 0.8225\sqrt{N} \tag{5.2b}$$

Values of X are the basis for the tables of minimum numbers of correct judgments required for significance, which can be found in any basic sensory evaluation text. A shortcut to these values can be expressed in terms of the size of the proportion that must exceed the chance probability, given by eqns 5.3a and 5.3b:

$$P_{obs} = p + \frac{1.55\sqrt{N}+1}{2N} \quad \text{for } p = 1/3 \tag{5.3a}$$

and

$$P_{obs} = p + \frac{1.655\sqrt{N}+1}{2N} \quad \text{for } p = 1/2 \tag{5.3b}$$

A quick worked example. For $N=100$ in a triangle test, the P_{obs} must exceed the chance probability of 1/3 by the following quantity:

$$\frac{1.55\sqrt{N}+1}{2N}=\frac{1.55(10)+1}{2(100)}=\frac{16.5}{200}=0.0825$$

Adding 0.0825 to 0.3333 gives us 0.4158. Multiplying by N (=100), we need 41.58 as a minimum number, so rounding up to the nearest whole number (panelist) we need 42 correct, which agrees with the tables of minimum numbers of correct judgments (for 100 panelists) in all the texts. Similarly, for $N=100$ in a duo–trio test, we must exceed the chance level by 17.55/200=0.08775, which when added to 0.5 gives us 0.58775 or 59 correct answers as a minimum for significance at $p<0.05$. For $p<0.01$, merely solve for a critical one-tailed value of $Z=2.33$, rather than $Z=1.645$.

5.2 Metrics for Degree of Difference

5.2.1 Proportions of Discriminators

Not long after the "invention" of discrimination testing, researchers began to look for measures of sensory difference or sensory performance, beyond the observed proportion correct. It is easy to see that, with a chance performance level, the raw percentage correct could not be a good measure of actual discrimination, because some people may have simply guessed correctly, without actually discriminating the product differences. Furthermore, statistical significance was not a good measure, because that is partly a function of the number of judges, a more-or-less arbitrary choice of the experimenter that has nothing to do with the discriminability of the products or the abilities of the judges in the test to see the difference. Remember that the standard error, in this case the standard error of the proportion is the denominator in the Z-score formula, is inversely related to the square root of N.

Corrections for guessing had been around for a long time, and the most common one was based on Abbott's formula, an adjustment originally used in entomology to find the true efficacy of an insecticide by taking into account the mortality in a control group (Finney, 1944, 1949). Finney's version is commonly stated as

$$P_{\text{corrected}}=\frac{P_{\text{observed}}-P_{\text{control}}}{1-P_{\text{control}}} \tag{5.4}$$

where $P_{\text{corrected}}$ is the corrected proportion, P_{observed} is the observed proportion of dead insects, and P_{control} is the mortality in the control group. In sensory testing, the values for the observed proportion correct and the chance proportion would be substituted for Finney's mortality proportions (Filipello, 1956). Flipping the equation around, we can ask what observed proportion would have to be attained to achieve a certain level of corrected (i.e., "true") discrimination. That produces the following relationship:

$$\begin{aligned}P_{\text{observed}}&=P_{\text{corrected}}+P_{\text{chance}}\left(1-P_{\text{corrected}}\right)\\&=P_{\text{chance}}+P_{\text{corrected}}\left(1-P_{\text{chance}}\right)\end{aligned} \tag{5.5}$$

Note that there are two ways to calculate the required level of P_{observed}. Soon, this correction factor was applied to threshold testing (Filipello, 1956), and it exists today in various threshold methods such as the ASTM method E-679 (ASTM 2008a, b). For threshold, we usually desire 50% true discrimination. In the ASTM method, a 3-AFC method is used, with a chance probability of 1/3. So substituting in eqn 5.5 with 0.5 for the $P_{corrected}$ and 1/3 for P_{chance}, we see that the required value of $P_{observed}$ is 2/3. An easy way to remember this is that the required proportion is halfway (0.5 of 50%) between the chance level and 100% correct. Halfway from 1/3 to 1.0 is 2/3. So there is a simple relationship between any observed proportion and the **proportion of discriminators**.

However, it is not necessary to use the 50% level as a benchmark, even in threshold testing. It might be useful to know, for example, if some levels below the group threshold would result in other levels of detection, such as 10% or 25% of the population. Using Abbott's formula's correction, it is easy to see that, for the 3-AFC ASTM procedure, we can get 10% discrimination at 40% correct and 25% detection at 50% correct, for example (Lawless, 2010). In difference testing, this benchmark became useful for tests against various levels of discrimination other than chance. That is, once we know an allowable proportion of discriminators, we can test for performance below that level as a test for sensory similarity or equivalence (Schlich, 1993; Meilgaard et al., 2006). This topic is discussed further in Chapter 6. Many sensory evaluation practitioners found this notion of proportions of discriminators to be useful, and it still forms a benchmark for decision making in some large food companies.

A quick worked example. For the 3-AFC, we would like to know what the correct performance must be to obtain 10% and 25% discriminators. Using the lower expression of eqn 5.5, we see that for 10% discriminators we get

$$P_{obtained} = 1/3 + 0.10(1 - 1/3) = 0.3333 + 0.0667 = 0.40 \text{ or } 40\%$$

And for 25% discriminators we get

$$P_{obtained} = 1/3 + 0.25(1 - 1/3) = 1/3 + (1/4)(2/3) = 1/3 + 1/6 = 3/6 = 0.5 \text{ or } 50\%$$

The concept of discriminators was noted in the beer literature by Fedinandus et al. (1970). They used the term "recognizers" for those true discriminators who could discern the difference, which is a somewhat unfortunate term as recognition has several different meanings in memory research and threshold testing, for example. Nonetheless the idea was simple, that there were some people who discerned the correct difference and others who did not, but some of whom guessed correctly. This is sometimes referred to as a "guessing model." Note that the status of being a discriminator is only a hypothetical entity and has nothing to do with the long-term performance of any specific judge on a standing discrimination testing panel. Using the notion of a "true discriminator" in a single test does not assume that any given judge has that ability as a consistent trait. That is, the differentiation we are talking about is momentary (and hypothetical), not a general condition of the people involved. It is calculated, but not observed anywhere in the data.

The calculation of an estimated proportion of discriminators provides a useful source of information in addition to the normal significance testing that is routinely done. It serves to remind management that, although we found a significant difference in a discrimination test, it is not the case that all consumers will discern the difference. Furthermore, if we have used a large enough sample size, that proportion could be very small indeed. The concept has one major drawback, however. Because it is based solely on the chance probability level, the idea does not take into account the difficulty or inherent variability in different test methods. For example, the 3-AFC and triangle tests would calculate the same proportion of discriminators for the same percentage correct, but we know from many observations that the triangle test is more difficult and involves a more variable (and thus noisy – i.e., error-full) decision process (Frijters, 1979; Ennis, 1990, 1993). Using the same products with the two tests, the 3-AFC procedure will invariably produce a higher proportion correct than the triangle (Byer & Abrams, 1953; Frijters, 1979).

Owing to this paradox, the concept of a proportion of discriminators is only meaningful within the context of a given sensory test method. Being a "discriminator" in a triangle test is much more difficult than being a discriminator in a 3-AFC procedure. This may not be a serious issue if your testing program only uses one kind of test and sticks with it, historically. But once someone uses a different test in a different research facility, different manufacturing site, or different subsidiary of your company, all bets are off. As discussed in Chapter 4, and as we will see Section 5.2.2, Thurstonian modeling provides a more sophisticated and more accurate yardstick for discrimination performance, and one that can reasonably compare results from different methods, a unifying principle.

5.2.2 Thurstonian Models

The Thurstonian models were discussed in Chapter 4, but a brief summary will be given here. Given two versions of a product, the theory assumes that each version gives rise to a distribution of sensations. This is most easily conceptualized as a pair of univariate normal distributions with equal variance, although those assumptions can be relaxed in more complicated versions of the theory. Think of this as if there were two sets of sensory experiences, differing in strength (perceived intensity; e.g., sweetness if sugar concentration was being varied). Sometimes the physically less concentrated item would be less sweet, but sometimes more sweet than the physically more concentrated item. Thus, the two distributions overlap, which leads to imperfect performance, which is just what you would expect in a discrimination test where the two items were not clearly that different. This theory is illustrated in Figure 5.1. The true sensory difference is based on the difference of the means of the two sampling distributions, based on standard deviation units. When obtained in a signal

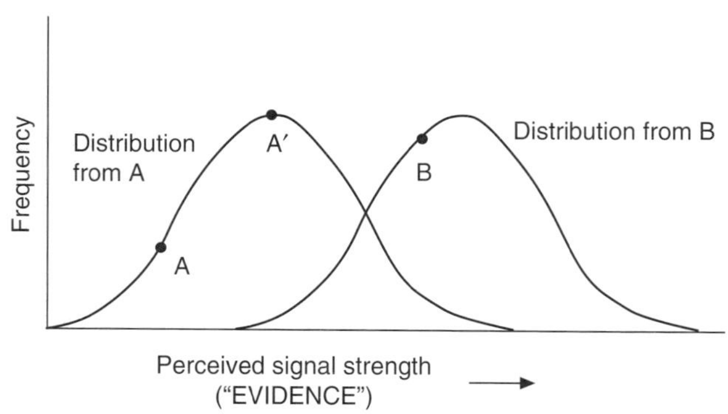

Figure 5.1 The Thurstone model for discrimination tests. Each product gives rise to a normally distributed set of sensations (over many trials), and on each particular trial, samples are drawn from these underlying distributions. The number of samplings on any trial is determined by the number of items in the test group; for example, two from one distribution and one from the other in a triangle procedure.

detection experiment, that distance quantity is called d' and when calculated theoretically in a Thurstonian model is commonly indicated by the Greek letter δ.

The theory then has to connect two factors: the number of items of each version that are given in a trial (i.e., the format of the test) and the cognitive strategy or decision rule that a person will use to get the correct answer. For the triangle test, for example, on any trial we will sample randomly twice from one distribution and once from the other. The cognitive rule that defines correct performance is the following: the sensory difference (think of it as a distance) between the duplicate sample must be smaller than both differences between each duplicate and the single test item. Thus, the correct answer depends upon three embedded paired comparisons from three randomly sampled sensations, a highly variable situation. This is called a **differencing strategy** or comparison of distances (COD). For the 3-AFC, on the other hand, the person must simply choose the strongest of the three sensations when the odd sample is the strongest. No paired comparisons need be done to figure sensory distances, just a rank ordering of intensities, a strategy sometimes called **skimming**. The equations defined by these sampling and comparison strategies are shown in Chapter 4. The important conclusion is that, for the same underlying sensory distance, the 3-AFC will lead to a higher percentage correct. Because they have the same chance performance level (1/3), then the 3-AFC will be statistically more powerful (Ennis, 1993).

Thurstonian δ values can be found for most of the common discrimination tests, along with tables for calculating variance (e.g., Ennis et al., 2011, and the sensR library from the R statistical platform). The variance factors are useful if one wishes to compare two obtained δ values from two sets of products, or to see if a δ value is nonzero, another measure of a statistically significant and perceivable difference. A third application is to compare the results from different test methods. For example, if one R&D facility is using a triangle test and another is using a duo–trio test, the δ values can now be compared. For any given pair of products, the δ values should be the same. In one version of a famous paradox, in a bitterness discrimination test, performance was 47% correct in the triangle but 71% correct under a 3-AFC instruction with the same individuals (Byer & Abrams, 1953). How could one get the triangle wrong but the 3-AFC correct? The paradox is resolved by the fact that judging three differences is much more variable than judging the strongest of three items. So the δ value obtained in both cases is about 1.3 (Frijters, 1979), showing that there is no paradox after all.

In conclusion, there are three pieces of information that can be gleaned from any discrimination test result: the statistical significance, the proportion of discriminators (but remember that value is method specific), and the Thurstonian δ value, a pure measure of sensory difference. Statistical significance is important when declaring a difference exists, but it not very meaningful when no significant difference is found. That is because there are many reasons for not attaining significance, such as too small a panel size or lack of replication. Also remember that statistical significance is an all-or-none event. That is, there is no real meaning to different levels of p-values ($p<0.05$, $p<0.01$, $p<0.001$, etc. ad nauseum) with superscript asterisks to boot. Journals should simply do away with this convention. Statements such as "marginally significant" or "highly significant" are not useful. After all, the p-value is only the probability of the result assuming a true null, which you are rejecting anyway! The result is either significant or it is not. Set a cutoff and stick with it. Delta values, on the other hand, provide a graded metric for the degree of difference, a useful measure in many circumstances. Examples include providing a benchmark or action standard for accepting or rejecting product reformulations, and in quality control standards.

5.3 Replication in Choice Tests

5.3.1 Traditional Approaches

Consider the common scenario of an in-house testing panel for simple difference testing. Usually, these people are pre-screened to have normal taste and smell function and are on call for tests, perhaps several times per week. Testing is usually done by invitation, or by some method of broadcasting the test schedule for the day to members of the qualified panel (e.g., email or hallway TV monitor announcements). So people will interrupt their normal workday to proceed to a sensory testing facility to help out, for which they may receive some kind of token reward like a food treat or snack. Given that they have taken the time to walk over to the test facility, it makes sense to have them participate in more than one test or more than one replicate, since the nature of a triangle test may make it shorter than the time needed to get there and back in the first place. So replication is a common occurrence in sensory testing, and the question arises as to how best to treat the replicated data set. Because the judgments are coming from the same individual, they are not theoretically or statistically independent, so treating the data for a duplicate test as if there were twice as many independent judgments was not generally recommended.

Several common sense approaches are reasonable (Lawless & Heymann, 2010: chapter 4). First, the replicates can be analyzed separately, as if they were unrelated tests. This can provide information on practice effects or warm-up, if the second test has a higher proportion correct, or about fatigue or adaptation if the second test is lower. Different products may or may not benefit from the additional testing; but, in general, a warm-up experience is a good thing. Second, the simple binomial model can still be used, but now by tabulation of those people who got both (or all) tests correct. So, for example, the chance probability now goes to 1/9 for a repeated triangle test or 1/8 for a duo–trio test done in triplicate. The new chance performance levels are simply substituted into the *Z*-score formula for the normal approximation to the binomial (eqn 5.1), or an exact binomial probability can also be calculated. This is a bit conservative, as it requires all tests to be correct in order to be counted as a correct judge, but it lowers the chance level and thus provides some protection from Type I error. A related alternative is to use a chi-square formula to take into account all the expected outcomes. For example, the expected outcomes for a duplicated triangle test are 1/9 with both correct, 4/9 with one of two correct, and 4/9 with neither correct. This affords some consideration of those people who got one of two tests correct (1/2 is higher than 1/3, after all) and might, in fact, have been discerning of the actual difference on one of the two trials.

5.3.2 Combining Replicates: Beta-Binomial Models

A valuable insight in considering replications is the notion that panelists or judges are not like random effects such as coin tosses or dice rolling. That is, some judges might be consistently discriminating, and others not. So why not take this into account, much like we do for the judge effect in analysis of variance? The **beta-binomial model** does just that. To the extent that judges are behaving randomly (i.e., no connection between their performance on the first and subsequent replicates), the data might actually be combined as if the judgments were for independent observers. To the extent that they are not random, some adjustments can be made in the binomial expectations based upon the degree of consistency. The model uses a beta distribution to describe and account for the judges' degree of consistency versus randomness. The combination of the beta and binomial distributions allows a test for significant discrimination, but using the larger overall N afforded by replication.

The consistency parameter is derived from two parameters of the beta distribution and is called **γ (gamma)**. γ can also be considered a measure of **overdispersion** (sort of the opposite of inconsistency or randomness), meaning there is an extra source of variance (i.e., panelists or assessors) that contributes to the overall pattern of variation. Using γ, rather than the original two parameters of the beta distribution, has several advantages. γ ranges from zero, signifying random behavior, to one, signifying completely consistent behavior from trial to trial. So it is intuitively understood. Second, there is a significance test for nonzero γ. This can indicate whether there is a degree of consistent discrimination among some panelists, and is thus an indication of both a perceivable difference (in addition to the usually significance test) and perhaps a segment of more discriminating individuals. γ is calculated as

$$\gamma = \frac{1}{r-1}\left[\frac{rS}{\mu(1-\mu)n} - 1\right] \tag{5.6}$$

where r is the number of replicates, n is the number of judges, S is a measure of dispersion, and μ is the mean proportion correct for the group as defined below. S is similar to the denominator of the standard deviation formula and thus becomes

$$S = \sum_{i=1}^{n}\left(\frac{x_i}{r} - \mu\right)^2 \tag{5.7}$$

and μ is

$$\mu = \frac{\sum_{i=1}^{n}\frac{x_i}{r}}{n} \tag{5.8}$$

where x_i is the total number correct for each panelist i.

The test for a significant overdispersion is given as follows (Bi, 2006):

$$Z = \frac{E - nr}{\sqrt{2nr(r-1)}} \tag{5.9}$$

where E is

$$E = \sum_{i=1}^{n}\frac{(x_i - rm)^2}{m(1-m)} \tag{5.10}$$

and m is the mean proportion correct from

$$m = \sum_{i=1}^{n}\frac{x_i}{nr} \tag{5.11}$$

If the Z-test is not significant, one can pool the replicates and use the overall N in the usual binomial tables or Z-formula (Liggett and Delwiche (2005) show examples). If not, tables adjusted for different levels of γ can be consulted for the minimum number of required judgments, such as table K in Lawless and Heymann (2010) for duplicated tests or those given in Bi (2006) for higher numbers of replicates.

Some authors have argued that the beta-binomial model is strictly not correct, because the beta distribution starts at zero, and zero correct judgments would be less than the chance level. So a chance-corrected beta-binomial model is favored by some theorists, with a left-censored beta distribution using the chance level as a lower bound (see Bi (2006) for a discussion and examples). To date, the advantages of the chance-corrected beta-binomial and the simple beta-binomial are not clear, and differences in the models appear to be small (Bi, 2006). The beta-binomial model can also be extended into tests with more than two choices, in which case the more general Dirichlet multinomial tests can be used (Bi, 2006; Gacula et al., 2009). See Appendix 5.B for some unusual properties of γ when it reaches extreme (versus minimal) levels.

5.4 Current Variations

5.4.1 Tetrad Tests

Recent publications on the **tetrad method** have generated interest in this kind of test as a substitute for the **triangle** procedure (Masuoka et al., 1995; Ennis et al., 1998; Ennis, 2012, Ennis & Rousseau, 2012). In the unspecified method of tetrads, four products are presented, two from the control treatment and two from the test items. The task is to sort them into two groups. There is only one of three ways to sort that is correct, and thus the chance probability is the same as the triangle test; that is, one-third (Figure 5.2). So the same statistical equations and the same lookup tables can be used for both tests. There is also a specified method of triads, in which the participant is instructed to pick the two strongest products, but since the nature of the difference is rarely known in product testing, further discussion here will only pertain to the unspecified method. An interesting variation on this is the "pick-2" method, which combines the tetrad procedure with a duo–trio-type instruction to match one of the two groups to a reference standard. Lee (2011) referred to pick-2 as an "extended duo–trio" and found no advantage to this test over the triangle procedure with a variety of products differing in discriminability. The chance probability for pick-2 drops to one out of six, because there is one of two groups to match once the sorting has been done

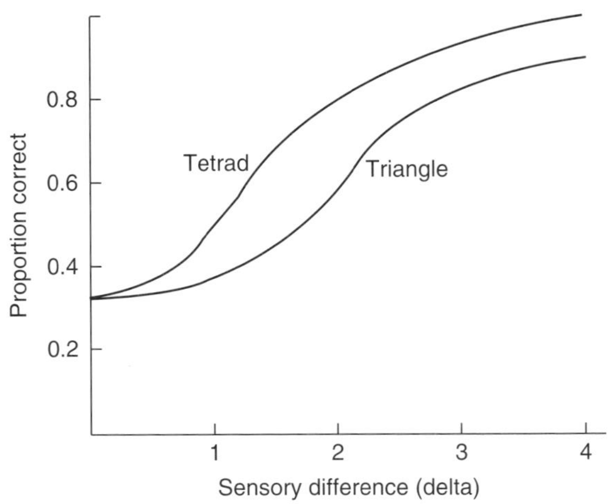

Figure 5.2 Psychometric functions for the tetrad method, compared with the triangle procedure. For a given sensory difference (δ value) the tetrad method will lead to a higher percentage correct. Because they have the same chance probability level (1/3), the tetrad test will be more statistically powerful.

correctly. At this time there is no psychometric model for that procedure and multiple cognitive strategies are conceivable.

Assume a unidimensional intensity continuum for the two product differences. In the tetrad, four sensations (s, s′, w, and w′) arise from the two products (s,w for strong and weak). Correct performance will be obtained when the two weakest sensations arise from the two presentations of the actually weaker products. Further assume that the weaker stimulus produces a sensation distribution with mean of zero and variance of one, and the stronger product produces a distribution of sensations with mean of δ, and also has a variance of one. Given these assumptions, the psychometric function is given by the following relationship:

$$P_c = 1 - 2\int_{-\infty}^{+\infty} \varphi(x)\left\{2\Phi(x)\Phi(x-\delta) - \left[\Phi(x-\delta)\right]^2\right\}\mathrm{d}x \tag{5.12}$$

where P_c is the proportion correct, and $\varphi(x)$ and $\Phi(x)$ are the density and cumulative distribution functions of the normal distribution. Recently, Bi and O'Mahony (2013) published variance tables for the tetrad method, along with comparisons with other discrimination procedures and several additional useful tables.

Where might the unidimensional Thurstonian model for the tetrad test break down? First, the model is based on a decision rule, a cognitive strategy that proposes that the person will answer correctly if the stimuli are correctly ordered (i.e., the sensations correspond to w/w/s/s in rank order), and more particularly that the two weakest sensations come from the two weakest stimuli. However, other cognitive strategies are conceivable. For example, a person might decide to group the two most similar items and ignore the other two. That could lead to an incorrect decision even if the sensations were ordered w/w/s/s. Second, the product sensory variation might not be unidimensional, but involve several attributes. Third, the varying attributes might be positively or negatively correlated; that is, there could be sources of redundant information. Fourth, the variation might not correspond to a simple intensity continuum, but could be a qualitative difference or even a temporal one (order of appearance of dominant sensations, for example). A qualitative difference might be color or hue, for example, or presence versus absence of some textural feature. Fifth, other factors might interfere, such as fatigue or sensory adaptation, due to the extra stimulus compared with the triangle test. Finally, retasting could change or alter the sensation pattern, as well as induce more fatigue or adaptation. For these reasons, the Thurstonian distance estimates δ might be inaccurate if based simply on translation from the percentage correct data (i.e., eqn 5.12).

Ennis (2012) provided a potentially useful strategy for assessing the deviations from Thurstonian predictions, combined with a strategy for replacement of triangle tests with tetrad tests in a sensory testing program. In order to do this, a series of difference tests is conducted on the same products using both triangle and tetrad tests. Owing to factors such as fatigue, adaptation, or memory load, the δ values may be smaller for the tetrad than for the triangle. If, in fact, they drop by no more than one-third, there is still a benefit in conducting the tetrad test in terms of the achievable test power. Specifically, the iso-power relationship holds when the tetrad δ obtained is 0.685 times the triangle δ. For example, if the triangle tests yield a mean δ value of 1.0, and the tetrad gives a mean value of 0.8, the tetrad is still more powerful. Conversely, you could achieve the same level of statistical power in the tetrad test with a smaller panel size or fewer replications, providing functional cost savings to the sensory testing program. See Ennis and Jesionka (2011) for useful sample size tables. A concise exposition of this relationship and some sample power tables can be found in Ennis and Rousseau (2012).

5.4.2 Variations on Duo–Trio

5.4.2.1 *Nonattribute 2-AFC and 2-AFC-R*

Hautus, Lee, and their colleagues have studied two further modifications of the duo–trio test and provided Thurstonian analyses for each. Results suggested that they could be more powerful than the duo–trio, although most experiments were conducted on very few subjects (e.g., three in Hautus and Irwin (1995) and seven in van Hout et al. (2011)) tested repeatedly, and thus the experimental situation does not correspond to the industrial testing situation with a large panel of persons (e.g., $N=50$) tested only once or twice per product session. These studies were primarily psychophysical investigations concerned with developing Thurstonian models for the tasks and studying different cognitive strategies. Furthermore, the comparisons were done as within-subject designs, and thus all the subjects received practice and exposure to all of the methods, a suboptimal design for teasing out differences between methods. In most of these publications, the modified duo–trio procedures were compared with the A–not-A test, usually with two other versions of each which include a "reminder" sample, a reference item repeated on each trial. In these procedures, sureness ratings (a kind of certainty judgment) were also collected on a six-point scale. The sureness rating permits a receiver operating characteristic (ROC) curve to be constructed, if desired, and allows for hit and false-alarm rates to be calculated, and thus a δ or d' estimate (see Chapter 4). Details of the procedures are a little fuzzy with some missing information, such as a script to show what were the verbatim instructions to subjects. However, the descriptions are as follows.

For the so-called 2-AFC method, one of two stimuli (products) was designated the reference, which we will call product A (the other, B). So the 2-AFC protocol resembles a modified duo–trio without a repeated reference. After a familiarization period, in which the reference item was tasted several times (e.g., at least four times in van Hout et al. (2011), with an *ad libitum* number allowed), then the trial would present both items in a pair. After tasting both, the subject designates one as the reference, the other not. At first glance, this would seem to be redundant information. If one sample matches the reference then the other logically must not, but recall that a sureness rating is collected after each single judgment. Thus, it is theoretically possible for someone to say "same, sure" as the first response and then "different, not sure" when pointing to the other item, if their certainty were greater for the one chosen as reference versus the one chosen as nonreference. However, it is difficult to see (and perhaps illogical) if the first item were most apparently and certainly the reference, why the other item wouldn't be "different, sure." The procedure is different from the traditional duo–trio in that (1) no reference is presented on each trial (Lawless & Heymann, 2010) but must be remembered from the familiarization period and (2) two (seemingly redundant) responses are required about each pair, rather than specifying simply a match to the reference item. Note that this is a different test than a true 2-AFC, in that no attribute is specified. Recall that in the 2-AFC an attribute (or overall strength) would be cued to the respondent (e.g., "pick the sweetest of these two items"). So the choice of this name is somewhat unfortunate. More descriptive choices might have been "unspecified 2-AFC with (redundant?) sureness ratings after each sample" or "modified duo–trio without trial-by-trial reference and sureness ratings after each sample."

The 2-AFC-R procedure is even more similar to the traditional duo–trio. The response scheme was the same as the so-called 2-AFC (unspecified), but the reference item was presented on each trial, just as in the normal duo–trio. In Lawless and Heymann (2010), this is called the "constant reference" version of the duo–trio test and their sample ballots clearly show that the reference item is given *on each trial*, just as in Peryam and Swartz (1950). As in the so-called 2-AFC, a sureness judgment is collected after each decision, two per trial. So perhaps a better name for this would be "constant reference duo–trio test with two sureness ratings per pair." Remember that for all four tasks (the A–not-A, A–not-A with reminder, 2-AFC, and 2-AFC-R) a sureness scale had responses after each stimulus, such as "same as reference, sure", "same as reference, not sure," "don't know but guess it's the same as the reference," "don't know but guess it's different from reference," "different, not sure," and "different, sure" (van Hout et al., 2011), thus producing a six-point category scale.

On a theoretical basis, the 2-AFC-R kind of test should allow either of two strategies. The first is a β strategy, a kind of classification of each stimulus as reference or not, based on an individual cutoff as in a yes/no signal detection task. The second is called a τ strategy, a comparison with a mental standard of how much of a perceived difference is necessary in order to consider the product as different from the reference. Adoption of one or the other strategy can be influenced by instructions to subjects and the kind of experience manipulated in the familiarization stage (Lee et al., 2007). Signal detection/Thurstonian models for these strategies (Hautus et al., 2009) suggest that (1) either strategy results in the same estimates of d' and (2) they should be more powerful than the traditional duo–trio test, which requires a "comparison of absolute distances" (Hautus et al., 2011: 434), also called a "COD" strategy and which is considered "suboptimal." The difference between a τ strategy for the 2-AFC-R and a COD strategy for the traditional duo–trio is a little murky at this time, with statements like the following appearing in the literature: "The commonly used cognitive decision strategy for [overall difference tests] is the comparison of differences strategy, which is sometimes called the tau strategy" (van Hout et al., 2011: 312). Furthermore, the decision-space diagrams of Hautus et al. (2009: figure 6) clearly show the evidence being weighed is the differences between each test product ("the evidence") and the reference. However, for some reason the COD and 2-AFC-R with τ strategy have different signal detection theoretic models (Hautus et al., 2011).

With regard to the power of the tests, at least one study has not shown any advantage to the reminder trial, with empirically determined *R*-indices for A–not-A-R and 2-AFC-R lower than those obtained for A–not-A and "2-AFC" (Lee et al., 2007). So the modifications by Hautus, Lee and colleagues seem like a potentially useful idea in improving the discriminative ability of unspecified overall difference tests. However, the claimed advantage to these tests rests mostly on theoretical grounds, with a need for further empirical testing in actual industrial testing situations, in which large panels of individuals are tested only once or a few times per session. Furthermore, better descriptive terminology should be developed to make the nature of these tests clear, and less confusable with commonly accepted test names such as the normal 2-AFC.

5.4.2.2 *Changing Position of the Reference*

A few studies have questioned the value of offering the reference product first in the series, and suggested that other positions in a duo–trio test might improve performance. There would be less memory load if the reference item was presented in the middle of the series, between the two test items for example. Repeating the reference at the end of the series could also be valuable, as an extra reminder about the sensory properties that one is attempting to match. Kim et al. (2010) and Lee and Kim (2008) found such an advantage for the duo–trio with a reference repeated at the end, as compared with placing the reference only at the beginning or in the middle of the series. Rousseau et al. (2002) found an advantage of the mid-placed reference duo–trio over the traditional order (a higher d' value), but this advantage unexpectedly disappeared when the question was rephrased to ask which change in sensation was larger, rather than which sample matched the reference. Also, Kim et al. (2010) did not see superiority of the middle position when similar stimulus sequences were compared, and in fact the advantage in that case went to the traditional reference-first version.

5.4.3 Same–Different and Dual Pair Tests

The same–different paired test has generated recent interest owing to its theoretically higher power/sensitivity than other traditional nonspecific difference tests. The Thurstonian models for this test were discussed in Chapter 4, as well as consideration of ROC curves, as a substantial literature has been devoted to this procedure and its signal detection analysis (Irwin et al., 1993, 1999; Hautus & Irwin, 1995; Stillman & Irwin, 1995; Irwin & Hautus, 1997). Tables for conversion of proportions observed to the corresponding $\delta\,(d')$ values and variance estimates can be found in Bi (2006). As in the A–not-A test, the variance estimates depend upon the specific proportions of responses on both test and control trials; that is, they are not a simple function of "percentage correct" as in the forced-choice procedures. The same–different test is usually thought to involve a **τ criterion**, a comparison with an internalized standard for the degree of (experienced) difference necessary to evoke a response of "different." It is worth noting that, on Thurstonian grounds, the same–different test is less powerful than the A–not-A or yes/no (single stimulus) methods, and generates different ROC curves. Statistical analysis of the method is straightforward if every participant receives both the test pair (A/B) and one version of a control pair (A/A or B/B). In that case, a 2×2 matrix can be constructed with the frequency counts for each combination of the two trials and a McNemar test can be conducted comparing the two discrepant cells. The critical cells are those in which the test pair was called "different" and the control pair "same" versus the cell in which the responses were reversed (test = same and control = different).

Comparisons of the triangle and same–different procedure have been made in terms of the d' values obtained. If the same–different procedure is in fact more powerful, it should generate a larger d' value, according to Rousseau and coworkers. In one study on yogurts, a small advantage was observed when the same–different was compared with the triangle, although not in all cases (Rousseau et al., 1998). When the same–different procedure was "modified" to include both test and control pairs (i.e., the normal recommended method), a superior d' value was obtained than was seen for the triangle. In terms of d' values, all methods produced results that were significantly different from zero, although confidence intervals tended to be smaller for the "modified" same–different procedure. In a second test with a more fatiguing product (Dijon-style mustard), no advantage was seen to the same–different test over the triangle, except when a familiarization period was instituted, perhaps due to a stabilization of the τ criterion (Rousseau et al., 1999). A further study in orange beverages found no differences in d' values for triangle, same–different, and dual pair tests (Rousseau & O'Mahony, 2001), echoing a lack of difference between triangle and same–different seen previously (Stillman & Irwin, 1995). It seems that, at this point, the purported advantage of the same–different test depends somewhat upon the specific procedure. A direct comparison on statistical grounds, using the standard McNemar test for the same–different versus the traditional binomial analysis of the triangle test, would be informative. A variation of the same–different test with sureness ratings in just such a comparison will be discussed below.

Another version of the same–different test that spurred some interest is the **dual pair** or four-interval AX (4IAX) test. In this method, two pairs are presented, one with identical samples (a control pair) and one with the two different samples, and the task of the judge is to specify which of the two pairs contains the different pair. This way of phrasing the question suggests a comparison of differences model, in which the magnitude of the experienced differences (i.e., sensory distance) from the two pairs is compared. This is probably different from the same–different strategy, which is usually to compare the experienced difference in a single pair to one's internalized τ criterion. So, at face value, the two tasks are actually quite different, and the expected decision rules are quite different. Of course, other decision strategies are possible, as discussed in the signal detection literature. A Thurstonian model for the dual pair test was presented by Rousseau and Ennis (2001) along with tables for conversion of percentage correct to δ values, and variance tables as well. The equations for δ, power, and sample size calculations are given in Appendix 5.A. The original paper contains simple worked examples. Extension of the model to accommodate multiple products was also published (Rousseau & Ennis, 2002a).

Empirical investigations of the method have focused on the d' values obtained, and whether they were comparable to or better than the d' values obtained from other methods, such as the triangle, duo–trio, and same–different test. Rousseau and O'Mahony (2000, 2001) observed no significant differences in d' values from the triangle, same–different, and dual pair. Similarly, Rousseau et al. (2002) found no differences among the triangle, duo–trio with middle reference, same–different, and dual pair, although the dual pair was expected to perhaps improve the d' value by minimizing response bias (τ) variations among the participants. The question arises as to whether the percentage correct obtained in a dual pair would be any different than that obtained in another general difference test, such as the duo–trio test. Given that they have the same chance probability under a binomial model, one could cross-reference through any given δ value in order to see what the percentage correct would be. The higher percentage correct would indicate a higher probability of statistical significance under a simple binomial model, which remains the most frequently used measure of significant differences, in spite of Thurstonian modeling. Based upon the tables of Rousseau and Ennis (2001) for the dual pair, a δ value of 1.0 corresponds to a percentage correct of 57.3%, while the percentage correct for a duo–trio would be 58.3% according to the table in Lawless and Heymann (2010). So there is little or no difference, and no apparent advantage to the dual-pair method from this example. In fact, power curves based on δ values show the method to be slightly less powerful than the duo–trio and triangle tests (Rousseau & Ennis, 2001). At this point, the dual-pair method cannot be justifiably recommended, unless other concerns are relevant.

5.4.4 Sureness Ratings

As discussed in Chapter 4, it is possible to extend the simple yes/no task in a signal detection procedure to include a rating scale. This affords the opportunity to assess several cutpoints for hit and false-alarm rates, thus permitting a better estimate of d', the calculation of an R-index, and the opportunity to plot an ROC curve. In the case of an R-index, the rating scale is usually formulated in terms of sureness or certainty ratings (signal, sure/signal, not sure/noise, not sure/noise, sure). As the signal detection task is functionally equivalent to the A–not-A test in sensory evaluation, that test can also readily include a rating scale (Thieme & O'Mahony, 1990;

Table 5.2 Examples of sureness rating scales for the A–not-A test and same–different test

Trial	Responses				
	Not-A, sure	**Not-A, unsure**	**A, unsure**	**A, sure**	**(Totals)**
Not-A (signal)	*A*	*B*	*C*	*D*	N_s
A (noise)	*E*	*F*	*G*	*H*	N_n
	Different, sure	**Different, not sure**	**Same, not sure**	**Same, sure**	
Test pair (different)	*A*	*B*	*C*	*D*	N_s
Control pair (identical)	*E*	*F*	*G*	*H*	N_n

Christensen et al., 2011). Statistical models for the A–not-A method were discussed by Bi and Ennis (2001), and the recent paper by Christensen et al. (2011) extends the statistical and Thurstonian models to include the effects of additional explanatory variables such as formulation variables or consumer sub-groups, a potentially valuable advance in analyzing designed experiments.

Sureness ratings can also be introduced into the same–different paradigm, also permitting the construction of an *R*-index (Delwiche & O'Mahony, 1997). Using a same–different method with sureness ratings is similar to simply using a degree-of-difference scale (Aust et al., 1985), although one can argue that certainty and degree of difference are conceptually two different things. The same–different test with sureness ratings was recently compared with the triangle test with a greater number of significant differences observed in cereal products (Kamerud & Larson, 2010), although Lee (2011) did not find that advantage in a simpler beverage system differing in sweetness level. Table 5.2 shows the response schemes for rating scales with both the A–not-A task and same–different. The letters *A*–*H* correspond to the frequency counts that would be in the data. The calculation of the *R*-index is the same for both tasks:

$$R = \frac{A(F+G+H)+B(G+H)+C(H)+0.5(AE+BF+CG+DH)}{N_s N_n} \tag{5.13}$$

However, the correspondence of the *R*-index to Thurstonian δ is not the same for the two tasks, so caution is needed in determining and interpreting statistical significance if the *R*-index is tested against its binomial expectation of "percentage correct" equal to 1/2. The A–not-A task produces a higher *R*-index (and is thus theoretically more sensitive) for a given δ value (Hautus & Irwin, 1995; Rousseau, 2006, 2007).

5.4.5 Difference Testing with Multiple Products

Kamerud and Larson (2010) noted that the same–different test could be more cost efficient than testing multiple triangles if a single control condition (i.e., an identical pair baseline) was used with three or more test items. The triangle test in that case would require nine samples (three triads) but the same–different test only eight; that is, four pairs such as T_1/C, T_2/C, T_2/C, and C/C, where T is a test item and C is the control item. This would be advantageous if several formulations were produced with graded amounts of some ingredient or graded differences in a processing variable.

The use of multiple standards in a difference test is important in quality control situations in which there may be batch-to-batch variation, or inherent differences from different manufacturing lines. Pecore et al. (2006) and Young et al. (2008) addressed this issue by suggesting that rating-scale degree-of-difference tests might include multiple control items in order to account for the normal production variability, an approach first considered by Aust et al. (1985). In the first paper, they described how multiple control lots could be incorporated, and in the second how multiple test lots could be added as well. This approach changes the critical decision-making comparison. For example, in the latter case, the critical comparisons are the mean differences between all test and control lots versus the mean differences between control lots compared amongst themselves and test lots compared amongst themselves. In order to facilitate this without too much of a testing burden on panelists, Young et al. used an incomplete block design.

Another approach to the issue of multiple control or multiple test lots is a variation on the three-stimulus choice procedures, known as methods of triads (Ennis et al., 1988). In the older psychometric literature, there were two versions of the triadic procedures. In Torgerson's "complete **method of triads**," the three items are presented in three separate trials, and in each the subject must state which of the first two items is most similar to the third, similar to an ABX task or to a duo–trio if the reference items to be matched are presented first. In Richardson's version, the three items are presented once and the subject must designate which two are most similar and which two are most different. Thus, the triangle test is a special case or variant of Richardson's method. With four products representing two versions each of two different formulations (A and B) there are 12 different combinations of the four items (A_1, A_2, B_1, and B_2) as the reference is rotated, as shown in Table 5.3. Note that there are 12 combinations, without rotation of the order of the two test items, which would produce 24 total sequences.

Ennis et al. (1988) first developed a Thurstonian model for triads. Rousseau and Ennis (2002b,c) further discussed this procedure within the context of batch-to-batch product variation and provided a hierarchical multivariate model for generating Thurstonian distances using this procedure (Rousseau & Ennis, 2002c). The modeling is hierarchical in the sense that different degrees of complexity (e.g., four versus two distributions for A and B) can be tested to see what is required to fit the data. Rousseau and Ennis (2002a) showed how multiple samples could be accommodated using the dual-pair method, with tetrads of various combinations. Fifteen combinations were required for four products, a considerable testing burden if a complete design is used (60 products to taste). So the practical utility of that approach remains to be demonstrated. Further information on multiproduct difference tests, including ranking, can be found in Bi (2006).

Table 5.3 Rotated reference combinations in a four-product triadic test design with two versions of each of two formulations (A and B)

Reference	Test sequence	Reference	Test sequence
A_1	A_1 A_2 B_1	B_1	B_1 A_1 A_2
A_1	A_1 A_2 B_2	B_1	B_1 A_1 B_2
A_1	A_1 B_1B_2	B_1	B_1 A_2 B_2
A_2	A_2 A_1B_1	B_2	B_2 A_1 A_2
A_2	A_2 A_1 B_2	B_2	B_2 A_1 B_1
A_2	A_2 B_1B_2	B_2	B_2 A_2 B_1

5.5 Summary and Conclusions

Simple difference tests constitute a large proportion of the sensory evaluation procedures involved in minor product modifications such as ingredient or supplier changes, processing improvements, and cost reductions. Their popularity is in part due to the simple nature of their statistical analysis and straightforward proven methods that can be standardized for any given product system. Furthermore, an ongoing program of routine testing can be instituted in most larger R&D facilities, making the tests highly cost efficient with in-house employee panels. Perhaps it is not surprising, then, that a large body of theoretical literature has evolved for modeling the psychological processes involved in discrimination. Yet it is appropriate to ask if this reliance on simple discrimination tests and the resulting concentration of resources is misplaced. After all, these tests only provide a yes/no answer about a statistically significant difference, in their simplest interpretation. We do not know in what ways the products are different, nor if that difference matters to consumers at all. So in a sense these tests are impoverished in information content, and in most cases only act as a starting point for further testing in a sequence once a difference has been established.

This chapter, as well as Chapter 4, has looked at ways to extend the information content of discrimination test results. One approach is to correct for the most likely guessing levels, then provide an adjusted estimate of the true proportion discriminating, using Abbott's formula. Of course, this is merely a mathematical adjustment that depends upon the chance performance level in any test, and is thus not an absolute measure of discriminability. Furthermore, it is method specific, and two methods with the same chance performance level will produce different adjusted proportions, depending upon the difficulty or inherent variability of the cognitive strategies required. A second measure is the theoretical Thurstonian difference, as modeled by simple univariate assumptions. However, this is also not without its pitfalls, as most foods are multidimensional, any ingredient or processing change is likely to involve multiple attributes, and the changes in these attributes may be correlated (positively or negatively), producing redundant sources of information. These complications are not considered by the Thurstonian models in their simplest form.

"Proportions of discriminators" is in some ways an unfortunate choice of terms. As the measure is method specific, it does not give us a general idea of how many consumers would discern the difference, only the likely proportion sitting in a test booth in a given test procedure. A better term might be "chance-corrected performance," which would not imply any amount of discrimination in the general population, only a better estimate of the actual percentage correct after adjustments for guessing. Nonetheless, this serves as a valuable reminder to management that a significant difference does not mean that all observers would discern the difference, and perhaps only a very small percentage indeed. Further testing is almost always required in order to determine the degree and nature of any consumer perception of the change. The major liability rides with the change being both detected by and disappointing to some segment of loyal or heavy users of the product, an issue in consumer "alienation."

The constant warning to management must be that a significant difference does not mean that the perceived change is necessarily important. It is also unfortunate that we have chosen the term "significant" to mean a result in which the observed events would occur less than 5% of the time under the assumptions of a true null hypothesis. After all, that null and its assumptions are beliefs you are rejecting anyway, making said assumptions wholly irrelevant from then on! "Confidence" as in "confidence levels" is equally uninformative and routinely misunderstood, as if they implied the stability or repeatability of the result. Perhaps in a more enlightened era, we will no longer need either of these terms in our everyday scientific communication. The key in discussing results with management and decision-makers is to insure proper interpretation of the result and avoid reliance on the face-value common-language interpretation of statistical terms such as "significant." Instead, the sensory professionals should work toward reasonable action standards and decision criteria, ones that take into account the fact that perception of differences is partial and never universal (100% correct performance is never attained) and that further testing is likely going to be needed to fill in the missing information about the nature of the change and its functional (i.e., consumer) impact.

What does the future hold? A growing body of literature suggests that discrmination tests may not be as sensitive as tests with an affective component. This was first described by MacRae and Geelhoed (1992), who showed that a preference test was more sensitive than a triangle in finding differences, although this was later challenged under Thurstonian analysis by Geelhoed et al. (1994). Recently, Booth et al. (2011) have offered some evidence that preference "discrimination" is more sensitive than bitterness discrimination, at least for some observers. Furthermore, an authenticity test, with an affective component, was found to be more sensitive than either descriptive testing or discrimination testing with consumers (Frandsen et al., 2003, 2007). It remains to be seen how the approach to the task may improve performance with the affective or emotional involvement of a consumer or panelist.

Appendix 5.A Psychometric Function for the Dual Pair Test, Power Equations, and Sample Size

Thurstonian models for the dual pair test are outlined in Rousseau and Ennis (2001). The paper includes the psychometric function relating proportion correct to the δ value, the power calculation, and sample size equations with simple worked examples. The psychometric function was modeled as a noncentral beta distribution, and tables for conversion from proportion correct to δ (the Thurstonian distance or sensory difference measure) were provided along with a variance table for hypothesis testing. Using a normal distribution, a somewhat simpler relationship is that proposed by Macmillan et al. (1977):

$$P_c = [\Phi(\delta/2)]^2 + [1 - \Phi(\delta/2)]^2 \tag{5.A.1}$$

where P_c is the proportion correct, δ is the sensory difference measure, and Φ is the cumulative normal distribution function. Another useful function is the opposite conversion, to get a δ value from the proportion correct in the data:

$$\delta = 2\Phi^{-1}\left[\sqrt{\frac{P_c}{2} - \frac{1}{4}} + \frac{1}{2}\right] \tag{5.A.2}$$

where Φ^{-1} is the inverse normal distribution function (converting from proportion to a Z-score).

The power of the test is given by two relationships (split up to simplify)

$$u = \frac{1}{2} + \frac{\Phi^{-1}(1-\alpha)}{2\sqrt{N}} \tag{5.A.3}$$

for a two-alternative task. N, of course, is the sample size, and we must also define α and the required degree of difference δ we are testing against. And u is substituted into

$$\text{Power} = 1 - \Phi\left[\frac{u-p}{\sqrt{p(1-p)/N}}\right] \tag{5.A.4}$$

where p is the proportion correct estimated from our required difference δ as in eqn 5.A.1 or the tables provided in Rousseau and Ennis (2001).

The sample size requirements can also be split into two functions as

$$y = \frac{\Phi^{-1}(1-\alpha)}{2} \tag{5.A.5}$$

and

$$N = \left\{\frac{\Phi^{-1}(\beta)\left[\sqrt{p(1-p)}\right] - y}{\frac{1}{2} - p}\right\}^2 \tag{5.A.6}$$

where p is defined from the required difference δ as above. A worked example can be found in Rousseau and Ennis (2001: appendix B).

Appendix 5.B Fun with γ

We conducted a replicated triangle test with 28 panelists. There are 14 correct judgments (mean correct proportion: 0.5) in each replicate. Considering them separately, this is not a significant difference. But the way the judgments are distributed across panelists matters greatly. Consider the following situations (two outcomes):

In Case A, the answers are distributed as follows:

- 14 panelists get zero correct
- 14 panelists get both correct.

Owing to the extreme consistency, $\gamma = 1.0$, no evidence for a binomial model, panelists are behaving very differently (evidence exists for "distinguishers" or "discriminators").

If we conduct a Z-test against $p = 1/9$ (for those getting both correct) we obtain $Z = 4.16$, $p < 0.001$, which is very strong evidence against the null.

In Case B, we have a random pattern of behavior on replication, but still maintain 14 correct in both reps, with answers distributed as follows:

- 7 panelists get zero correct
- 14 panelists get one correct
- 7 panelists get both correct.

In other words, a symmetric distribution similar to what one would expect by chance. Now $\gamma = 0$, which provides justification for combining reps and using a binomial model. Now N is increased to 56, and is now significant at $p < 0.05$ ($28 > 26$, with 26 being the critical minimum number correct. What if we look at the Z-test against $p = 1/9$ for both correct? Now it yields $Z = 1.35$ (not significant) and by that method we have insufficient evidence against the null.

A paradox(?): as γ increases, the minimum number correct must increase in the beta-binomial model. So Case A gives *less evidence* (than Case B) against the null in the beta-binomial model, but more evidence in a method that considers $p(\text{chance}) = 1/9$.

References

ASTM 2008a. Standard practice for determining odor and taste thresholds by a forced-choice ascending concentration series method of limits, E-679-04, Annual Book of Standards, vol. 15.08, ASTM International, Conshocken, PA, pp. 36–42.

ASTM 2008b. Standard practice for defining and calculating individual and group sensory thresholds from forced-choice data sets of intermediate size, E-1432-04, ASTM International Book of Standards, vol. 15.08, ASTM International, Conshocken, PA, pp. 82–9.

Aust, L.B., Gacula, M.C., Beard, S.A., and Washam, R.W. 1985. Degree of difference test method in sensory evaluation of heterogeneous product types. Journal of Food Science, 50, 511–13.

Bi, J. 2006. Sensory Discrimination Tests and Measurements. Blackwell Publishing., Ames, IA.

Bi, J. and Ennis, D.M. 2001. Statistical models for the A–not A method. Journal of Sensory Studies, 16, 215–37.

Bi, J. and O'Mahony, M. 2013. Variance of d' for the tetrad test and comparisons with other forced-choice methods. Journal of Sensory Studies, doi: 10.1111/joss.12004.

Booth, D.A., Sharpe, O., and Conner, M.T. 2011. Gustatory discriminative norms for caffeine in normal use point to supertasters, tasters and nontasters. Chemosensory Perception, 4, 154–62.

Byer, A.J. and Abrams, D. 1953. A comparison of the triangle and two-sample taste test methods. Food Technology, 7, 183–7.

Christensen, R.H.B., Cleaver, G. and Brockhoff, P.B. 2011. Statistical and Thurstonian models for the A–not A protocol with and without sureness. Food Quality and Preference, 22, 542–9.

Dacremont, C., Sauvegeot, F. and Ha Duyen, T. 2000. Effect of assessors expertise level on efficiency of warm-up for triangle tests. Journal of Sensory Studies, 15, 151–62.

Delwiche, J. and O'Mahony, M. 1997. Changes in secreted salivary sodium are sufficient to alter taste sensitivity: use of signal detection measures with continuous monitoring of the oral environment. Physiology and Behavior, 59, 605–11.

Ennis, D.M. 1990. Relative power of difference testing methods in sensory evaluation. Food Technology, 44(4), 114, 116–17.

Ennis, D.M. 1993. The power of sensory discrimination methods. Journal of Sensory Studies, 8, 353–70.

Ennis, J.M. 2012. Guiding the switch from triangle testing to tetrad testing. Journal of Sensory Studies, 27, 223–31.

Ennis, J.M. and Jesionka, V. 2011. The power of sensory discrimination methods revisited. Journal of Sensory Studies, 26, 371–82.

Ennis, J.M. and Rousseau, B. 2012. Reducing costs with tetrad testing. IFPress, 15(1), 3–4.

Ennis, D.M., Mullen, K., and Frijters, J.E.R. 1988. Variants of the method of triads: unidimensional Thurstonian models. British Journal of Mathematical and Statistical Psychology, 41, 25–36.

Ennis, J.M., Ennis, D.M., Yip, D., and O'Mahony, M. 1998. Thurstonian models for variants of the method of tetrads. British Journal of Mathematical and Statistical Psychology, 51, 205–15.

Ennis, D.M., Rousseau, B., and Ennis, J.M. 2011. Short Stories in Sensory and Consumer Science. IFPress, Richmond, VA.

Ferdinandus, A., Oosterom-Kleijngeld, I., and Runneboom, A.J.M. 1970. Taste testing. MBAA Technical Quarterly, 7(4): 210–27.

Filipello, F. 1956. A critical comparison of the two-sample and triangular binomial designs. Journal of Food Science, 21, 235–41.

Finney, D.J. 1944. The application of the probit method to toxicity test data adjusted for mortality in the controls. Annals of Applied Biology, 31, 68–74.

Finney, D.J. 1949. The adjustment for a natural response rate in probit analysis. Annals of Applied Biology, 36, 187–95.

Frandsen, L.W., Dijksterhuis, G., Brockoff, P., Nielsen, J., and Martens, M. 2003. Subtle differences in milk: comparison of an analytical and an affective test. Food Quality and Preference, 14, 515–26.

Frandsen, L.W., Dijksterhjuis, G.B., Brockhoff, P.B., Nielsen, J.H., and Martens, M. 2007. Feelings as a basis for discrimination: comparison of a modified authenticity test with the same–different test for slightly different types of milk. Food Quality and Preference, 18, 97–105.

Frijters, J.E.R. 1979. The paradox of the discriminatory nondiscriminators resolved. Chemical Senses, 4, 355–8.

Gacula, M., Singh, J., Bi, J., and Altan, S. 2009. Statistical Methods in Food and Consumer Research. Second edition. Elsevier/Academic Press, Amsterdam.

Geelhoed, E.N., MacRae, A.W., and Ennis, D.M. 1994. Preference gives more consistent judgments than oddity only if the task can be modeled as forced choice. Perception & Psychophysics, 55, 473–7.

Hautus, M.J. and Irwin, R.J. 1995. Two models for estimating the discriminability of foods and beverages. Journal of Sensory Studies, 10, 203–15.

Hautus, M.J., van Hout, D., and Lee, H.-S. 2009. Variants of A not-A and 2-AFC tests: signal detection models. Food Quality and Preference, 20, 222–9.

Hautus, M.J., Shepherd, D., and Peng, M. 2011. Decision strategies for the A not-A, 2AFC and 2AFC-reminder tasks: empirical tests. Food Quality and Preference, 22, 433–4.

Irwin, R.J. and Hautus, M.J. 1997. Likelihood-ratio strategy for independent observations in the same–different task: an approximation to detection-theoretic model. Perception & Psychophysics, 59, 313–16.

Irwin, R.J., Stillman, J.A., Hautus, M.J., and Huddleston, L.M. 1993. The measurement of taste discrimination with a same–different task: A detection theory analysis. Journal of Sensory Studies, 8, 229–39.

Irwin, R.J., Hautus, M.J., and Butcher, J.C. 1999. An area theorem for the same–different experiment. Perception & Psychophysics, 61, 766–9.

Kamerud, J. and Larson, G. 2010. Use of same–different test and *R*-index to efficiently compare multiple product differences. Poster presented and the Society of Sensory Professionals Meeting, Napa, CA, October 27–29.

Kim, M.-A., Lee, Y.-M., and Lee, H.-S. 2010. Comparison of d' estimates produced by three versions of a duo–trio test for discriminating tomato juices with varying salt concentrations: the effects of the number and position of the reference stimulus. Food Quality and Preference, 21, 504–11.

Lawless, H.T. 2010. A simple alternative analysis for threshold data determined by ascending forced-choice method of limits. Journal of Sensory Studies, 25, 332–46.

Lawless, H.T. and Heymann, H. 2010. Sensory Evaluation of Foods, Principles and Practices. Second edition. Springer, New York, NY.

Lee, H.-S. and Kim, K.-O. 2008. Difference test sensitivity: comparison of three versions of the duo–trio method requiring different memory schemes and taste sequences. Food Quality and Preference, 19, 97–102.

Lee, H.-S., van Hout, D., and Hautus, M.J. 2007. Comparison of performance in the A–not A, 2-AFC and same–different tests for the flavor discrimination of margarines: the effect of cognitive discrimination strategies. Food Quality and Preference, 18, 920–8.

Lee, J. 2011. Comparison of discrimination test methods: triangle test, same–different sureness test and extended duo–trio test. MPS project paper, Cornell University, Department of Food Science.

Liggett, R.A. and Delwiche, J.F. 2005. The beta-binomial model: variability in overdispersion across methods and over time. Journal of Sensory Studies, 20, 48–61.

Macmillan, N.A., Kaplan, H.L., and Creelman, C.D. 1977. The psychophysics of categorical perception. Psychological Review, 84, 452–71.
MacRae, A.W. and Geelhoed, E.N. 1992. Preference can be more powerful than detection of oddity as a test of discriminability. Perception & Psychohysics, 51, 179–81.
Masuoka, S., Hatjopoulos, D., and O'Mahony, M. 1995. Beer bitterness detection: testing Thurstonian and sequential sensitivity analysis models for triad and tetrad methods. Journal of Sensory Studies, 10, 295–306.
McClure, S.T. and Lawless, H.T. 2010. Comparison of the triangle and a self-defined two alternative forced choice test. Food Quality and Preference, 21, 547–552.
Meilgaard, M., Civille, G.V., and Carr, B.T. 2006. Sensory Evaluation Techniques. Fourth edition. CRC Press, Boca Raton, FL.
O'Mahony, M. 1995. Sensory measurement in food science: fitting methods to goals. Food Technology, 29, 72–82.
O'Mahony, M. and Goldstein, L.R. 1986. Effectiveness of sensory difference tests: sequential sensitivity analysis for liquid food stimuli. Journal of Food Science, 51, 1550–3.
O'Mahony, M. and Rousseau, B. 2002. Discrimination testing: a few ideas, old and new. Food Quality and Preference, 14, 157–64.
O'Mahony, M.A.P.D.E. and Odbert, N. 1985. A comparison of sensory difference testing procedures: sequential sensitivity analysis and aspects of taste adaptation. Journal of Food Science, 50, 1055–8.
Pecore, S., Stoer, N., Hooge, S., Holschuh, N., Hulting, F., and Case, F. 2006. Degree of difference testing: a new approach incorporating control lot variability. Food Quality and Preference, 17, 552–5.
Peryam, D.R. and Swartz, V.W. 1950. Measurement of sensory differences. Food Technology, 4, 390–5.
Rousseau, B. 2006. Indices of sensory difference: *R*-index and *d′*. IFPress, 9(3), 2–3.
Rousseau, B. 2007. Measuring product similarities: are two indices, *R*-index and *d′*, interchangeable? Poster presented at the 7th Pangborn Sensory Science Syposium, Minneapolis, MN.
Rousseau, B. and Ennis, D.M. 2001. A Thurstonian model for the dual pair (4IAX) discrimination method. Perception & Psychophysics, 63, 1083–90.
Rousseau, B. and Ennis, D.M. 2002a. The multiple dual-pair method. Perception & Psychophysics, 64, 1008–14.
Rousseau, B. and Ennis, D.M. 2002b. Discrimination testing with multiple samples. IFPress, 5(1), 2–3.
Rousseau, B. and Ennis, D.M. 2002c. Multivariate difference testing with multiple samples. IFPress, 5, 2–3.
Rousseau, B. and O'Mahony, M. 2000. Investigation of the effect of within-trial retasting and comparison of the dual-pair, same–different and triangle paradigms. Food Quality and Preference, 11, 457–64.
Rousseau, B. and O'Mahony, M. 2001. Investigation of the dual-pair method as a possible alternative to the triangle and same–different tests. Journal of Sensory Studies, 16, 161–78.
Rousseau, B., Meyer, A., and O'Mahony, M. 1998. Power and sensitivity of the same–different test: comparison with triangle and duo–trio procedures. Journal of Sensory Studies, 13, 149–73.
Rousseau, B., Rogeaux, M., and O'Mahony, M. 1999. Mustard discrimination by same–different and triangle tests: aspects of irritation, memory and tau criteria. Food Quality and Preference, 10, 173–84.
Rousseau, B., Stroh, S., and O'Mahony, M. 2002. Investigating more powerful discrimination tests with consumers: effects of memory and response biases. Food Quality and Preference, 13, 39–45.
Santosa, M. and O'Mahony, M. 2008. Sequential sensitivity analysis for same–different tests: some further insights. Journal of Sensory Studies, 23, 267–83.
Schlich, P. 1993. Risk tables for discrimination tests. Food Quality and Preference, 4, 141–51.
Stillman, J.A. and Irwin, R.J. 1995. Advantages of the same–different method over the triangular method for the measurement of taste discrimination. Journal of Sensory Studies, 10, 261–72.
Thieme, U. and O'Mahony, M. 1990. Modifications to sensory difference test protocols: the warmed up paired comparison, the single standard duo–trio and the A–not A test modified for response bias. Journal of Sensory Studies, 5, 159–76.
Van Hout, D., Hautus, M.J., and Lee, H.-S. 2011. Investigation of test performance over repeated sessions using signal detection theory: comparison of three nonattribute-specified difference tests 2-AFCR, A–not A and 2-AFC. Journal of Sensory Studies, 26, 311–21.
Young, T.A., Pecore, S., Stoer, N., Hulting, F., Holschuh, N., and Case, F. 2008. Incorporating test and control product variability in degree of difference tests. Food Quality and Prefernce, 19, 734–6.

6 Similarity and Equivalence Testing

Science Cannot Prove a Negative.
Joseph H. Hotchkiss

Power is People.
Hildegarde Heymann

6.1 Introduction: Issues in Type II Error

The desirable outcome in many simple difference tests is a conclusion that the test product or modified product is sensorially equivalent to a standard, a control product, or some existing version of a current product. This is true of minor ingredient and processing changes, as well as of cost reductions, nutritional modifications, and shelf-life evaluations. Thus, one can argue that a major part of product research is concerned more with sameness than with differences. Of course, this is not necessarily true for an innovative new product development scenario. But it is generally true for established, successful brands where formulation or processing changes are done for various reasons.

The liability in an equivalence conclusion is that there really is a perceivable difference, at least to some individuals, and that the sensory test has missed this fact. This is the familiar problem of a type II error – missing a difference that is really there; that is, failing to reject the null hypothesis

Quantitative Sensory Analysis: Psychophysics, Models and Intelligent Design, First Edition. Harry T. Lawless.

when it is false. Most of normal science and, indeed, most of our statistical training is aimed at preventing type I error; that is, declaring a significant effect when none really occurred. This can have devastating consequences in a basic research program. However, in applied product research in the foods and consumer products industries, the type II error is often much more troublesome. It can create problems in missed opportunities in product improvements, or franchise risk when loyal consumers are alienated by a product change that they do not like.

One can look at equivalence in two general categories. The first is functional equivalence, in which a sensory test result is considered "good enough" to move the project ahead with acceptable risk levels. The second type of equivalence is one that is proven beyond a reasonable doubt by airtight statistical criteria. This would be required in an advertising claim situation, for example, where one needed to prove parity or support an "unsurpassed" claim. In the first scenario, the testing might involve an internal discrimination or descriptive panel, with a known track record, and a finding of "no significant difference" is considered adequate evidence for moving the project ahead by adopting the formula change. Often this is done under severe time constraints, so the luxury of a full consumer test is not an option, although a small central location test could sometimes be conducted as a disaster check. In the second situation, a large consumer sample is probably needed, and if the advertising claim is made on a national basis, a geographically distributed sampling (e.g., in 8 to 12 cities in four or more regions) might be necessary (ASTM, 2008). In this case, the usual approach is to be able to prove that the result obtained has a 95% confidence interval that does not overlap the cutpoint for declaring "sameness." This can be a difficult and strict rule that mostly applies to consumer preference equivalence for ad claim purposes, rather than for a simple formulation change where the goal is primarily to avoid alienation of the loyal or heavy-user groups.

One might ask why all equivalence testing should not involve a statistical criterion of airtight proof, consideration of confidence interval boundaries, and extensive consumer testing validation. An old saying has it that "the enemy of good enough is perfect." Often, the internal discrimination panel – which has been selected and screened, operates under controlled laboratory conditions, and has a known track record of detecting differences – is sufficient to substantiate a "safety net" logic. That is, if the internal panel with all of its discriminative sensitivity does not see a difference, then it is unlikely that consumers will see a difference either. Of course, there is always a chance of type II error, and no test is perfect. But that can be said of a large-scale consumer test as well. The same reasoning can be applied to a descriptive analysis result in which the analysis of variance (ANOVA) and planned comparisons (e.g. Duncan, Tukey tests) show no difference between means on various attribute scales. That may be sufficient for research purposes with minor formula changes and in situations where the risk of consumer alienation is low, and/or time pressure to get the new product to market is an overriding concern.

Note that no two products are every truly the same, and even identical products may be considered different, as predicted by Thurstonian models and signal detection theories. So we are really dealing with an issue of sufficient perceived similarity to get by. Zero difference is not provable (science cannot prove a negative). Once it is viewed this way, the issue of equivalence becomes one of how close is close enough? At this point management has to weigh in, and there is no substitute for a history of product knowledge and a record of changes that have been attempted in the past. Of course, the consumer's tolerance for product variability has to play a part in these decisions. A product that has been consumed since childhood, like a nationally established brand of breakfast cereal, may have tight limits on what consumers will react to. On the other hand, with a product like wine that has customary and expected seasonal variation the consumer may be more tolerant of changes.

This chapter will first examine qualitative rule-based approaches to similarity testing, and then proceed to formal statistical tests. A finding of no significant difference is often actionable, although those who prefer a strict statistical treatment of equivalence will undoubtedly view this as a step backward. Yet many research programs in successful companies have used a nonsignificant statistical result with an established trustworthy instrument as evidence for equivalence, and many continue to do so to this day. Further discussion on equivalence testing can be found in US FDA (2001), Bi (2005, 2006, 2007), Meilgaard et al. (2006), Gacula et al. (2009), and Lawless and Heymann (2010: chapter 5). Consumer preference equivalence, as a two-tailed situation, will be discussed briefly. The interested reader is referred to the excellent ASTM document on advertising claim substantiation, E-1958, for further information and guidelines for large-scale preference consumer tests. Formal statistical treatment of equivalence testing can be found in Welleck (2003). A recent review is found in Bi (2005) and an alternative approach in Ennis and Ennis (2010).

6.2 Commonsense Approaches to Equivalence

6.2.1 When is a Nonsignificant Difference Meaningful?

Finding no significant difference in a statistical test is usually considered an ambiguous result. Rather than saying that we accept the null hypothesis, statisticians like to phrase the nonsignificant result as something like "there was insufficient evidence against the null." This is more intellectually honest than claiming a difference does not exist. There are many reasons why a significant statistical result might not be obtained. First, the sample size N could be too low for the size of the effect you are trying to detect. Second, you may have done a sloppy experiment, with uncontrolled sources of variation, thus increasing the sample standard deviations. In sensory testing, a host of methodological goofs are possible from people who think sensory testing is easy and straightforward and do not recognize the complexity of any situation using human observers. For example, we might have used the "wrong" people in the test. It would not make sense to have a test for boar taint in pork using people who were not screened for anosmia (smell-blindness) to the critical aroma compound, androstenone.

Given this ambiguity, we can ask whether any nonsignificant result is ever useful in decision-making. Under what conditions is a nonsignificant test evidence for sensory equivalence or similarity? Perhaps a better way to phrase this is when is a nonsignificant result likely to be unreliable (i.e., not trustworthy)? The previous paragraph gives several hints about the nature of a test we would not trust. The sensory professional can ask several questions: First, did I test enough people or use sufficient replication? A test with low N will lack horsepower. Second, did I use a good sensory method? A test conducted without attention to details or one that uses unjustified modifications of standard practices is suspect. There are lots of sensory tests done informally, such as benchtop "cuttings" with discussion of samples that are not blind-coded and decision-making based on consensus rather than independent judgments and hard data. Third, did I use the right people? A group of convenient volunteers who are not familiar with the test protocol and who have not been screened to insure a basic level of sensory discrimination talent is not a trustworthy source of information. Finally, did I do my part to insure control of unwanted sources of variation? Were there proper sample controls with regard to serving size, cooking/preparation, and temperature? Was the test conducted in a proper situation without extraneous odors and distractions? What about the people administering and conducting the test? Do they know what they are doing? Are they properly trained? Do they know the pitfalls and liabilities in taking any shortcuts? Do they follow a written standard operating procedure or lab manual? A critical set of considerations concerns the known record, reliability, and accuracy of the method. Is this a new procedure without a track record, or do I know and trust that it will detect true differences (and avoid false alarms) based on longstanding experience with the panel and procedure? Do I know how a significant difference will translate into consumer perception? What size of a difference, when detected by this panel, is likely to be meaningful to a significant proportion of consumers?

The opposite of the flaws suggested above will all contribute to a test whose result is believable and trustworthy, whether the outcome indicates a significant difference or not. In mathematical terms, we have decreased the β risk and increased the sensitivity and power of the test by using an appropriate size panel and controlling factors that add to the unwanted variability between samples. The latter is an effective lowering of the standard deviation, which, as we will see below, is an important factor in power calculations. The main points are to have good methods, standardized procedures, and evidence, usually based on a record that the panel, when acting as an analytical instrument, will detect differences when they are present and will avoid false alarms. Accuracy, of course, is largely a matter of translating the result of an internal discrimination panel into consumer perception. Such "calibration" studies are infrequent, although they are immensely valuable for an ongoing testing program with a standing panel for discrimination or descriptive analysis.

A final consideration in deciding whether a no-difference result is trustworthy is the question of how close your result is to the significance cutoff. If you are one judge short, and the p-value is 0.06, the result is much less convincing for a no-difference conclusion than if the number of correct answers is right at the chance-predicted head count. Of course, the p-value is not a simple metric. It depends both on the proportion correct and the sample size N, so both must be considered if you are going to use this criterion. An older approach to discrimination used a sequential testing strategy (Wald, 1945) in which testing would continue if the result was ambiguous; that is, neither significant nor indicating low β risk. By increasing the number of observations, one would eventually cross over either the α or β cutoff levels (Amerine et al., 1965). A graphical and worked example can be found in Meilgaard et al. (2006: chapter 6), and is illustrated in Figure 6.1. This approach seems to have been lost in current practice, perhaps because it is logistically cumbersome to keep testing and analyzing results as you go. However, it may be worth reconsideration if your testing program is capable of adding additional panelists after the first results are collected.

6.2.2 A "Double Control" Approach

Up to this point, we have discussed accepting a null result as evidence of sensory equivalence under some conditions. Those consist primarily of using the results from a panel with known discrimination abilities (i.e., a known detection rate). But suppose you do not have such a panel. Is there any alternative? One idea is to build the positive and negative control situations into the test itself. That is, one can include products with a known difference and products with known equivalence in order to see whether the panel is appropriately accurate and to serve as a kind of calibration check. Then the null result becomes meaningful when compared with the positive control pair (samples with a known difference) as long as said difference is detected. The negative control pair should probably be a small difference,

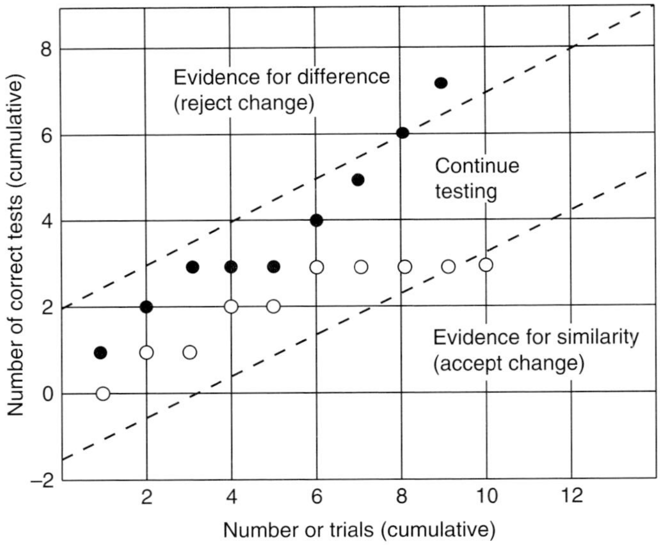

Figure 6.1 The sequential testing approach attributed to Abraham Wald (1945) which was in early sensory texts (Amerine et al., 1965) and currently in Meilgaard et al. (2006). As testing progresses, the cumulative record may cross the lower line, indicating evidence for equivalence (open circles), or the upper line indicating a significant difference (filled circles). If the cumulative record remains between the two boundaries, testing continues. The lower limit is given by

$$y = \frac{\log\beta - \log(1-\alpha) - n\log(1-p_a) + n\log(1-p_o)}{\log p_a - \log p_o - \log(1-p_a) + \log(1-p_o)}$$

where p_o is the probability associated with chance and p_a is the probability associated with the alternative hypothesis. In the example above, p_a is assigned as 2/3 or 0.667, given the 50% detection rate in a triangle test with $p_o = 1/3$. The upper boundary is given by

$$y = \frac{\log(1-\beta) - \log\alpha - n\log(1-p_a) + n\log(1-p_o)}{\log p_a - \log p_o - \log(1-p_a) + \log(1-p_o)}$$

rather than identical samples. This serves primarily as a check on false alarms, or type I error. Of course, if the test results show a missed difference in the positive control, then the result is not actionable. Also, if there is a difference found for the negative control, then the result is similarly ambiguous. The actionable result only occurs when the panel detects a difference in the positive control pair and no difference in the negative control.

An example of this logic can be found in a sweetener **cross-adaptation** study in the sense of taste (Lawless & Stevens, 1983). Before the sweet receptor was known and characterized, psychophysical workers used the method of cross-adaptation to find evidence for overlapping versus different receptors in taste and olfaction. The idea was simple. You would adapt the tongue to one substance, until the taste sensation disappeared. Typically, this was done in a flow system, which provided a stabilized area of stimulation and a fair amount of stimulus control. Then the experimenter would switch the flow to a second taste substance, to see whether there was no taste or a perceivable taste sensation. If no taste was perceived, it was considered evidence that the second chemical used the same (adapted) pathway as the first substance. It had been "cross-adapted." If a taste was perceived, then the conclusion would be that the first and second chemicals must be binding or transduced by difference mechanisms or pathways. This logic had been successfully applied to bitterness studies showing evidence for multiple receptor mechanisms, as is now confirmed by genetic studies (McBurney et al., 1972; Lawless, 1979; Bufe et al., 2005).

But the problem is that the conclusion from cross-adaptation is that the critical result is a null finding. That is, the cross-adapted item is not different from zero taste, or at least not different from the degree of adaptation (or taste intensity) seen when the first substance follows itself. In other words, a finding of "no significant difference." The finding of negative cross-adaptation (lack thereof) also raises issues. Lawless and Stevens realized that the pattern of results could only be meaningful if three conditions were met: First, the second substance must in fact be capable of adaptation (i.e., it should "cross-adapt" itself). This is actually a case of self-adaptation, but one that should be found in the same protocol and same apparatus as the two-chemical scenario. Second, the first substance must have the power to have an adapting effect. This is another case of self-adaptation, but focusing on the first chemical in the series, rather than on the second. These first two conditions are fundamentally a demonstration that your protocol in fact works. This is not a trivial demonstration, as some researchers had found that taste adaptation is sometimes difficult to obtain, not present in all subjects, and/or incomplete (Meiselman & Halpern, 1973). Third, some pair in the study must not cross-adapt, so that the effective demonstration of a positive control is also present.

Having effectively demonstrated that the procedure "works" in some predictable situations (self-adaptation) and does not find cross-adaptation universally (i.e., in others), the study's conclusions become scientifically acceptable. These control conditions were not always evident in some previous taste studies.

The application of this kind of logic is similar to the kinds of procedures recommended for quality control and shelf-life studies (see Lawless and Heymann (2010: chapter 17)). Often, a quality control panel is a small panel and the data may not be treated statistically, but by heuristics or "rules of thumb" for rapid decision-making. Do we bottle these batches of product we have made on third shift in the middle of the night? Sometimes there is no room or time for the luxury of a statistically analyzable panel result. When does the decision to go ahead and "pass" the production batches become acceptable? When the panel has known discriminative abilities as shown by a record of detecting bad samples, and/or when they have encountered spiked samples or control items that are known to be rejectable and they make the proper decision.

6.2.3 Power and Sample Size Considerations

6.2.3.1 *Sample Size*

Section 6.2.2 described approaches to equivalence testing that are primarily qualitative and rule driven, rather than truly statistical. This section will examine two traditional aspects of equivalence, namely power calculations and sample size considerations. Once again, we will be looking at ways to use a null result (i.e., a finding of no significant difference) as an actionable finding. However, the decision rule now takes into account the known or calculated power of the test, and considerations in experimental design of how many observations we should make to have confidence in the result.

The logical starting point for the experimental design considerations is the sample size, usually the number of judges or panelists or consumers in the test. For a basic two-sample test without replication, the general form for the sample size equation is (Amerine et al., 1965; Gacula et al., 2009)

$$N = \left(\frac{Z_\alpha \sigma_\alpha + Z_\beta \sigma_b}{m_1 - m_a} \right)^2 \tag{6.1}$$

for scaled data, where Z_α and Z_β are the Z-scores associated with our predetermined (i.e., acceptable) levels of α and β risk, the σ-values are the standard deviations associated with the null and alternative hypotheses, and $m_1 - m_a$ is the difference between means of the null (or observed) mean and the mean expected under the alternative hypothesis. The denominator is thus the effect size one is testing against. This last item may seem like an arbitrary decision, but most sensory professionals dealing with an established product line will have some general idea of what size of a difference is likely to be meaningful to consumers. If the standard deviations are the same, the equation simplifies somewhat to

$$N = \frac{(Z_\alpha + Z_\beta)^2 \sigma^2}{(m_1 - m_a)^2} \tag{6.2}$$

And the equation can be further simplified if one expresses the difference between the means in standard deviation units.

A quick worked example. If we have a standard deviation of one unit on a nine-point hedonic scale (a common observation) and we wish to test for half that as a difference between means (0.5 units), we let the Z-scores equal 1.96 for a two-tailed α of 0.05 and 1.645 for a two-tailed β risk of 0.10, we get

$$N = \frac{(1.96 + 1.645)^2 1^2}{(0.5)^2} \cong 52$$

Of course, if we have only a one-tailed direction (e.g., we are testing to make sure the result is *not lower than* the current benchmark score), then the Z-scores will change accordingly.

For a method involving proportions or frequency counts, the formulas are very similar:

$$N = \left(\frac{Z_\alpha \sqrt{p_o q_o} + Z_\beta \sqrt{p_a q_a}}{p_o - p_a} \right)^2 \tag{6.3}$$

where p_o is the probability associated with the null, p_a is the probability associated with the alternative hypothesis, and $q = 1 - p$.

A quick worked example. Suppose we conduct a triangle test and want to be 90% sure that we do not miss a difference in which the correct performance was actually 0.50. This is a reasonable proportion, as it corresponds to 25% detection after using the correction for guessing; that is, Abbott's formula (Ferdinandus et al., 1970; Finney, 1971) as discussed in Chapter 5. Plugging into Eq. 6.3 we get the following:

$$N = \left[\frac{1.645\sqrt{(1/3)(2/3)} + 1.645(0.5)}{(1/3) - 0.5} \right]^2 \cong 92$$

So, to be 90% sure that we are able to detect a difference that is only 17% above the chance level, we need about 92 judges. It is the small difference in the denominator here that is driving the requirement for a rather substantial sample size. If the requirement were to detect a performance level of 66.7% (corresponding to 50% true detection after correction for guessing) then the sample size requirement falls to about 22. Although 50% detection might be considered a cutoff for "threshold," this is a rather gross difference, by discrimination testing standards.

For scaled data and more complex designs, an extensive discussion of sample size in consumer testing was provided by Hough et al. (2006). They realized that simple two-product tests are not the only experimental designs in consumer work, nor in research and development of new products. A useful table was presented with different levels of α, β, the effect size, and the experimental error. For scaled data, the experimental error was expressed as the root-mean-square error (presumably from the ANOVA of a multifactor experiment), which is after all a kind of pooled variance estimate. Because it is a variance estimate, we need to take the square root to get back to the analogous standard deviation for formulae such as eqn 6.1 or 6.3. A further insight was to express both the effect size and error term as a proportion of scale length, because different researchers may choose to use different rating scales. For an error term of 0.23 (about 23% of scale length for a pooled standard deviation estimate), an effect size of 0.1 (10% of scale), α at 0.05 and β at 0.10, we get a sample size requirement of 112. This is very close to the general sample size recommended in various sensory evaluation texts for consumer tests (Meilgaard et al., 2006; Lawless & Heymann, 2010).

In summary, power, sample size, the α level, the variance, and one additional factor all interact to determine the statistical "system." The final factor is the size of the difference one wishes to test against; that is, the denominator of the sample-size equation. This factor can be explicitly stated in the form of the alternative hypothesis. Let us parse this relationship to see what happens to the sample size N as various factors are changed. First, it should be obvious that the sample size requirements go down as the standard deviations decrease. Next, if we take on more β risk (and thus lower the power) we require fewer people. Third, if we decide to test for a larger critical difference, the required N goes down. One way to think of this is that it is easier to detect a large difference than a small difference. Being sure not to miss a small difference requires a more powerful test and larger N. Finally, if we let α float up, perhaps changing our traditional critical p-value from say 0.05 to 0.10, the sample size required is less. That is simply because, as α floats up, Z_α will become smaller, in this case from 1.645 to 1.28. This may seem somewhat paradoxical, but you can think of it this way: if I have failed to detect a difference with a relaxed α of 0.10 (or higher), I have made the hurdle for significance lower, and yet not seen a difference anyway. So the difference is even less likely to be truly present. All other factors being equal, α and β risks will be inversely related.

6.2.3.2 *Power Calculations*

By rearranging eqn 6.1, one can solve for Z_β and then the power of the test can be found by converting from Z to the corresponding proportion and subtracting from one. The corresponding proportion is simply the cumulative value of the normal distribution function. Assuming the same standard deviation for the null and alternative distributions, letting μ_D be the critical difference between means, and taking the square root of eqn 6.1 gives

$$\sqrt{N} = \frac{\sigma\left(Z_\alpha + Z_\beta\right)}{\mu_D} \tag{6.4}$$

Moving terms from left to right gives

$$\frac{\sqrt{N}\mu_{\mathrm{D}}}{\sigma} = \frac{\mu_{\mathrm{D}}}{\sigma / \sqrt{N}} = \frac{\mu_{\mathrm{D}}}{\mathrm{SE}} = Z_{\alpha} + Z_{\beta} \tag{6.5}$$

Thus:

$$Z_{\alpha} - \frac{\mu_{\mathrm{D}}}{\mathrm{SE}} = \frac{Z_{\alpha}\mathrm{SE} - \mu_{\mathrm{D}}}{\mathrm{SE}} = Z_{\beta} \tag{6.6}$$

The quantity $Z_{\alpha}\mathrm{SE}$ gives a cutoff on the measured scale X_{c} in Gacula's (1991) notation, which can be a useful shortcut. It helps obtain the value of the alternative hypothesis parameters, and thus our obtained β risk.

As $\Phi(Z_{\beta}) = \beta$, where Φ is the cumulative normal distribution function, we can get the power from

$$\text{Power} = 1 - \Phi\left(\frac{Z_{\alpha}\mathrm{SE} - \mu_{\mathrm{D}}}{\mathrm{SE}}\right) \tag{6.7}$$

The equivalent formula for a test on proportions is

$$\text{Power} = 1 - \Phi\left[\frac{Z_{\alpha}\sqrt{p_{\mathrm{o}}q_{\mathrm{o}}/N} - (|p_{\mathrm{o}} - p_{\mathrm{a}}|)}{\sqrt{p_{\mathrm{a}}q_{\mathrm{a}}/N}}\right] \tag{6.8}$$

A quick worked example, from Gacula (1991). Suppose we conduct a consumer study with 92 participants and a product which scores a 6.10 on a nine-point scale. The control product scores a 5.9, and we are tempted to conclude they are equivalent. Both products had a standard deviation of 1.1, giving a standard error of 0.11. However, a score of 6.12 would have been statistically significant, so more information is needed before we draw any conclusions. Fortunately, management has decided that any difference smaller than 0.5 standard deviations is "good enough" to be considered sensorially equivalent. That makes the mean for the alternative hypothesis 5.9+0.5(1.1)=6.45. Looking backward at an action cutoff of 6.12, we can see that the value of 6.12 is about three standard errors lower than 6.45. So a Z_{β} of −3 produces a p-value of about 0.01, or 99% power. Plugging directly into the equation gives us [1.96(0.11)−0.55]/0.11=3.22 for Z_{β}. All is well.

These general relationships will be revisited in Section 6.3.2 when we consider Ennis's equations for power based on Thurstonian modeling considerations. Summarizing, one common and traditional strategy for using a null result as evidence for equivalence is to make sure that the test has sufficient statistical power to detect a difference. Although this strategy does not produce airtight evidence, it is often sufficient for purposes of cost reductions and other small product changes, where product development and maintenance are the issues, rather than advertising claims or a need to withstand litigation. In those two circumstances, more evidence is generally needed, and strategies for parity testing for ad claims will be discussed later. For now, we can consider a test that is not significantly different and a conclusion of equivalence to be "functionally valid," as suggested in this quote from Cohen (1988), who published widely on power issues (italics added):

> Research reports in the literature are frequently flawed by conclusions that state or imply that the null hypothesis is true. For example, following the finding that the difference between two

sample means is not statistically significant, instead of properly concluding from this failure to reject the null hypothesis that the data do not warrant the conclusion that the population means differ, the writer concludes, at least implicitly that there is no difference. The latter conclusion is always strictly invalid, and it is functionally invalid *unless power is high.*

6.3 Allowable Differences and Effect Size

6.3.1 Chance-Corrected Performance Cutoffs

The use of a nonsignificant effect as evidence for equivalence, with known power, has largely been abandoned by the larger scientific community in areas such as drug delivery and bioequivalence (US FDA, 2001), who are using more specific statistical tests to prove that a given result lies within some range considered "equivalent." In sensory discrimination testing, the first statistical approaches to proving equivalence by actually constructing confidence intervals and/or rejecting a hypothesis were given by MacRae (1995), Schlich (1993), and Carr (Meilgaard et al., 2006) at about the same time in the early 1990s. The concept was simple. First, one had to specify what level of difference would be considered an upper bound on functional equivalence, much like specifying the effect size we are testing against in a power calculation. In other words, how much of a difference can we tolerate? Zero difference is not possible, although many managers and executives would probably like to impose that kind of limit. All three of the aforementioned authors chose to use some version of the concept of proportions of discriminators as their criterion for similarity/difference. In other words, one could use Abbott's formula to specify a proportion correct in the data that would correspond to an acceptable proportion of discriminators.

Once that limit is specified, it can be used to construct a functional null hypothesis capable of rejection. Another way to view this is that the level chosen constitutes a functional upper bound, and one can then test to see that the 95% confidence interval in our actual data does not overlap that cutoff. This is the same as testing to see that you are above a chance performance level with 95% confidence, and thus rejecting the null and declaring a difference, except that it turns the test on its head. Rather than looking downward at a floor to see that we are significantly above it, we are looking up at a ceiling to see that we are significantly below it. Simple binomial statistics suffice in both cases. The approach is illustrated in Figure 6.2.

The method was intuitively appealing, but it had several drawbacks. First, it required a larger sample size than most people were using for discrimination panels. So a program adjustment was often required to increase the size of any standing discrimination panel. Second, the chance performance level provides a floor that does not go away just because we are doing a one-tailed test in the other direction. Having both a ceiling and a floor can produce a narrow window within which the proportion obtained and its confidence interval must fit. This is obviously a problem if (1) the proportion of allowable discriminators is very low (lowering the ceiling towards the chance level) or (2) the sample size is too low, and thus the confidence interval too large to fit inside the "window." Third, the notion of proportions of discriminators is method specific, and thus a limit set for a triangle actually represents a different sensory difference in a 3-AFC test, for example. Calculations would arrive at the same proportion or discriminators for any given proportion correct observed in the data. Nonetheless, many companies found the concept sufficiently useful to retain it in decision-making. This is generally acceptable if you are only using one kind of test and you know

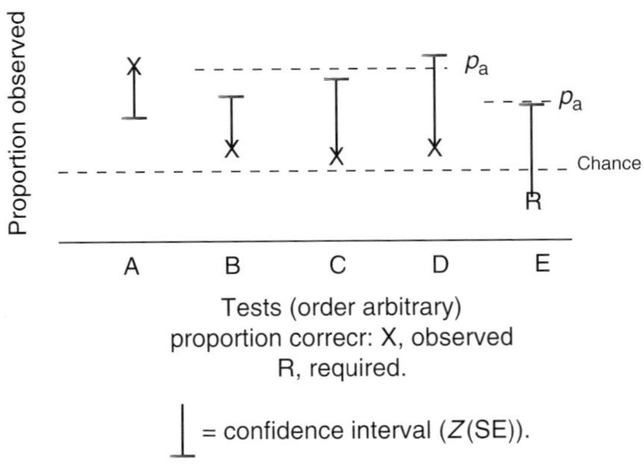

Figure 6.2 The test for significant similarity, based upon proportions of allowable discriminators, as compared with the test for significant differences in a choice (e.g., 3-AFC) or sorting (e.g., triangle) discrimination test with a known chance level. The similarity limit, p_a can be determined from Abbott's formula for converting from the allowable proportion of discriminators to the expected proportion correct. In situation A, there is the normal test for significant differences, in which the observed proportion correct minus the downside confidence interval, $\mathrm{CI}=Z(\mathrm{SE})=Z\sqrt{pq/N}$, must be greater than chance. In the similarity test, example B, the upside confidence interval must be less than the allowable boundary defined by p_a. In example C, the panel size is less, lowering N and thus increasing the standard error and confidence interval, but still is able to fit in the window between the upper boundary and chance. In example D, the sample size is still smaller and performance too high to justify a significant similarity. In example E, the allowable value for p_a has been lowered. Also, the required proportion correct, R, is below chance, and therefore this test is not practical.

both its track record and consumer implications. The alternative would be to use a Thurstonian difference measure, which cuts across different methods and can provide a common yardstick for sensory equivalence. Thurstonian measures will be discussed below.

The three authors expressed the approach somewhat differently. MacRae (1995) phrased it in terms of traditional confidence intervals. He astutely observed that a fairly large panel size would be required for many of the common parameters that one would adopt for declaring similarity/equivalence. Schlich's tables gave an entire strategy. First, specify your α, β, and proportion of true discrimination. Then his tables provide a recommended sample size and a crossover point for the critical number correct. Below this number, you can declare a significant similarity, with your specified β level, and above this number you could declare a significant difference. This is useful because it is possible (paradoxically) to have both a significant difference and significant similarity, depending upon the limits you set and your sample size N. Carr's tables required no prespecification of a fixed sample size, but rather let that be a parameter in the model (Meilgaard et al., 2006). This is perhaps a bit more useful for sensory programs with a constrained number of recruits for any given test day. However, the tables do have some "holes" where the parametric window becomes too close to the chance level. Requiring performance below chance is not a useful strategy.

The tables can be found in the original publications, and parts of them are also reproduced in Lawless and Heymann (2010). For a quick calculation, remember that the standard error of a proportion is given by

$$\mathrm{SE} = \sqrt{\frac{pq}{N}} \tag{6.9}$$

where p is the chance proportion, $q = 1 - p$, and N is the sample size. A confidence interval then is given by

$$CI = P_{obs} \pm Z(SE) \tag{6.10}$$

And for a one-sided upper confidence interval we can use 1.645 as the value for Z.

One further adjustment is needed, owing to the fact that we are testing against a proportion of discriminators, although the cutoff in the tables will be expressed as a raw proportion correct. For the triangle test, for example, the proportion of discriminators is given by $1.5P_{obs} - 0.5$. Also, the standard error will be increased by the same multiplicative factor of 1.5. So the full expression for the upper 95% confidence interval limit becomes

$$\text{Upper CI} = (1.5P_{obs} - 0.5) + 1.645(1.5)\sqrt{\frac{pq}{N}} \tag{6.11}$$

as found in Meilgaard et al. (2006) (and more recent editions) and used to construct the tables for Carr's test for significant similarity. One merely has to solve for the largest $P_{critical}$ such that $P_{critical} + \text{Upper CI} < P_{discriminators}$. If $P_{obs} < P_{critical}$, then you can declare significant similarity under this model.

6.3.2 Use of δ as a Cutoff

Power and sample size estimates can also be obtained from the use of Thurstonian δ values, rather than the chance-corrected proportions of discriminators. This approach has the advantage of using a universal measure of sensory discriminability, one that is not tied to any specific method (Ennis, 1993). The δ value is one that can be applied to any choice test in which the cognitive strategy for obtaining a correct judgment is known and thus the psychometric function can be calculated. Once the psychometric function is known, one can translate any δ value taken as a criterion and translate that into a proportion correct. Having done so, the general calculations for sample size and power are given as follows, broken into two parts for easier notation (after Rousseau and Ennis (2001)):

$$y = \frac{\Phi(1-\alpha)\sqrt{m-1}}{m} \tag{6.12}$$

where m is the number of alternatives in the choice test, and

$$N = \left[\frac{\Phi^{-1}(\beta)\sqrt{p(1-p)} - y}{1/m - p}\right]^2 \tag{6.13}$$

where p is the value determined from δ in the psychometric function for that particular method. The power of the test is then given by

$$\text{Power} = 1 - \Phi\frac{u - p}{\sqrt{p(1-p)/N}} \tag{6.14}$$

where

$$u = \frac{1}{m} + \frac{\Phi^{-1}(1-\alpha)\sqrt{m-1}}{m\sqrt{N}} \tag{6.15}$$

and p is derived from the tables of the δ values as above.

A quick worked example. Let us say we want to detect a δ value of 1.0, in a triangle test ($m=3$) of 100 judges with α at the usual 0.05. The triangle p corresponding to $\delta=1.0$ is 0.414 so the value of u becomes

$$u = \frac{1}{3} + \frac{1.645\sqrt{2}}{3\sqrt{100}} = 0.4109$$

And power is thus

$$1 - \Phi\frac{0.4109 - 0.414}{\sqrt{0.414(0.586)/100}} \approx 0.524$$

This might not be considered very good power because we stand about a one-in-two chance of missing the difference with this sample size.

One might be tempted to ask, why not use a test against $d'=0$? Suppose we find that the d' obtained is not significantly different from zero. Can one conclude that the experiment has given evidence for sensory equivalence? Of course, this is much like basing a decision on a failure to reject the null, which has all of the limitations and pitfalls outlined above. What about setting a d' value as an upper limit and then testing to see if we are significantly below that level? Unfortunately, the tables for the B-factors, from which the variance values for d' are calculated, pass through a minimum near $d'=0.61$ for the triangle test and $d'=0.71$ for the duo–trio. This creates a kind of paradox, as the variance increases as d' approaches zero. Thus, it is easier to test for a value significantly below a d' of 1.0 but more difficult to find a significant difference below 0.5. Furthermore, the confidence intervals are increasing in size as d' gets small. So, neither of these criteria is recommended.

A solution to this problem was proposed by Bi (2011). One can still use a d' value as a criterion for the upper bound on similarity, but rather than testing using the variance of d', return to the binomial formula to get your confidence interval or exact binomial limits. The method has several steps for a one-tailed test based on the choice method, such as the triangle, duo–trio, or n-AFC procedures. First, set your d' limit (d'_0) and set up a null hypothesis that says the population estimate of d' is equal to or greater than the limit. In other words:

$$H_0: d' \geq d'_0 \quad \text{versus} \quad H_a: d' < d'_0 \tag{6.16}$$

This has the advantage of a statistical test with a rejectable null hypothesis. When H_0 is rejected, we can declare evidence for equivalence. Next, we convert our critical d' limit to a proportion (P_0) correct that corresponds to that d', based on the psychometric function for the test in question. Tables or software for the conversion are readily available. Third, we find the lower confidence bound, such that an observed number correct forms the largest value in the tail of the binomial distribution such that the tail area corresponds to our α-level cutoff. Then we can be sure that the observed number would only occur, say, 5% of the time (or less) given a true null. Another way to think about this is that there is a 5% chance or less

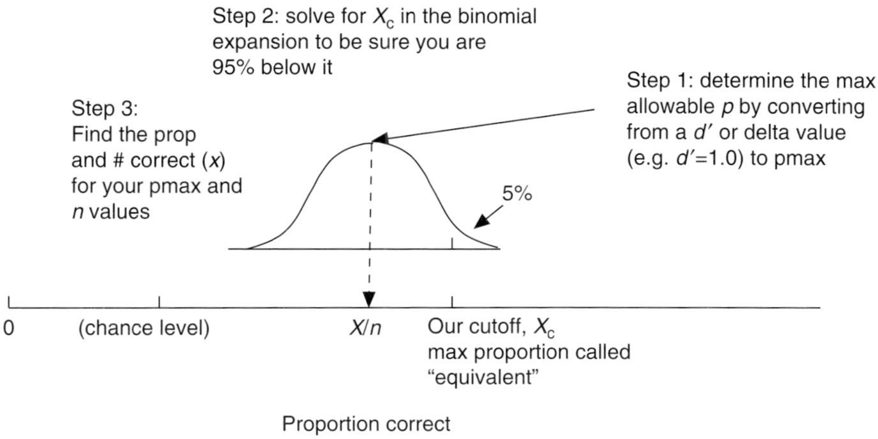

Figure 6.3 An illustration of the one-tailed binomial test approach for a 95% confidence nonoverlapping with the cutoff for declaring similarity.

of seeing this result, if d' in the true population were equal to or greater than the critical d'_o. If the true d' were even greater, then we would see this result even less than 5% of the time.

Mathematically, we are solving for the greatest value of X_c (number correct) out of N judgments such that

$$\sum_{X=0}^{X_c} \binom{N}{X} P_o^X \left(1-P_o\right)^{N-X} < \alpha \tag{6.17}$$

So, this is equivalent to summing the binomial probabilities (areas) in the tail of the distribution from zero to X such that the sum is less than α. This approach is illustrated in Figure 6.3.

Power calculations are equally straightforward. The power of any test for an obtained d' value d'_c, corresponding to a proportion correct of P_c, with $d'_c < d'_o$ is given by

$$\text{Power} = \sum_{X=0}^{X_c} \binom{N}{X} P_c^X \left(1-P_c\right)^{N-X} \tag{6.18}$$

A quick worked example (from Bi (2011)). Suppose we have set our cutoff d' value at $d'=0.2$. Note that this is a small difference. For a triangle test, the corresponding proportion correct from the tables of the psychometric function is 0.337. Putting this into eqn 6.18, we find that the upper bound for a triangle test where $N=100$ is 25.

Looking at a normal distribution approximation, the critical one-tailed value for rejecting H_0 is −1.645. This is negative because we are looking for values below the assumed proportion P_o; that is, we are now looking south rather than north for a test of equivalence, rather than difference. Using the Z-score test on proportions and solving for X where X/N is the largest observed proportion correct, we can use and still reject the null using the following expression:

$$X \le NP_o - 1.645\sqrt{NP_o\left(1-P_o\right)} \tag{6.19}$$

Using our worked example from Bi (2011) and this inequality, we get an upper bound of 25.92, so once again we need no more than 25 correct judges in order to reject the null.

Note: this is below chance because, for 100 judges in a triangle, 33 or more correct is expected! So we have actually set a limit here that is probably unattainable. The primary reason is that our d' limit of 0.2 is very small indeed, placing an impossible constraint on the requirement. In Carr's tables, for example, the value would be blank for a situation in which it is not possible to be below the critical limit and still above chance. So you should be careful to understand the practical limits of this approach and make sure your requirements are actually attainable given δ and N.

6.4 Further Significance Testing

6.4.1 Two One-Sided Tests for Scaled Data

A standard method for testing equivalence in scaled data in biological and pharmacological studies is the two one-sided test (TOST) approach (US FDA, 2001). This has been criticized on theoretical grounds, primarily by Ennis and co-workers, who provided an alternative approach that is outlined below. If we have some scaled values of a variable X representing the difference score between two sets of observations, with mean μ, a value that defines equivalence τ, and some allowable difference θ, using the notation of Ennis and Ennis (2009, 2010), the null and alternative hypotheses take the following forms:

$$H_0: \mu \leq \tau - \theta \quad \text{or} \quad \mu \geq \tau + \theta \quad \text{and} \quad H_a: \tau - \theta < \mu < \tau + \theta \tag{6.20}$$

So, we are asking whether the population value of the mean difference is likely to fall within or outside our required value plus or minus its acceptable interval. This boils down to two t-tests, with the requirement that the usual null be rejected in both cases. You must be significantly above the lower limit and significantly below the upper limit.

Bi (2005, 2007) recommended a modified version of this test, a similarity test for two means, as might come from some scaled data such as acceptability ratings, descriptive panel data, or quality control panel data. The critical test statistic is T_{AH} after the original authors of the test, Anderson and Hauck. If we have two means M_1 and M_2 from two groups of panelists with N panelists per group and a variance estimate S, the test proceeds as follows:

$$T_{AH} = \frac{M_1 - M_2}{S\sqrt{2/N}} \tag{6.21}$$

The variance estimate S can be based on the two samples, where

$$S^2 = \frac{S_1^2 + S_2^2}{N} \tag{6.22}$$

and we must also estimate a noncentrality parameter δ:[1]

$$\delta = \frac{\Delta_0}{S\sqrt{2/N}} \tag{6.23}$$

where Δ_0 is the allowable difference interval.

[1] Note that this is not the Thurstonian δ.

The calculated p-value is then

$$p = t_v(|T_{AH}| - \delta) - t_v(-|T_{AH}| - \delta) \quad (6.24)$$

where t_v is the common (central) t distribution value for $v=2(N-1)$ degrees of freedom. If p is less than our cutoff, usually 0.05, then we can conclude that our difference is within the acceptable interval and we have equivalence. A worked example can be found in Bi (2005).

6.4.2 Parity Testing for Preference

The binomial two-tailed test for equivalence is given as follows (Ennis & Ennis, 2009, 2010), which is very similar to the one-sided equation of Bi in eqn 6.17. Let t once again be our true value, say 0.5 for a preference test, and θ be the allowable interval. We now need the difference in two binomial tails as defined by

$$p = \sum_{k=0}^{N-m} \binom{N}{k} (0.5-\theta)^k (0.5+\theta)^{N-k} - \sum_{k=0}^{m-1} \binom{N}{k} (0.5-\theta)^k (0.5+\theta)^{N-k} \quad (6.25)$$

where m is the smaller of the two frequency counts in the preference study.

A quick worked example, given by Ennis and Ennis (2009). Suppose we conduct a large-scale consumer test with 800 consumers, 410 of whom prefer the winning product. Can we conclude that there is really no preference based upon this information? Plugging the value of 390 in for m, using a boundary of 0.45 to 0.55 in the preference proportions as the defining interval for "equivalence," we can use eqn 6.25 as follows:

$$p = \sum_{k=0}^{410} \binom{800}{k} (0.45)^k (0.55)^{N-k} - \sum_{k=0}^{389} \binom{800}{k} (0.45)^k (0.55)^{N-k} = 0.018$$

thus rejecting the null and declaring significant parity. This approach forms the basis for parity tests on preference, with tables published by Ennis (2008a).

The normal approximation to the binomial is even easier to use, as it is much less computationally intensive than summing all the terms in the tail of the exact binomial. Of course, if one has a program to do that or can work it up in a spreadsheet, this is a trivial concern. The normal distribution test is done as follows:

$$p = \Phi\left(\frac{|x| - \theta}{\sigma}\right) - \Phi\left[\frac{-(|x| + \theta)}{\sigma}\right] \quad (6.26)$$

A quick worked example. Using the example of Ennis and Ennis again, with 800 consumers and 390 choosing one alternative. They apply a continuity correction to adjust to 389.5 and then divide by 800 to get a proportion of 0.4869. We then take the difference from the null value of 0.5 to get a value for x of –0.131. Using a sigma estimate of $\sqrt{(0.45)(0.55)/800} = 0.0176$, the p-value is calculated as

$$p = \Phi\left(\frac{0.0131 - 0.05}{0.0176}\right) - \Phi\left[\frac{-(0.0131 + 0.05)}{0.0176}\right] = 0.01791$$

which agrees very well with the exact binomial value of 0.018 found above.

When the variance or standard deviation is not known, Ennis and Ennis (2009) have argued for an adjustment factor in order to protect type I error. Remember that the null in this case is a statement about nonequivalence, so rejecting that too often would mean concluding equivalence too often (Ennis uses the term "too liberal" for these occurrences). In order to protect the α level, eqn 6.26 is modified to become

$$p = \Phi\left(\frac{|x| - \sqrt{c}\theta}{s}\right) - \Phi\left[\frac{\left(-|x| + \sqrt{c}\theta\right)}{s}\right] \tag{6.27}$$

where s is an estimate of variance based upon the sample standard deviation and c is the adjustment factor applied to the noncentrality parameter.

6.5 Summary and Conclusions

This chapter has discussed various approaches to help the sensory practitioner justify a decision of sensory equivalence, similarity, or parity. These range from qualitative decisions to formal statistical approaches with a rejectable null hypothesis of nonequivalence. Qualitative decisions involve an acceptance of the null (i.e., failure to reject) as an action standard based upon the known record of the measuring instrument (i.e., the sensory panel) in terms of its ability to detect differences. This is unsophisticated, but reasonable for most research decisions involving minor ingredient, processing, or cost reduction changes, where there is low risk of consumer alienation. The next, more sophisticated step is to do formal sample size and power calculations so that when no difference is detected it is under circumstances that validate the equivalence decision. Knowing the power of your test is probably a good idea in any event. The final and most scientifically sound approach is to set up a null hypothesis of nonequivalence, based upon an acceptable interval of agreed-upon similarity, and then to amass evidence to reject the nonequivalence null in favor of the alternative.

Exactly how to go about the latter approach is a matter of some debate. For scaled data, an industry standard has been the TOST approach. Various modifications of that have been proposed to adjust for its shortcomings. Notably, in the sensory community, Ennis and colleagues have argued that the TOST has some odd properties that make it suboptimal (e.g., Ennis, 2008b). For one thing, the two tests are dealing with correlated information, and thus their covariance is not zero. They are not independent in that sense. Second, for situations with a small difference but high variance, the TOST will never reject the null. This is a kind of paradox that is avoided by Ennis and Ennis when they proposed an alternative based on a noncentral chi-square distribution. In their 2009 paper, they do state that the TOST may be a useful approximation under some conditions and that it is easy to conduct. They also critique the approach of Anderson and Hauck, which is more powerful than the TOST, but in their opinion also too liberal. The equations outline in Section 6.4.2 are consistent with their approach, and the reader is referred to the original publications for further details on situations with known versus unknown variance, and adjustments made with a noncentrality parameter. Third, the exact binomial presented above produces more accurate rejection regions that the TOST (Ennis, 2008b). More detailed and basic information can be found in the article by Berger and Hsu (1996). An interesting and spirited debate about the value of TOST versus alternatives can be found in Volume 19(3) of *Food Quality and Preference*.

In summary, one can take more comfort in a decision that is based on rejection of the null rather than one based on a failure to reject. That being said, it is important to realize that the p-value in the significance test, using a null of nonequivalence, does not translate directly into evidence for the alternative hypothesis, at least not on a quantitative level. Statements such as "Equivalency hypothesis testing, like all hypothesis testing, involves a statement of the null hypothesis in a form that can be tested and if it is rejected, there is evidence in favor of the alternative" (Ennis, 2008b: 347) are common. This is not strictly true. It is tempting to think that the quantity $1-p$ gives the probability of the alternative hypothesis, but it does not, as that depends upon the incidence (prior probability in Bayesian terms) of the null and alternative situations. The p-value, remember, only shows the *rarity of the result obtained* given the assumption of a true null, which you are rejecting anyway.

Whatever the approach, researchers should generally be prepared for a significant cost for equivalency testing in terms of the sample size required. More observations are generally needed than are required for finding a result of a significant difference, and protecting α. As stated at the outset by Heymann, power resides primarily in the number of people you test.

References

Amerine, M.A., Pangborn, R.M., and Roessler, E.B. 1965. Principles of Sensory Evaluation of Food. Academic Press, New York, NY.

ASTM. 2008. Standard guide for sensory claim substantiation. Designation E-1958-07. Annual Book of Standards, vol. 15.08. ASTM International, West Conshohocken, PA, pp. 186–212.

Berger, R.L. and Hsu, J.C. 1996. Bioequivalency trials, intersection–union tests and equivalency confidence sets. Statistical Science, 11, 283–302.

Bi, J. 2005. Similarity testing in sensory and consumer research. Food Quality and Preference, 16, 139–49.

Bi, J. 2006. Sensory Discrimination Tests and Measurements. Blackwell Publishing, Ames, IA.

Bi, J. 2007. Similarity testing using paired comparison method. Food Quality and Preference, 18, 500–7.

Bi, J. 2011. Similarity testing using forced choice methods in terms of Thurstonian discriminal distance, d'. Journal of Sensory Studies, 26, 151–7.

Bufe, B. Breslin, P.A.S., Kuhn, C., Reed, D.R., Tharp, C.D., Slack, J.P., Kim, U.-K., Drayna, D., and Meyerhof, W. 2005. The molecular basis of individual differences in phenylthiocarbamide and propylthiouracil bitterness perception. Current Biology, 15, 322–7.

Cohen, J. 1988. Statistical Power Analysis for the Behavioral Sciences. Second edition. Lawrence Erlbaum Associates, Hillsdale, NJ.

Ennis, D.M. 1993. The power of sensory discrimination methods. Journal of Sensory Studies, 8, 353–70.

Ennis, D.M. 2008a. Tables for parity testing. Journal of Sensory Studies, 23, 80–91.

Ennis, D.M. 2008b. Rejoinder to Bi and Meyners. Food Quality and Preference, 19, 347–8.

Ennis, D.M. and Ennis, J.M. 2009. Hypothesis testing for equivalence defined on symmetric open intervals. Communications in Statistics, 31, 1792–803.

Ennis, D.M. and Ennis, J.M. 2010. Equivalence hypothesis testing. Food Quality and Preference, 21, 253–6.

Ferdinandus, A., Oosterom-Kleijngeld, I., and Runneboom, A.J.M. 1970. Taste Testing, MBAA Technical Quarterly, 7(4), 210–27.

Finney, D.J. 1971. Probit Analysis. Third edition. Cambridge University Press.

Gacula, M.C., Jr. 1991. Claim substantiation for sensory equivalence and superiority. In: Sensory Science Theory and Applications in Foods. H.T. Lawless and B.P. Klein (Eds). Marcel Dekker, New York, NY, pp. 413–36.

Gacula, M.C., Singh, J., Altan, S., and Bi, J. 2009. Statistical Methods in Food and Consumer Research. Academic Press (Elsevier), Burlington, MA.

Hough, G., Wakeling, I., Mucci, A., Chambers, E.C., IV, Gallardo, I.M., and Alves, L.R. 2006. Number of consumers necessary for sensory acceptability tests. Food Quality and Preference, 17, 522–6.

Lawless, H.T. 1979. The taste of creatine and creatinine. Chemical Senses, 4, 249–52.

Lawless, H.T. and Stevens, D.A. 1983. Cross-adaptation of sucrose and intensive sweeteners. Chemical Senses, 7, 309–15.
Lawless, H.T. and Heymann, H. 2010. Sensory Evaluation of Foods, Principles and Practices. Springer Publishing, New York, NY.
MacRae, A.W. 1995. Confidence intervals for the triangle test can give reassurance that products are similar. Food Quality and Preference, 6, 61–7.
McBurney, D.H., Smith, D.V., and Shick, T.R. 1972. Gustatory cross adaptation: sourness and bitterness. Perception & Psychophysics, 11, 228–32.
Meilgaard, M., Civille, G.V., and Carr, B.T. 2006. Sensory Evaluation Techniques. Fourth edition. CRC Press, Boca Raton, FL.
Meiselman, H.L. and Halpern, B.P. 1973. Enhancement of taste intensity through pulsatile stimulation. Physiology and Behavior, 11, 713–16.
Rousseau, B. and Ennis, D.M. 2001. A Thurstonian model for the dual pair (4IAX) discrimination method. Perception & Psychophysics, 63, 1083–90.
Schlich, P. 1993. Risk tables for discrimination tests. Food Quality and Preference, 4, 141–51.
US FDA. 2001. Guidance for Industry. Statistical Approaches to Bioequivalence. US Department of Health and Human Services, Food and Drug Administration, Center for Drug Evaluation and Research (CDER), www.fda.gov/downloads/Drugs/Guidances/ucm070244.pdf (accessed 27 March 2012).
Wald, A. 1945. Sequential tests of statistical hypotheses. Annals of Mathematical Statistics, 16, 117–86.
Welleck, S. 2003. Testing Statistical Hypotheses of Equivalence. CRC Press (Chapman and Hall), Boca Raton, FL.

7 Progress in Scaling

> *The concept of measurement is basic to an understanding of the various methods devised to quantify sensory attributes. Measurement is the assigning of numbers by rules to represent properties of objects or events. The numbers, as symbols of properties in the world of objects and events, can be manipulated in accordance with the rules of mathematics. If the properties of the number system reflect the properties of objects or events, new information about the measured properties may be revealed from such symbolic manipulations.*
>
> G.A. Gescheider (1997: 186).

7.1 Introduction

7.1.1 Common Scales in Current Usage

Like most of the methods in this book, scaling is a system of numerification. It is fundamentally a process of matching between two continua. One continuum is anchored to the perception of the product, usually on a single unidimensional attribute. Three major types are the perceived

Quantitative Sensory Analysis: Psychophysics, Models and Intelligent Design, First Edition. Harry T. Lawless.

sense of intensity (How sweet is this?), extent (How many chocolate chips do I see on this cookie: none, few, or a lot?), or appeal (How much do I like this product?). It is also possible to have people scale overall difference (or similarity), which tends to be multidimensional. The second continuum is the response continuum that is eventually translated into numerical data. The continuum can be numbers themselves or some physical (usually visual) continuum such as the length of a line that can be measured or a series of check boxes that can be numbered from left to right. In the case of perceived intensity, the matching process is a fundamental part of the science of psychophysics, as discussed in Chapters 1 and 2.

This chapter will deal with response options and how they may be treated as data. Common response options are shown in Figure 7.1. They can broadly be grouped as categorical or continuous, although the boundary between the two becomes blurred at some point. Examples of discrete, categorical scales include integer category scales, verbally labeled response options, and a series of check boxes. Examples of continuous scales include the line scales, also called visual analog scales. Other physical continua may be matched in a more or less continuous fashion, such as finger span (distance between two fingers matches intensity) or handgrip force, but these are rare in industrial practice. Line scales themselves

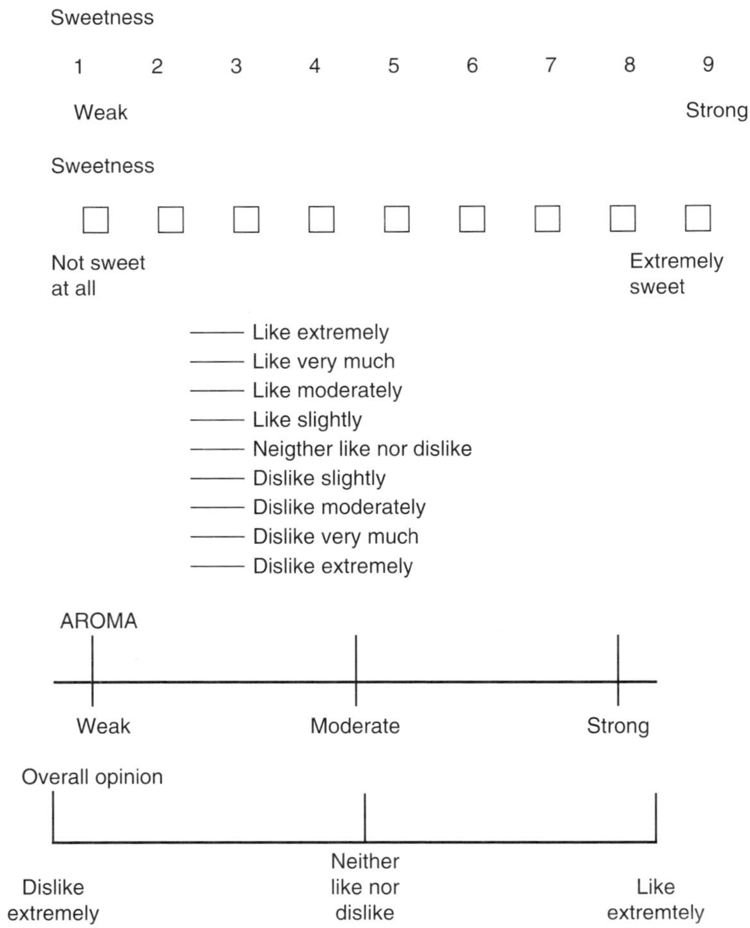

Figure 7.1 Common types of scales in current usage.

take on many variations, such as the addition of tick marks or verbal anchor labels at the endpoints and sometimes at various locations along the line itself. Numbers can be either continuous, as in the case of magnitude estimation that allows fractions and decimals, or discrete, as in the case of integer category scales. The distinction between discrete and continuous scales becomes blurred when one considers the limits of resolution possible in the data, such as the number of decimal places allowed in a numerical scale or the number of pixels on a computer screen showing a line scale.

Magnitude estimation is sometimes considered a separate category within scaling methods because there is no visual display and the number generation process is unique. In this method, the respondent is instructed to generate numbers to represent ratios or proportions among the stimuli or products, on the continuum of interest (e.g., sweetness). Thus, if one item is perceived to be twice as sweet as another, it should be given a number twice as large as that given to the other item. In most versions, any numbers are allowed, including fractions and decimals, as well as zero for things that are not perceived at all. Negatives are obviously not appropriate for rating perceived intensity, but they could be used for disliking in hedonic versions of this method. The technique can be done with or without an anchoring example, called a modulus. In the modulus procedure, a person will typically be given a standard or reference sample and be instructed to use a fixed number such as 10. Then all subsequent ratings are made relative to this number; for example, if the second item is twice as intense it gets the number 20. In spite of a flurry of activity concerning magnitude estimation in the 1970s, the method has not caught on in applied sensory testing. However, there are vocal proponents of various hybrid versions of the technique, called **labeled magnitude scales** (LMSs), discussed in Section 7.2.

It is also possible to derive a measure of intensity or difference indirectly from discrimination tests or other choice procedures. This is called indirect scaling and is largely based on the theories of Thurstone, as discussed in Chapter 4. An introduction to the practical aspects of scaling methods can be found in Lawless and Heymann (2010: chapter 7). Scaling is also prone to a host of context effects, which are discussed in Lawless and Heymann (2010: chapter 9) and in Poulton (1989). The psychophysical theory and history of scaling is covered in Gescheider (1997). Various approaches to dealing with panelist differences are discussed in Naes et al. (2010).

7.1.2 Empirical Scaling Comparisons

Different scaling methods were developed in different laboratories over the years, and it was common for some of these research groups or food commodity groups to use a particular method without much consideration of what else was out there. The burgeoning growth of professional societies in the mid-20th century, such as the Institute of Food Technologists and the American Society for Testing and Materials, brought people into contact who were using different scaling techniques. Debates arose about the best or correct method to use, and these sometimes turned acrimonious. Category scales were common in the food research community and magnitude scales in psychophysics and psychology. Line scales were used for pain research (Huskisson, 1983; Sriwatanakul et al., 1983) and in food research (Baten, 1946; Stone et al., 1974), and they had their proponents in both psychology and descriptive analysis.

Some arguments were presented on theoretical grounds. An example is the curvilinear relationship that is seen between magnitude estimation data, which tends to fit a power function, and category scaling data, which often tended to fit a log function as follows:

$$\mathrm{ME} = kI^n \text{ and thus } \log(\mathrm{ME}) = n\log I + \log k \tag{7.1}$$

where ME is the magnitude estimation datum, I is the physical stimulus intensity, n is a characteristic exponent for that sensory continuum, and k is a constant that depends upon the units of measurement;

$$\mathrm{CR} = k' \log I \tag{7.2}$$

where CR is the category rating, I is once again the stimlus intensity in physical units, and k' is another constant (different from the k in eqn 7.1). As predicted by these relationships, the plot of a data set showing category ratings and magnitude estimates for the same data formed a curve (Stevens & Galanter, 1957). This curvilinear relationship implied that only one could be a linear transformation of the actual perceived sensation intensity. This "linear translation" requirement was taken as a test of the validity of the scaling method (e.g., Anderson, 1974). These arguments are presented in Chapter 2.

Workers in sensory evaluation and consumer research are a pragmatic lot. So the question often arose of which scaling method was practically better (i.e., more useful). The theoretical arguments could be put aside and the scaling methods studied and evaluated just like any other analytical measurement tool. The fundamental goals of any measurement technique are precision and accuracy. Precision is basically a matter of low variability and the repeatability of any measurement. When we ask the question a second time, we would like to get the same number back. In statistical terms, standard deviations should be low. Adjusting for differences in scale range, this can be rephrased as having a lower coefficient of variation. Because lower variability means better discrimination among products, this could be viewed as finding more significant differences when all other factors such as sample size (number of judges) were held equal. Consider the simple t-test, which in its most generic form looks like this:

$$t = \frac{M_1 - M_2}{\sigma / \sqrt{N}} \tag{7.3}$$

where M_1 and M_2 are means, σ is a measure of variability, and N is the sample size. The denominator is called the standard error, and when we take the sample size out of the picture we get a true signal-to-noise ratio. That is, multiplying t times $\sqrt{N}$ yields a pure difference measure analogous to d' in signal detection terms (see Chapters 3 and 4), the difference between means divided by the standard deviation. Given the same products scaled by two methods, we assume the mean experiences are the same, so the main issue becomes the size of the standard deviation. Even if the methods have different numerical ranges, the standard deviation takes this into account. Alternatively, you can rescale the data from the two methods to bring them into a common range, such as percentage of total scale range or percentage of actual numerical range used in the study.

These comparisons are critical in applied sensory testing because a primary goal is to provide the most sensitive, discriminating test method and one that minimizes the occurrence of type II error – missing a difference that is really there. Type II error leads to franchise risk when the product change is a liability for consumers (but goes undetected in your research) and opportunity risk when the product difference is positive for consumers (but once again, not detected and thus not acted upon in your research outcome). So many researchers compared two or more scaling methods using the same sets of products and asked the question of whether the methods produce more product differences or higher statistical differentiation.

Accuracy is a little harder to judge, as we are essentially dealing with subjective events. Validity, as it is known in the behavioral sciences, is a bit more tractable. Is there evidence that the method is measuring what it is supposed to measure? Are there converging lines of evidence that indicate one scaling method is more valid? This can bring us back to the arguments about linear response translations as discussed in Chapter 2. But other criteria can be brought to bear. If a method is giving accurate valid results, one might suppose that it is not prone to contextual biases or that it does not show variable results as a function of product context. Unfortunately, all scaling methods studied so far seem to be affected by product context (e.g., Lawless et al., 2000). Other criteria might include the correlation with product choice or consumption (Lawless et al., 2010a,b) and the ability of the scaling method to uncover consumer segments with different preferences (Villanueva & Da Silva, 2009). Both of these criteria speak to the utility of the technique. A related issue is the ability of the method to make valid between-group comparisons, and to differentiate different genetic groups (Bartoshuk et al., 2003, 2004a,b). This idea will be revisited in the discussion of LMSs (Section 7.2).

Rather than individually review the large number of studies done to compare scaling methods, we can ask about the overall trend. In terms of product discrimination and statistical differentiation, most methods (e.g., line scales, category scales, and magnitude estimation) tend to give similar results (Giovanni & Pangborn, 1983; Pearce et al., 1986). Examples are the two studies done by Lawless and Malone (1986a, b) that looked at over 20,000 judgments from consumers on a variety of different product attributes. Line scales, category scales, and magnitude estimations all produced remarkably similar data, over both wide and narrow stimulus ranges, and about equal levels of product differentiation. Magnitude estimation tended to produce slightly more variable results than the other methods when used by consumers as opposed to college students, but perhaps that is not surprising given the students' more recent and/or frequent exposure to math. The overall conclusion is that all three scaling methods work well to differentiate products when used sensibly.

Of course, any method can give poor results if used inappropriately. Scales must allow respondents sufficient response alternatives to differentiate items (Bendig & Hughes, 1953), the attribute being rated must be understood, and any anchor words (none, extremely, etc.) must be sensible for what is being scaled. In most cases, the continuum should be simple and unidimensional. Quality grading scales that combine various features are really scoring schemes, and are not considered true scaling methods as used in this chapter. An example would be a shrimp grading scheme where increasing defect "numbers" are applied to reflect both increasing density of black spotting and increasing darkness of color. Exceptions to the unidimensional rule are overall degree of difference (DOD) scales and, of course, hedonics (like/dislike). For both of these, respondents may integrate multiple attributes in coming up with a decision on a numerical judgment. But the fact that different people may consider different attributes (and give them different weight) is still problematic for those techniques.

7.2 Labeled Magnitude Scales for Intensity

7.2.1 Desiderata

Another goal in developing a valid scaling method has been the desire to find a method that would reflect true ratio properties in the results. Presumably, that would allow one to make a statement such as "this product is twice as sweet as that one." For this reason, some scientists still cling to the notion that magnitude scales are the best to use because they give ratio-level data. However, one should be careful with the logical jump in claiming that because I have given the subject *ratio instructions*, they have actually produced ratio-level data. As seen repeatedly in contextual studies, ratio scaling methods produce variable results based on simple changes like stimulus range, choice of modulus, and numerical examples, and in showing contrast effects (Baird & Noma, 1978; Gescheider, 1997). If the ratios were accurate, one would not expect them to move all over the place as a function of arbitrary methodological variables.

Another problem with magnitude scales is that they do not tell us anything about the actual level of the stimulus intensity. Is it weak, moderate, or strong? We only know that item A is purported to be twice as sweet as item B. For this reason, various workers have committed the "heresy" (in the ratio scaling community) of putting verbal category-style labels onto magnitude scales. The hybrid could either resemble a numerical category scale (Borg, 1982) or a line scale (Green et al., 1993). Inherent in this work is the notion that words themselves can be scaled; that is, they can be given numerical ratings reflecting what they generally mean to a person in terms of relative intensities of experiences (Gracely et al., 1978a,b). This notion is not so radical as it sounds. Scaling words was at the basis for the development of the nine-point hedonic scale by Thurstone and colleagues (Jones & Thurstone, 1955; Jones et al., 1955), who used a simple category scale in their derivation of values for hedonic terms. Researchers have gone about this in different ways, either scaling the words by themselves (I suppose we can call that "naked" scaling) or with the words embedded in a list of common items that are also scaled (a good term might be "contextual" or "grounded" scaling).

7.2.2 Labeled Magnitude Scales for Intensity

The first prominent scientist to commit the heresy was Gunnar Borg. In studying perceived exertion, Borg realized that using a sensory standard was unwise. In training a panel for descriptive analysis, it is common to provide reference standards for different intensity levels such as examples of products that are weak, moderate, or strong on a given attribute (Meilgaard et al., 2006). However, for perceived exertion, this would make no sense. Riding an exercise bicycle under a certain load for a certain period of time would produce different experiences in different individuals due to differences in physical conditioning. However, if they rode the bicycle to the point of exhaustion, that experience would in theory be quite similar for different individuals. So Borg used a high end-anchor such as "maximal" in his scales to represent the top of the scale. He also scaled the words, and superimposed them on a simple category scale such as the CR10 scale (for **category ratio scale** with 10 "points") for perceived exertion, as shown in Figure 7.2. Note that the spacing of the verbal category labels is not equal, but approximately logarithmic.

Historically, the CR10 scale was found useful for continua outside perceived exertion, such as pain. Nowadays, it is common to rate perceived exertion on a 6 to 20 scale, called the Borg RPE scale. Why a scale would start at six, rather than zero, is a curious choice. However, if we subtract 5 from each scale point we end up with a 1 to 15 scale, which is very similar to some of the category scales in current use in sensory descriptive analysis. Fifteen categories may be a sort of "magic number" (Miller, 1956). Various versions of the CR scale appeared as Borg and colleagues tinkered away (Marks et al., 1983; Borg, 1990). One important assumption was that the range of sensations was not only the same for all individuals, but that the range was about the same across different sensory modalities. If true, this would allow the scale to function as a general ruler for all senses and

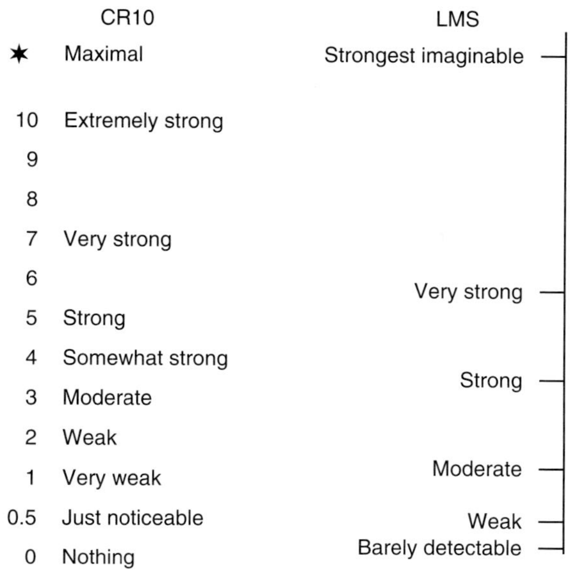

Figure 7.2 The CR scale of Borg and the LMS of Green et al. (1993).

attributes, a potentially highly useful characteristic. However, this assumption appeared questionable (Marks et al., 1983), and the alternative was to construct specific scales relative to the common context of experiences in that specific modality. More on this in Section 7.2.3.

The next important step was an attempt to construct a general tool for measuring oral sensations, using the Borg scale as a model. Green and colleagues gave subjects verbal category labels to rate (including barely detectable, weak, moderate, strong, very strong, and strongest imaginable) and embedded this list in a series of common oral experiences ranging in intensity levels. When spaced according to their geometric means, a roughly logarithmic spacing was achieved, as shown in Figure 7.2, and this came to be known as the LMS (Green et al., 1993). The uneven spacing was consistent with other workers who had scaled verbal intensity descriptors, such as Gracely (Gracely et al., 1978a,b). However, there were some differences between the spacings; for example, Borg's "moderate" was at 26% of maximal, whereas Green et al.'s scale had that descriptor at 17% of maximal. Also, the ranges from lowest to highest label were somewhat different, with a range of 1.84 log units for the LMS versus 1.36 log units for the CR scale. Gracely's pain scaling showed a range of 1.8 to 1.9 log units. The implication of this range was that the ratio of strongest to weakest sensations was less than a factor of 100.

Note that the response instrument takes the form of a line scale with labels. Subjects were sometimes given explicit instructions to choose areas between the labels. For example, they could be told to focus on the most appropriate descriptor, and then place their mark on the line at a point between that label and the next most appropriate descriptor. Data from the scale were found to coincide well with data from magnitude estimation on the same stimuli (Green et al., 1993). However, this finding was not universal. Zheng et al. (1998) found the LMS to give consistently higher power function exponents for five tastants. For perceived exertion and loudness, the CR scale gave lower exponents than did magnitude estimation (Marks et al., 1983). In that same study of perceived exertion, the CR scale was found to correlate better with heart rate ($r=0.50$ to 0.66) than magnitude estimation did ($r=0.10$ to 24). So the methods were not always congruent.

7.2.3 Context Effects and Generalized Scales

The importance of context on the form and use of the scale was soon recognized. In a second series of studies on the LMS, Green et al. (1996) examined the explicit context for the high end-anchor. The question concerned the interpretation of "strongest imaginable" and whether the results would change if that phrase were taken to mean just a specific taste, all oral sensations, or oral sensations excluding pain. Direct comparisons of these conditions were not performed, but were compared once again with magnitude estimation (ME) data. When the anchors were specific to a taste quality, or to oral sensations excluding pain, the LMS data produced steeper psychophysical functions than the ME results did; that is, they spanned a wider range. The contrast is perhaps not surprising. The strongest sweetness of a candy is nowhere near the strongest taste of a hot pepper.

This result was completely consistent with previous psychophysical literature on range effects. The general rule is that when stimuli span a small range they produce steeper functions, and when they span a wide range the functions are shallower (Teghtsoonian & Teghtsoonian, 1978). The effective range of the scale in this case was the mental frame of reference. In the context of a narrow range of imagined taste sensations, the stimuli presented caused an expanded response continuum. Relative to the strongest taste imaginable, a strong taste would be given a higher rating than when it was compared with the strongest imaginable oral pain. This was the first hint that the LMS might not be producing data with absolute ratio properties, but like other scaling methods was subject to context effects, including the self-imposed frame of reference of the subjects. Further research on context effects confirmed that the LMS was as liable to contextual shifting, such as contrast effects, as any other scaling method was (Lawless et al., 2000). It seemed that the search for a valid scaling method that would produce ratio-level data regardless of context was still elusive.

Meanwhile, the LMS was found to be useful for another purpose. The literature on propylthiouracil (PROP) taste sensitivity groups suggested that a specific version of the LMS was a better tool for inter-group comparisons of taste and other responses. This research also spawned the question of what context was most appropriate. Would an LMS anchored to the most general context possible produce a valid scale? Many thought so. The critical reasoning hinged on the idea that PROP **supertasters** lived in a world that simply contained more intense oral sensations than other people (Bartoshuk et al., 2003, 2004a,b). This was easy to demonstrate by matching intensities of tastes to other continua such as loudness of sounds. Supertasters matched tastes to louder sounds than nontasters did. Supertasters were found to have more fungiform papillae, and reacted much more strongly to PROP than even normally sensitive individuals. The corollary to this notion was that this was a genetic effect, since the well-known taster/nontaster dichotomy to phenylthiocarbamide (also known as phenylthiourea), a compound chemically related to PROP, was shown to have a genetic component (see Bufe et al. (2005)). PROP tasters and supertasters had been shown to be more responsive to a variety of taste and tactile stimuli than less responsive individuals were, but the effects were not always consistent (Tepper, 2008; Hayes & Keast, 2011).

The key to this line of thought was the following argument. Since supertasters lived in a world of more intense oral sensations, their response to any actual stimulus, compared with their frame of reference for the "greatest imaginable," could actually be marked lower on the scale than the rating given by a less sensitive individual (Dionne et al., 2005). The less

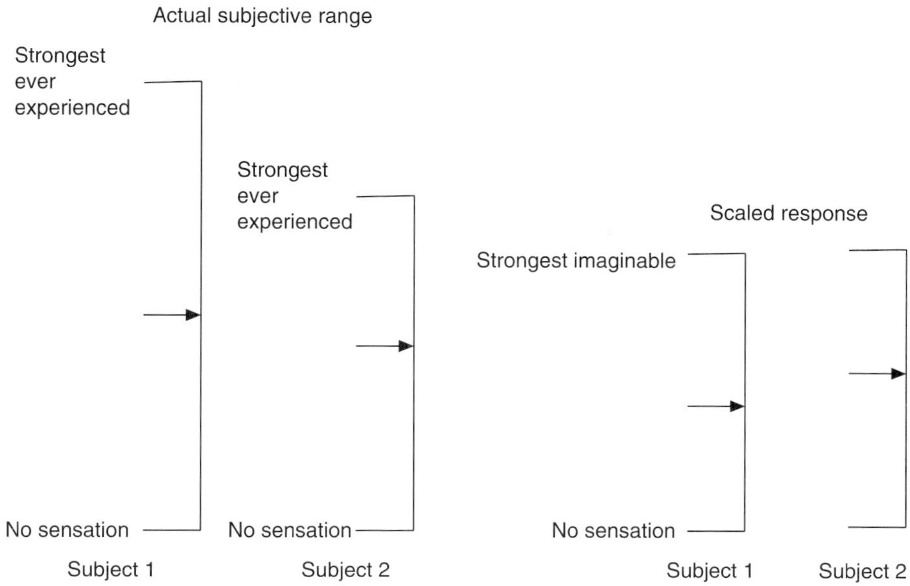

Figure 7.3 The actual sensory experience of a highly sensitive individual could be more intense than that of a less sensitive person, as shown in the left panel. Assume that Subject 1 is rating pain relative to her experience in childbirth. Subject 2 has led a sheltered life and never experienced anything worse than a stubbed toe. They now rate a stimulus in an experimental session. Paradoxically, when rated relative to their maximum experience, the rating of the sensitive person is lower on the actual response scale, due to contrast with the imagined upper bound.

sensitive person might get a similar experience, but since their frame of reference was truncated relative to the supertaster, the response choice (rating) could actually be higher. Figure 7.3 shows how this can occur. Scales that presumed the same meaning of adjective labels to people, and presumed to have the same semantic upper bound, were prone to making inaccurate comparisons (Bartoshuk et al., 2003, 2004a,b). Clearly, there needed to be a way to try to get all the individuals into the same mental frame of reference, in order to provide a valid comparison across people and groups. One solution might be to use the most general possible frame of reference, but making the upper anchor on the LMS "the greatest imaginable sensation *of any kind*" (Bartoshuk et al., 2004a). This was called the **generalized labeled magnitude scale (gLMS)**. Results with comparison of the generalized scale to magnitude matching were encouraging, and the psychophysical community embraced the more general version of the LMS.

Whether a common frame of reference that provides a comparable subjective anchor is actually achievable remains open to debate. Research on pain showed that, relative to the brightest light ever seen, the strongest pain ever experienced was about equal for male subjects, but about 20% higher for females who had selected childbirth as their most intense experience (Bartoshuk et al., 2004a). So basing things on experience is variable; people have different contexts. Using the term "imaginable" does not seem to solve this problem. Comparing the "strongest imaginable experience of any kind" to the strongest sensation ever experienced produce a value 40% higher (Dionne et al., 2005). Logic dictates that if greatest imaginable is simply 1.4 times as great as strongest experienced, then "*it varies across subjects just as the most intense sensation experienced also varies*" (Dionne et al, 2005: 127; italics added). Dionne et al. went on to suggest that multiple standards based on common experiences might help solve this problem, but evidence for that approach as a solution is still wanting.

7.2.4 Labeled Magnitude Scales: Summary and Conclusions

LMSs have not achieved very widespread use in industrial sensory evaluation. They have vocal proponents in the psychophysical community, especially in taste research. However, the goals of the two groups are often quite different. In sensory testing of foods and consumer products, the goal is to have a discriminating scaling method, one that can detect perceived differences among products. Inter-group comparisons are sometimes made in studies of market segmentation, but are less common than in taste research. Cost efficiency is another concern. As in the case of other line scales, the response choice must be measured, and this process can be laborious unless this is done by computer-assisted data collection. A third criterion concerns whether the method is user friendly. With a trained panel, almost anything is possible. With consumers, extensive instructions and/or practice examples are not common, but they may be necessary in order to get proper usage of the LMS-type of scales by untrained persons (Lim & Fujimara, 2010). Ideally, a response method should be easily grasped by consumers. Research thus far indicates that the hedonic versions of the LMS-type scale are prone to consumers making categorical choices at or near the verbal labels (Cardello et al., 2008; Lawless et al., 2010a,b) , unless given explicit instructions and examples to use the space on the lines in between the verbal anchors (Lim & Fujimara, 2010). Usage of only the verbal labels would negate the property of the scale as a continuous rating technique and bring into question whether the data still have ratio properties. Of course, one can argue that as long as the word spacing is accurate the resulting data will still have ratio properties. But whether the data produced have ratio-level properties is still unproven, as it is for magnitude estimation itself. We simply know that the two methods are correlated. To date, the only method showing reasonable evidence of ratio properties is best–worst scaling, as discussed by Finn and Louviere (1992).

Another important issue with the LMS is the fact that, for most foods, only the bottom one-fourth of the available scale is likely to be used. The vast majority of sensations generated by foods and consumer products are in the range of weak to moderate intensity. The verbal label "moderate" only occurs at 17% of full scale range. Thus, the largest portion of the response scale is not being used in any practical application, unless one is dealing with very intense products such as a spicy hot pepper sauce. Whether this kind of response compression is a liability or not remains to be seen, although one recent study seems to indicate so. Ludy and Mattes (2011) did direct comparisons of the LMS and line scales for oral burn, oral thermal stimulation, oral tactile stimulation, and auditory sensations. Although both methods produced highly significant stimulus effects in ANOVA (common in scaling studies), the line scale produced higher F-ratios for all continua. As expected, the psychophysical functions from the line-scale data had steeper slopes, consistent with the fact that, for "everyday" sensations, only a small part of the LMS gets used.

Furthermore, from a theoretical perspective the compression at the bottom of the scale is potentially counterproductive. Weber's law (see Chapter 1) tells us that the physical size of the just-noticeable difference (JND) is smaller for weaker stimuli; thus, there are more JNDs at the low end of the stimulus range. If people are more discriminating with weaker sensations, why would one want to compress that part of the response range? The tradeoff is the alleged ratio properties of the scale. However, in applied work there is a significant liability in missing a real difference if the response method is not discriminating enough. The question arises as to whether the LMS type of scale allows sufficient space and alternatives for users to express their perceptions of finer differences at the low end. Note that Borg's original CR10 scale inserts an extra category between zero and one.

In summary, the LMS and gLMS do not appear to provide any magic bullet to solve all the problems inherent in scaling. They are as prone to context effects and contextual shifting as any other method is. The search for a common frame of reference on which to base inter-individual comparisons of subjective magnitude continues. In areas like taste and pain, there is good evidence that people are not living in the same sensory worlds, and so the validity of any choice for the high end-anchor remains a subject of debate.

7.3 Adjustable and Relative Scales

7.3.1 Three Categories of Relative Scales

Section 7.2 described several situations where the results of scaling were affected by the context or frame of reference. One can argue that humans are natural comparators and that they are always basing their judgments on a frame of reference. If all judgments are relative in nature, why not take advantage of this trait, rather than trying to work around it? This is the approach taken in several scaling methods in which explicitly relative judgments are made. Three general techniques will be discussed in this section: ratings relative to a standard, scaling methods that anchor themselves with the high and low stimuli, and adjustable methods in which judgments can be altered as more items are sampled.

7.3.2 Scaling Relative to a Reference

One technique that explicitly involves a comparison is rating a product as higher or lower than a reference product in some scalable attribute. This method appears in the literature from time to time (Mahoney et al., 1957; Land & Shepard, 1984) and is thought to provide a simpler kind of scaling task than placing a product in some absolute position on a scale, perhaps without explicit reference to anything else. Of course, many descriptive analysis techniques involve physical reference standards during training, and one function of these standards is to anchor the intensity scale for that attribute. In other words, examples are used to show the panelist trainee what is a little and what is a lot of that attribute. These techniques will be discussed in Section 7.4. In this section, we are referring to scales that generally have the reference marked in the center of a line or category scale. Ratings to the right generally mean a sensation was more intense or greater than the reference, and ratings to the left signify less of that sensation. Examples are shown in Figure 7.4. These scales are very similar to just-about-right (JAR) scales, but in JAR scales the center of the scale is the imagined ideal level of the attribute in that particular product.

These kinds of scales were found to produce more significant differences than a normal category or line scale in two studies (Larson-Powers & Pangborn, 1978; Stoer & Lawless, 1993). However, the advantage was small, and in the first study the participants were more practiced with the relative scale. The obvious disadvantage of this kind of method is that it transmits no information about the absolute levels of each attribute. That is, we do not know if they are weak, moderate, or strong, unless we have scaled the reference item using other procedures, like a complete descriptive profile. A suitable application for the procedure would be in a quality control setting, in which the test products are compared with a standard

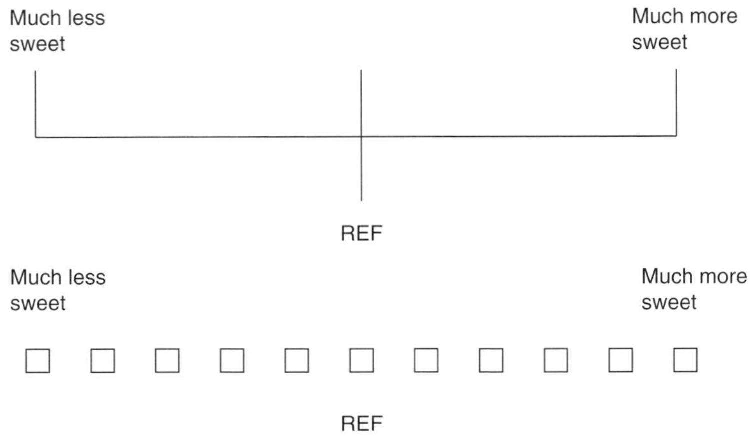

Figure 7.4 Line scale (upper) and category scale (lower) versions of a relative-to-reference scale.

or in-spec version (Gillette & Beckley, 1992; Beckley & Kroll, 1996). If combined with hedonic ratings for liking and disliking, the ratings could be subject to a penalty/benefit analysis (see Chapter 8) where the advantage or liability for being different from the reference could be scaled. However, this has not appeared in the literature at the time of publication.

7.3.3 Rank-Rating

Another type of relative positioning appears in a technique called **rank-rating** (Kim & O'Mahony, 1998; Park et al., 2003; O'Mahony et al. 2004) or positional relative rating (Cordonnier & Delwiche, 2008). In this method, respondents are asked to place the cups or vessels holding the product on a paper marked with the response scale in front of them. As subsequent items are tasted, the previous items can be re-tasted and repositioned. So the method not only takes advantage of the comparative nature of the human judge, but also allows for adjustment in previous ratings. If a person is unfamiliar with the scale or with the range of products to be rated, they can adjust for this lack of knowledge as the experiment progresses.

The method was evaluated for how well subjects could track increases in salt concentration when rating saltiness of water solutions (Kim & O'Mahony, 1998; Park et al., 2003). Fewer reversals were noted with this method than with nonadjustable ratings. The major disadvantage of the technique is that it has not been evaluated in a setting with multiple attributes, such as a descriptive procedure. Presumably, the procedure would start over for each new attribute. However, this would not be an issue for a simple evaluation of consumer acceptability, for example. Note that the procedure allows both re-tasting and repositioning, and the relative contributions of each aspect have not been thoroughly evaluated (Koo et al., 2002). Also, the common efficiency criterion of producing significant product differentiation has not been thoroughly evaluated relative to other common scaling techniques. The procedure is lengthy, so cost efficiency is not achieved. An open question is whether people take advantage of the opportunity to amend previous decisions, and how often. Additionally, because the ratings are more relative than common scaling, they could be more prone to contextual effects such as shifting due to contrast (e.g., "since this next item is really weak, I'll push the last more intense item up on the scale").

7.3.4 Self-Anchoring, Fully Relative Scales

Perhaps the most completely relative scaling method is the procedure of Gay and Mead (Gay & Mead, 1992; Mead & Gay, 1995). In this method, the entire product set is inspected, and the lowest and highest items placed at the lowest and highest points on the response scale. Others are then distributed between these endpoints. Villanueva and colleagues have studied a hedonic version of this scale with consumers (Villanueva et al., 2005; Villanueva & Da Silva, 2009). Apparently, the method was used in the 1970s for wine evaluation at the Long Ashton Research Station in the UK (Villaneuva & Da Silva, 2009). This is logically similar to the use of line and category scales by Anderson (1974) in his functional measurement studies. In those experiments, he would often show the subject examples of the greatest and least items. However, they would be anchored not at the endpoints, but slightly higher than the bottom and lower than the top of the response scale. Presumably, this use of an indented scale would provide a safety zone for subjects in case something even more or less intense was presented, and in theory should get rid of end-of-scale avoidance problems. Some of the early use of indented line scales in descriptive analysis may be traceable to Anderson's methods. This method shares some of the liabilities of other relative scaling techniques in that (1) absolute information is lost and (2) only one feature can be scaled at a time and, thus, it is not suitable for descriptive profiling. The ability of the method to provide enhanced discrimination or its proclivity for context effects is unknown.

7.4 Explicit Anchoring

7.4.1 Examples: Pepper Heat and Texture Scales

In work with trained panels, it is often the practice to provide reference standards for specific intensity levels in order to calibrate the panelists and to try to get them to use the scale in a similar fashion across the range of sensations that are likely to be encountered. A good example is the ASTM (2008) standard for assessing pepper heat, which provides examples of weak, moderate, and strong sensations in training, and sometimes a reminder stimulus for the "weak" anchor point in actual product evaluation sessions. The response scale is a 15 cm line scale with cross-hatching at the endpoints and three intermediate points spaced as follows: no heat (left endpoint), threshold (at 1.25 cm), weak (at 5 cm), moderate (at 10 cm), and strong (at the right endpoint, 15 cm), as shown in Figure 7.5.

This kind of **anchoring** has a long history in texture analysis. In the original General Foods Texture Profile, reference standards were provided for a number of scales, with

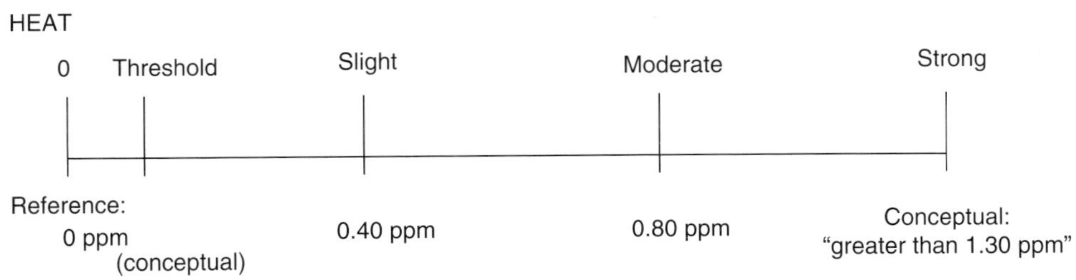

Figure 7.5 The ASTM E-1083 15 cm line scale for pepper heat, with the reference standards for various concentrations of vanillyl nonamide, a synthetic capsaicinoid.

Table 7.1 Examples of reference materials in texture profiling

Category scale point	Product	Sample size
Hardness scale		
1	Philadelphia brand cream cheese	1.2 cm (orig. ½ inch)
2	Egg white, hard boiled	1.2 cm from tip
3	Uncooked skinless kosher frankfurter	1.2 cm
4	American processed cheese	1.2 cm
5	Olives, giant, stuffed	1 olive
6	Cocktail peanuts	1 nut
7	Uncooked fresh carrot	1.2 cm
8	Peanut brittle, candy part	(unspecified)
9	Rock candy	(unspecified
Gumminess scale		
1	40% flour paste	15 ml*
2	45% flour paste	15 ml
3	50% flour paste	15 ml
4	55% flour paste	15 ml
5	60% flour paste	15 ml

*Originally one tablespoon.

specific examples for individual scale points, usually on a one to nine scale (Brandt et al., 1963). Examples are shown in Table 7.1. Note that some of the scales use common commercial products that are consistently reliable in their properties and can be obtained nationwide. In other cases, such as the gumminess scale, reference standards were constructed in model systems. The list of reference standards was updated and modified from time to time with improvements in the terms (Civille & Szczesniak, 1973; Civille & Liska, 1975). Muñoz (1986) also updated the reference list, and also switched to a 15 cm line scale, with anchoring products reflecting low, medium, and high levels of each attribute. The time commitment to be trained and to internalize a mental image of all the reference points was substantial, with 2 weeks of intensive training (2–3 hours a day) followed by up to 6 weeks of daily practice (Lawless & Heymann, 2010).

7.4.2 Anchoring Options in Descriptive Analysis

In a landmark paper, Muñoz and Civille (1998) delineated three options for intensity anchoring in descriptive analysis. The key is how the top of the scale gets defined. One option they defined as an attribute specific scale. In a descriptive profile, each attribute has its own reference for the highest scale point: that being the strongest product for that specific attribute. The full scale range is used for each and every attribute, since a product may be "high for that attribute in that product category." However, the absolute meaning of the data can change from attribute to attribute, and thus the data from different attributes are not comparable. For example, a vanilla cookie might be very sweet, but have only a mild aromatic vanilla note. However, if that vanilla note is the highest in the product set, it gets rated at the top of the vanilla scale. So even though the sweetness and vanilla flavor get the same rating, they are not meant to imply a strong sensation in both cases.

A second option is the product-specific scale. In this case the high end-anchor (e.g., "very strong") is defined as the strongest sensation experienced in that product category. For common experiences, most people have no problem with these kinds of judgments. For example, we can visualize "a large mouse running up the trunk of a small elephant" and make no mistake about which animal is actually larger. As long as we have some familiarity with the attribute (size) and a frame of reference for the category (mice or elephants) we can adjust our estimates accordingly. In this case, ratings from different attributes are now comparable as they are being made relative to that strongest attribute in the most intense product. However, the data across different product categories cannot be compared. What is highly sweet for an orange juice panel may be quite different than what is rated as highly sweet for a confectionary panel. So a different training regimen and perhaps a different set of panelists must be instituted for each product category.

A third option is a **universal intensity scale** that has reference standards for absolute levels of intensity across all product categories and attributes. Training for this kind of scale can be laborious, but once the panel is calibrated, all the data can meaningfully be compared both across attributes and products. On a 15-point scale, what is a 5 for the sweetness of orange juice is the same sensation intensity as a 5 on a sweetness scale for brownies. Furthermore, panelists can switch product categories and use the same response instrument. Also, any new formulations that might cause a need for recalibration of the attribute-specific or product-specific scales are already encompassed by the universal scale. There is no need for retraining simply because something more intense has been evaluated. The only potential liability is that only a small section of the scale is being used when the product category is one of very mild sensations. One approach to this problem is to allow the use of decimal fractions. That functionally transforms the common 15-point scale to a 150-point scale.

Three problems arise with this level of calibration. First, a universal scale makes sense for sensations that are in modalities, such as taste and smell, that encompass about the same subjective range from low to high. Once we add something from a different modality, like pain, the upper end of the scale may seem too restrictive. Of course, with a numerical 15-point scale instead of a line scale, there is nothing to prohibit the panel from using larger numbers, as in magnitude estimation. Some researchers have also allowed panelists to extend their line scales if needed (Lawless, 1977). The second issue concerns whether all scales are similarly intensive versus extensive. Common references for intensity make sense if the range is similar. However, some kinds of continua are more extensive, such as the density of chocolate chips visible on the surface of a cookie or the size of a brownie. Some may be more qualitative than intensive, such as the apparent pitch of fizzing effervescence in a beverage, the color of orange juice, or the browning of an apple slice. So the universal scaling approach may be limited to only perceived intensity. Many other kinds of continua can be quantitatively evaluated. The third issue concerns actual physiological individual differences. Some people may be taste-blind or smell-blind (anosmic) to some taste or flavor compounds. It would seem fruitless to try to get them into agreement with the rest of the panel when they are actually experiencing something quite a bit less intense. Of course, there are other strategies for dealing with those kinds of individual differences, some of which are discussed later in this chapter.

7.4.3 Anchoring as Calibration; Feedback and Training

Inherent in this approach is the notion that descriptive panelists are like measuring instruments. So they should be calibrated just as any physical instrument in order to be accurate. This is quite foreign to psychophysical thinking, which uses only untrained observers to report their impressions and simple predictable stimuli to evoke those responses. In a psychophysical study, it would be almost unheard of to try to force a calibrated response, although ratings relative to a reference standard are common (Gescheider, 1988). This is not so in sensory evaluation, which treats trained panelists as instruments used to measure the (unknown) sensory properties of products. In sensory evaluation practices, it is then important to both train (i.e., calibrate) and monitor the performance of such a panel. Measuring and monitoring the performance of panels, as well as individual assessors, is a large topic, involving both univariate and multivariate statistical techniques.

Although intensity anchoring and training have received a lot of attention in the texture literature (Civille & Szczesniak, 1973), it is less often discussed in other forms of descriptive analysis, except for those techniques using a universal intensity scale (Meilgaard et al., 2006). Much more attention has been paid to the development of the attribute list or lexicon and how that may be developed through consensus or through ballot training with reference standards (Lawless & Heymann, 2010: chapter 10). Several criteria can be brought to bear to determine that a panel is well trained and in-sync regarding the use of an intensity scale. One measure, of course, is simply the standard deviations for any particular product, which are expected to decrease as training progresses (Cliff et al., 2000). Another criterion is the product by panelist interaction in ANOVA, or rather the lack thereof. Assuming the panelists have roughly the same rank order of the products, if they are also using the scale in the same fashion, their judgments should cover about the same range. Thus, there will be no interaction effect (i.e., no significant differences in slope) if the products were to be arranged in rank and a line graph constructed.

Recent efforts at providing systematic and immediate feedback to panelists on intensity targets have taken advantage of computer-aided systems for data acquisition. Based upon previous ratings from the same or previously trained individuals, an ellipse can be shown on a line scale indicating an acceptable range, or the 90% confidence interval from those previous evaluations (Findlay et al., 2006, 2007; also discussed in Meullenet et al. (2007)). In these studies, it was thought to be too disruptive to provide feedback after each individual judgment, but the ellipses would appear when a complete "screen" of five attributes was completed. Judges were allowed to re-taste, but not change their ratings, permitting "discrete self-correction." In comparisons with more conventional types of training with post-test debriefing, the immediate feedback technique was able to reduce training time by about half.

7.5 Post Hoc Adjustments

7.5.1 Nonmodulus Magnitude Estimation

As discussed in Chapter 2, the technique of scaling using magnitude estimation without a modulus leaves the raw data in different ranges for different subjects. One person might use a choice of numbers between 1 and 100, while another might choose to work from 1 to 1000. Of course, in most psychophysical studies there is no attempt to train the participants or calibrate them to use the same scale range. In order to bring these two people into the same range before any further analysis, it would be necessary to multiply the data from the first person by 10, or divide the second by 10. So a post hoc adjustment was often performed on the data (Lane et al., 1961). A multiplicative constant would be derived for each subject, based on the ratio of their overall mean to the overall mean of the group as a whole. Geometric means were commonly used in magnitude estimation studies (antilog of the mean of the log data), as well as log transforming the data due to the occurrence of positively skewed (almost lognormal) data and high outliers.

Data transformation is not a new idea. A common approach with quantitative data is to convert the person's scores to Z-values by subtracting the score from the mean and dividing by the standard deviation for that particular attribute and person. Other kinds of additive and multiplicative adjustments are possible, as discussed in Section 7.5.2.

7.5.2 Rescaling, MSC, and Brockoff-type Adjustments

The kind of data transformation one is prepared to make depends upon the model one adopts for the panelist differences. The two most common differences are a change in the overall level of the data on the scale and a change in the range or span of scale that is used for a particular product set and attribute. If one adopts the model shown, it would make sense to transform panelist data to take out those "nuisance" effects. Various approaches are concisely discussed by Naes et al. (2010: chapters 3 and 4), as well as graphical methods for visualizing different panelist patterns. For each product, attribute, and panelist, a simple model would look like this (Brockoff & Skovgaard, 1994):

$$y_{ijk} = \alpha_{ik} + \gamma_{ik} v_{jk} + \varepsilon_{ijk} \tag{7.4}$$

where y is the score for panelist i, product j, and attribute k. Replicates can also be added to the model. The panelist level effect, then, is the additive constant α, and the panelist scaling effect is the multiplicative constant γ. ε represents the residual variation, so this is very similar to an ANOVA model with a panelist main effect and a panelist by product interaction term. Figure 7.6 shows some examples of panelists who would have different constants in this model.

A practical model for dealing with this structure is the MSC model proposed by Martens (discussed in Naes et al. (2010)), used with spectroscopic data. Once again, we have a level effect and a span or range effect; so

$$y_{ijk} = a_{ik} + b_{ik}\overline{y}_{jk} + \varepsilon_{ijk} \tag{7.5}$$

which is very similar to the Brockoff model, except that now we can estimate the constants a and b by least squares methods, across the products. $\overline{y}$ represents the product means, and once again we have i panelists, k attributes, and j products. The data adjustment consists of the following transformation:

$$y_{ijk(\text{new})} = \frac{y_{ijk(\text{old})} - \hat{a}_{ik}}{\hat{b}_{ik}} \tag{7.6}$$

where $\hat{a}$ and $\hat{b}$ are the fitted constants for that panelist and attribute. The overall effect of this transformation is to make the panelist's new scores line up with the panel averages, and only residual deviations such as rank order violations are left over.

The statistical adjustment performed by a two-way ANOVA with a panelist effect does half that job. Because the a-effect in the Brockoff model represents the panelist level difference, it is functionally equivalent to the panelist main effect in ANOVA. Thus, any ANOVA model which partitions out the panelist effect from the error term is effectively setting the a-effect to zero (as far as the product differences are concerned), leaving only slope

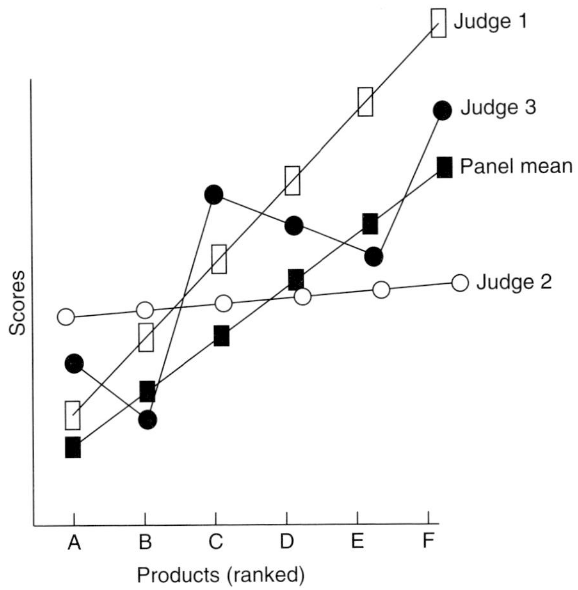

Figure 7.6 A plot of hypothetical panelists showing different range and level effects, as found in the Brockoff and MSC models. Note that the panel mean scores are shown by the filled rectangles. Judge 1 (open rectangles) has both a level effect (as the line is generally higher than the panel means) and a range effect (as the line is steeper than the slope of the panel mean function). Judge 2 (open circles) has a level effect and range effect, showing poorer discrimination. Judge 3 (filled circles) has little or no range effect and no level effect, but a lot of residual variation. This kind of "loose cannon" can feed the error term in ANOVA, but the other judges would not contribute much to error after adjustments for their levels and range.

differences. As long as panelists are still ranking products in the same order, the ANOVA model has a good chance of finding significant product differences.

Whether range and level effects are truly nuisance variables is partly a philosophical question. They could be due to actual sensitivity or discriminative differences among the panelists. Do you want to leave that information out of the picture? Perhaps it represents something about the real human variability and is represented in the population at large. Or are these just differences due to some kind of scale usage bias? If so, perhaps they should be taken out of the picture in order to provide a more sensitive test to better detect perceived product differences.

7.5.3 A Visual Ranking Test

As we have seen with the repeated measures partitioning and rescaling types of adjustments, the critical question for finding significant product differences often boils down to a question of consensus on rank orders. That is, if all panelists are ranking products in the same order, the ANOVA will detect the pattern in repeated measures analysis, or after rescaling takes out the panelist main effect. Naes and colleagues developed a graphical procedure for viewing the degree of panelist agreement, called the eggshell plot (Naes et al., 1994; Naes, 1998). This graph allows rapid recognition of whether a panelist is following the consensus rankings or is showing any substantial deviations. The original data do not have to be from actual rankings; scaled data will do, as long as they can be converted

Table 7.2 Data for the eggshell plot of Figure 7.7

A: consensus rank	B: cumulative rank	C: "null rank"	D: cumulative null rank	E: plotted value (B – D)	
1	1	5	5	–4	
2	3	5	10	–7	
3	6	5	15	–9	
4	10	5	20	–10	
5	15	5	25	–10	
6	21	5	30	–9	
7	28	5	35	–7	
8	36	5	40	–4	
9	45	5	45	0	
F: Judge 1 rankings	**G: Judge 1 cumulative**	**H: Judge 1 plotted (G – D)**	**I: Judge 2 rankings**	**J: Judge 2 cumulative**	**K: Judge 2 plotted (K – D)**
1	1	–4	2	2	–3
2	3	–7	4	6	–4
3	6	–9	6	12	–3
6	12	–8	8	20	0
4	16	–9	3	23	–2
5	21	–9	7	30	0
7	28	–7	1	32	–3
8	36	–4	9	41	1
9	45	0	5	45	0

to ranks. The plot works well with larger numbers of products, and the example below will show the procedure for nine products.

The plot works as follows. First, find a consensus ranking. This can be done by taking the mean scale values and ranking them, for example. Calculate the cumulative rank value for each product. Next, calculate the cumulative rank values for each individual judge. Then calculate the cumulative rank value that would be obtained if all products were ranked the same. For example, with nine products ranked the same, each would have a rank of 5. So the cumulative values would be 5, 10, 15, and so on. Subtract these cumulative null values from the consensus accumulation and from each assessor's accumulation.

Table 7.2 shows this process for nine products, and two hypothetical judges, one who largely agrees with the consensus and one who is clearly daft. The consensus line resembles a parabola with a minimum halfway through the ranks and with a higher right (than left) shoulder in these plots. Figure 7.7 shows the **eggshell plot**. Judges, like Judge 1, who agree with the consensus will tend to parallel or coincide with the lower curve of the consensus. Judges who are nondiscriminating or very discrepant will tend to float toward the top of the graph, as in the case of Judge 2. If many judges are plotted, the graph will appear somewhat like a cracked eggshell. Note that this is only one measure of a panelist's competence. As Naes (1998) noted, the overall variability of an individual, and consistency of the replicates are also important factors.

7.6 Summary and Conclusions

Many different types of scales have been used for reporting perceived intensity and perceptions along other sensory continua. Like–dislike or **hedonic scales** will be discussed in Chapter 8. Most common options work well when used sensibly.

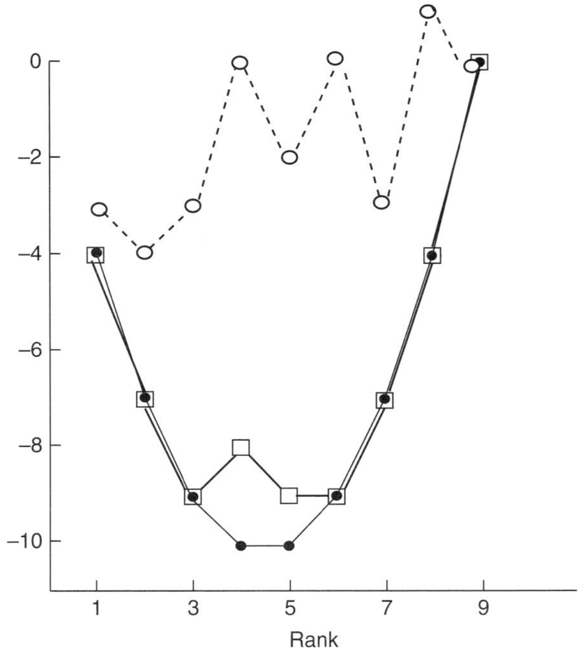

Figure 7.7 A sample eggshell plot for two hypothetical judges, one who mostly agrees with the panel rank order and one who does not.

New techniques continue to be developed. Whether they are true improvements or not remains a question for objective research. The criteria for a good scale in sensory evaluation are all related to reducing variability, and thus increasing the ability of the instrument to detect differences among products. Secondary criteria are whether the scale is user friendly and whether it is cost efficient from a data collection perspective. These are in contrast to the primary criterion in psychophysics, which is to produce a scale with at least interval and preferably ratio properties. What works well for some products and situations may not be the best approach for others. For example, the relative to reference scales seem to work well for quality control and shelf-life studies, but would not provide sufficient information for a complete descriptive profile of a product.

Appendix 7.A Examples of Individual Rescaling for Magnitude Estimation

Table 7.A.1 shows the data (plotted in Figure 7.A.1) from three hypothetical subjects in a scaling study. The second section shows the data after the individual multiplicative transformations are applied as in Lane et al. (1961). The bottom section shows the data after the MSC-type of transformation. Subject 1 shows a linear response with a low range of number usage. Subject 2 shows a higher range and some variability in rank orders. Subject 3 uses a much wider range of numbers and deviations from linearity.

Table 7.A.1 Data from three hypothetical subjects

Stimulus	Subject 1 data	Subject 2 data	Subject 3 data
10	2	5	10
20	4	22	15
30	6	20	50
40	8	30	28
50	10	40	100
Transformed	data	data	data
10	4.83	3.75	4.910
20	9.66	16.49	7.360
30	14.50	14.99	24.560
40	19.33	22.49	13.750
50	24.16	29.98	49.120
MSC transformed	−20.00	−23.59	−15.85
	−10.00	−1.79	−13.26
	0.00	−4.36	4.87
	10.00	8.46	−6.53
	20.00	21.28	30.78

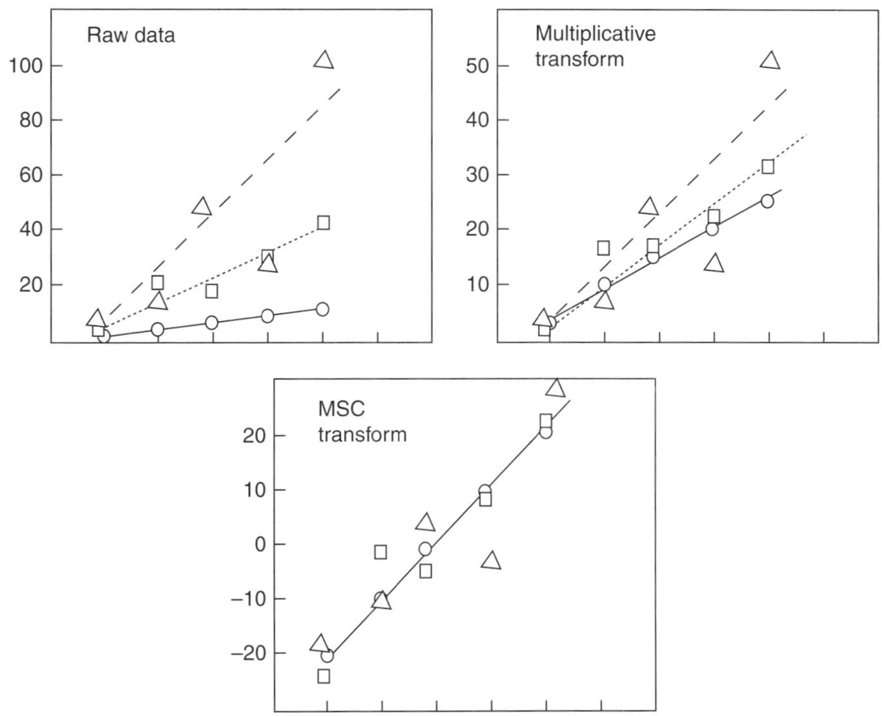

Figure 7.A.1 A plot of the data in Table 7.A.1, circles for Subject 1, squares for Subject 2, and triangles for Subject 3. Approximate best-fitting lines are drawn. Note that the MSC transformation sets all three subjects to the same fit line, a common slope and intercept.

References

Anderson, N.H. 1974. Algebraic models in perception. In: Handbook of Perception, Volume 2, Psychophysical Judgment and Measurement. E.C. Carterette and M.P. Friedman (Eds). Academic Press, New York, NY, pp. 215–98.

ASTM. 2008. Standard test method for sensory evaluation of red pepper heat. Designation E 1083-00. In: Annual Book of ASTM Standards, Volume 15.08, End Use Products. American Society for Testing and Materials, Conshohocken, PA, pp. 49–53.

Baird, J.C. and Noma. E. 1978. Fundamentals of Scaling and Psychophysics. John Wiley & Sons, Inc., New York, NY.

Bartoshuk, L.M., Duffy, V.B., Fast, K., Green, B.G., Prutkin, J., and Snyder, D.J. 2003. Labeled scales (e.g. category, Likert, VAS) and invalid across-group comparisons: what we have learned from genetic variation in taste. Food Quality and Preference, 14, 125–38.

Bartoshuk, L.M., Duffy, V.B., Green, B.G., Hoffman, H.J., Ko, C.-W., Lucchina, L.A., Marks, L.E., Snyder, D.J., and Weiffenbach, J.M. 2004a. Valid across-group comparisons with labeled scales: the gLMS versus magnitude matching. Physiology & Behavior, 82, 109–14.

Bartoshuk, L.M., Duffy, V.B., Chapo, A.K., Fast, K., Yiee, J.H., Hoffman, H.J., Ko, C.-W., and Snyder, D.J. 2004b. From psychophysics to the clinic: missteps and advances. Food Quality and Preference, 15, 617–32.

Baten, W.D. 1946. Organoleptic tests pertaining to apples and pears. Food Research, 11, 84–94.

Beckley, J.P. and Kroll, D.R. 1996. Searching for sensory research excellence. Food Technology, 50(2), 61–3.

Bendig, A.W. and Hughes, J.B. 1953. Effect of number of verbal anchoring and number of rating scale categories upon transmitted information. Journal of Experimental Psychology, 46(2), 87–90.

Borg, G. 1982. A category scale with ratio properties for intermodal and interindividual comparisons. In: Psychophysical Judgment and the Process of Perception. H.-G. Geissler and P. Petzold (Eds). VEB Deutscher Verlag der Wissenschaften, Berlin, pp. 25–34.

Borg, G. 1990. Psychophysical scaling with applications in physical work and the perception of exertion. Scandinavian Journal of Work and Environmental Health, 16, 55–8.

Brandt, M.A., Skinner, E.Z., and Coleman, J.A. 1963. Texture profile method. Journal of Food Science, 28, 404–9.

Brockoff, P.B. and Skovgaard, I. 1994. Modeling individual differences between assessors in sensory evaluation. Food Quality and Preference, 5, 215–24.

Bufe, B., Breslin, P.A.S., Kuhn, C., Reed, D., Tharp, C.D., Slack, J.P., Kim, U.-K., Drayna, D., and Meyerhof, W. 2005. The molecular basis of individual differences in phenylthiocarbamide and propylthiouracil bitterness perception. Current Biology, 15, 322–7.

Cardello, A., Lawless, H.T. and Schutz, H.G. 2008. Effects of extreme anchors and interior label spacing on labeled magnitude scales. Food Quality and Preference, 21, 323–34.

Civille, G.V. and Liska, I.H. 1975. Modifications and applications to foods of the General Foods sensory texture profile technique. Journal of Texture Studies, 6, 19–31.

Civille, G.V. and Szczesniak, A.S. 1973. Guidelines to training a texture profile panel. Journal of Texture Studies, 4, 204–23.

Cliff, M.A., Wall, K., Edwards, B.J., and King, M.C. 2000. Development of a vocabulary for profiling apple juices. Journal of Food Quality, 23, 73–86.

Cordonnier, S.M. and Delwiche, J.F. 2008. An alternative method for assessing liking: positional relative rating versus the 9-point hedonic scale. Journal of Sensory Studies, 23, 284–92.

Dionne, R.A., Bartoshuk, L., Mogil, J., and Witter, J. 2005. Individual responder analyses for pain: does one pain scale fit all? Trends in Pharmacological Sciences, 26, 125–30.

Findlay, C.J., Castura, J.C., Schlich, P., and Lesschaevew, I. 2006. Use of feedback calibration to reduce the training time for wine panels. Food Quality and Preference, 17, 266–76.

Findlay, C.J., Castura, J.C., and Lesschaeve, I. 2007. Feedback calibration: a training method for descriptive analysis. Food Quality and Prefernce, 18, 321–8.

Finn, A. and Louviere, J.J. 1992. Determining the appropriate response to evidence of public concern: the case of food safety. Journal of Public Policy and Marketing, 11, 12–25.

Gay, C., and Mead, R. 1992 A statistical appraisal of the problem of sensory measurement. Journal of Sensory Studies, 7, 205–28.

Gescheider, G.A. 1988. Psychophysical scaling. Annual Review of Psychology, 39, 169–200.
Gescheider, G.A. 1997. Psychophysics. The Fundamentals. Third edition. Lawrence Erlbaum, Mahwah, NJ.
Gillette, M.H. and Beckley, J.H. 1992. In-plant sensory quality assurance. Paper presented at the Annual Meeting, Institute of Food Technologists, New Orleans, LA, June.
Giovanni, M.E. and Pangborn, R.M. 1983. Measurement of taste intensity and degree of liking of beverages by graphic scaling and magnitude estimation. Journal of Food Science, 48, 1175–82.
Gracely, R.H., McGrath, P., and Dubner, R. 1978a. Ratio scales of sensory and affective verbal-pain descriptors. Pain, 5, 5–18.
Gracely, R.H., McGrath, P., and Dubner, R. 1978b. Validity and sensitivity of ratio scales of sensory and affective verbal-pain descriptors: manipulation of affect by Diazepam. Pain, 5, 19–29.
Green, B.G., Shaffer, G.S., and Gilmore, M.M. 1993. Derivation and evaluation of a semantic scale of oral sensation magnitude with apparent ratio properties. Chemical Senses, 18, 683–702.
Green, B.G., Dalton, P., Cowart, B., Shaffer, G., Rankin, K., and Higgins, J. 1996. Evaluating the "labeled magnitude scale" for measuring sensations of taste and smell. Chemical Senses, 21, 323–34.
Hayes, J.E. and Keast, R.S.J. 2011. Two decades of supertasting: where do we stand? Physiology and Behavior, 104, 1072–4.
Huskisson, E.C. 1983. Visual analogue scales. In: Pain Measurement and Assessment. R. Melzack, (Ed.). Raven Press, New York, NY, pp. 34–7.
Jones, L.V. and Thurstone, L.L. 1955. The psychophysics of semantics: an experimental investigation. Journal of Applied Psychology, 39, 31–6.
Jones, L.V., Peryam, D.R., and Thurstone, L.L. 1955. Development of a scale for measuring soldier's food preferences. Food Research, 20, 512–20.
Kim, K.-O. and O'Mahony, M. 1998. A new approach to category scales of intensity I: traditional versus rank-rating. Journal of Sensory Studies, 13, 241–9.
Koo, T.-Y., Kim, K.-O., and O'Mahony, M. 2002. Effects of forgetting on performance on various intensity scaling protocols: magnitude estimation and labeled magnitude scale (Green scale). Journal of Sensory Studies, 17, 177–92.
Land, D.G. and Shepard, R. 1984. Scaling and ranking methods. In: Sensory Analysis of Foods. J.R. Piggott (Ed.). Elsevier, London, pp. 141–77.
Lane, H.L., Catania, A.C., and Stevens, S.S. 1961. Voice level: autophonic scale, perceived loudness and effect of side tone. Journal of the Acoustical Society of America, 33, 160–7.
Larson-Powers, N. and Pangborn, R.M. 1978. Descriptive analysis of the sensory properties of beverages and gelatins containing sucrose or synthetic sweeteners. Journal of Food Science, 43, 47–51.
Lawless, H.T. 1977. The pleasantness of mixtures in taste and olfaction. Sensory Processes, 1, 227–37.
Lawless, H.T. and Heymann, H. 2010. Sensory Evaluation of Foods, Principles and Practices, Second edition. Springer, New York, NY.
Lawless, H.T. and Malone, J.G. 1986a. The discriminative efficiency of common scaling methods. Journal of Sensory Studies, 1, 85–96.
Lawless, H.T. and Malone, G.J. 1986b. A comparison of scaling methods: sensitivity, replicates and relative measurement. Journal of Sensory Studies, 1, 155–74.
Lawless, H.T., Horne. J., and Speirs, W. 2000. Contrast and range effects for category, magnitude and labeled magnitude scales. Chemical Senses, 25, 85–92.
Lawless, H.T., Popper, R., and Kroll, B. 2010a. Comparison of the labeled affective magnitude (LAM) scale, an 11-point category scale and the traditional nine-point hedonic scale. Food Quality and Preference, 21, 4–12.
Lawless, H.T., Sinopoli, D., and Chapman, K.W. 2010b. A comparison of the labeled affective magnitude scale and the nine point hedonic scale and examination of categorical behavior. Journal of Sensory Studies, 25(S1), 54–66.
Lim, J. and Fujimara, T. 2010. Evaluation of the labeled hedonic scale under different experimental conditions. Food Quality and Preference, 21, 521–30.
Ludy, M.-J. and Mattes, R.D. 2011. Noxious stimuli sensitivity in regular spicy food users and non-users: comparisons of visual analog and general labeled magnitude scaling. Chemosensory Perception, 4, 123–33.
Mahoney, C.H., Stier, H.L., and Crosby, E.A. 1957. Evaluating flavor differences in canned foods. II. Fundamentals of the simplified procedure. Food Technology, 11(Supplemental Symposium Proceedings), 37–42.
Marks, L.E., Borg, G., and Ljunggren, G. 1983. Individual differences in perceived exertion assessed by two new methods. Perception & Psychophysics, 34, 280–8.

Mead, R. and Gay, C. 1995. Sequential design of sensory trials. Food Quality and Preference, 6, 271–80.
Meilgaard, M., Civille, G.V., and Carr, B.T. 2006. Sensory Evaluation Techniques. Fourth edition. CRC Press, Boca Raton, FL.
Meullenet, J.-F., Xiong, R., and Findlay, C.J. 2007. Multivariate and Probabilistic Analyses of Sensory Science Problems. IFT Press/Blackwell Publishing, Ames, IA.
Miller, G.A. 1956. The magical number seven, plus or minus two: some limits on our capacity for processing information. Psychological Review, 62, 81–97.
Muñoz, A.M. 1986. Development and application of texture reference scales. Journal of Sensory Studies, 1, 55–83.
Muñoz, A.M. and Civille, G.V. 1998. Universal, product and attribute specific scaling and the development of common lexicons in descriptive analysis. Journal of Sensory Studies, 13, 57–75.
Naes, T. 1998. Detecting individual differences among assessors and differences among replicates in sensory profiling. Food Quality and Preference, 9, 107–10.
Naes, T., Hirst, D., and Baardseth, P. 1994. Using cumulative ranks to detect individual differences in sensory profiling. Journal of Sensory Studies, 9, 87–99.
Naes, T., Brockoff, P.B., and Tomic, O. 2010. Statistics for Sensory and Consumer Science. John Wiley & Sons, Ltd, Chichester, UK.
O'Mahony, M., Park, H., Park, J.Y., and Kim, K.-O. 2004. Comparison of the statistical analysis of hedonic data using analysis of variance and multiple comparisons versus an *R*-index analysis of the ranked data. Journal of Sensory Studies 19, 519–29.
Park, J.Y., Jeon, S.Y., O'Mahony, M., and Kim, K.-O. 2003. Induction of scaling errors. Journal of Sensory Studies 19, 261–71.
Pearce, J.H., Korth, B., and Warren, C.B. 1986. Evaluation of three scaling methods for hedonics. Journal of Sensory Studies, 1, 27–46.
Poulton, E.C. 1989. Bias in Quantifying Judgments. Lawrence Erlbaum Associates, Hillsdale, NJ.
Sriwatanakul, K., Kelvie, W., Lasagna, L., Calimlim, J.F., Wels, O.F., and Mehta, G. 1983. Studies with different types of visual analog scales for measurement of pain. Clinical Pharmacology and Therapeutics, 34, 234–9.
Stevens, S.S. and Galanter, E.H. 1957. Ratio scales and category scales for a dozen perceptual continua. Journal of Experimental Psychology, 54, 377–411.
Stoer, N.L. and Lawless, H.T. 1993. Comparison of single product scaling and relative-to-reference scaling in sensory evaluation of dairy products. Journal of Sensory Studies, 8, 257–70.
Stone, H., Sidel, J., Oliver, S., Woolsey, A., and Singleton, R.C. 1974. Sensory evaluation by quantitative descriptive analysis. Food Technology, 28, 24–9, 32, 34.
Teghtsoonian, R. and Teghtsoonian, M. 1978. Range and regression effects in magnitude scaling. Perception & Psychophysics, 24, 305–14.
Tepper, B.J. 2008. Nutritional implications of genetic taste variation: the role of PROP sensitivity and other taste phenotypes. Annual Review of Nutrition, 28, 367–88.
Villanueva, N.D.M. and Da Silva, M.A.A.P. 2009. Performance of the nine-point hedonic, hybrid and self-adjusting scales in the generation of internal preference maps. Food Quality and Preference, 20, 1–12.
Villanueva, N.D.M., Petenate, A.J., and Da Silva, M.A.A.P. 2005. Comparative performance of the hybrid hedonic scale as compared to the traditional hedonic, self-adjusting and ranking scales. Food Quality and Preference, 16, 691–703.
Zheng, C.-J., Hoffman, H.J., Lucchina, L.A., Bartoshuk, L.M., and Weiffenbach, J.M. 1998. Comparison of the Green scale versus magnitude estimation for taste perception. Annals of the New York Academy of Sciences, 855, 820–2.

8 Progress in Affective Testing: Preference/Choice and Hedonic Scaling

> *... there is a true search for variation in our appreciation of foods. Usually we do not eat the same food every day, when the possibility of variation exists. And even within a meal we like variation. Certainly rice, potatoes, bread or pasta are eaten every day in different parts of the world, but they are consumed with different side dishes on different days. It seems that repeatability in our food choices is an exception rather than a rule.*
>
> Köster et al. (2002)

8.1 Introduction

A variety of techniques are available in sensory testing with consumers that are used to evaluate the affective or hedonic reactions to products. They can be divided into two categories: choice methods and scaling methods. In the choice methods, a consumer is asked to choose which of a pair of products is better liked, or to rank a set of products from most to least liked. The simple paired test can be thought of as a ranking test of N products where $N=2$. Scaling methods, on the other hand, involve response choices that indicate degrees of liking or disliking. The classic example is the **nine-point hedonic scale** commonly used in the sensory evaluation of foods. It is a balanced scale with the adverbs "slightly,"

Quantitative Sensory Analysis: Psychophysics, Models and Intelligent Design, First Edition. Harry T. Lawless.

"moderately," "very much," and "extremely" applied to the words "like" or "dislike" and has a neutral point in the middle labeled "neither like nor dislike."

The key to either form of consumer testing lies in the screening and recruitment of the appropriate sample. In general, participants must be users of the product or product category, or sometimes simply purchasers of the product or product category. Most testing would also screen for frequency of usage, as a consumer who uses the product only rarely may not be in the target population to which you wish to project the results. The second key characteristic of consumer sensory testing is that it is **blind-labeled**; that is, the products are presented with only the minimal concept necessary for the consumer to have a proper frame of reference when judging their reactions. This is in contrast to a great deal of market research testing, which typically includes a test of the product concept, in all its detail, as well as a taste test.

This chapter will discuss various techniques, design options, models, and analyses for paired preference first, and then for affective scaling methods. The paired preference section will focus on use of the no preference option and the potential use of replication in preference testing. The little-known Ferris model for replicated nonforced preference will be presented. The section on scaling will examine some new hedonic scales such as the labeled affective magnitude (LAM) scale and its relatives. Best–worst scaling will be introduced as it has received recent attention and may be a candidate for achieving an actual ratio scale for hedonics. Topics related to product optimization, such as the just-about-right scales and their analyses, will be covered in Chapter 13 on product optimization and ideal profiling. It is assumed that the reader has at least a passing familiarity with paired preference and its binomial-based statistical analysis, and the use of *t*-tests and analysis of variance for scaled hedonic data. Basic information and further background can be found in Lawless and Heymann (2010: chapters 13 and 14) and other sensory evaluation textbooks. The terms "affective" and "hedonic" will be used as synonyms to indicate the valence or liking/disliking reactions to products, as opposed to more objective attributes like sweetness.

8.2 Preference Testing Options

8.2.1 Forced Choice

The most common form of the preference test is one in which a choice is forced between the two items. The analysis of this is straightforward, using a null hypothesis of a population proportion choosing each option equally, and a two-tailed test because the winner is not predicted ahead of time. The common analysis is via a *Z*-score approximation to the binomial distribution (see eqn 8.1) or its χ^2 equivalent (an exact (cumulative) binomial calculation can also be done as shown in eqn 8.3):

$$Z = \frac{(P_w - p) - (1/2N)}{\sqrt{pq/N}} = \frac{(X - Np) - 0.5}{\sqrt{Npq}} \tag{8.1}$$

where P_w is the proportion for the larger (winning) choice, equal to X votes divided by N total judgments, p is the chance proportion under the null and $q = 1 - p$. Because $p = q = 0.5$, these reduce to the following simpler formulae:

$$Z = \frac{(P_w - 0.5) - (1/2N)}{0.5/\sqrt{N}} = \frac{[X - (N/2)] - 0.5}{0.5\sqrt{N}} \tag{8.2}$$

where $1/2N$ is the continuity correction, which becomes negligible as N gets large, as occurs in most consumer tests with typically $N > 100$.

The exact binomial probability is given by a simplification of the binomial expansion as

$$p_{\mathrm{L}} = 2\sum_{x=0}^{L}\binom{N}{x}0.5^{N} \tag{8.3}$$

where p_{L} is the probability based on the proportion for the smaller (losing) item and L is the total count for that item; and once again there are N consumers. This finds the probability in the tail of the binomial distribution for L out of N votes and the probability is doubled because the test is two tailed. The parenthetical quantity $\binom{N}{x}$ is simply the combinatorial coefficient for N items taken x at a time, or $N!/[(N-x)!x!]$. This calculation can be laborious for a large consumer test, so in most cases the Z-score approximation will do. Various lookup tables are found in many sensory textbooks for the minimum required count for a significant winning product (e.g., Lawless & Heymann, 2010: tables 13.2 and F.M.).

8.2.2 Dealing With No-Preference Votes

A forced choice in preference testing is not always used, and the most common alternative is to have a "no preference" option available. Sometimes, this may be required for purposes of advertising claim substantiation (ASTM, 2008). Alternative responses include "like both equally," "dislike both equally," and "don't care" (Lawless & Heymann, 2010). The preference scale can also be expanded to indicate degrees of preference, using adverbs such as "slightly," "moderately," "very much," or "extremely" to modify statements such as "I prefer sample 459 to sample 873 ______________" (Filipello, 1957). Because choice is no longer forced, a simple binomial analysis is not appropriate. Analysis either requires certain strategies to return to the binomial test, or alternatives such as a multinomial analysis. Because the analysis becomes more complex, the no-preference option is not generally favored by sensory professionals.

There are several strategies for dealing with no-preference votes, and each has its advantages and disadvantages. The six options are summarized in Table 8.1.

Table 8.1 Options for handling no preference responses

Strategy	Comment	Advantages	Disadvantages
Drop	Lowers N	Simplicity	Loss of information; increased type II error from lower sample size
Split 50/50	Conservative, adds noise based on null	Decreases type I	Increased type II risk?
Split proportionally	Justification unclear	Decreases type II	May be too lenient; increased type I error risk
Thurstonian model	Model is 2-AC	Compromise in type I versus type II risk; uses all information; provides τ and δ estimates	Depends upon validity of model; may need additional software
Confidence interval estimation	Not common	Simplicity	Requires $N > 100$; no preference counts less than 20%
Allot to competitor	Very conservative	Used in some advertising claim substantiation	Could obscure a true win.

The no-preference votes may simply be dropped from the analysis, lowering the total sample size accordingly. In making advertising claims, this approach is sometimes used by making qualified statements such as "*Among those expressing a preference*, Product X beat Product Y" (ASTM, 2008). A lower sample size (decreased *N*) results in increased risk of type II error – missing a win that is really present in the underlying population. Remember that the balance between type I and type II risk is inverse, all other things being equal. So when the probability of missing a difference goes up, the probability of declaring a difference that is false usually decreases and vice versa.

The no-preference responses may be apportioned in three different ways. They may be split evenly (50/50) and counted into the existing preference groups. This maintains the sample size but adds noise to the system because a 50/50 split is what is expected by chance alone. Thus, you are favoring the null hypothesis and making it harder to declare a win. You can also split the no-preference votes in proportion to those who did express a preference. For example, if those expressing a preference showed a 60/40 split, the no-preference votes would be apportioned to those products 60/40 as well. This is based on research from Odesky (1967). He observed that the proportional split when forced was about the same ratio as when the no-preference votes were allowed, implying that the nonpreferring consumers would have voted in the same proportions as the rest of the sample if they had been forced to choose one product over another. Whether this finding is valid for all kinds of products has, however, been challenged (Angulo & O'Mahony, 2005). A third method for apportionment is to allot the no-preference votes to the competitor's product. This is rare, and only done in some cases for advertising claims of product superiority. Obviously, this is a very conservative approach, and one that could obscure an actual preference win.

A confidence interval may be constructed around each proportion of those who did express a preference and tested for overlap. If the confidence intervals do not overlap, there is a significant preference win for the product with the larger proportion. This requires a large sample size and low proportion of no-preference votes (Quesenbery & Hurst, 1964). The formula for the confidence interval is

$$\mathrm{CI} = \frac{\chi^2 + 2X \pm \sqrt{\chi^2\left[\dfrac{\chi^2 + 4X(N - X)}{N}\right]}}{2(N + \chi^2)} \tag{8.4}$$

where X is the number of panelists preferring one of the two products, N is the total number of panelists, and χ^2 is the value for two degrees of freedom or 5.99.

A quick example. Suppose the data show preference for product $X_1 = 83$, $X_2 = 65$ and no preference equals 14. We need to construct two confidence intervals to see if they overlap. The confidence interval for product X_1 (83/162 = 51.2% preference) is given by

$$\mathrm{CI} = \frac{5.99 + 2(83) \pm \sqrt{5.99\left[\dfrac{5.99 + 4(83)(162 - 83)}{162}\right]}}{2(162 + 5.99)} = \frac{171.99 \pm 31.15}{335.98}$$

which gives an interval from 42% to 60% for product X_1.

The confidence interval for product X_2 (65/162 = 40%) is given by

$$\mathrm{CI} = \frac{5.99 + 2(65) \pm \sqrt{5.99\left[\dfrac{5.99 + 4(65)(162 - 65)}{162}\right]}}{2(162 + 5.99)} = \frac{135.99 \pm 30.39}{335.98}$$

This gives an interval from 31% to 50% for product X_2. The lower bound of the higher proportion (42%) overlaps with the upper bound of the lower proportion (50%); there is no evidence for a significant win using this approach.

8.2.3 A Thurstonian Model

A more recent approach is based on a Thurstonian model (see Section 4.5). The no-preference votes are modeled as a function of τ, a zone of indifference, and is based on an extension of the paired comparison test for difference with an "equal" option (Braun et al., 2004). The degree of preference/difference is called d', and in some literature the Greek letter δ is used. We can assign a Z-value to each proportion. These proportions will correspond to areas under the difference distribution, as shown in the upper part of Figure 8.1. Next, we transform this distribution by subtracting d' from everything and by dividing by the square root of two, getting back to our unit variance. This produces the transformed distribution shown in the

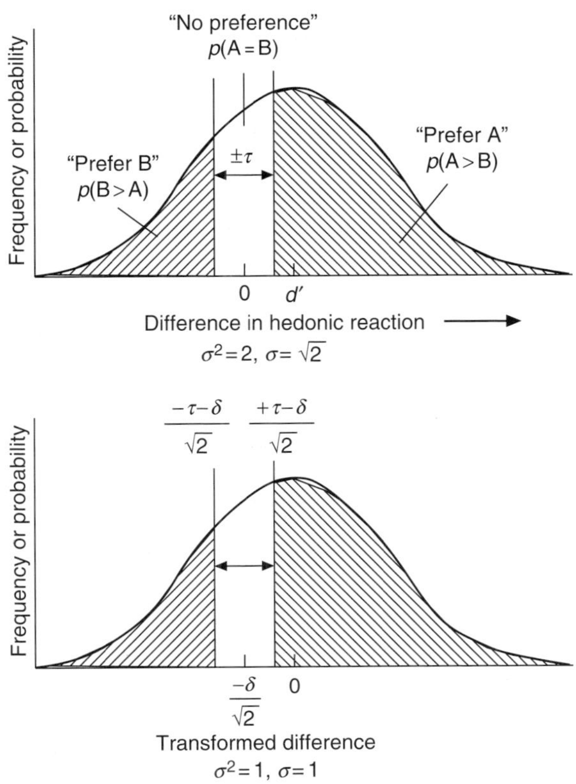

Figure 8.1 A Thurstonian model for the 2-AC paradigm (nonforced choice).

lower panel of Figure 8.1. The assignment of Z-scores produces two equations in two unknowns, and so can be solved for τ and δ as follows:

$$Z_1 = \frac{-\tau - \delta}{\sqrt{2}} \tag{8.5a}$$

and

$$Z_2 = \frac{+\tau - \delta}{\sqrt{2}} \tag{8.5b}$$

where Z_1 is the Z-score for the proportion preferring product A, and Z_2 is the Z-score for the sum of the proportion for A *and* the no-preference votes.

A quick example, where 20% prefer B, 20% have no preference, and 60% prefer A. The left boundary corresponds to the Z-score for 0.2 or –0.84. The next boundary corresponds to the Z-score the sum of 0.2 and 0.2, or 0.4, which equals –0.25. Solve for τ and δ as

$$Z_1 = \frac{-\tau - \delta}{\sqrt{2}} = -0.84 \quad \text{and} \quad Z_2 = \frac{+\tau - \delta}{\sqrt{2}} = -0.20$$

giving $\tau = 0.42$ and $d' = 0.77$.

Testing for a significant preference is not straightforward, however. Two approaches have recently been published. The problem is to find the variance of the δ estimates. According to Christensen et al. (2012), the variance–covariance matrix of the parameters can be obtained as the inverse of the negative Hessian of the log-likelihood function evaluated at the maximum likelihood estimates (i.e., of the second partial derivatives at the solution). Standard errors are then obtained as the square roots of the diagonal elements of the matrix. The Hessian can be found by algebraic derivation or numerical evaluation. Examples and discussion are found in Christensen et al. (2012). Fortunately, they have programmed the numerical evaluation in the R package SENSR and the function TWOAC. So it is possible to obtain the standard error estimates from this freeware. A somewhat different approach was taken by Ennis and Ennis (2012). They chose to evaluate the difference in –2 log (likelihood) values between a model with $\delta = 0$ (with τ estimated) and a model with both δ and τ estimated. A significant gain in the model including both τ and δ would provide evidence for a significant preference (i.e., nonzero δ). Whichever approach is taken, the pure effect size δ is valuable to know for future comparisons of similar products or alternative methodologies.

8.2.4 No-Preference Avoidance and the Dufus Factor

What kind of response would one expect if physically identical products were presented in a paired preference test? If the no-preference response option is available, it would be reasonable to expect most consumers to say they have no preference, possibly because they cannot tell the difference. Somewhat surprisingly, this does not occur, or occurs only infrequently. When presented with identical products, 70–80% of consumers regularly express a preference and only 20–30% will take the no-preference option. This observation was first published by Marchisano et al. (2003), although Ennis mentioned observing this kind of behavior earlier with tobacco products. The effect is important, because it challenges an assertion attributed to Gridgeman (1959), that the meaning of a 50/50 preference split could be disambiguated by allowing the no-preference option. Gridgeman reasoned that if persons truly had no preference, they should avail themselves of this option, but if there were in fact stable segments of consumers that truly favored one of the two products, they should continue to express a preference even though the no-preference option was available. However, if the Marchisano effect is real and consistent, Gridgeman's assertion is incorrect, and only replicated testing can determine if there are stable consumer segments of approximately equal size.

The Marchisano et al. (2003) finding spurred a flurry of research to further investigate this effect and perhaps explain it. Chapman and Lawless (2005) replicated the effect in a series of studies with milk and cottage cheese, with a remarkably consistent use of the no-preference option of only 30% with identical products. A study of discrimination errors did not clarify the reason for the effect. When given a same–different test, about 40–50% of respondents called the identical milks different, but that does not match a 70% false-alarm rate. However, momentary perceived differences, as predicted by Thurstonian theory, could certainly play a part. Among those who thought the milks were different, the avoidance of no preference increased to over 90% (Chapman et al., 2006). The effect proved resistant to manipulation. Showing examples of clearly identical color swatch samples before the taste test did not change the behavior. Consumers' expectations of the degree of difference anticipated did not predict the no-preference avoidance either (Chapman et al., 2006), and explicit instructions that some of the product pairs might be indiscriminable did not help the situation either (Chapman et al., 2010). The effect remains something of a mystery, although task demands (the tendency to want to please the experimenter) could play a role if a consumer thinks they are expected to express a preference.

Another curious finding in the Chapman and Lawless (2005) study was that some consumers were unable to discriminate skim (nonfat) milk from milk with 2% fat (a low-fat milk). When given a dual standard test (matching the two test items to two previously inspected reference samples), and with the reference items left in plain sight, only 88% correct performance was observed. After correction for guessing, this suggests only 75% of the consumer sample was discriminating. This shocking result may underscore the finding that 100% correct discrimination is rarely observed in threshold tests. It was noted in the experience of Stone and Sidel, who in their textbook remarked, "… about 30% of the consumer population does not satisfactorily discriminate among products …" (Stone & Sidel, 1993: 13). So this upper limit on discrimination and/or the presence of a nondiscriminating proportion of the public may produce a constant level of background noise or unavoidable error variance in any consumer study. We nicknamed this nondiscrimination the "dufus factor."

8.3 Replication

8.3.1 Simple χ^2, McNemar, and Reverse McNemar

Replication in a preference test could be considered a risky business. What might happen and what consequences would arise if the proportions of choices on the first presentation do not match the subsequent results? There are many reasons why choices might not be totally stable and repeatable. This was discussed at length by Köster et al. (2002), who found less than 50% consistency from trial to trial in the most liked (chosen) food in a set of novel foods for children and only somewhat higher levels with adults. Chapman and Lawless (2005) found 44% changing of responses in a test of milks that allowed the no-preference option, although the marginal proportions of the groups that preferred

Table 8.2 The four outcomes for a repeated forced preference test

	First trial	
Second trial	**Prefer product A**	**Prefer product B**
Prefer product A	Choice: A/A	Choice: B/A
Prefer product B	Choice: A/B	Choice: B/B

each type of milk remained stable. Wilke et al. (2006) argued in favor of replicated preference testing, and found stable preferences for colas over four replicates, and an increasing preference split for cereals, as if preferences became clearer or stronger over repeated exposures. However, this pattern is not universal. Hauswirth et al. (2010) attempted to replicate the preference splits in the famous "Pepsi challenge" of Pepsi versus Coke. On the first trial, the historical finding of a 57% preference for Pepsi (among those expressing a preference) was duplicated, but on the second trial the preference splits were equal (i.e., the effect washed out).

As we saw in Section 8.2, one of the few ways to discover whether a 50/50 split (or some other nonsignificant preference split) is indicative of true preferring segments is to replicate the test with the same individuals. Another option is to ask for degree of preference, but that is only indicative of the strength of the momentary feeling and does not necessarily tell you about the loyalty or stability of that choice. So the question arises as to how a replicated test should be analyzed. The simplest first choice is to do a χ^2 test on the expected frequencies under a true null. That is, for a duplicate test, one would expect one-fourth of the sample to choose product A twice (A/A), one-fourth to choose product B twice (B/B) and one-half to change from A to B or B to A (Meyners, 2007; Harker et al., 2008). However, the resulting χ^2 value only tells you if the counts obtained differ from what is expected, and not specifically whether there is a significant proportion of A/A choices or more A/A choices than B/B choices.

At this point a strategic evaluation should be conducted. Management should decide which of the following two comparisons are more important, or perhaps both may be. We have the data listed in Table 8.2 to deal with.

If management decides that the consistent cells are the only important ones, then a comparison of cell A/A with B/B is appropriate. If management decides that it is important that people switch to our product on the second trial, then a comparison of B/A with A/B is important. The simplest test for comparing the diagonal cells is the **McNemar test** for changes, and a kind of reverse McNemar can be done for the comparison of cells A/A with B/B. This is legitimate because the McNemar test for changes is simply another version of the binomial test on proportions (see Appendix 8.A for the proof). So it really does not matter whether you are testing the changing cells or the consistent cells. The McNemar test simply tells you if one cell is larger than another and ignores the other pair.

The McNemar test is simple. If we label the cells A, B, C, and D for A/A, B/A, A/B, and B/B, respectively, it takes the following form:

$$\chi^2 \frac{(|B-C|-1)}{B+C} \tag{8.6}$$

This is a χ^2 test with one degree of freedom (critical value: 3.84). In other words, it is a test of the difference of opposite corners, and the same test can be applied for cell A versus D.

8.3.2 Beta Binomial

A variety of other statistical models are available for simple replicated preference tests with a forced choice. These are discussed and simulated in Cochrane et al. (2005). One of these is the **beta binomial model** used for analysis of replicated discrimination tests that is also applicable to replicated preference tests. In theory, it would allow combination of the replications and count them as *X* times the sample *N*, for *X* replicates. However, this is only useful to the extent that you are willing to consider the replicate tests as if they had been done on different individuals. Nonetheless, it will allow a count of the grand total of votes for your product, including both the consistent choosers (counted twice in a duplicate test) and those who switch to or from your product. To the best of my knowledge this is not common in practice, but the basic ideas will be shown here as an option that could be tried.

The beta distribution is used to model the behavior of individuals, and provides an estimate of whether they are behaving randomly, or in some other significant pattern. The consistency parameter is called γ, which varies from zero (random behavior) to one (completely consistent behavior). So in essence it provides a test of the size of the AA, BB cells versus a random distribution (equal proportions in all four cases). Logically, that would appear to satisfy the requirements for a test of consistent preference segments, which would be valuable information. If γ is not significantly different from zero, then you appear to have some random behavior that justifies using the individuals as if they were producing a new independent judgment on each replicate, and thus a simple binomial analysis on the total count (i.e., for the winning product) would be appropriate.

Let us assume we are interested in one of the two products, product A. For r replicates and n panelists, γ is

$$\gamma = \frac{1}{r-1}\left[\frac{rS}{\mu(1-\mu)n} - 1\right] \tag{8.7}$$

where

$$\mu = \frac{\sum_{i=1}^{n} x_i / r}{n} \tag{8.8}$$

is the mean proportion preferring product A and x_i is the number of preference choices for the winning product summed across replicates for that panelist i.

The S-value is part of a variance estimate:

$$S = \sum_{i=1}^{n}\left(\frac{x_i}{r} - \mu\right)^2 \tag{8.9}$$

In other words, S is the sum of squared deviations of each subject from the mean, where x is the number of choices of product A for a given subject, analogous to the numerator of a standard deviation. The test for nonzero γ (Bi, 2006) is based on Tarone's test using a Z-statistic as follows:

$$Z = \frac{E - rn}{\sqrt{2rn(r-1)}} \tag{8.10a}$$

and

$$E = \frac{\sum\left[X - (r\mu)\right]^2}{\mu(1-\mu)} \tag{8.10b}$$

where μ is the sample mean choosing A and X is now the total choices of A for each consumer.

Once again, the practice of combining replicates for a total count depends upon how comfortable the analyst is with treating N consumers as $N \times r$ separate "individuals" for r replicates. In my opinion, counting consumers twice seems like a considerable break with the reality of the situation. Computing γ is still a good idea, but what you do from there should be carefully justified. If the test for nonzero γ is not significant, then you have evidence of random behavior anyway, and so the chances are greatly reduced that there is a significant win for either product.

8.4 Alternative Models: Ferris *k*-visit, Dirichlet Multinomial

8.4.1 The Ferris *k-visit Model*

For replicated tests with a no-preference option, there are several analyses available. One nearly forgotten model is the ***k*-visit model** attributed to George Ferris, who worked as a statistician for General Foods in the 1950s. Ferris had some keen insights, and reasoned that there might be good reason to replicate a preference test, even in a home-use situation, hence the "*k*-visit" designation for *k* replicates. Ferris realized there was a certain amount of apparently random switching that could occur in the test, because some people could not discriminate the products, did not really have much of a preference, or simply changed their minds. These situations created a certain amount of noise in the system, which could be estimated and used as a correction factor. That is, he hypothesized that the proportion of people choosing product A consistently (A/A) could be contaminated by some of the random behavior, and thus the proportion in cell A/A was not a true estimate of the actual proportion that had allegiance to product A. However, there was an underlying proportion whose true value and variance could be estimated.

First, the data would be tabulated as shown in Table 8.3, with frequency counts.

The model requires the calculation of a consistency parameter p by the following means:

$$p = \frac{M - \sqrt{\left[M^2 - \left(N_{oo} + N_y/2\right)\left(2N_x + N_y\right)\right]}}{2N_{oo} + N_y} \tag{8.11}$$

Table 8.3 Notation for the nonforced replicated preference test in the Ferris *k*-visit model

	Trial 1 response		
Trial 2 response	**Prefer A**	**No preference**	**Prefer B**
Prefer A	N_{aa}	N_{oa}	N_{ba}
No preference	N_{ao}	N_{oo}	N_{bo}
Prefer B	N_{ab}	N_{ob}	N_{bb}

where, for convenience, the quantities M, N_y, and N_x were first calculated as follows:

$$N_y = N_{ao} + N_{oa} + N_{bo} + N_{ob}$$
$$\text{(some people switching to or from no preference)} \quad (8.12a)$$

$$N_x = N_{ab} + N_{ba}$$
$$\text{(some people switching products, A to B or B to A)} \quad (8.12b)$$

$$M = N - N_{aa} - N_{bb}$$
$$\text{(all those showing no consistent preference plus the consistent no-preference votes or } N_{oo}\text{)} \quad (8.12c)$$

Ferris called the underlying true proportions of those who prefer A as π_a and those who prefer B as π_b. Thus, a test for the size of π_a versus π_b would tell us if we had a consistent win for product A or B.

So the underlying proportions are corrected for the nonpreferring, nondiscriminating, and random behavior as follows:

$$\pi_A = \frac{\left[N_{aa}\left(1-p^2\right)\right]-\left[\left(N-N_{bb}\right)p^2\right]}{N\left(1-2p^2\right)} \quad (8.13)$$

$$\pi_B = \frac{\left[N_{bb}\left(1-p^2\right)\right]-\left[\left(N-N_{aa}\right)p^2\right]}{N\left(1-2p^2\right)} \quad (8.14)$$

And their variance and covariance estimates are

$$\text{Var}\left(\pi_A\right) = \frac{\pi_A\left(1-\pi_A\right)+\left(3\pi_o p^2\right)/2}{N} \quad (8.15)$$

$$\text{Var}\left(\pi_B\right) = \frac{\pi_B\left(1-\pi_B\right)+\left(3\pi_o p^2\right)/2}{N} \quad (8.16)$$

$$\text{COV}\left(\pi_A,\pi_B\right) = \frac{\left(\pi_o p^2/2\right)-\left(\pi_A\pi_B\right)}{N} \quad (8.17)$$

where π_o is $N-\pi_a-\pi_b$.

The covariance estimate becomes important in the following significance test (Bi, 2006):

$$Z = \frac{\pi_A - \pi_B}{\sqrt{\text{Var}\left(\pi_A\right)+\text{Var}\left(\pi_B\right)-2\text{COV}\left(\pi_A,\pi_B\right)}} \quad (8.18)$$

Ferris suggested the model is applicable for a test with $N > 100$ and preferably $N > 200$. Because the recommended sample size for claim substantiation is about 300 (for a superiority claim),

Table 8.4 Data for the Ferris *k*-visit example

	Prefer A	**No preference**	**Prefer B**
Prefer A	$N_{aa}=457$	$N_{oa}=12$	$N_{ba}=14$
No preference	$N_{ao}=14$	$N_{oo}=24$	$N_{ob}=17$
Prefer B	$N_{ab}=8$	$N_{ob}=11$	$N_{bb}=343$

this should be common in actual practice. Bi (2006) also suggested a test for a proportion against a baseline, as one might do to support an "unsurpassed" type of advertising claim:

$$Z=\frac{\pi_{\mathrm{A}}-P_{\mathrm{b}}}{\sqrt{\mathrm{Var}\left(\pi_{\mathrm{A}}\right)}} \tag{8.19}$$

where P_{b} represents the baseline proportion one might want to test against (e.g., 0.45).

A quick example. Suppose there is a replicated test (i.e., duplicate, two "visits") with 900 consumers and the data are cast as in Table 8.4.

Questions: (1) Is there a significantly higher preference for product A or product B? (2) Is the preference for the winning product higher than 45%? (Example from Ferris (1958) and Bi (2006: 72–6).)

($N=900$.) The basic equations we need are

$$N_y=N_{\mathrm{ao}}+N_{\mathrm{oa}}+N_{\mathrm{bo}}+N_{\mathrm{ob}}=14+12+17+11=54$$

$$N_x=N_{\mathrm{ab}}+N_{\mathrm{ba}}=8+14=22$$

and

$$M=N-N_{\mathrm{aa}}-N_{\mathrm{bb}}=100$$

And we need to retrieve N_{oo} from the data table, or 24.

$$p=\frac{M-\sqrt{M^2-\left(N_{\mathrm{oo}}+N_y/2\right)\left(2N_x+N_y\right)}}{2N_{\mathrm{oo}}+N_y}=0.257$$

And the best estimates of each segment/proportion become:

$$\pi_{\mathrm{A}}=\frac{\left[N_{\mathrm{aa}}\left(1-p^2\right)\right]-\left[\left(N-N_{\mathrm{bb}}\right)p^2\right]}{N\left(1-2p^2\right)}=0.497$$

$$\pi_{\mathrm{B}}=\frac{\left[N_{\mathrm{bb}}\left(1-p^2\right)\right]-\left[\left(N-N_{\mathrm{aa}}\right)p^2\right]}{N\left(1-2p^2\right)}=0.370$$

Next, we need the variability and covariance estimates for the Z-tests:

$$\mathrm{Var}\left(\pi_{\mathrm{A}}\right)=\frac{\pi_{\mathrm{A}}\left(1-\pi_{\mathrm{A}}\right)+\left(3\pi_{\mathrm{o}}p^2\right)/2}{N}=0.000296 \text{ or } \pm 0.017\ (1.7\%)$$

as $0.017 = \sqrt{0.000296}$ (converting variance to the standard deviation),

$$\mathrm{Var}(\pi_B) = \frac{\pi_B(1-\pi_B) + (3\pi_o p^2)/2}{N} = 0.000297$$

$$2\mathrm{COV}(\pi_A, \pi_B) = \frac{(\pi_o p^2/2) - (\pi_A \pi_B)}{N} = -0.000198$$

Now for the hypothesis test:

$$Z = \frac{0.497 - 0.370}{\sqrt{0.000297 + 0.000296 - (-0.000198)}} = 4.038$$

which is obviously a significant win for product A (Z>1.96).

8.4.2 Dirichlet Multinomial

An alternative to the Ferris model is the **Dirichlet multinomial**. This model extends the reasoning of the beta-binomial approach to the situation where there are more than two alternatives (Gacula et al., 2009). Let x_1 be the sum for product A over all choices, x_2 the sum of no preference, and x_3 the sum for product B. Let there be n panelists, r replicates, and m response choices (in this case three). We have N total observations ($= n \times r$). The model uses a simple χ^2 test against the expected values for a null result. This approach requires two estimations. First, what is the expected value of the no preference proportion? Once that proportion, P_n, is found, estimated, or assumed, the expected frequencies for each product become $1 - 2P_n$. For example, if we assume that the baseline of no preferences is 20%, then we are testing against expected frequencies of 0.4 for product A, 0.2 for no preferences, and 0.4 for product B. Some authors call this an identicality norm (Ennis & Ennis, 2012) (in difference or equivalence testing), and in theory it could be estimated by doing a preference test with identical products. The second parameter to be estimated is an overdispersion or consistency parameter that is analogous to γ in the beta binomial model. In keeping with the notation of Gacula et al. (2009), we will call this factor C. Then the χ^2 formula is adjusted by this factor as follows:

$$\chi^2 = \frac{nr}{C} \sum_{j=1}^{m} \frac{(p_j - p_{\exp})^2}{p_{\exp}} \tag{8.20}$$

The C-factor itself is found in a longer equation, that once again looks at the consistency or variation among the consumers, relative to the means for the whole group. So, once again, it is somewhat analogous to a variance estimate of the panelist's patterns of individual behavior. C is given by the rather longish expression

$$C = \frac{r}{(n-1)(m-1)} \sum_{j=1}^{m} \frac{1}{p_j} \sum_{j=1}^{n} \left(\frac{x_{ij}}{r} - p_j \right)^2 \tag{8.21}$$

And the p_j are simply the mean proportions for each of the response options. In other words:

$$p_j = \frac{\sum_{i=1}^{n} x_{ij}}{nr} \quad (8.22)$$

So, in order to get the C-value, we are taking each individual panelist's vote count for that reponse option, subtracting the group mean, and then squaring and summing these squared deviations for the whole group, just as in the numerator of a standard deviation. We are then summing these, weighted by $1/p_j$ for each of the three (in this case) response options.

Is there a pattern of responding analogous to a nonzero γ in Tarone's Z-test? This is another Z-statistic, given by eqn 8.23. This assumes there are an equal number of replicates for each consumer, which would be the most common design in a consumer test. If there are different numbers of replicates for different consumers, a more general formula is given in Gacula et al. (2009: 545).

$$Z = \frac{\left[N\sum_{j-1}^{m}\frac{1}{x_{oj}}\sum_{i=1}^{n}x_{ij}\left(x_{ij}-1\right)\right]-N\left(r-1\right)}{\sqrt{2\left(m-1\right)N\left(r-1\right)}} \quad (8.23)$$

where x_{ij} is the total number of that choice j for panelist i, multiplied by $x_{ij} - 1$, then summed across all panelists, then weighted by $1/x_{oj}$, which is the count for choice j. Repeat for each choice j. Once again, $N = r \times n$, the product of replicates times consumers.

The approach outlined above will let us know whether there is an overall deviation from the expected response pattern, but it does not show whether one product was preferred to another. For example, a significant χ^2 could arise simply from an overabundance of no preference votes. A test of the difference between proportions to let us know if there is a product win can be executed by the expression

$$Z = \frac{p_A - p_B}{\sqrt{V\left(p_A\right)+V\left(p_B\right)-2\mathrm{COV}\left(p_A, p_B\right)}} \quad (8.24)$$

where $V(x)$ is the variance estimate and COV(x,y) is the covariance estimate. So there is a little more work to do to get to the bottom line to see if there is a product win. Note that this formula is very conservative relative to the common test for the difference of two proportions. A more liberal formula would use the standard error of the proportions and ignore the covariance, giving

$$Z = \frac{p_A - p_B}{\sqrt{(\hat{p})(1-\hat{p})(1/n_1 + 1/n_2)}} \quad (8.25)$$

where $\hat{p}$ is the average of the two proportions, a kind of null estimate.

A quick example. Table 8.5 shows the data from 25 hypothetical consumers, 15 of which prefer product A, 5 of which do not care and only 5 consistently like product B. Of course, any real consumer test would have far more individuals in the pool. So we have $N = 75 = 3(25)$ judgments. The overall proportions are 0.507 for A, 0.293 no preference, and 0.20 for B. When tested against an equality norm of (0.4,0.2,0.4) we get a $\chi^2 = 8.06$, which is significantly higher than the critical value for two degrees of freedom of 5.99. So something is going on. But there is considerable overdispersion; that is, nonrandomness or consistency. The C-factor is 1.59, which corresponds to a γ value of 0.33 and a Tarone's

Table 8.5 Data from the Dirichlet multinomial example

Consumer	Votes for product A	No preference	Votes for product B
1	3	0	0
2	3	0	0
3	3	0	0
4	3	0	0
5	3	0	0
6	2	1	0
7	2	1	0
8	2	1	0
9	2	1	0
10	2	1	0
11	2	1	0
12	2	1	0
13	2	1	0
14	2	1	0
15	2	1	0
16	1	1	1
17	1	1	1
18	1	1	1
19	0	3	0
20	0	3	0
21	0	1	2
22	0	1	2
23	0	1	2
24	0	0	3
25	0	0	3
Sum	38	22	15
Mean	1.52	0.88	0.6
Proportion	0.5067	0.2933	0.2000
Variance	0.14		0.11
Covariance A,B	−0.088		
C-factor	1.59		

Z-test of 3.25, so there is significant overdispersion; that is, a pattern of at least some consumers acting consistently. Unfortunately, in spite of the large difference between p_A and p_B, the *Z*-score returns a nonsignificant value ($Z<1$) partly due to the high (negative) covariance. Using the test for two proportions, we get a more liberal $Z=2.26$. But this is ignoring the covariance. Whether one wishes to accept the more conservative or more liberal test is a judgment call. Certainly, the overall pattern seems to justify a conclusion of a difference, although the small sample size would suggest a good degree of caution until further consumers could be tested. The reader may wish to see what happens when the sample size is doubled, with the same pattern of results.

8.5 Affective Scales

8.5.1 Category Scales and Line Marking Techniques

A variety of scaling techniques have been used to measure liking or pleasantness of food flavors and other experiences from products, including magnitude estimation, line or visual analog scales (VASs), and the popular nine-point category scale attributed to Peryam and colleagues. For a review see Lim (2011), and for general information on scaled acceptability testing see Lawless and Heymann (2010: chapter 14). Related techniques include behaviorally oriented scales (Schutz, 1965), appropriateness scales (Schutz, 1988, 1994; Cardello & Schutz, 1996), satisfaction scales, scales that allow for adjustment or repositioning of previous judgments (Kim & O'Mahony, 1998; Cordonnier & Delwiche, 2008), various pictorial scales suitable for use with children (Kroll, 1990), and counting the number of persons who are accepting of the product (Lagrange & Norback, 1987). These different techniques are discussed in Lawless and Heymann (2010).

By far the most common scale for assessing liking is the nine-point hedonic scale. The nine-point scale uses the words "like" and "dislike," modified by the adverbs "slightly," "moderately," "very much," and "extremely," to indicate increasing degrees of response and contains a neutral phrase, "neither like nor dislike." This is the most common tool for measuring consumer opinion in applied sensory testing and has a long history and good track record (Peryam & Girardot, 1952; Peryam & Pilgrim, 1957). Attempts to measure the psychological spacing of the terms using a Thurstonian analysis of category ratings of different phrases showed that the terms are roughly evenly spaced, with the exception that the phrases "slightly" are nearer to the neutral point than is the spacing of any of the other adjacent phrases (Jones & Thurstone, 1955; Jones et al., 1955). Thus, the numbers 1 through 9 are usually assigned to the ratings for purposes of statistical analysis.

However, the scale has been criticized for uneven spacing, for the tendency of consumers to avoid the end categories that tends to truncate the scale, and for deviations from normality in the data distributions. The lack of interval-scale properties and deviations from normality suggest that the use of nonparametric statistics would be more appropriate for data generated by this scale. To avoid these problems, a variety of other scaling techniques have been used, such as line scales, also known as VASs (Lawless, 1977; Rohm & Raaber, 1991; Hough et al., 1992). The VAS has a history of use in pain measurement (Huskisson, 1983). Usually, the ends are labeled with a phrase such as "like extremely" at one end and "dislike extremely" at the other and a neutral phrase is placed at the center marker (Wright, 2007). A version of the line scale with pips or markers equally spaced along the line was found to compare favorably against the nine-point scale in terms of product differentiation and identification of consumer segments (Villanueva et al., 2005; Villanueva & Da Silva, 2009), although advantages were slight (Lawless, 2009). A simplified version of the LAM scale (discussed in Section 8.5.2) was used by Wright (2007) in which the end anchors "greatest imaginable like(/dislike)" were used instead of the usual "like (dislike) extremely."

8.5.2 Magnitude Estimation and Category-Ratio Scales

Magnitude estimation has been used for hedonic scaling, in both unipolar and bipolar forms (Moskowitz & Sidel, 1971; Moskowitz, 1980, 1986; McDaniel & Sawyer, 1981; Pearce et al., 1986, Lavanka & Kamen, 1994). However, its use in applied sensory evaluation has been limited, in part because of the difficulty of making ratio judgments for some consumers as well as the need for data transformation to adjust for scale range differences and deviations from normality. A related approach is the magnitude scaling of verbal phrases, which are then placed on a line using the mean scale values as spacing indicators for the appropriate relative positions of the words. These scales are variously referred to as labeled affective

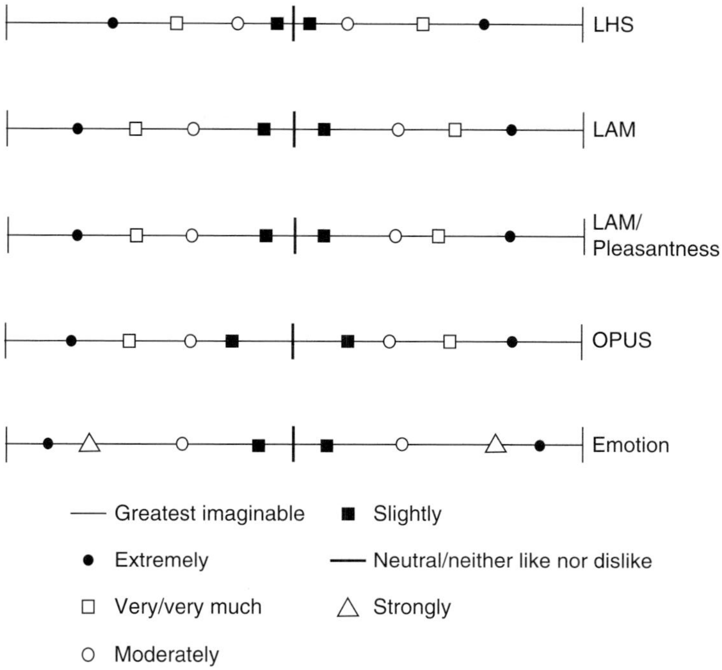

Figure 8.2 Labeled affective scales for liking and pleasantness. Note that the spacing of the various modifiers is somewhat different depending upon the method used. The LHS scale has the most compressed spacing, the OPUS scale positions the "slightly" phrases a bit farther from neutral, and the emotion scale places the "extremely" phrases closer to the ends.

or labeled hedonic scales (LHSs) and have been used for both degrees of liking/disliking and for pleasantness/ unpleasantness.

Because the word phrases are generally scaled using magnitude estimation, it has been proposed that the resulting data have ratio properties. Whether this is true or not, and whether affective reactions can stand in comparison with one another in ratios, remains to be determined, although there are supporters of this view (Lim, 2011). One must remember that, just because there have been ratio instructions given to subjects, this does not mean the data actually have ratio properties. This issue was discussed further in Chapter 2. Nonetheless, there are proponents of the idea that being able to say, for example, that this product was liked "twice as much" as that product would be a valuable kind of statement (Prescott, 2009).

Some of these scales are shown in Figure 8.2. They differ in several key properties. First, were the word phrases scaled by themselves (e.g., Schutz & Cardello, 2001), or in the context of other experiences (e.g., Lim et al., 2009)? Next, does the scale refer to liking (and disliking) or to pleasantness (and unpleasantness) of experiences? Third, what does the high end-anchor refer to in terms of the frame of reference? Is it specified? Usually, the modifier at the high end is the phrase "greatest imaginable." Does it refer to all experiences or just some related set of products? All of these scales derive from the work of Borg (1982) and Green et al. (1993, 1996) on category-ratio scales for measurement of sensory intensity, but extend that type of scale into the realm of hedonics. Five of these scales are discussed here.

The first attempt was made by Schutz and Cardello, who gave magnitude estimation instructions to subjects and then had them label the judgment with a valence indicator (plus or minus for liking versus disliking) (Schutz & Cardello, 2001; Cardello & Schutz, 2004). No other context or set of familiar experiences was used in the scale development sessions. The words consisted of the common nine-point hedonic terms and the high end-anchors "greatest imaginable like" and "greatest imaginable dislike." The scale was called the LAM scale. Similar procedures were used to develop scales for clothing comfort (Cardello et al., 2003) and perceived satiety (Cardello et al., 2005). This scale was modified by Keskitalo et al. (2007) to change the like/dislike to pleasant/unpleasant, due to a lack of direct translation into Finnish for the word "dislike." However, the word spacing was similar. Guest et al. (2007) also developed a pleasantness scale for oral experiences, using magnitude estimation and embedding the word phrases into a set of common oral experiences such as "the sourness of a slice of lemon." This scale was called the OPUS for oral pleasantness/unpleasantness scale. A similar scale was developed by Lishner et al. (2008) for scaling emotional responses. They embedded their phrases in the context of various emotion-inducing pictures (e.g., injured children, naked women) but used a −100 to +100 scale without ratio instructions. They also included modifiers at the low end such as "barely," "a little," and "mildly" and one additional higher modifier "strongly." (These extra scale points are not shown in Figure 8.2.) Next, Lim et al. (2009), working in Green's laboratory, used the magnitude procedure for liking/disliking, but in contrast to the LAM scale development process, they included a wide-ranging set of common experiences in addition to the word phrases. They termed the scale an LHS. These five scales and the resulting word spacings are shown in Figure 8.2.

The context within which the phrases are scaled obviously can affect the spacing. An argument was presented that the best context was the most wide, in an analogy to the argument presented for labeled magnitude scales of intensity. The argument was that using the context of greatest imaginable "experiences of any kind" rather than no explicit context or a limited context would produce data that provided more valid comparison of ratings among different persons. This proposition was discussed in Chapter 7 and will not be discussed further here. Of note, studies of the "any kind" versus more limited contexts such as "foods and beverages" showed that the inclusion of a more extreme end-anchor compressed both the range of ratings (Cardello et al., 2008) and the range of values for the phrases, making the spacing more compact and towards the center of the line (Lawless et al., 2010c). In the latter study, the compressed spacing led to poorer product differentiation in a food preference survey. However, in Cardello et al. (2008), no detrimental effect was observed. Note that in Figure 8.2, the spacings of the LHS words are more compressed than the spacings of the LAM scale, as one might expect when the phrases are scaled in the context of a wide range of common experiences.

Are the LHSs really better than the traditional nine-point scale? Turning aside the issue of whether the data are truly ratio level, we can ask whether the scales provide any better product differentiation than the nine-point scale. Several studies have found such an advantage for the LAM scale (Schutz and Cardello, 2001; Greene et al, 2006, El Dine & Olabi, 2009). An advantage might be expected on the basis of allowing more room at the top and bottom of the scale, and thus mitigating end-avoidance tendencies as well as providing a response option for extremely well-liked or disliked products. However, the advantage is not always seen. In an extensive consumer study of well-liked products, Lawless et al. (2010a) found that the LAM scale was sometimes superior and the nine-point scale was sometimes superior in both product differentiation and identification of consumer segments. A similar result was found for potato chips by Lawless et al. (2010b). Hein et al. (2008) found about equal product differentiation for snack bars on the first of two replicates, but superior differentiation by the nine-point scale on the second go-round. So, in terms of discriminatory ability, the advantage is questionable. In addition, the line scale requires measurement of the response in data collection. Of course, this is less of an issue in recent times with the advent of widespread computer-assisted data collection software.

Another concern with these scales has been the tendency of consumers to use the scale categorically; that is, by making marks primarily where the verbal anchors occur (Cardello et al., 2008; Lawless et al., 2010a,b). Such behavior would tend to work against the notion that people are making ratio judgments, although one can argue that, if the spacing is correct, the data still have ratio properties; they are just more variable ("noisy") due to this odd consumer tendency. The effect can be reduced by reading a detailed set of verbal instructions of about 250 words with examples given (Lim & Fujimaru, 2010) or by using a physically longer line with more visible room between verbal phrases (Lawless et al., 2010b).

8.6 Ranking and Partial Ranking

8.6.1 Standard Rank Analysis.

Ranked data are usually analyzed by the Friedman test for ranks (Lawless & Heymann, 2010) or by the Kramer rank sum analysis, which provides simple lookup tables for significant differences in rank sums (Newell & MacFarlane, 1987). In consumer tests, it would be unusual to have respondents rank more than a few products. However, there are some circumstances, like wine judging, where larger numbers of items may be ranked or scored for purposes of quality grading or assigning awards in medal competitions. One such scheme, "Borda count" is discussed in Section 8.6.2. Dealing with ranked data from larger numbers of samples presents some challenges. One approach is to use Durbin's rank test on incomplete block designs (Bi, 2009).

A common approach is to perform a Friedman test to see if there is an overall difference, and then follow it with paired comparisons to see if one of two products is ranked consistently higher. The equation for the Friedman test is

$$\chi^2 = \left[\frac{12}{K(J)(J+1)}\sum_{j=1}^{J} T_j^2\right] - 3K(J+1) \tag{8.26}$$

where there are K consumers ranking J products and T represents the rank totals for that product. It is distributed as a χ^2 variable with $J-1$ degrees of freedom. Individual products can then be compared by a least significant difference test as follows:

$$\text{LSD} = 1.96\sqrt{\frac{K(J)(J+1)}{6}} \tag{8.27}$$

8.6.2 Borda Counts

A potential application of rank sum techniques occurs in wine judging. In wine judging for competitive medals, the question is debated about how best to combine the scores of various judges. Following the painful discoveries that many so-called expert wine judges do not replicate their ratings and that medal awards are largely due to random chance (Hodgson,

2008, 2009), various strategies have been suggested. One is to weight the scores from judges based upon their reliability before summing or averaging. But this begs the question of why unreliable judges are not vetted and disqualified in the first place. Another issue concerns the contribution of judges who may be more critical, and thus provide lower scores. The contribution of such judges may be underweighted in a simple average, especially if scores are positively skewed, which brings up the simple average due to the high outliers.

As an alternative to the typical 100-point scoring schemes with some kind of averaging, Hulkower (2012) suggested a procedure called a **Borda count** for combining the rankings (rather than ratings) of wine judges. He also had the insight that, in the rankings, the judges could provide benchmarks for their suggested medal award cutoffs and treat those data as if they were the ranks of actual products. The Borda count is named after the mathematician Jean-Charles de Borda who published the scheme in 1781 for electing new members to the Paris Academy of Sciences (Hulkower, 2012). It is based on an inverse rank sum, where, for N products, the scores become the value N minus the rank. So in 10 products, the top ranked gets a score of 9. A worked example will follow.

The Borda scheme has some useful mathematical properties that are desirable in any voting method, as outline by Hulkower. First, any strictly transitive ranking can be chosen by a judge (if A>B and B>C then A>C). The transitivity of the individual rankings is preserved in the final scores. If every voter ranks one item over another, the overall outcome will preserve this. Finally, the ranking of two products in the outcome is determined by the relative ranking of each judge and the number of alternatives between them. This is the only combinatorial scheme that does not result in any paradoxes (Hulkower, 2012). A useful variation is to insert a cutoff (such as no candidate below this rank is acceptable) into the product rankings. Such a cutoff or benchmark could clearly be used in quality judging or even sensory quality control panels. In Hulkower's scheme for wine awards, the judges insert benchmarks for gold, silver, and bronze medals, along with the rankings of the actual wine samples.

Table 8.6 shows an example for six judges, nine wines, and the three benchmarks. Only wines tied or above the benchmark are awarded that medal. Note that ties are allowed.

Table 8.6 Example of Borda counts for nine wines and six judges (after Hulkower (2012))

	Wines (A–I), benchmarks and rankings											
Judge	**A**	**B**	**C**	**D**	**E**	**F**	**G**	**H**	**I**	**Gold**	**Silver**	**Bronze**
1	12	3	9	8	2	11	7	4	6	1	5	10
2	11	3	10	7	1	12	8	2	5	4	6	9
3	10	1	9	6	3.5	11	12	3.5	7	2	5	8
4	10	2	7	6	1	12	9	3	4	5	8	11
5	11	4	10	8	2	12	7	1	5	3	6	9
6	5	3	6	10	2	8	12	9	11	1	4	7
Borda scores (= 12 – rank)												
1	0	9	3	4	10	1	5	8	6	11	7	2
2	1	9	2	5	11	0	4	10	7	8	6	3
3	2	11	3	6	8.5	1	0	8.5	5	10	7	4
4	2	10	5	6	11	0	3	9	8	7	4	1
5	1	8	2	4	10	0	5	11	7	9	6	3
6	7	9	6	2	10	4	0	3	1	11	8	5
Sum	13	56	21	27	60.5	6	17	49.5	34	56	38	18

In this example, the final rankings are as follows:

E > B = gold > H > silver > I > D > C > bronze > G > A > F

In other words, wines E and B receive a gold medal, H gets silver, I, D, and C get bronze, and G, A, and F get no award. Hulkower also suggested a superior category for wines that all judges ranked higher than the gold benchmark, called "double gold" (why not platinum?). None are so ranked in this example.

8.6.3 Best–Worst (Max-Diff) Data Handling and Models: A True Ratio Scale?

A technique for assessing consumers' relative preferences that has enjoyed some popularity in marketing research is called **best–worst scaling** (Finn & Louviere, 1992). In this technique, the consumer views a set of products, usually three or more, and chooses the most liked and least liked from the set. Obviously, with only three products this is equivalent to simple ranking. However, with a large product set the consumer can be asked to evaluate different combinations, and each individual test set is not limited to only three items. The method is sometimes applied to intensity scaling, wherein it is referred to as max-diff for maximum difference (although in that case it would be more accurate to call it strongest–weakest). It can be viewed as an extension of the method of paired comparisons (in this case paired preference), for which there are many appropriate models, including Thurstonian analysis (see Chaper 4) and logistic modeling (Luce, 1959; Baird & Noma, 1978).

There are several claimed advantages to this method when used for consumer choice. First, it is alleged to have a superior discriminatory power than simple scaling. Second, using two of the analysis options, either an interval or ratio scale can be obtained. These are discussed below. Third, it is a simple, easy, and natural task for consumers to express their most- and least-favored products. Consumer choice is often clearest when asked about the extremes (Marley & Louviere, 2005). Owing to these potential benefits, best–worst scaling has been studied for food preference tests by Jaeger and colleagues (Hein et al., 2008; Jaeger et al., 2008; Jaeger & Cardello, 2009). In their first study, Jaeger et al. (2008) found better product discrimination by best–worst scaling than an unstructured line scale (as shown by *F*-ratios and significant differences in analysis of variance (ANOVA)), although the mean scores were highly correlated. However, the best–worst scaling required each consumer to view 18 samples in six triadic sets, and so the exposure and data collection were quite different. A more direct comparison would perhaps have replicated the ratings to provide an equal number of judgments per product for each method. In their second study (Hein et al., 2008), best–worst scaling produced higher *F*-ratios and better product discrimination than the nine-point hedonic scale, LAM scale, unstructured line scale, and preference ranking (but no *F*-ratio comparison could be made for ranking as Friedman's test was used). However, on the second replicate, the nine-point hedonic scale fared much better than any other method (a second replicate of the best–worst was not done owing to the large number of samples presented). The dramatic increase in the nine-point ratings suggests that consumers would use this scale much more effectively once they have been exposed to the product set. If so, one could make a strong argument for replication in food acceptance tests. In a third study of tasted fruit juices and a food preference survey, the LAM scale was compared with best–worst scaling (Jaeger & Cardello, 2009). For the taste test, the LAM scale differentiated the products better, but the reverse was true for the food preference survey, so there was no apparent advantage to either method. Given the larger number of comparisons, the best–worst method would appear to be well suited to paper-and-pencil kinds of surveys, as opposed to taste tests in which the fatigue factor can be significant (Mueller et al., 2009).

Different approaches have been taken for analysis of these data. With a set of multiple products taken three at a time, the design is effectively an incomplete block design with simple ranking, and thus Durbin's rank test can be applied (Bi, 2009). A univariate Thurstonian model can be applied (Ennis et al., 2011), presumably by maximum likelihood methods, although details are sparse. Proprietary software is available for this purpose (Ennis et al., 2011). The simplest analysis is to produce totals for each product of when it was graded best and when it was graded worst, and then subtract these two quantities to arrive at a score (Jaeger et al., 2008). These results are claimed to have interval scale properties, and thus can be appropriately analyzed by ANOVA or general linear models. Another analysis is a multinomial logistic analysis, available in various commercial software packages and in the MLOGIT library of the R software platform. This is claimed to produce ratio-level data (Finn & Louviere, 1992). However, because the multinomial logistic analysis scales everything relative to a reference product, the zero point would seem to be arbitrary (as in Thurstonian models) rather than meaningful. If so, the ratio-scale claim seems questionable. The output of the multinomial analysis will produce a set of weights or utilities, which can be converted to a probability scale by taking the utility of any item and dividing it by the sum of all items, for easier interpretation (see Jaeger et al. (2008)).

8.7 Conclusions

Simple paired preference tests and nine-point scales for acceptability ratings remain the two most common tools for assessing consumer appeal. However, many variations exist, such as replicated preference and nonforced preference with a no-preference response option. These variations generally provide additional information, but require different analysis and careful interpretation. For example, what is the management strategy when the first replicate shows that a certain product wins but the second does not? For acceptance scaling, newer techniques have evolved, such as labeled magnitude scales and best–worst rankings. The advantages of these newer methods are at this time unclear, although it would make sense, for example, to expand the nine-point scale (as in the LAM scale and LHS) to mitigate the tendency of consumers to avoid end-categories and thus effectively truncate any scale. In general, more response options allow for more information to be transmitted, up to a point (Bendig & Hughes, 1953). This is an active area of research in sensory evaluation, applied psychophysics, and sensometrics, and new developments in methods and analyses are likely to continue. Comparisons of models and approaches need to be made on the same data sets to provide information on the relative risks of type I and type II errors for different analyses.

Appendix 8.A Proof that the McNemar Test is Equivalent to the Binomial Approximation *Z*-Test (AKA Sign Test)

The McNemar test takes the following form:

$$\chi^2 = \frac{(|B-C|-1)}{B+C} \quad \text{and} \quad \chi^2 = \frac{(|B-C|)}{B+C} \tag{8.A.1}$$

if the continuity correction is ignored.

B and C are diagonal entries for frequency counts in a 2×2 matrix, usually the cells that represent the counts of responses that differed on the first and second assessments (hence the "McNemar test for changes").

The Z-score approximation to the binomial takes the form

$$Z = \frac{P_{\text{obs}} - \frac{1}{2}}{\frac{1}{2}\sqrt{1/N}} \quad \text{And} \quad Z = \frac{P_{\text{obs}} - P_{\text{chance}}}{\sqrt{p(1-p)/N}} \qquad (8.A.2)$$

(ignoring continuity correction).

Since $P_{\text{chance}} = p = \frac{1}{2}$ under the null hypothesis (i.e., equal population proportions), eqn 8.A.2 simplifies to

$$Z = \frac{P_{\text{obs}} - \frac{1}{2}}{\frac{1}{2}\sqrt{1/N}} \quad \text{and} \quad Z = \frac{2P_{\text{obs}} - 1}{\sqrt{1/N}} \qquad (8.A.3)$$

A few useful identities will simplify matters:

$$Z^2 = \chi^2$$

$$B + C = N$$

Dividing by N gives

$$\frac{B}{N} + \frac{C}{N} = 1 \quad \text{and} \quad \frac{C}{N} = 1 - \frac{B}{N}$$

If $B > C$, then $P_{\text{obs}} = B/N$. (If $C > B$, $P_{\text{obs}} = C/N$, but the proof remains the same with that substitution.)

Taking the square root of the χ^2 formula we get

$$\sqrt{\chi^2} = \frac{B-C}{\sqrt{B+C}} = \frac{B-C}{\sqrt{N}} = Z \qquad (8.A.4)$$

Dividing both the numerator and denominator of eqn 8.A.4 by N gives us

$$Z = \frac{\frac{B}{N} - \frac{C}{N}}{\sqrt{1/N}}, \quad \text{since} \quad \frac{\sqrt{N}}{N} = \sqrt{\frac{1}{N}} \qquad (8.A.5)$$

Taking C/N out by substituting $1 - \frac{B}{N}$ gives

$$Z = \frac{\frac{B}{N} - \left(1 - \frac{B}{N}\right)}{\sqrt{1/N}} \qquad (8.A.6)$$

and thus

$$Z = \frac{2\frac{B}{N} - 1}{\sqrt{1/N}} \tag{8.A.7}$$

Since $B/N = P_{obs}$, this is the same expression as eqn 8.A.3. QED.

References

Angulo, O. and O'Mahony, M. 2005. The paired preference test and the no preference option: was Odesky correct? Food Quality and Preference, 16, 425–34.

ASTM. 2008. Standard Guide for Sensory Claim Substantiation. Designation E 1958-07. Vol. 15.08 Annual Book of ASTM Standards. ASTM International, Conshohocken, PA, pp. 186–212.

Baird, J.C. and Noma. E. 1978. Fundamentals of Scaling and Psychophysics. John Wiley & Sons, Inc., New York, NY.

Bi, J. 2006. Sensory Discrimination Tests and Measurements. Blackwell Publishing, Ames, IA.

Bi, J. 2009. Computer-intensive methods for sensory data analysis, exemplified by Durbin's rank test. Food Quality and Preference, 20, 195–202.

Bendig, A.W. and Hughes, J.B. 1953. Effect of number of verbal anchoring and number of rating scale categories upon transmitted information. Journal of Experimental Psychology, 46(2), 87–90.

Borg, G. 1982. A category scale with ratio properties for intermodal and interindividual comparisons. In: Psychophysical Judgment and the Process of Perception. H.-G. Geissler and P.Petzold (Eds). VEB Deutscherverlag der Wissenschaften, Berlin, pp. 25–34.

Braun, V., Rogeaux, M., Schneid, N., O'Mahony, M., and Rousseau, B. 2004. Corroborating the 2-AFC and 2-AC Thurstonian models using both a model system and sparkling water. Food Quality and Preference, 15, 501–7.

Cardello, A.V. and Schutz, H.G. 1996. Food appropriateness measures as an adjunct to consumer preference/acceptability evaluation. Food Quality and Preference 7, 239–49.

Cardello, A.V. and Schutz, H.G. 2004. Research Note. Numerical scale-point locations for constructing the LAM (labeled affective magnitude) scale. Journal of Sensory Studies, 19, 341–6.

Cardello, A.V., Winterhaler, C., and Schutz, H.G. 2003. Predicting the handle and comfort of military clothing fabrics from sensory and instrumental data: development and application of new psychophysical methods. Textile Research Journal, 73, 221–37.

Cardello, A.V., Schutz, H.G., Lesher, L.L., and Merrill, E. 2005. Development and testing of a labeled magnitude scale of perceived satiety. Appetite, 44, 1–13.

Cardello, A., Lawless, H.T., and Schutz, H.G. 2008. Effects of extreme anchors and interior label spacing on labeled magnitude scales. Food Quality and Preference, 21, 323–34.

Chapman, K.W. and Lawless, H.T. 2005. Sources of error and the no-preference option in dairy product testing. Journal of Sensory Studies 20, 454–68.

Chapman, K.W., Grace-Martin, K., and Lawless, H.T. 2006. Expectations and stability of preference choice. Journal of Sensory Studies 21, 441–55.

Chapman, K.W., Lovelace, E., Cardello, A., and Lawless, H.T. 2010. Preference for one of two identical stimuli: explicit instructions and personality traits. Journal of Sensory Studies, 25, 35–53.

Christensen, R.H.B., Lee, H.-Y., and Brockoff, P.B. 2012. Estimation of the Thurstonian model for the 2-AC protocol. Food Quality and Preference, 24, 119–28.

Cochrane, C.-Y.C., Dubnicka, S., and Loughin, T. 2005. Comparison of methods for analyzing replicated preference tests. Journal of Sensory Studies, 20, 484–502.

Cordonnier, S.M. and Delwiche, J.F. 2008. An alternative method for assessing liking: positional relative rating versus the 9-point hedonic scale. Journal of Sensory Studies, 23, 284–92.

El Dine, A.N. and Olabi, A. 2009. Effect of reference foods in repeated acceptability tests: testing familiar and novel foods using 2 acceptability scales. Journal of Food Science, 74, S97–S106.

Ennis, D.M. and Ennis, J.M. 2012. Accounting for the no difference/preference responses or ties in choice experiments. Food Quality and Preference, 23, 13–17.

Ennis, D.M., Rousseau, B., and Ennis, J.M. 2011. Short Stories in Sensory and Consumer Science. The Institute for Perception, Richmond, VA.

Ferris, G.E. 1958. The *k*-visit method of consumer testing. Biometrics, 14, 39–49.

Filipello, F. 1957. Organoleptic wine-quality evaluation. I. Standards of quality and scoring vs. rating scales. Food Technology, 11, 47–51.
Finn, A. and Louviere, J.J. 1992. Determining the appropriate response to evidence of public concern: the case of food safety. Journal of Public Policy and Marketing, 11, 12–25.
Gacula, M., Singh, J., Bi, J., and Altan, S. 2009. Statistical Methods in Food and Consumer Research. Elsevier/Academic Press, Amsterdam.
Green, B.G., Shaffer, G.S., and Gilmore, M.M. 1993. Derivation and evaluation of a semantic scale of oral sensation magnitude with apparent ratio properties. Chemical Senses, 18, 683–702.
Green, B.G., Dalton, P., Cowart, B., Shaffer, G., Rankin, K., and Higgins, J. 1996. Evaluating the "labeled magnitude scale" for measuring sensations of taste and smell. Chemical Senses, 21, 323–34.
Greene, J.L., Bratka, K.J., Drake, M.A., and Sanders, T.H. 2006. Effective of category and line scales to characterize consumer perception of fruity fermented flavors in peanuts. Journal of Sensory Studies, 21, 146–54.
Gridgeman, N.T. 1959. Pair comparison, with and without ties. Biometrics, 15, 382–8.
Guest, S., Essick, G., Patel, A., Prajapati, R., and McGlone, F. 2007. Labeled magnitude scales for oral sensations of wetness, dryness, pleasantness and unpleasantness. Food Quality and Preference, 18, 342–52.
Harker, F.R., Amos, R.L., White, A., Petley, M.B., and Wohlers, M. 2008. Flavor differences in heterogeneous foods can be detected using repeated measures of consumer preferences. Journal of Sensory Studies, 23, 52–64.
Hauswirth, J., Sinopoli, D., and Lawless, H.T. 2010. Does loyalty dictate blind preference? Poster presented at the Society of Sensory Professionals, Napa, CA, 28 October.
Hein, K.A., Jaeger, S.R., Carr, B.T., and Delahunty, C.M. 2008. Comparison of five common acceptance and preference methods. Food Quality and Preference, 19, 651–61.
Hodgson, R.T. 2008. An examination of judge reliability at a major U. S. wine competition. Journal of Wine Economics, 3, 105–13.
Hodgson, R.T. 2009. An analysis of the concordance among 13 U.S. wine competitions. Journal of Wine Economics, 4, 1–9.
Hough, G., Bratchell, N., and Wakeling, I. 1992. Consumer preference of Dulce de Leche among students in the United Kingdom. Journal of Sensory Studies, 7, 119–32.
Hulkower, N. 2012. A mathematician meddles with medals. AAWE Working Paper, No. 97, pp. 1–9.
Huskisson, E.C. 1983. Visual analogue scales. In: Pain Measurement and Assessment. R. Melzack, (Ed.). Raven Press, New York, NY, pp. 34–7.
Jaeger, S.R. and Cardello, A.V. 2009. Direct and indirect hedonic scaling methods: a comparison of the labeled affective magnitude (LAM) scale and best-worst scaling. Food Quality and Preference, 20, 249–58.
Jaeger, S.R., Jørgensen, A.S., Aaslying, M.D., and Bredie, W.L.P. 2008. Best-worst scaling: an introduction and initial comparison with monadic rating for preference elicitation with food products. Food Quality and Preference, 19, 579–88.
Jones, L.V. and Thurstone, L.L. 1955. The psychophysics of semantics: an experimental investigation. Journal of Applied Psychology, 39, 31–6.
Jones, L.V., Peryam, D.R., and Thurstone, L.L. 1955. Development of a scale for measuring soldiers' food preferences. Food Research, 20, 512–20.
Keskitalo, K., Knaapila, A., Kallela, M., Palotie, A., Wessman, M., Sammalisto, S., Peltonen, L., Tuorila, H., and Perola, M. 2007. Sweet taste preferences are partly genetically determined: identification of a trait locus on chromosome 16[1–3]. American Journal of Clinical Nutrition, 86, 55–63.
Kim, K.-O. and O'Mahony, M. 1998. A new approach to category scales of intensity I: traditional versus rank-rating. Journal of Sensory Studies, 13, 241–9.
Köster, E.P., Couronne, T., Leon, F., Levy, C., and Marcelino, A.S. 2002. Repeatability in hedonic sensory measurement: a conceptual exploration. Food Quality and Preference, 14, 165–76.
Kroll, B.J. 1990. Evaluating rating scales for sensory testing with children. Food Technology, 44(11), 78–80, 82, 84, 86.
Lagrange, V. and Norback, J.P. 1987. Product optimization and the acceptor set size. Journal of Sensory Studies, 2, 119–36.
Lavanaka, N. and Kamen, J. 1994. Magnitude estimation of food acceptance. Journal of Food Science, 59, 1322–4.
Lawless, H.T. 1977. The pleasantness of mixtures in taste and olfaction. Sensory Processes, 1, 227–37.

Lawless, H.T. 2009. Commentary on comparative performance of the nine-point hedonic, hybrid and self-adjusting scales in the generation of internal preference maps. Food Quality and Preference, 21, 165–6.

Lawless, H.T. and Heymann, H. 2010. Sensory Evaluation of Foods, Principles and Practices, Second edition. Springer, New York, NY.

Lawless, H.T., Popper, R., and Kroll, B.J. 2010a. Comparison of the labeled magnitude (LAM) scale, an 11-point category scale and the nine-point hedonic scale in a multiproduct consumer field study. Food Quality and Preference, 21, 4–12.

Lawless, H.T., Sinopoli, D., and Chapman, K.W. 2010b. A comparison of the labeled affective magnitude scale and the nine point hedonic scale and examination of categorical behavior. Journal of Sensory Studies, 25(S1), 54–66.

Lawless, H.T., Cardello, A.V., Chapman, K.W., Lesher, L.L., Given, Z., and Schutz, H.G. 2010c. A comparison of the effectiveness of hedonic scales and end-anchor compression effects. Journal of Sensory Studies, 25, 18–34.

Lim, J. 2011. Hedonic scaling: a review of methods and theory. Food Quality and Preference, 22, 733–47.

Lim, J. and Fujimaru, T. 2010. Evaluation of the labeled hedonic scale under different experimental conditions. Food Quality and Preference, 21, 521–30.

Lim, J., Wood, A., and Green, B.G. 2009. Derivation and evaluation of a labeled hedonic scale. Chemical Senses, 34, 739–51.

Lishner, D.A., Cooter, A.B., and Zald, D.H. 2008. Addressing measurement limitations in affective rating scales: development of an empirical valence scale. Cognition and Emotion, 22, 180–92.

Luce, D. 1959. Individual Choice Behavior. John Wiley & Sons, Inc., New York, NY.

Marchisano, C., Lim, J., Cho, H.S., Suh, D.S., Jeon, S.Y., Kim, K.O., and O'Mahony, M. 2003. Consumers report preference when they should not: a cross-cultural study. Journal of Sensory Studies, 18, 487–516.

Marley, A.A.J. and Louviere, J.J. 2005. Some probabilistic models of best, worst and best–worst choices. Journal of Mathematical Psychology, 49, 464–80.

McDaniel, M.R. and Sawyer, F.M. 1981. Preference testing and sensory evaluation: magnitude estimation vs. the 9-point hedonic scale. Journal of Food Science, 46, 182–5.

Meyners, M. 2007. Easy and powerful analysis of replicated paired preference tests using the χ^2 test. Food Quality and Preference, 18, 938–48.

Moskowitz, H.R. 1980. Psychometric evaluation of food preferences. Journal of Foodservice Systems, 1, 149–67.

Moskowitz, H.R. 1986. New Directions for Product Testing and Sensory Analysis of Foods. Food and Nutrition Press, Westport, CT.

Moskowitz, H.R. and Sidel, J.L. 1971. Magnitude and hedonic scales of food acceptability. Journal of Food Science, 36, 677–80.

Mueller, S., Francis, I.L., and Lockshin, L. 2009. Comparison of best–worst and hedonic scaling for the measurement of consumer wine preferences. Australian Journal of Grape and Wine Research, 15, 1–11.

Newell, G.J. and MacFarlane, J.D. 1987. Expanded tables for multiple comparison procedures in the analysis of ranked data. Journal of Food Science, 52, 1721–5.

Odesky, S.H. 1967. Handling the neutral vote in paired comparison product testing. Journal of Marketing Research, 4, 199–201.

Pearce, J.H., Korth, B., and Warren, C.B. 1986. Evalutaion of three scaling methods for hedonics. Journal of Sensory Studies, 1, 27–46.

Peryam, D.R. and Girardot, N.F. 1952. Advanced taste test method. Food Engineering, 24, 58–61, 194.

Peryam, D.R. and Pilgrim, F.J. 1957. Hedonic scale method of measuring food preferences. Food Technology, (September), 9–14.

Prescott, J. 2009. Rating a new hedonic scale: a commentary on "Derivation and evaluation of a new hedonic scale" by Lim, Wood and Green. Chemical Senses, 34, 735–7.

Quesenberry, C.P. and Hurst, D.C. 1964. Large sample simultaneous confidence intervals for multinomial proportions. Technometrics, 6, 191–5.

Rohm, H. and Raaber, S. 1991. Hedonic spreadability optima of selected edible fats. Journal of Sensory Studies, 6, 81–8.

Schutz, H.G. 1965. A food action rating scale for measuring food acceptance. Journal of Food Science, 30, 365–74.

Schutz, H.G. 1988. Beyond preference: appropriateness as a measure of contextual acceptance. In: Food Acceptability. D.M.H. Thomson (Ed.). Elsevier, London, pp. 115–34.

Schutz, H.G. 1994. Appropriateness as a measure of the cognitive-contextual aspects of food acceptance. In: Measurement of Food Preferences. H.J.H. MacFie and D.M.H. Thomson (Eds). Chapman and Hall, pp. 25–50.

Schutz, H.G. and Cardello, A.V. 2001. A labeled affective magnitude (LAM) scale for assessing food liking/disliking. Journal of Sensory Studies, 16, 117–59.

Stone, H. and Sidel, J. 1993. Sensory Evaluation Practices. Second edition, Academic Press, New York, NY.

Villanueva, N.D.M. and Da Silva, M.A.A.P. 2009. Performance of the nine-point hedonic, hybrid and self-adjusting scales in the generation of internal preference maps. Food Quality and Preference, 20, 1–12.

Villanueva, N.D.M., Petenate, A.J., and Da Silva, M.A.A. P. 2005. Comparative performance of the hybrid hedonic scale as compared to the traditional hedonic, self-adjusting and ranking scales. Food Quality and Preference, 16, 691–703.

Wilke, K.D. Cochrane, C.-Y.C., and Chambers, E., IV. 2006. Multiple preference tests can provide more information on consumer preferences. Journal of Sensory Studies, 21, 612–25.

Wright, A.O. 2007. Comparison of hedonic, LAM, and other scaling methods to determine Warfighter visual liking of MRE packaging labels, includes web-based challenges, experiences and data. Presentation at the 7th Pangborn Sensory Science Symposium, Minneapolis, MN, 12 August. Supplement to Abstract Book/Delegate Manual.

9 Using Subjects as Their Own Controls

And Jacob said to Rebekah his mother, Behold, Esau my brother is a hairy man, and I am a smooth man. My father peradventure will feel me, and I shall seem to him as a deceiver; and I shall bring a curse upon me, and not a blessing. And his mother said unto him, upon me be thy curse, my son. Only obey my voice, and go fetch me them. And he went, and fetched, and brought them to his mother, and his mother made savoury meat, such as his father loved. And Rebekah took goodly raiment of her eldest son Esau, which were with her in the house, and put them upon Jacob her younger son and she put the skins of the kids of the goats upon his hands, and upon the smooth of his neck and she gave the savoury meat and the bread, which she had prepared, into the hand of her son Jacob.

Quantitative Sensory Analysis: Psychophysics, Models and Intelligent Design, First Edition. Harry T. Lawless.

And Jacob went near unto Isaac his father; and he felt him, and said, The voice is Jacob's voice, but the hands are the hands of Esau. And he discerned him not, because his hands were hairy, as his brother Esau's hands: so he blessed him.

Genesis 27: 11–17, 22–3 (KJV)

Strategy 1. Product 1 is served to assessor 1 and product 2 is served to assessor 2. This is a strategy with an obvious weakness. Assume, for instance, that clear differences between the two measurements are observed. Without additional information it is impossible to tell whether this is caused by differences between the assessors or differences between the products. This is called 'confounding of effects' and represents a classical mistake in experimentation.

Lea et al. (1998: 7)

Part I: Designs using Parametric Statistics

9.1 Introduction to Part I

Important considerations in intelligent experimental design are the variability that exists between people and the tendency of a single individual to behave in a consistent manner. That is, people may differ, but individuals have certain consistent styles of responding; for example, in the way they use a rating scale. Any individual's data, then, tend to be correlated depending upon these personal biases or tendencies. This presents both a problem and an opportunity for sensory analysis to deal with error variability and attempt to minimize it. Those personal tendencies can add to the error variance in a study, so any design or analysis that minimizes them or accounts for them is useful.

A decrease in error variability will lead to a stronger statistical test, all other things being equal. One can think of this as enhanced signal-to-noise ratio. Many of our statistics take the form of a ratio, such as a t-test (difference between means divided by an estimate of the standard error) or an F-test from analysis of variance (ANOVA). The F-statistic for a product difference is the product variance divided by the error variance. Lower error variance leads to a more powerful test, and thus a decrease in β risk or the probability of a **type II error**.

Type II errors are especially troublesome in sensory evaluation. They are committed when we miss a difference between products that is really present. Most of basic science is concerned with avoiding **type I error**; that is, avoiding a false alarm (i.e., rejecting the null hypothesis when it is true). This occurs when there really is no difference but we declare one anyway. In scientific research, it would be disastrous to initiate a research program on the basis of a false finding that leads one down a blind alley. So we do our best to rule out the probability that the result was due to chance alone. The normal concerns with α risk, type I errors and p-values do a reasonable job of helping avoid that situation. But in the real world of product testing, type II errors can be even more important. Missing a real difference can lead to franchise risk and loss of opportunity. **Franchise risk** occurs when we offend our loyal consumers or heavy users of our product by making a change that went unnoticed in our test, but was perceived (and possibly negatively) by these long-time users of our brand. This is sometimes called **alienation**, and is something to be avoided. Alienation is especially dangerous for longstanding consistent brands that have a loyal following. This is an important consideration in ingredient substitution, processing, or packaging changes, as

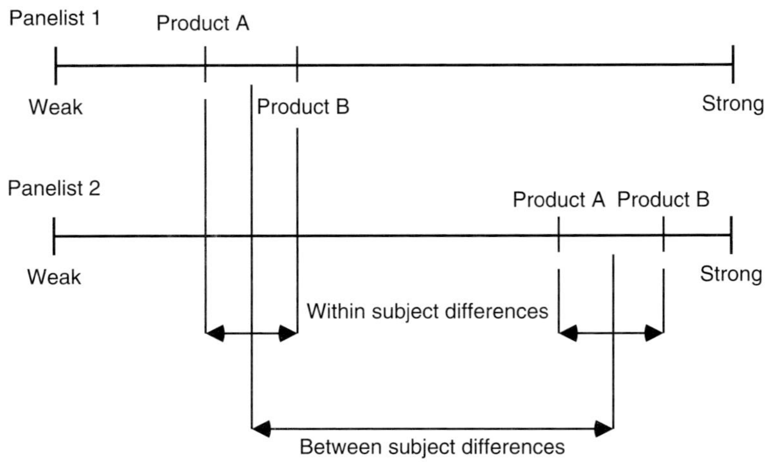

Figure 9.1 Two products rated by two judges. Both judges agree that product B is higher in this attribute than product A and by about the same amount. However, they disagree on the absolute strength of the products, perhaps because Judge 1 is less sensitive to this characteristic than Judge 2.

well as cost-reduction programs. **Opportunity risk** can occur in several situations. One unfortunate scenario is when there has been a sensory improvement in our product, but the sensory test fails to detect the change (another form of type II error). So we lose the opportunity to please or even delight the consumer with a product improvement. Opportunity risk can also involve type I error, if we have made a change in the product (e.g., a cost reduction) that would have gone unnoticed by the consumer. If, instead, our test gave a false positive result, then we could miss the chance to initiate the change in production and reap the benefits of the cost savings.

If we can separate the between-person variability from the within-person variability, we can often take the former out of the estimate of error variance. An important concept here is **partitioning** of variance. The more variance we can attribute to a specific cause, the less likely it will contribute to and appear in our estimate of overall error. As the error or unexplained variance decreases, our chances of finding an important difference are improved. The partitioning of individual variation can be very beneficial in sensory testing, because people differ in the range of the scale they like to use, or they may differ in their physiological sensitivity on an absolute basis. Consider the data in Figure 9.1 from two individuals rating two products on a line scale. For convenience, the ratings have been placed on the same lines, although in common practice the person would probably use a separate scale for each item. What do we know from these responses? It is obvious that the two persons differ, and by quite a lot. However, their ratings of the two items show some consistent patterns. Product B is always rated higher than product A. Furthermore, they differ by about the same amount on the scale. In statistical terms, there is a large between-subject variation. The within-subject variation, however, is consistent between the two persons and tells us something about the two products and how they differ.

A second important consideration is philosophical. What do we adopt as our overarching philosophy in assessing human psychological processes? One philosophical point of view is that human mental processes and sensory experiences are essentially private and unknowable. This is contrary to the psychology of introspectionism and contrary to some psychophysical theories. What we can and do observe, and what is undeniable irrefutable public data, is a person's behavior. So all we really have as data are the overt responses they produce. Certainly, we can concoct theories about their perceptions and sensations from this behavior, but it is only the overt response that we can really count on.

Consider two stimuli that produce two different responses. If I have given each stimulus to a different person, I really do not know if they had different experiences or not. The people might have given different responses due to different response strategies, response biases, a different frame of reference, or any number of other factors. If the two stimuli produced the same responses, the converse problem holds. I cannot really know if the two persons had the same experience or not. Any conclusion is based on the totally subjective responses of two individuals. Now consider the situation in which the same person is given the two stimuli in rapid succession. Our test subject changes response from stimulus one to stimulus two. They have been given close in time, so no contextual factors have changed, nor has our observer's physiological state. The person has their own biases and response tendencies, so those are not varying. We stand on much firmer ground now, if we wish to argue that the two stimuli have in fact produced different sensations. Even if we cannot know that for certain, we can still argue that the data show a change in behavior. If this change persists, for example with different stimulus orders, we can put forth a strong argument that there has been a change in sensations.

The second scenario has been called "using a subject as their own control." That is, the first response serves as a kind of baseline, and what we know for sure is that there was a difference on the second trial, one that we can quantify. Suppose that the first stimulus was a salt solution that was rated on a 15-point scale at a 5, which corresponds to a verbal anchor of "weak." Suppose the second stimulus is rated as a 10, which corresponds to "moderate." Perhaps I cannot really know whether these two stimuli are weak or moderate in any absolute sense. Perhaps this person is insensitive to salt, or just does not like to use big numbers. What I do know is that the rating changed by five units on a 15-point scale. So that change from baseline becomes a data point. It is an objective verifiable observation. There is nothing subjective about it, nor do we need to worry about whether the person had an experience that was actually weak, moderate, strong, near threshold, or whatever. Even if we do not care about the absolute subjective intensity of the experience, we have a data point that says the two things were not equal and that they changed in a certain direction and by a certain amount.

Whether you are comfortable with a psychophysical system that equates responses with sensations or not, this objectivist approach has some appeal. I may never convince my scientific colleagues that I can say one stimulus is twice as strong as another, or even that such a statement could be true with any degree of scientific certainty. But even the most hard-nosed empiricist cannot deny that the data show what the data show – that behavior was different when I changed the stimulus. Of course, one can argue that the mental, contextual, or physiological state of the individual might also have changed, as in the case of sensory adaptation, where the same stimulus produces a different response in the same person. But such phenomena are still usually open to investigation, hypothesis testing, and objectively based theories.

In the sections that follow, we will examine the statistical advantages in using subjects as their own controls in experimental designs. This chapter will deal with more or less continuous data that are analyzed by parametric statistics such as the *t*-test on means or analysis of variance (ANOVA). Part II of this chapter will examine some of the nonparametric statistical tests that can be done when there are multiple observations from the same subjects, panelists, or consumers. In that case we will deal with frequency counts of responses from the same people, or counting responses from the same groups of individuals tested twice or more. This chapter assumes that the reader has some familiarity with basic statistics, *t*-tests and ANOVA. For a refresher, you can review the materials in Lawless and Heymann (2010: appendices A and C) or in the basic statistics book for sensory evaluation by O'Mahony (1986). A comprehensive statistical text for sensory and consumer data is the updated book from Gacula et al. (2009). A specific treatment of ANOVA for sensory data is found in Lea et al. (1998) as well as Naes et al. (2010), who deal extensively with the issue of adjusting for scale usage differences among panelists.

9.2 Dependent Versus Independent *t*-Tests

An example of the dilemma discussed above arises in the rated **degree of difference (DOD)** test. In this test a pair of items is rated on a scale, such as a line scale or category scale, usually anchored with phrases like "exactly the same" and "completely different" at opposite ends of the scale. Suppose we give a person a pair of items that includes a control product and a new or altered test product. They choose a rating category or make a mark on a line scale to indicate the perceived degree of difference. But what does that mark mean? How much of a difference is meaningful? Two items will rarely be perceived as identical, even if they are physically the same or from the same batch of product. So there is almost always a nonzero difference rating. On its own, that single rating is virtually meaningless!

But the way out of this jam is to provide the judge with a control pair of identical samples as well as the test pair. Now we have a baseline, or a false-alarm rate in the terms of signal detection theory (see Chapter 3). The rating for the test items can now be compared with the rating given for the control pair. When both kinds of pairs are given to the same judge, we have an intelligent design, and the data can be treated by a paired or dependent *t*-test. The person's own criterion for how much a difference matches the scale category or distance on the line scale is taken out of the picture, as we have a difference score to deal with from the two trials – a degree of difference in two difference ratings, not all that complicated.

Probably the most common statistical test on two products when there are scaled data is the *t*-test. Student's *t*-test was named after William Sealy Gosset, who published under the pseudonym "Student" while working on barley varieties for the Guinness breweries. The test has several forms, all used to test the differences between two mean values with a small sample of observations. In setting up any sensory experiment, we have the choice of making our measurements on different groups of people, or making measurements on the same individuals but multiple times. In this section we will compare the results from data sets that are numerically similar, but originate from these two different experimental designs. The examples will show that when the data are correlated (i.e., when each individual's responses are related to the others when they are asked to evaluate two or more items), this correlation will provide an advantage in the statistical analysis and the sensitivity of the test to differences. Section 9.3 will consider the same issue, but from the situation when there are three or more items, and ANOVA is used instead.

The general formula for a *t*-test is that *t* is equal to the difference between means divided by the standard error of the mean(s). The standard error of the mean is the standard deviation divided by the square root of *N*, the number of judges or observations. The exact numerical calculations for the *t*-statistic are a little different depending upon whether we are comparing the mean value with a population value or standard, whether we are comparing two independent groups, or whether we have the two means from two products evaluated by the same group of individuals. The latter is sometimes called a monadic sequential design, whereas having the products viewed by different groups is called a monadic design in consumer testing (Lawless & Heymann, 2010).

The type of t-test that is conducted when there are different groups is often called an **independent groups *t*-test**. Sometimes the experimental constraints might dictate a situation where we have two groups which taste only one product each. The data are not paired or related in any way. A set of calculations is needed to estimate the standard error for the denominator of the t-test, since two groups were involved and they have to be combined somehow to get a common or pooled estimate of the standard deviations. We also have degrees of freedom given by the sum of the two group sizes minus 2, or $(N_{\text{Group1}} + N_{\text{Group2}} - 2)$. The t-value is given by

$$t = \frac{M_1 - M_2}{SE_{\text{pooled}}} \tag{9.1}$$

where M_1 and M_2 are the means of the two groups and $\text{SE}_{\text{pooled}}$ is the pooled standard error. For the independent t-test, the pooled error requires some extra work and gives an estimate of the error combining the error levels of the two groups. Recall that the standard deviation is given by

$$S = \sqrt{\frac{\sum_{i=1}^{N}(X_i - M)^2}{N-1}} \tag{9.2}$$

or its computational equivalent

$$S = \sqrt{\frac{\sum_{i=1}^{N} X_i^2 - \frac{(\sum X)^2}{N}}{N-1}} \tag{9.3}$$

and that the standard error is the standard deviation divided by the square root of N. To get this pooled estimate, the following formula applies:

$$\text{SE}_{\text{pooled}} = \sqrt{\frac{S_1^2}{N_1} + \frac{S_2^2}{N_2}} \tag{9.4}$$

where the subscripts refer to the two groups. A shortcut method (or at least computationally equivalent) is given by the following formula:

$$\text{SE}_{\text{pooled}} = \sqrt{\left(\frac{1}{N_1} + \frac{1}{N_2}\right) \frac{\sum x^2 - \frac{\left(\sum x\right)^2}{N_1} + \sum y^2 - \frac{\left(\sum y\right)^2}{N_2}}{N_1 + N_2 - 1}} \tag{9.5}$$

where x and y refer to the data points from the two groups. These pooled estimates will be used in the example that follows.

Here is a worked example of an independent groups t-test. There are two products evaluated by two different groups of 12 judges each. The raw scores are given in the first two columns of Table 9.1.

First, we should ask what the data are telling us. Always look over your data. Product B seems to get higher ratings than product A, and on the average they differ by one scale point.

Table 9.1 Hypothetical data set for ratings of two products by two groups of 12 judges each

Product A	**Product B**	**A²**	**B²**	
3	5	9	25	
3	5	9	25	
4	6	16	36	
5	4	25	16	
4	6	16	36	
5	7	25	49	
6	8	36	64	
5	6	25	36	
4	3	16	9	
4	6	16	36	
5	5	25	25	
4	4	16	16	
52	65	234	373	Sum
4.3333	5.4167			Mean
0.8876	1.3790	0.7879	1.9015	SD-variance
12	12			*N*
meandiff	−1.0833			
t	−2.2884			

The standard deviations are reasonably small (0.89 and 1.38), with the data centered on a value of about 5.0. Is this pattern different and consistent enough to give us a significant difference between the means? The critical value for t for 20 degrees of freedom is 2.09. Because the absolute value of our t-value obtained is greater than the critical value, we have a significant difference between the means at a p-value less than 0.05.

Note that the last sets of four judges (the last two rows of the table) gave similar ratings to both products, either a 4 or a 5. These four people do not seem to be able to differentiate the products. They are not transmitting any information of any real value. Let us substitute four new values and see what happens to the data. Let us give product A ratings of 1 and 7, and product B values of 2 and 8. You might think this will help find a difference, because now we have two sets of observations, which, if they were paired, would differ by one scale point (1 versus 2, and 7 versus 8) just like the means. However, because these sets of judges were either very low or very high on the scale, they have increased the standard deviations and the error variance of both groups. So even though our mean difference has been reinforced, the effect on the overall variance is such that it *decreases* the value of the t-statistic. Remember, as the standard deviations go up, the denominator goes up, resulting in a smaller t-value. The data and analysis is shown in Table 9.2.

Because of the added variability from having judges who are high and low on the scale, the difference is no longer significant, with a t-value of 1.68, even though these four people are now showing a pattern of product B being higher than product A. But suppose these values had been contributed by paired observations from the same group of 12 judges? The next analysis shows a drastic difference in the outcome once we have paired observations.

Another kind of t-test is the test of paired observations, also called the dependent t-test. To calculate this value of t, we first arrange the pairs of observations in two columns and then subtract each one from the other member of the pair to create a difference score. The difference scores then become the numbers used in further calculations. The null hypothesis is that the mean of the difference scores is zero. We also need to calculate a standard

Table 9.2 Hypothetical data set for ratings of two products by two groups of 12 judges each

Product A	Product B	A²	B²	
3	5	9	25	
3	5	9	25	
4	6	16	36	
5	4	25	16	
4	6	16	36	
5	7	25	49	
6	8	36	64	
5	6	25	36	
4	3	16	9	
4	6	16	36	
1	2	1	4	
7	8	49	64	
51	66	243	400	Sum
4.2500	5.5000			Mean
1.5448	1.8340	2.3864	3.3636	SD-variance
12	12			*N*
meandiff	−1.2500			
t	−1.8058			

deviation of these difference scores, and a standard error by dividing this standard deviation by the square root of *N*, the number of panelists:

$$t = \frac{M_{\text{diff}}}{S_{\text{diff}} / \sqrt{N}} \tag{9.6}$$

where M_{diff} is the mean of the difference scores and S_{diff} is the standard deviation of the difference scores.

Let us take the data from Table 9.2 and now treat it as repeated judgments from 12 judges. This is an example of a design in which each panelist tasted both products, and we can perform a **dependent *t*-test**, also called a paired *t*-test. In the following data set, there is a fair degree of variance in the actual ratings being given by subjects. This might be due to individual differences in sensitivity, as would be found with bitter tastes, for example. On the other hand, most panelists agreed that one product was stronger in this attribute and, therefore, rated higher than the other. This kind of situation is very common in sensory work, and often leads to significant paired *t*-test results. However, if the data were treated with the between-groups *t*-test, the resulting *t* would be lower that that necessary to reject the null hypothesis, and no statistically significant difference would be reported, as we saw in Table 9.2. Table 9.3 shows the new analysis using difference scores.

Note that there are fewer degrees of freedom than in the independent groups *t*-test because our calculations are now based on only 12 difference scores. So $N = 12$ and the degrees of freedom $df = 11$. The critical *t*-value is a little higher, at 2.20. However, owing to the consistent pattern in the difference scores, there is now a highly significant result with a *t*-value of 3.022. Referring back to Figure 9.1, we have taken the between-subject difference out of the calculations and are left with only the scores reflecting the product differences, which show a consistent pattern. Figure 9.2 shows the pattern of correlation that "helps" the *t*-test find a significant difference.

Another way to look at this is to examine the consistency of the changes. Ten of the twelve judges raised their ratings for Product B versus A, and only two went in the other direction. So it is not only the pattern of correlation that is important, but the consistency of

Table 9.3 Hypothetical data set for two products rated by a panel of 12 judges

Product A	Product B	Difference D	D²	
3	5	−2	4	
3	5	−2	4	
4	6	−2	4	
5	4	1	1	
4	6	−2	4	
5	7	−2	4	
6	8	−2	4	
5	6	−1	1	
4	3	1	1	
4	6	−2	4	
1	2	−1	1	
7	8	−1	1	
51	66	−15	33	Sum
4.2500	5.5000			Mean
		1.1382		SD
12	12			*N*
meandiff	−1.2500			
t	−3.8044			

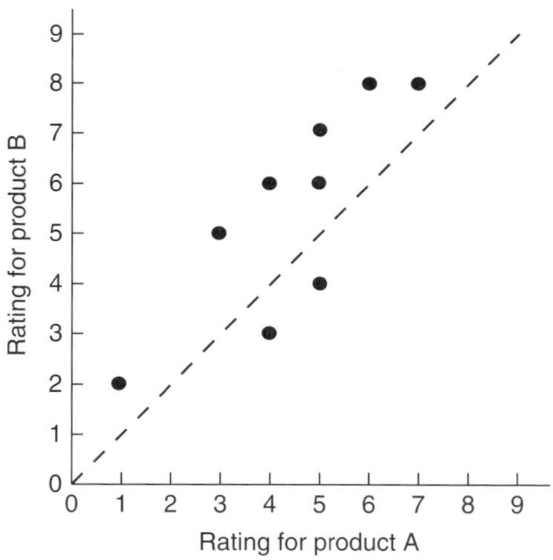

Figure 9.2 Scatterplot showing the pattern of correlation for the dependent *t*-test on the data from Table 9.3. Points above the dashed line show ratings that were higher for product B. Points below the line show ratings that were higher for product A. The correlation and the consistent pattern of change both help create a significant *t*-test in the dependent or paired *t*-test design.

the direction of the change that will yield a significant difference in the paired or dependent *t*-test. Note that 10 of the 12 judges fall above the diagonal dashed line ($y = x$) in Figure 9.2. In this case, having the paired or repeated design provides an advantage in cancelling out the inter-subject variability and helping us see the pattern of differences perceived by the judges. In a real product development scenario, we might have avoided missing an important difference (avoided type II error) and avoided potential franchise or opportunity risks.

Section 9.3 will examine a similar situation in a three-product test and show how partitioning the panelist variance will improve the error term in ANOVA.

Table 9.4 Data set and calculations for a one-way (between groups, monadic design) ANOVA*

Judge	Product A	Product B	Product C	A^2	B^2	C^2
1	6	8	9	36	64	81
2	6	7	8	36	49	64
3	7	10	12	49	100	144
4	5	6	6	25	36	36
5	6	5	7	36	25	49
6	5	6	9	25	36	81
7	7	7	8	49	49	64
8	4	6	8	16	36	64
9	7	6	5	49	36	25
10	**8**	**8**	**8**	64	64	64
Sum	61	69	80	385	495	672
SUM^2	3721	4761	6400			
N	30		Sum X^2	1552		
SST	82.00		Total *T*	210		
NP	10		T^2/N	1470		
SSP	18.20					
SSE	63.80					
Source	SS	df	MS	*F*		
Product	18.200	2	9.100	3.851		
Error	63.800	27	2.363			

*SST: total sums of squares; SSP: product sums of squares; SSE: sums of squares for error, which are the total minus the product sums of squares; MS: mean squares or variance estimates.

9.3 Within-Subjects ANOVA ("Repeated Measures")

As in the case of the *t*-test discussed in Section 9.2, we will start with an analysis based on the experiment in which different groups of judges evaluate each product. That is, we have a monadic design, sometimes called a between-groups or between-subjects experiment. Table 9.4 shows the data from 10 judges for each of three groups, each evaluating one product. This is the classic one-way ANOVA, without having panelists evaluate more than one product. Obviously, this is not a very efficient experiment, but it might be necessary in some cases where the product was very fatiguing or had a lot of carry-over, aftertaste, or some other reason that would preclude multiple tastings.

The data set shows that the ratings from these groups tend to increase from product A to B to C, and the ANOVA results help confirm this pattern. Once again, it is assumed that the reader is familiar with basic ANOVA. The ratio of $MS_{products}$ to MS_{error} gives the *F*-value of 3.85. At 2 and 29 degrees of freedom, this exceeds the tabled critical value of 3.36, so there is a significant difference between at least two of the product means.

Next, let us alter the data set slightly. The final row of the data set shows values of 8, 8, and 8, which does not contribute to the overall pattern of product differences noted in the mean values. In the data set of Table 9.5, the values of 1, 2, and 3 have been substituted. Note that these are consistent with the overall pattern of increasing values for the three products from A to C, but are somewhat low on the scale compared with the overall mean value. Table 9.5 shows the ANOVA calculations for this data set, once again using a simple one-way analysis as before, as if we had three separate groups of judges, one group testing each product.

Unfortunately, the substitution does not help the ANOVA find any significant trend, even though the data are more consistent with that trend than the original data set with its constant

Table 9.5 Data set and calculations for a second one-way (between groups, monadic design) ANOVA with the data from Table 9.4 altered in the bottom row

Judge	Product A	Product B	Product C	A^2	B^2	C^2
1	6	8	9	36	64	81
2	6	7	8	36	49	64
3	7	10	12	49	100	144
4	5	6	6	25	36	36
5	6	5	7	36	25	49
6	5	6	9	25	36	81
7	7	7	8	49	49	64
8	4	6	8	16	36	64
9	7	6	5	49	36	25
10	**1**	**2**	**3**	1	4	9
Sum	54	63	75	322	435	617
SUM^2	2916	3969	5625			
N	30		Sum X^2	1374		
SST	145.20		Total T	192		
NP	10		T^2/N	1228.8		
SSP	22.20					
SSE	123.00					
Source	SS	df	MS	F		
Product	22.200	2	11.100	2.437		
Error	123.000	27	4.556			

values in the bottom row. The F-value obtained (2.437) is now below the critical value of 3.36, and we can no longer reject the null hypothesis of any difference among the three mean values. This is because the values that were substituted were low on the scale. So, even though they have helped the product sums of squares a little, they have added a lot to the pooled error term.

Table 9.6 shows what happens when a two-way ANOVA is performed so that we can partition out the judge variability. To do this, we accumulate sums across rows and square those values, just as we accumulated totals down columns for the product variance estimates. The row totals are used to provide an estimate of the judge main effect, which is how much the judges differ in their overall levels on the scale. Once we calculate it, the judge effect can be subtracted from the error term. Anything that makes the error term smaller will improve the F-value we obtain for the product effect, because the denominator of the F-ratio is the error variance estimate.

Notice that because the error term is smaller, the F-ratio is larger (9.95!) and now the result shows a significant difference. So the two-way or repeated measures ANOVA is not hurt by the low outlier amongst the judges, as long as the trend for that judge is parallel to the overall trend in the data set. You can think of the partitioning out of the judge effect as setting all the judges to the same mean value. The only difference we have left is a difference in slope, by which we mean the overall trend across products if it is plotted as a line graph. But there is no free lunch. By estimating the judge effects and taking them out of the error, there is a cost of a few degrees of freedom. Had we set all the judges to the same mean value, we would no longer allow their overall height of level to vary, so there are degrees of freedom lost. But the loss of degrees of freedom in the error term is not so great a penalty as to counteract the decrease in error variance, which is substantial.

Figure 9.3 shows the judge trends across the three products as a line graph. Judge number 10 is the low outlier but has a consistent trend with most of the other judges. The overall

Table 9.6 Data set and calculations for a two-way (within subjects, monadic – sequential design or "repeated measures") ANOVA with the same data from Table 9.5

Judge	Product A	Product B	Product C	A^2	B^2	C^2	Judge sum	$Jsum^2$
1	6	8	9	36	64	81	23	529
2	6	7	8	36	49	64	21	441
3	7	10	12	49	100	144	29	841
4	5	6	6	25	36	36	17	289
5	6	5	7	36	25	49	18	324
6	5	6	9	25	36	81	20	400
7	7	7	8	49	49	64	22	484
8	4	6	8	16	36	64	18	324
9	7	6	5	49	36	25	18	324
10	1	2	3	1	4	9	6	36
Sum	54	63	75	322	435	617	192	3992
SUM^2	2916	3969	5625					
N	30		Sum X^2	1374		SSJudges	101.87	
SST	145.20		Total *T*	192				
NP	10		T^2/N	1228.8				
SSP	22.20							
SSE	21.13							
Source	SS	df	MS	*F*				
Product	22.200	2	11.100	9.954				
Panelists	101.870	9						
Error	21.133	18	1.174					

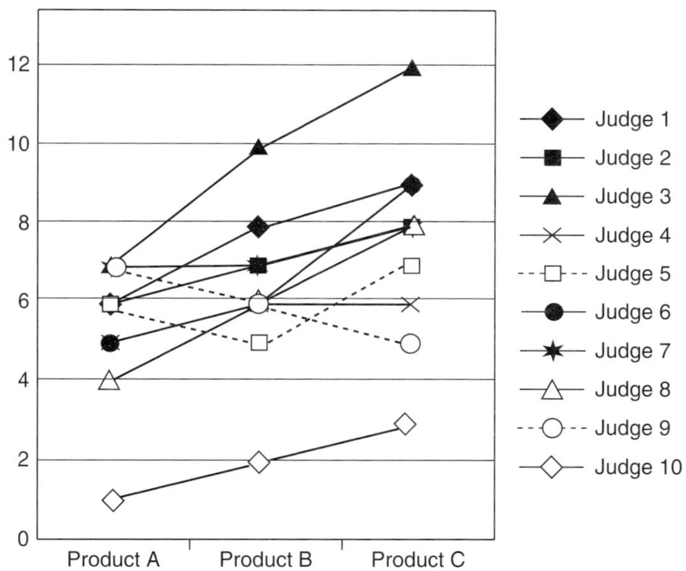

Figure 9.3 The overall trends for the 10 judges in Table 9.6 plotted as a line graph to see the trends. Most judges were increasing from product A to product B to product C. This was also true of the low outlier at the bottom of the graph, Judge number 10. So the overall variance is increased by this judge (as opposed to the values of 8, 8, and 8 in the first data set), but because the trend, as shown by the slope, is consistent with the group, the low values do not matter to the two-way ANOVA, which partitions away the overall height of the trendlines, as if they were all set to the same mean value.

variance is increased, but the two-way ANOVA takes out the overall height differences among the judges, thus negating any negative effect the low outlier would have on the estimation of error, and therefore any negative effect on the product differences.

9.4 Issues

9.4.1 A Word on Using the Correct Error Term in the *F*-Test

Note that the residual error term in the two-way ANOVA is only tapping into differences in slope, as it is portrayed in Figure 9.3. Another term for this is that the residual variance consists of the interaction effect of panelists (judges) with the product effect. An interaction is a change within a change or the effect of one variable modifying the effect of a second variable. In this case there is the product-related change (C > B > A), and then the individual panelist modification of that trend. As a general rule, the test for any main effect is that variable's MS (mean squares, our variance estimate) divided by the error estimate. The error estimate is the interaction effect of that source with the panelists or judges.

This concept is straightforward in the above example because, after we partition out the product effect and judge effect, all that is left over for error is the judge-by-product interaction term. But suppose there were three or more variables in the design? A common design is one where there is a product test, replication, and judges. So now there are three main effects, three two-way interactions and one three-way interaction. One could easily add other factors, like time, giving a four-factor design. No matter what the design, the correct error term consists of the interaction term of that factor with judges. So the product *F*-test is the product MS divided by the product-by-judge interaction MS. The replication *F*-test is the replicates MS divided by the MS for the replication by judge interaction.

The reason for this choice of terms is that the *F*-ratio denominator and numerator must only differ in their variance components by the one variance source in question. That is, the numerator must contain sources of all the variance except for the one being tested. Let us say that is the product effect. The product variance has embedded in it both the judge-by-product interaction source and the residual error source; in other words

$$\text{MS}_{\text{prod}} = \text{Var}_{\text{prod}} + \text{Var}_{\text{prod}\times\text{judges}} + \text{Var}_{\text{error}} \tag{9.7}$$

Another way to think about this is that the product data points arise due to the change in the products and also the specific ways each panelist evaluates those changes. So the denominator has to consist of the following:

$$\text{MS}_{\text{prod}\times\text{judges}} = \text{Var}_{\text{prod}\times\text{judges}} + \text{Var}_{\text{error}} \tag{9.8}$$

where Var_{X} are the theoretical variance components feeding into the mean square obtained. In order to construct a proper *F*-ratio, the numerator must only differ by the theoretical variance component of interest in the test; so, for the product test we need to use the $\text{MS}_{\text{prod}\times\text{judge}}$ interaction term, so that

$$F = \frac{\text{MS}_{\text{prod}}}{\text{MS}_{\text{prod}\times\text{judges}}} = \frac{\text{Var}_{\text{prod}} + \text{Var}_{\text{prod}\times\text{judges}} + \text{Var}_{\text{error}}}{\text{Var}_{\text{prod}\times\text{judges}} + \text{Var}_{\text{error}}} \tag{9.9}$$

Note that VAR_{prod} is the only difference in the two parts of the ratio and, thus, this is a legitimate *F*-test.

It is altogether commonplace, especially in fixed-effect designs, to use the smallest within-cell error term as the divisor, which consists only of Var_{error}. For the model given above, this would be incorrect. The fact that the interaction term is embedded in the product effect is dictated by the mixed-effects model, in which the product effect is a fixed effect but the judge effect is a random effect. What is the difference between a fixed and random effect? The products were chosen to represent specific items (say three concentrations of a flavor), rather than being a random sample of a larger population, for example three random choices from a large bank or library of flavor components. Judges, on the other hand, are a random sample of the population of all such judges (trained or not) and we wish to generalize our results to all such samples and the general population from which they were chosen. Note that it does not matter whether or not the judges were trained, they still represent a sample of that larger population. They are not any specific fixed values like 4, 5, and 6% sucrose in a beverage.

So the reader should use caution in setting up the *F*-test and be certain to use the correct error term for a mixed model. For further details on fixed versus random effects models, see Lawless and Heymann (2010), Lea et al. (1998), or Winer (1971). Winer's book is an excellent introduction to ANOVA and explains the variance component models used for correct *F*-tests.

One alternative line of thought in ANOVA modeling is to test whether the product by panelist interaction is statistically significant first, and then drop it from the model if it is not significant. This would theoretically allow you to use the smallest within-cell error term in a multifactor ANOVA, and skip having to worry about the interaction term as the error divisor. However, this is a risky approach. You are basing your choice on a nonsignificant effect. Nonsignificance means a failure to reject the null hypothesis, which is usually ambiguous. So it is safer to use the correct (interaction) error term unless you have some very good reasons for dropping the judge-by-product interaction from your model.

9.4.2 Limitations to Repeated Measures Models

The situation in which there are judges who are high or low on a scale (compared with a panel mean) is a common occurrence in descriptive analysis and almost ubiquitous in consumers' use of scales. The examples in Sections 9.2 and 9.3 show how the situation can be dealt with statistically by having an intelligent experimental design and the correct statistical test. However, in the case of ANOVA, the situation is a little more complicated. One assumption of the repeated-measures mixed-model ANOVA is that the pattern of covariance is equal or homogeneous. Suppose the design had five time periods in which the products were evaluated, and the time variable was a factor in the design and analysis. The repeated measures model depends upon the assumption that the responses from first time block are correlated (or not) with the second time block *to the same degree* as the first and last time blocks. That is, the pattern of covariance must be homogeneous.

This is clearly unrealistic for some repeated measures like the time block example given above, or any set of replications that extends over multiple sessions. There are several classic approaches to dealing with the covariance problem. One is to estimate the degree to which the assumption of homogeneity of covariance is violated and then make some adjustments in the ANOVA for this error. Most statistical programs will do this, using adjustments named for their statistical inventors, the Greenhouse–Geisser adjustment or the Huynh–Feldt (Greenhouse & Geisser, 1959; Huynh & Feltd, 1976). Both of these techniques adjust your degrees of freedom in a conservative manner to try to account for a violation of the assumptions and yet protect you from making a type I error. The corrections are shown by an "epsilon" value shown in the ANOVA results, and adjusted *p*-values, often abbreviated G–G or H–F. Another solution is to use a multivariate ANOVA (MANOVA) approach, which does not labor under the heterogeneity of covariance assumption of repeated measures. One can then compare the results of the repeated measures and MANOVA results to see whether they draw the same conclusion, which is common.

Regardless of this problem, there is still great value of partitioning out the judge effect or taking the judge's scale usage tendencies out of the picture. A different approach to scale usage differences is to use an individual fudge factor to bring each of the judges into the same scale range, mathematically setting them at the same overall mean value. Such an adjustment was discussed for magnitude estimation data in Chapter 2. A further possible adjustment is a scaling factor that takes into account the range of values used by that judge (and not just the mean).

This is functionally equivalent to also setting their slope values to be the same if we plotted the product trends as a line graph and if all judges were ranking the products the same way. Another alternative is to make an adjustment based on the overall standard deviation of that panelist for each attribute evaluated. Various methods of this type are discussed in Naes et al. (2010: section 4.2). It is not clear whether such adjustments are valid or necessary when one has already partitioned the panelist effect, or one could simply convert the data to ranks and do a Friedman test. Of course, statistical partitioning and numerical fudge factors are two approaches to try to get the panel data to be more orderly or homogeneous. Screening and training a panel is the other side of the coin, and is an up-front strategy that should be considered in any descriptive analysis or other trained panel situation such as a quality control panel.

Part II: Nonparametric Statistics

9.5 Introduction to Part II

As we saw in Part I, there are often advantages to a within-subjects design in terms of the statistical advantages it brings and the enhanced ability to avoid missing a difference and committing type II errors. Part I of this chapter dealt with common statistics used for continuous data, such as an intensity scale. In Part II we will discuss a few common situations in which a person views two or more products and makes a response from some set of limited choices, as opposed to a line scale or magnitude estimation in which the choices are, at least in theory, unlimited. Thus, the data in this chapter will consist of frequency counts in a table, rather than a quantitatively measured data point from each product and every individual. If each product is viewed by a different individual, the common statistical test would be a χ^2 test (for two or more response options), its binomial equivalent or the *Z*-score approximation to the binomial, the latter two being used when there are only two choices.

However, the everyday χ^2 test assumes that the data for each product are independent; that is, they are not assumed to come from the same participants. Most panel tests or consumer tests would not simply give one product to each of the participants; it would be a wasteful experimental design in terms of resources. So a common practice is to use monadic-sequential tests, in which each person views both (or all) of the products. Fortunately, we have a number of statistical tests that can be used with multiple observations from the same people. In the remainder of this chapter we will look at a few common examples, and suggest approaches that may be potentially more powerful than simply treating the data (i.e., falsely) as if they had been independent observations. Further examples can be found in Chapter 10 on check-all-that-apply (CATA) methods.

These methods consist primarily of the Stuart–Maxwell test for pairs of products (Stuart, 1955). For multiple products, repeated designs, or extra variables there are extensions and alternatives such as the Cochran–Mantel–Haenszel test (Fritz, 2009). A special case of the Stuart test is the McNemar test, when there are only two possible responses. This test is used when the data are cast into a 2×2 matrix, with each panelist counted in only one cell. The Stuart tests are not often taught in basic sensory evaluation training, so they are presented here with several examples and applications.

9.6 Applications of the McNemar Test: A–not-A and Same–Different Methods

Two difference tests that fit the conditions for a within-subjects or paired samples design are the **A–not-A** test and the **same–different test**. The A–not-A test presents one product at a time. Usually, there is some inspection period in which the panelist familiarizes themselves with the standard product; that is, the one designated as "A." The panelist or judge must respond with either "A" if they think the test item is an example of the standard, or "not A" if they think it does not match or is perceived to be something different. In the same–different test, pairs of products are presented, one pair at a time, and the judge must respond "same" or "different."

Note that, in both kinds of tests, if only one judgment is taken from each person, the test is not bias free. That is, the person in the test must assess the sensory properties relative to some criterion they set for how much of a difference will have to appear to say "not A" or "different." This is sometimes called a τ criterion (see Chapter 4 for further details). However, the situation becomes a bit clearer if a control product or an identical pair is given. In the case of the A–not-A test, then, the test would also include bona fide examples of the control or standard product (i.e., a true example of A) in the test phase. Similarly, in the same–different test, a control pair of identical products (or products from the same batch) can be given. This provides the opportunity for a comparison against a baseline. In signal detection terms, we can now estimate a false-alarm rate (see Chapter 3). The same situation applies to the rated DOD test, a scaled version of the same–different test. In the DOD test, it is preferable to provide ratings to a control pair of identical products. Those ratings then provide a baseline for a t-test comparing the mean ratings of the test pair with the mean ratings of the control pair.

A common design for these tests would be to present the control item (in A–not-A) or control pair (in same–different) to the same panelists that are evaluating the test item or test pair. Thus, there is a set of paired observations from the same individual. In the case of the rating scale DOD test, this permits a paired (or dependent) t-test, as discussed in Part I of this chapter. When we are counting responses, rather than using scaled or continuous data, there are nonparametric equivalents to provide us with comparisons of the paired observations from the same individual. A good example is the McNemar test, also called the McNemar test for the significance of changes (Siegel, 1956). The test is well suited to before-and-after test designs where a person is providing data after some treatment. An example would be voting preferences before and after a campaign advertisement.

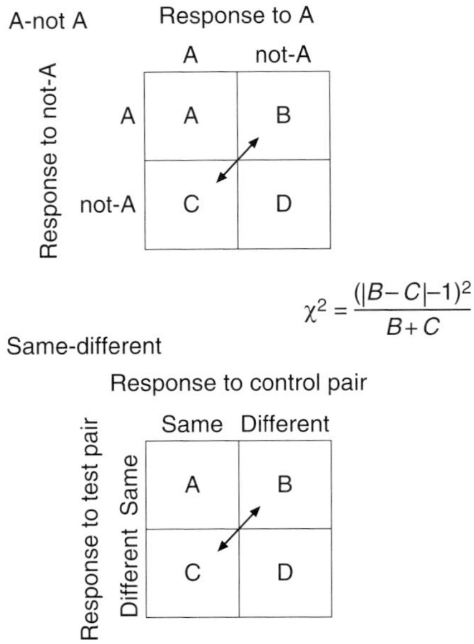

Figure 9.4 Statistical analysis using the McNemar analysis for the A–not-A test and the same–different test.

$$\chi^2 = \frac{\left(|B-C|-1\right)^2}{B+C} \tag{9.10}$$

Figure 9.4 shows the setup for the McNemar test for both the A–not-A test and the same–different test and the formula is shown in eqn 9.10. This is a χ^2 distributed statistic with one degree of freedom, so the critical value is 3.84. The critical cells are designated as *B* and *C*, where the response was different on the two trials.[1] Note that the people who give the same response on both trials are effectively discounted from the analysis. Some might say that is a legitimate choice in data handling, because they have not transmitted any information. In the case of the A–not-A test, they have experienced both samples as examples of the standard, or neither sample as an example of the standard. The McNemar test statistically rephrases the important question as "Did more people see the A samples as A, *and* the not-A samples as not-A, than the reverse?" The reverse, of course, would be to call the not-A sample "A" and the A sample "not-A," erroneously. In a similar fashion, the McNemar test is looking at the same–different pairs and asking, "Were there more people who called the control pair 'same' and the test pair 'different' than the reverse?

A sensory professional should look carefully at this situation and ask whether discounting the nonchanging votes is losing important information or not. There are several ways to view this. One position is that these are important data, even though they seem to be part of the background noise. The data suggest that there is a group of people who truly cannot tell the difference. That in itself could be relevant information in the decision process. However, bear in mind that, in any consumer test, there are likely to be about 25% of the participants who cannot distinguish even an obvious difference. An example is the inability to discriminate skim milk from milk with 2% fat, an obvious visible difference to most people (Chapman & Lawless, 2005). An ongoing argument is whether such non-discriminators should be

[1] This is mathematically equivalent to a sign test on the frequency counts in cells *B* and *C*.

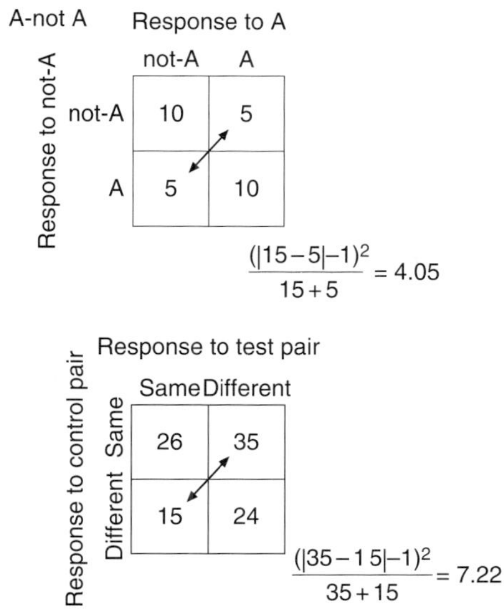

Figure 9.5 Worked examples of the McNemar analysis for the A–not-A test and the same–different test.

discounted (or screened out) in a simple difference test. Another view of the nonchanging responders is that they are not telling you anything important. The product samples might not be different after all, or these individuals might be a sampling of just those people with "tin tongues" who cannot tell the difference between any samples whatsoever. So, there is a strategic choice involved here in adopting the McNemar approach, as is true of the Stuart tests in general. In matrix terms, we are dealing with the pattern in the off-diagonal cells, and ignoring what happens along the diagonal cells that indicate no change.

Figure 9.5 shows two worked examples from the A–not-A and same–different designs. In the first case, there were 40 judges in the A–not-A test and the critical cells showed that 15 of them distinguished the A sample as "A" and the not-A samples as "not-A" as opposed to five who did the reverse. With only 40 testers, and 20 in the off-diagonal or changing cells, this was just enough to support a conclusion of a significant (i.e., perceivable) difference. In the second example, 100 consumers were given a same–different test. Thirty-five "correctly" called the identical pair "same" and the test pair "different," with 15 showing the reverse pattern. Note that with the bigger *N* compared with the first example, we no longer need a three-to-one margin to obtain significance. The sensory professional should also note that, in this example, 50% of the test group did not show any difference in response, suggesting that although the difference appears perceivable to some people, not everyone gets it and it is likely that the difference is small.[2]

What would happen if we used the commonplace χ^2 test for independent samples and, thus, used all four cells in the design? Owing to the substantial proportion of judges in the nonchanging diagonal cells, there is no significant effect (from a simple χ^2 test) in either example in Figure 9.5. Not even close! In these cases, looking for the difference in *changes* was a valuable insight that might help us avoid type II error, and the risk of missing a potentially important perceivable difference, at least to some of the population.

[2] In both of these examples, an alternative analysis would be to subject them to a Thurstonian model, and determine a d' value and its variance estimate. This value then could be tested against zero, statistically. A nonzero d' would be another legitimate test for statistical significance.

Such a difference might be a negative product change that we would want to know about to avoid consumer alienation. Or it could be a positive product change that might potentially bring better customer satisfaction or a larger consumer base into the franchise. Of course, the difference test does not tell us about that, only that the change is likely to be perceived by some, albeit small, percentage of panelists or consumers.

9.7 Examples of the Stuart–Maxwell

9.7.1 A Quick Look at the Stuart Test

In Section 9.6 the simple 2×2 design with two products and two categories of responses was shown to fit the McNemar test for changes. However, many common sensory tests have more than two response categories or more than two products. If there are two response categories (i.e., binary data), and more than two products, the Cochran Q statistic is appropriate and is discussed in many basic statistical texts. In this section, examples of the Stuart test and the Stuart–Maxwell statistic will be shown, which deals with the situation when there are two products to compare but more than two response categories. The Stuart–Maxwell statistic is distributed as a χ^2 statistic. The calculations do not assume any order to the categories. This could be a shortcoming if you have a rating scale such as the degree of difference scale. Why ignore the fact that the response categories possess meaningful rank orders? Extending the Stuart test for ordered categories will be discussed at the end of this chapter. The Stuart test is also known as a test of marginal homogeneity.

Looking back at the McNemar test, we can see that it derives its horsepower from cases in which the two cells representing changes in response had different frequency counts. The cells in which subjects do not change their responses do not even enter the picture. The Stuart test is somewhat different, in that all cells will contribute to the calculations. However, as in the case of the McNemar, it will tend toward statistical significance if the cells n_{ij} and n_{ji} differ. In this notation, i and j refer to the ith and jth response categories for the two products, respectively, and the data will be entered into a symmetric matrix with n_{ij} representing the frequency counts in a given cell and with $N_{.j}$ and $N_{i.}$ representing the row and column totals, respectively. That is, the period symbol will represent a sum across that variable. Consider a 3×3 matrix. The Stuart test will calculate some differences between $N_{.j}$ and $N_{i.}$. Let us take the first row and first column. Part of the calculation will be a difference score, D_1, taken from the difference of $N_{.1}$ and $N_{1.}$. For the cell n_{11} the contribution is the same for both marginal sums – it is in both the first row and the first column. But if cell n_{12} differs from cell n_{21}, then their marginal totals in that row and that column will also differ. So the Stuart test gains strength when the off-diagonal cells with opposite subscripts are different, as shown in Figure 9.6.

Here is the formal mathematical statement. The general form for the Stuart–Maxwell statistic is given by the test statistic S, in the notation of Best and Rayner (2001), and is a special case of the Cochran–Mantel–Haenszel (CMH) statistic (sometimes called Q_{CMH}). Bi (2002) discussed the application of this to sensory data as well. This is approximately a χ^2-distributed variable that is equal to a row vector of the marginal sum differences, times the inverse of a covariance matrix, times the column vector of the marginal differences. That is, two of the matrices consist of the marginal differences and its transpose. In matrix notation, as used by Bi, the calculation is as follows:

$$\chi^2 = S = d'V^{-1}d \tag{9.11}$$

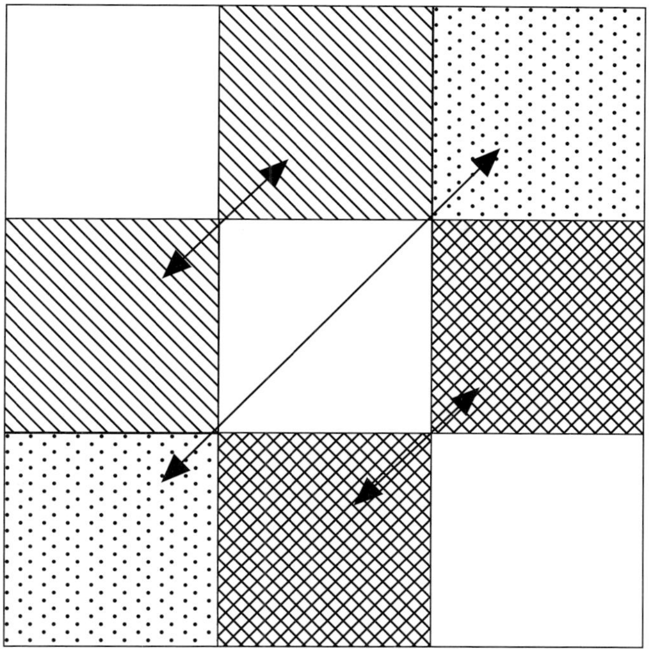

Figure 9.6 Schematic of cells that contribute to a significant Stuart–Maxwell statistic when they differ. Cells with similar shading or cross-hatching are critical comparisons. Note that they are "off-diagonal" in the sense that their subscripts would be opposite if the rows are numbered from top to bottom and columns from right to left. Also note that if the marginal totals compare the first row with first column and the second row with the second column, all of the critical comparisons have been included. Thus, it is a general property of the Stuart test that the calculations need not include the final row and column – no degrees of freedom are left.

where d' is the row vector of the marginal differences, $N_{\cdot i} - N_{i\cdot}$ (row sum minus column sum), *omitting the final last pairing*, d is the corresponding column vector (transpose of d'), and V^{-1} is the inverse of the estimated covariance matrix V.

The covariance matrix is simple to find, and is given as

$$V = \left\| v_{ij} \right\|$$

When $i = j$, the (diagonal) entries v_{ij} find the sum of the marginal totals minus two times the value of their intersecting interior cell. Once again, let a period symbolize the totals over the other variable; then

$$v_{ii} = n_{i\cdot} + n_{\cdot i} - 2n_{ii} \quad (9.12)$$

and when $i \neq j$ (off-diagonal), then

$$v_{ij} = -n_{ij} - n_{ji} = -(n_{ij} + n_{ji}) \quad (9.13)$$

Equations 9.12 and 9.13 produce the entire matrix V, which then needs to be inverted (V^{-1}), then multiplied times the difference vectors *in the proper order*. How the inversion is done for a 3×3 matrix is shown at the Wolfram Mathworld website (http://mathworld.wolfram.com/MatrixInverse.html). The Wolfram Mathworld site (http://mathworld.wolfram.com/Determinant.html) also gives an algebraic solution for the determinant of a 3×3 matrix, which must be found when calculating the matrix inverse.

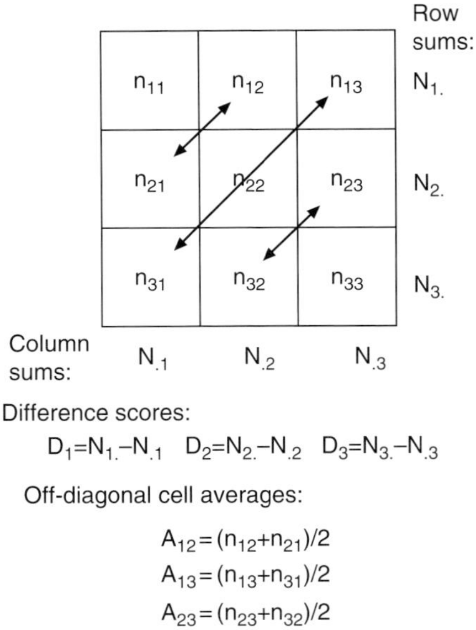

Figure 9.7 The calculations for the Stuart test for a 3×3 matrix, using the notation of Section 9.7.1.

9.7.2 A 3×3 Example: JAR data

A good application of the Stuart test is found for the just-about-right (JAR) scale. In this case we will truncate or combine cells so there are only three categories: too much of the attribute, too little, or about right. For this situation, an algebraic solution exists, so there is no need to invoke the matrix inversion and matrix multiplication at this level. Figure 9.7 shows the calculations needed, using the notation for cells, rows, and column totals given in Section 9.7.1.

The first step is to calculate the three difference scores D_x for the row and column totals, where $x = 1$, 2, and 3; so $D_1 = N_{.1} - N_{1.}$. Next, we get three off-diagonal cell averages, called A_{12}, A_{23}, and A_{13}, where $A_{12} = (n_{12} + n_{21})/2$, for example. The difference scores D_x are squared and multiplied by the average A_{ij} corresponding to the cells in which they did *not* participate (this seems counterintuitive, so be careful). These values are summed and form the denominator of the Stuart–Maxwell χ^2 ratio. Finally, the numerator consists of the sum of the three pairs of products of the three averages, multiplied by 2. The equation that results is

$$\chi^2 = \frac{A_{23}D_1^2 + A_{13}D_2^2 + A_{12}D_{31}^2}{2\left(A_{12}A_{23} + A_{13}A_{23} + A_{12}A_{13}\right)} \tag{9.14}$$

Figure 9.7 illustrates these calculations.

A worked example. Table 9.7 shows a sample data set from 135 consumers evaluating two products. It appears that product 2 has more judgments in the "not sweet enough" category and product 1 is a bit high in the "too sweet" response count. Is there a significant difference here? Perhaps product 2 is less expensive and there is really no reason to change to product 1 if there is no difference. The table shows the calculations for the row sums, column sums, difference scores, and off-diagonal cell averages. The χ^2 calculation then becomes

Table 9.7 Worked example of 3×3 Stuart test on JAR data

		Product 1			
		Not sweet enough	**Just about right**	**Too sweet**	**Row totals**
Product 2	**Not sweet enough**	5	22	33	60
	Just about right	16	14	11	41
	Too sweet	11	16	7	34
	Column totals	32	52	51	
Marginal total differences		D^2	Off-diagonal cell averages		
D_1	28	784	A_{12}	19	
D_2	−11	121	A_{23}	22	
D_3	−17	289	A_{13}	13.5	
Denominator products			Numerator products		
$A_{12}\times A_{13}$	418	$D_1^2 \times A_{23}$	17,248		
$A_{13}\times A_{23}$	297	$D_2^2 \times A_{13}$	1,633.5		
$A_{12}\times A_{23}$	256.5	$D_3^2 \times A_{12}$	5,491		
Sum	971.5	Sum = 12.54	24,372.5		
χ^2	24,372.5/(2×971.5)				

$$\chi^2 = \frac{22(784)+13.5(121)+19(289)}{2\left[22(19)+13.5(22)+19(13.5)\right]} = 12.54$$

This value (12.54) exceeds the critical value of 5.99 for two degrees of freedom (the degrees of freedom are the number of response categories minus one), so we can conclude that the pattern for the two products was in fact different. What would happen if we had treated the data as arising from independent groups and done a common χ^2 test? Generally, the independent-groups χ^2 will give a slightly higher value than the Stewart–Maxwell test for a 3×3 matrix. However, it is simply the incorrect analysis, as the data are paired, not independent.

Another useful application of this statistic is in nonforced preference when consumers evaluate a duplicate of the product pair, or are given some kind of before-and-after treatment such as presentation of an advertisement. Sometimes the preference result seen in the first pair will not replicate (Kim and Setser, 1980). Whether there is a significant shift or a washout of the result could be a useful piece of information. If some kind of treatment or condition is imposed between the two preference choices, then the Stuart test would tell you whether there is a significant pattern in the shifting of preference choices and, thus, suggest an effect of said treatment. Because there is a no-preference option, the results can be cast as a 3×3 table, as in the case of JAR data. Figure 9.8 shows the critical comparisons.

9.7.3 Some 4×4 examples for two products

A number of common product comparisons are done with four-category response options. Examples include the *R*-index, the same–different test with sureness ratings, and the degree

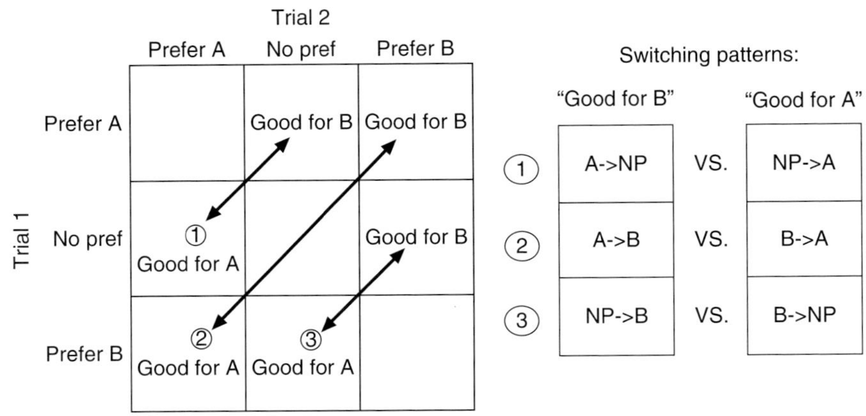

Figure 9.8 An example of how the Stuart test can be applied to replicated nonforced preference data. A significant χ^2 value would indicate an asymmetry in the switching behavior. The three critical comparisons are shown by the arrows numbered 1, 2, and 3. The right-hand panel shows which direction of change favors product A or product B.

of difference rated on a category scale. This type of analysis can be extended to any number of scale points, so it is equally applicable to five or more categories. Also note that there are alternative methods to analyze categorized ratings data. An example is Thurstonian analysis (Ennis et al., 2011) that yields both an overall degree of difference measure (δ or d'), a test of statistical difference, as well as estimates of category boundary locations. The advantage of the Thurstonian analysis is that it uses the fact that the categories are meaningfully ordered. Stuart tests do not. However, a d' analysis does not take into account the fact that the data are paired (dependent or repeated measures on the same individuals) or that there is covariance structure in the data set.

9.7.3.1 *R-index Data*

An appropriate application of the Stuart test occurs when data are gathered in an *R*-index study, and each participant evaluates one test item ("new") and a second control item ("control"). The rating scale would involve four responses as follows: "new, sure," "new, unsure," "control, unsure," and "control, sure," symbolized as NS, NU, CU, and CS, respectively. Table 9.8 shows a sample data set, and the *R*-index calculation. The Stuart test of marginal homogeneity returns a χ^2 value of 8.65 ($p=0.034$). In this case there was rather close agreement between the *R*-index binomial test and the Stuart test results, but the Stuart test takes into account the paired nature of the data and, therefore, is more appropriate.

The *R*-index binomial test can take one of the two forms shown below (see Bi (2006) for further discussion). The left-hand expression assumes the maximal value of the standard error under the null ($p=0.5$), while the right-hand expression uses the value of R obtained, instead of $p=0.5$. Be sure to express R as a decimal fraction in these equations, not as a percentage as is commonly done (e.g., $R=0.584$ not 58.4).

$$Z=\frac{R-0.5}{0.5\sqrt{\dfrac{1}{N-1}}} \quad \textit{or} \quad Z=\frac{R-0.5}{\sqrt{\dfrac{R(1-R)}{N-1}}} \tag{9.15}$$

Table 9.8 *R*-index data and submission to the Stuart test analysis

		Response to test product				
		NS	**NU**	**CU**	**CS**	**Totals**
Response to control product	NS	5	10	2	5	22
	NU	5	7	8	2	22
	CU	8	12	8	3	31
	CS	8	8	7	2	25
	Total	26	37	25	12	100
		NS	NU	CU	CS	
R-index table	Test product	26	37	25	12	
	Control product	22	22	31	25	
R	0.596		MH CHIsq	8.6516		
Z	1.937		*p*	0.0343		
p	0.0264					

Table 9.9 Same-different data, *R*-index value and Stuart test results

		Response to control pair				
		SS	**SU**	**DU**	**DS**	**Totals**
Response to test pair	SS	15	7	3	5	30
	SU	8	20	8	2	38
	DU	5	12	20	3	40
	DS	8	8	8	10	34
	Total	36	47	39	20	142
		SS	SU	DU	DS	
R-index table	Control pair	36	47	39	20	
	Test pair	30	38	40	34	
R	0.565		MH CHIsq	6.686		
Z	1.552		*p*	0.0826		
p	0.0603					

A rating scale can also be applied to same–different or degree of difference measures. One option is to include a sureness rating, sometimes called a certainty judgment, similar to the *R*-index scale (Delwiche & O'Mahony, 1997). Table 9.9 shows some hypothetical data from a same–different test with a sureness rating scale. This approach was recently studied by Kamerud and Larson (2010) and found to be superior to triangle tests on three important criteria. It did better than the triangle in terms of obtaining statistical significance, it was better able to track ingredient changes, and it was more cost efficient because multiple test pairs (three in their case) could be compared with a single control pair. Thus, there were eight samples given in each of five separate tests in their study (four pairs per test) rather than nine samples required by the three repeated triangles. If more than three samples (versus control) are tested, the efficiency increases further.

The same–different test with a rating scale is very similar to the *R*-index method, except that the observations are made on paired samples, rather than individual products. As in any good test, there is a baseline – a control pair of identical items. The question is whether a test pair is judged as "more different" than the control pair, based on the distribution in the data matrix. Note that it is possible to compute an *R*-index from these data and to conduct a simple binomial test on the proportion obtained versus 0.5, using one of the formulae shown in eqn 9.15. However, this does not take into account that the data are paired and that there is covariance structure in the data set. As we have seen in this chapter and Chapter 8, it is often advantageous to take into account the paired observations from the same individual, rather than treat them as if they were independent judgments.

Table 9.9 shows the data set, with 142 consumers taking part in the difference test. The results from the Stuart test, as well as the *R*-value that would be obtained, are shown. The *R*-value would be obtained had we simply treated the data as arising from independent samples. The responses are "same, sure," "same, unsure," "different, unsure," and "different, sure," coded as SS, SU, DU, and DS, respectively. In this case, neither approach showed a significant difference.

9.7.3.2 *Degree of Difference on a Category Scale*

The same analysis can be applied to any rating scale where there are repeated measures on the same judges. Instead of a sureness or certainty scale, for example, the analysis would also be applicable to any degree of difference scale, as long as it has categorical responses, or responses can be binned into categories, as might be done with a line scale or magnitude estimation data. Many such category scales are commonly used in consumer surveys, such as scales for satisfaction, appropriateness, and purchase intent. Rather than treat such data as continuous (e.g., assigning the values 1 through 4 to the categories) and using a *t*-test, it seems more reasonable, or at least honest, to admit that the scale is not interval-level data, and treat it as truly categorical. There are fewer statistical assumptions involved and less chance of violating those assumptions.

9.8 Further Extensions of the Stuart Test Comparisons

Replicated 2×2 comparisons. Occasionally,it may be advantageous to look at more than one sample in the kinds of 2×2 comparisons discussed for analysis with the McNemar approach. An example would be repeated measures on the same group or examining different groups of consumers and their changing attitudes before and after being given some nutritional information about a product. In these cases the CMH statistic may be used. An example is given in McDonald (2009a, b). The associated web site provides an Excel spreadsheet for repeated 2×2 tables that provides the CMH statistic and a probability value. Another worked example is given in Gacula et al. (2009). The statistic is simple to compute and distributed as a χ^2 with one degree of freedom. First, cast each of the tables as four cells with entries *a*, *b*, *c*, and *d*. Now choose one of the four corners, for example cell *a*. Strangely, it does not matter which one you choose, but for comfort's sake one might pick the cell with the most information; that is, one that shows a significant shifting pattern. The CMH statistic becomes

$$\chi^2_{\text{CMH}} = \frac{\left\{\left|\sum_{i=1}^{k}\left[a - \frac{(a+b)(a+c)}{N}\right]\right| - 0.5\right\}^2}{\sum_{i=1}^{k}\frac{(a+b)(a+c)(b+d)(c+d)}{N^3 - N^2}} \tag{9.16}$$

where there are k treatments, each one of the 2×2 matrices. The subscripts that correspond to the kth matrix have been omitted from the formula for readability. This formula looks daunting, but it is merely a version of the familiar χ^2 in which the denominator is formed by the sum of each observed value minus an expected value, and each sum squared. Thus, the numerator is simply the cell minus its expected value (row total times column total, divided by N) and the denominator is a variance estimate. Note there is a continuity correction, which is not always used (see McDonald (2009a)).

Best and Rayner (2001) discussed the extension of the Stuart test to designs with more than two products. They showed an example in which three products are rated on a five-point scale by eight panelists. If we let r be the number of products and s the number of rating scale points, then the derived statistic S is χ^2 distributed with $(r-1)(s-1)$ degrees of freedom. The calculation involves a summation of the individual Stuart-type comparisons, multiplied by the quantity $(r-1)/r$ as follows:

$$S = \frac{r-1}{r}\sum_{i=1}^{r} d_i^{\mathrm{T}} V^{-1} d_i \tag{9.17}$$

in which the d_i represent the difference vectors for the marginal sums and V^{-1} is the transposed covariance matrix as discussed above. As noted above, the Stuart test does not take into account the fact that the response categories may be meaningfully ordered. Best and Rayner (2001) also discussed alternatives that use this information, deriving them as special cases of the CMH statistic. The interested reader is referred to that paper and also to the treatment in Landis et al. (1979) and the book by Agresti (1990).

A simple alternative to the CMH statistic is to assign difference scores to each pair of judgments. One can then assume that the scores are treatable with parametric statistics; in the case of two products, a simple t-test will suffice. Suppose we have a same–different test with a rating scale for sureness. As in the example above, this produces four categories, "same, sure," "same, unsure," "different, unsure," and "different, sure." In this design there are two pairs: an identical control pair and one test pair consisting of the standard product and the test product (a "test" pair). The optimally discriminating panelist would rate the control pair as "same, sure" and the test pair as "different, sure." If we assign the numbers 1 through 4 to the response categories and subtract to get a difference score, this panelist would receive the maximum possible score of +3. A panelist whose ratings differed by only one category, and in the expected direction, would receive a score of +1. Panelists who mistook the test pair as a control and gave it "same, sure" and also mistook the control pair as "different, sure" would receive a difference score of −1, and so on. Panelists in the diagonal cells that did not change their choices would receive a score of zero. Table 9.10 shows the complete scoring scheme.

Table 9.10 Sample scoring system for difference scores assigned to rating scale data

		Response to control pair			
		Same, sure	**Same, unsure**	**Different, unsure**	**Different, sure**
Response to test pair	Same, sure	0	−1	−2	−3
	Same, unsure	+1	0	−1	−2
	Different, unsure	+2	+1	0	−1
	Different, sure	+3	+2	+1	0

The t-test will be against an expected value of zero, and the test should be one-tailed because there is an a priori prediction that the test pair should be judged more "different" than the control pair. If we apply this scheme to the data in Table 9.9, we obtain a t-value of 2.397 which is significant at 141 degrees of freedom.

A word on ranked data. A variety of options are available to the sensory professional when dealing with multiple observations from the same individuals. If the data are binary, the matrix is amenable to a simple Cochran's Q test. Sometimes, another option to response categories is ranking. Thus, the familiar Friedman test or Kramer rank sum tests are applicable. Note that, in the case of the Kramer test, the structure of the response options virtually dictates that the data are in essence repeated measures. That is, in ranking, the panelist must consider each product relative to the others in the set, unless some kind of incomplete design has been used. In that case, Durbin's test is appropriate (Bi, 2009). The Friedman test looks for consistency among the rankings across the panelists or consumers. So the χ^2 value it returns is a function of the consistency of the pattern, just as in the case of the paired (dependent) t-test, repeated measures ANOVA, and the Stuart and McNemar tests. All of these are allowing a view of the data sets with regard to how *individuals* do or do not show similar patterns of response across multiple products or product pairs. This is in contrast to being forced to consider the absolute responses from different individuals as if they were viewing a product only once, without any product comparison, implied or overt.

9.9 Summary and Conclusions

Part I made a case for experimental designs in which assessors can be used as their own baseline or controls. In these intelligent designs, each person evaluates each and every product, so their personal tendencies in scale usage or other inter-person differences can be partitioned and accounted for. This is simply a fundamental principle of psychological and psychophysical testing. In statistical terms, it avoids the problem of confounding product differences with assessor differences. Part II showed how some statistical tests can be applied to complete block designs for choice data and simple rating scales, where averaging and t-tests are not appropriate.

Although this was not meant to be a book about sensory statistics, it is difficult to discuss experimental design without referring to the appropriate analysis. They go hand in hand. Table 9.11 summarizes some of the tests that can be done with related-samples choice data. Further discussion of the strategies for handling JAR data with Stuart-type analyses and CHM statistics can be found in Fritz (2009). The article by Best and Rayner (2001) is an excellent treatment of the Stuart test applied to sensory data, and gives extensions for multiple products and consideration of ordered categories.

Alternatives to these simple tests are available. One option is the G-statistic, a likelihood ratio statistic, that is a general alternative to χ^2 tests with frequency data (MacDonald,

Table 9.11 Tests for related-samples choice data

		Number of products	
		2	**>2**
Number of response categories	2	McNemar	Cochran Q
	>2	Stuart–Maxwell	CMH

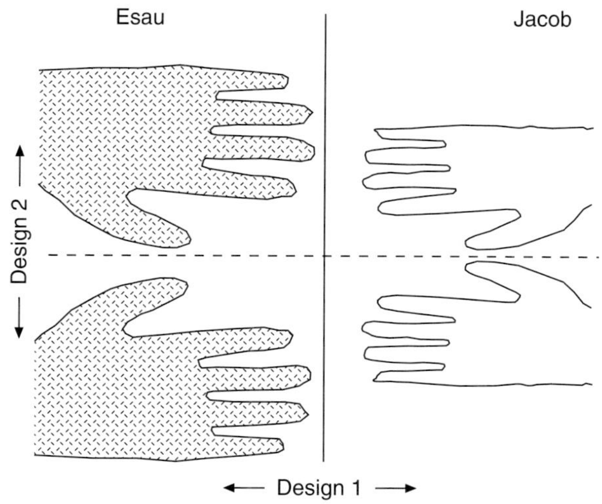

Figure 9.9 Two possible experimental designs for testing a skin care product on Esau and Jacob. Design 1 gives each product to a different person, but Design 2 compares the products on the left and right hands of each brother.

2009b). For very low frequency counts, Fisher's exact test is appropriate. Another alternative is to submit the data to a Thurstonian analysis and find a d' value and its variance estimate. If the 95% lower confidence interval does not overlap zero, there is evidence of a significant d' and, thus, a perceivable difference or preference. However, as noted above, the current common Thurstonian models do not take into account the repeated measures nature of the designs discussed here. The sensory professional should also be careful not to go on "fishing expeditions" and search through multiple analyses until one that yields a significant result is found. Such a strategy might help avoid type II errors (missing a difference), but remember that both type I and type II errors can be committed and either one may be trouble.

Returning to the biblical brothers Esau and Jacob, suppose one were designing a skin care product and wanted to test it on these two fellows. Figure 9.9 show two possible experimental designs. In Design 1, the products are given, one to each man, for testing on his hands. In Design 2, the products are both given to each man, for a comparison of their left and right hands. Given the difference in hair growth, it would make sense to factor out the natural variation in skin character by doing a direct comparison within each person (Gonik & Smith, 1993). The second design would confound the inter-person difference with the product difference, as suggested in the opening quote by Lea et al. (1998). This kind of thinking may be second nature to most experimentalists dealing with human variation, but this chapter should give extra ammunition to those dealing with experimental designs dictated by management preference or time constraints.

Appendix 9.A R Code for the Stuart Test

Assumes you have loaded R and can run it.
Assumes you have loaded package named COIN into your R collection of available libraries

Entries following the "greater than" sign are what you type after the R prompt ">"

Entries after the double hashtag "##" are just commentary.

```
>library (coin)

>stuart_data<-c(d1, d2, d3....dn)
##Assigns the data matrix

>Stuart_table<-as.table(matrix(stuart_data, nrow=4)
##Assigns a table format and tells it you have 4 rows & columns

>Mh_test(stuart_table)
## Returns chi-square value.
## Degrees of freedom are number of responses minus one (not rows times columns).
```

References

Agresti, A. 1990. Categorical Data Analysis. John Wiley & Sons, Ltd, New York, NY.

Best, F.J. and Rayner, J.C.W. 2001. Application of the Stuart test to sensory evaluation data. Food Quality and Preference, 12, 353–7.

Bi, J. 2002. Statistical models for the degree of difference test. Food Quality and Preference, 13, 31–7.

Bi, J. 2006. Statistical analyses for *R*-index. Journal of Sensory Studies, 21, 584–600.

Bi, J. 2009. Computer-intensive methods for sensory data analysis, exemplified by Durbin's rank test. Food Quality and Preference, 20, 195–202.

Chapman, K.W. and Lawless, H.T. 2005. Sources of error and the no preference option in dairy product testing. Journal of Sensory Studies, 20, 454–68.

Delwiche, J. and O'Mahony, M. 1997. Changes in secreted salivary sodium are sufficient to alter taste sensitivity: use of signal detection measures with continuous monitoring of the oral environment. Physiology and Behavior, 59, 605–11.

Ennis, D.M., Rousseau, B., and Ennis, J.M. 2011. Short Stories in Sensory and Consumer Science. IFPress, Richmond, VA.

Fritz, C. 2009. Appendix G: Methods for determining whether JAR distributions are similar among products (chi-square, Cochran–Mantel–Haenszel (CMH), Stuart–Maxwell, McNemar). In: Just-About-Right Scales: Design, Usage, Benefits, and Risks. L. Rothman and M.J. Parker (Eds). ASTM Manual MNL63. ASTM International, Conshohocken, PA, pp. 29–37.

Gacula, M., Singh, J., Bi, J., and Altan, S. 2009. Statistical methods in Food and Consumer Research. Elsevier/Academic Press, Amsterdam.

Gonik, L. and Smith, W. 1993. The Cartoon Guide to Statistics. HarperCollins Publishers, New York, NY.

Greenhouse, S.W. and Geisser, S. 1959. On methods in the analysis of profile data. Psychometrika, 24, 95–112.

Huynh, H. and Feldt, L.S. 1976. Estimation of the Box correction for degrees of freedom in the randomized block and split plot designs. Journal of Educational Statistics, 1, 69–82.

Kamerud, J. and Larson, G. 2010. Use of same–different test and *R*-index to efficiently compare multiple product differences. Poster presented and the Society of Sensory Professionals Meeting, Napa, CA, 27–29 October.

Kim, K. and Setser, C.S. 1980. Presentation order bias in consumer preference studies on sponge cakes. Journal of Food Science, 45, 1073–4.

Landis, R.J., Cooper, M.M., Kennedy, T., and Koch, G.C. 1979. A computer program for testing average partial association in three-way contingency tables (PARCAT). Computer Programs in Biomedicine, 9, 223–46.

Lawless, H.T. and Heymann, H. 2010. Sensory Evaluation of Foods, Principles and Practices. Second edition. Springer, New York, NY.

Lea, P., Naes, T., and Rodbotten, M. 1998. Analysis of Variance for Sensory Data. John Wiley & Sons, Ltd, Chichester, UK.

McDonald, J.H. 2009a. Cochran–Mantel–Haenszel test for repeated tests of independence. In: Handbook of Biological Statistics. Second edition. Sparky House Publishing, Baltimore, MD, pp. 88–94. See also http://udel.edu/~mcdonald/statcmh.html (accessed 24 December 2011).

McDonald, J.H. 2009b. *G*-tests for independence. In: Handbook of Biological Statistics. Second edition. Sparky House Publishing, Baltimore, MD, pp. 64–9. See also http://udel.edu/~mcdonald/statgtestind.html (accessed 26 Decembrer 2011).

Naes, T., Brockoff, P.B., and Tomic, O. 2010. Statistics for Sensory and Consumer Science. John Wiley & Sons, Ltd, Chichester, UK.

O'Mahony, M. 1986. Sensory Evaluation of Food. Statistical Methods and Procedures. Marcel Dekker, New York, NY.

Siegel, S. 1956. Nonparametric Statistics for the Behavioral Sciences. McGraw-Hill, New York, NY.

Stuart, A. 1955. A test for homogeneity of the marginal distribution in a two-way classification. Biometrika, 42, 412–16.

Winer, B.J. 1971. Statistical Principles in Experimental Design. Second edition. McGraw-Hill, New York, NY.

10 Frequency Counts and Check-All-That-Apply (CATA)

Check all that apply:

____ *I would like to use CATA questions more often.*
____ *CATA questions help me understand my consumers.*
____ *I would like to understand statistical analysis of CATA.*
____ *I would not use this in place of JAR questions, which I truly love.*
____ *This method seems like a waste of time.*

After Cowden et al. (2009)

10.1 Frequency Count Data: Situations — Open Ends, CATA

Frequency counts of responses in consumer research can be found in a number of situations. Probably two of the most common are counts of the responses to open-ended questions and checklist data such as **check-all-that-apply (CATA)**. Open-ended questions are often used to evoke responses from consumers in their own words, without any prompting or

Quantitative Sensory Analysis: Psychophysics, Models and Intelligent Design, First Edition. Harry T. Lawless.

suggestions as to possible answers. An example would be "Why did you like the product?" following a positive hedonic response or "Why did you dislike the product?" following a negative response. Typically, a good interviewer will continue to probe the respondent with "what else?" after they stop speaking or responding. This continues until the person runs out of things to say.

Open-ended responses offer some challenges. One is the bias inherent to this kind of question that favors people who have high verbal ability or just like to talk. By far the biggest challenge is how to code and group responses. If with three different people one says the product was sour, one tart, and one acidic, do you count these as three of the same response? Obviously, there is a fair amount of professional judgment in deciding what to throw into the same category. A consumer or sensory specialist may be required to produce a code sheet or coding system that specifies what responses will be counted together. Good pilot data, information from focus groups, or knowledge of previous test results can be very valuable for that purpose.

A somewhat more structured approach is to use a checklist. Consumers can be instructed to check all that apply (commonly called CATA) or to check the X most important or salient characteristics of the product. Typically, this would be limited to three or four items if there is such a limitation. Some have referred to these counts as "citation frequency" (Campo et al., 2010). CATA lists are typically constructed through pilot work or focus group information (Hooge, 2008). An interesting twist on the method is to have consumers check all the attributes that would appear in their ideal product. This entices the respondent to think about what is really important; that is, what the best example of this product should possess in terms of attributes. To the best of my knowledge, no one has been asked to check what the product should *not* have (a "negative must-have" in the Kano scheme), but it would seem reasonable to offer a choice like "low bitterness" as a response option if one was testing beer. Using these ideal product checklists will be discussed in Section 10.4.

Frequency count data are seen in many other situations. The most common are discrimination testing choices (correct odd sample choice in a triangle test, for example) or preference test choices. Both of these examples have exact mathematical expectations for chance-level responding and a corresponding numerical statement of the null hypothesis. In a triangle test, for example, the null hypothesis states that the population proportion correct is equal to one-third (and not that there is "no difference" as is sometimes assumed). The alternative hypothesis is that the population proportion correct is greater than one-third (i.e., one-tailed). For frequency counts or word choices, we have no such baseline to test against. In any test of two products, one can of course construct a null that states that the proportions of choices for the two products are equal. In a χ^2 test, this is like assuming the expected frequency is the average across cases. So this is a different situation than the null hypothesis with a known or expected baseline distribution, such as the binomial as it is applied to forced-choice discrimination tests.

Another sticky statistical issue is that different consumers will be contributing different amounts of information to the data set, depending upon how many responses they give. Thus, there is unequal weighting and an advantage to those high verbal responders. Perhaps this is not so important when one looks at a single attribute and compares the frequencies for two products. But across the entire data set we are getting more information and an unequal contribution from different people. To the best of my knowledge this issue has not been dealt with statistically, although one could easily envision some kind of weighting scheme that adjusted for people who like to check lots of options in a CATA question.

Of course, with a limit on the number of options (check your top four, for example) this problem disappears.

One of the earliest applications of frequency counts was found in the *Atlas of Odor Character Profiles* (Dravnieks, 1985), in which several hundred fragrance and aroma compounds and materials were profiled on 147 odor descriptors. In addition to average intensity ratings, Dravnieks (1985) tabulated a "percent applicability" score to indicate how often a term was selected for that particular substance. More recently, researchers have been looking at citation frequency as a potential alternative to traditional descriptive profiling, often with complex products such as wine (McCloskey et al., 1996; Le Fur et al. 2003; Campo et al., 2008). In one surprising result, Campo et al. (2010) found that the citation frequency approach showed more differences and more accurate separation of wines regarding their aging potential than did traditional descriptive analysis. They attributed this result to the larger set of descriptors that was available on the CATA list. It should also be noted that all of their CATA panelists received training with reference standards for each aroma attribute. This is somewhat unusual as most sensory evaluation practitioners would be using CATA with untrained consumers. It does, however, speak to the potential versatility of the method.

Sinopoli and Lawless (2012) used a CATA method to evaluate the characteristics of potassium chloride (KCl) and sodium chloride (NaCl) mixtures in water. Following the methods of Hooge (2008), who did a similar study with soups, they first used focus groups to generate the list of terms used in the later consumer evaluations. Results showed lower saltiness impact from KCl, and increasing unpleasant tastes (bitter, metallic) with increasing KCl concentration, and reduction in the off-tastes with increasing NaCl. These results were parallel to those observed with classical psychophysical methods; that is, intensity ratings (Murphy et al., 1981). The CATA approach was able to uncover additional terms for off-tastes (e.g., metallic) that were not part of traditional psychophysical approaches, which usually assumed only four tastes.

In another application, Ares et al. (2010b) applied a traditional descriptive analysis to chocolate milk desserts and then had 70 untrained consumers perform a CATA evaluation and give hedonic ratings. External preference maps were then constructed by **principal components analysis (PCA)** on the trained panel data and **multiple factor analysis (MFA)** on the frequency counts from consumers. The maps were similar and indicated similar optima and preferences when hedonic data were superimposed. A similar result was found by Dooley et al. (2010) with the external preference maps from commercial vanilla ice cream products. The maps constructed from a trained panel evaluation and CATA counts from consumers were very similar. If this result proves to be more general, it may provide a shortcut to external preference mapping, in that consumer data can be used for both phases. That is, the consumer data can be used for both map construction and the correlation of hedonic scores with the positions in the map, in order to plot vectors for increasing liking or ideal points (see Lawless and Heymann (2010, chapter 19)).

Open-ended counts have recently been subject to similar comparisons. Bécue-Bertaut et al. (2008) were able to relate liking scores to open-end frequency counts for wine description using MFA. The descriptions were culled from only two individuals and published in a popular wine magazine and the scores were generated from a panel of wine experts. Ares et al. (2010a) used open-ended comments and liking scores to discover drivers of liking for milk desserts. The consumer data (frequencies and hedonics) were later combined with trained panel descriptive data using MFA to produce a composite map.

10.2 Simple Data Handling

How should two products be compared, using frequency counts from CATA questions? The simplest, easiest, and maybe quickest analysis is a simple comparison of two proportions or frequencies using a binomial test on proportions. Note that the binomial-based test on proportions is mathematically equivalent to a χ^2 test (proof in Appendix 10.A) as long as both tests use or both tests do not use the continuity correction.

Siegel (1956) referred to a one-sample χ^2 test when comparing two independent groups. The normal χ^2 formula can be used:

$$\chi^2 = \sum_{i=1}^{k} \frac{(O_i - E_i)}{E_i} \tag{10.1}$$

where O is the observed count, E is the average count across k products; in this case $k=2$. This version is without Yate's continuity correction and must have expected observations greater than five. There is one degree of freedom for a two-product test (df $=k-1$), so the critical value for χ^2 is 3.84. In order to use this test, the groups must be of the same size.

A quick example. Suppose we do a consumer test with two groups of 100 consumers each. For product A, 27 consumers checked a box for "crisp" and 39 did so for product B. Is this a significant difference? Omitting the continuity correction, we get an average of $(27+39)/2=33$, and a χ^2 value of

$$\chi^2 = \frac{(27-33)^2}{33} + \frac{(39-33)^2}{33} = \frac{36+36}{33} = 2.18$$

This fails to meet our critical value, so we cannot conclude there is a significant difference in the rate of choosing "crisp" for the two products (insufficient evidence).

If the groups are not the same size, and/or there is no clear baseline, one can use the following formula suggested by Kanji (1993), for sample sizes of n_1 and n_2 and observed proportions p_1 and p_2:

$$Z = \frac{p_1 - p_2}{\sqrt{P(1-P)\left(\frac{1}{n_1} + \frac{1}{n_2}\right)}} \tag{10.2}$$

where P is a weighted average of the two proportions

$$P = \frac{p_1 n_1 + p_2 n_2}{n_1 + n_2} \tag{10.3}$$

Kanji recommends that the sample size for each group be greater than 30 to justify the approximation to the normal distribution. Note that the denominator is a version of the familiar standard error of a proportion, $(pq/n)^{1/2}$. For a two-tailed test with no a priori prediction, the critical value of Z is ± 1.96.

A quick example. Suppose we have consumer groups of 196 and 205 who choose the term "satisfying" for product A 138 times and product B 97 times, respectively. Our proportions are $138/196 = 0.704$ and $97/205 = 0.473$, and our weighted average P comes out to 0.586 and $1-P$ is 0.414. Crunching through our formula, eqn 10.2, we get

$$Z = \frac{0.704 - 0.743}{0.586(0.414)\left(\frac{1}{235} + \frac{1}{401}\right)} = \frac{0.231}{\sqrt{0.0016372}} = 5.70$$

This exceeds our critical Z-score of 1.96 at α=0.5, so there is sufficient evidence to reject the null and conclude that the frequencies of choice were significantly different for the two products. The hefty sample size in this hypothetical study did not hurt one's chances, either.

At the end of this chapter we will discuss testing against a fixed benchmark or single proportion, as one might do for substantiating an advertising claim based on a fixed percentage ("most women agree ...," which implies over 50%) or a ratio ("dentists would recommend our product over the market leader by a 2 to 1 margin," which implies a 2/3 to 1/3 split).

10.3 Repeated or Within-Subjects Designs

Many consumer tests involve a monadic sequential design, in which two or more products are viewed by all participants. For two products this is a paired test, and for more than two it is a complete block design. The binomial and χ^2 tests for simple frequency counts assume independent observations, which the paired test or repeated test is not. For a two-product test or comparison, the appropriate solution is to cross-tabulate their responses for each product; that is, whether or not they checked a box for one product, both, or neither. Then the appropriate statistic becomes the McNemar tests for changes. An example is shown in Table 10.1.

The McNemar test compares the frequency of box *B* versus *C*; that is, more people should have changed in one direction than the other. These are "discordant" cells – people answered differently for the two products. Note that people who checked both boxes or neither box are not transmitting any useful information, and are essentially discounted (although this is debatable).

$$\chi^2 = \frac{(|B-C|-1)^2}{B+C} \quad \text{or} \quad \chi^2 = \frac{(B-C)^2}{B+C} \tag{10.4}$$

The second version is taken from Gacula et al. (2009), who omit the continuity correction. They point out that this is also equivalent to a Z-test:

$$Z = \frac{|B-C|}{\sqrt{B+C}} \tag{10.5}$$

Although we are only using the cells where there was a change in response, there is nothing wrong with reporting the marginal frequency counts in your data summary. The McNemar

Table 10.1 A 2×2 breakdown for McNemar test comparing two products tested by the same consumers

	Product 1	
Product 2	**Checked box**	**Did not check**
Checked box	A	B
Did not check	C	D

is a generalization of the Stuart test, which is in fact a test on the equivalence of marginal totals. Moving from a binomial or χ^2 test on the totals to the McNemar changes the analysis to a potentially more sensitive test that takes into the account the dependent/related-samples nature of the data. There is some potential decrease in power because the effective N is somewhat lower by using only the discordant cells. Note that $B+C<25$ is suspect. So you need a hefty sample size, and most consumer tests would qualify. If the total of both cells is less than 25 then an exact binomial should be performed, once again comparing cells B and C, and not the marginal totals. The test is the mathematical equivalent of a sign test and is a χ^2 test with df=1 (critical χ^2 is 3.84).

For a data set with more than two products, an appropriate statistic is **Cochran's *Q* test**, not to be confused with Cochran's C test for outliers. Some researchers have used Friedman's "analysis of variance on ranks" as a test statistic (e.g., Ares et al. 2010b), but as that is designed for ranked data, and these are binomial (checked or not checked), Cochran's Q seems a better fit. You cast the data in a table with products as columns and consumers as rows. Every entry is thus either zero or one, for not checked versus checked. If we have k products and N consumers, we need column totals C_j, row totals R_i, and the overall total of checked boxes for this particular word or attribute. The test statistic Q is approximately distributed as a χ^2 variable with $k-1$ degrees of freedom. The computations may be done in several ways as follows:

$$Q = \frac{k(k-1)\sum_{j=1}^{k}\left(C_j - \bar{C}\right)^2}{k\sum_{i=1}^{N} R_i - \sum_{i=1}^{N} R_i^2} \tag{10.6}$$

where $\bar{C}$ is the average count for columns (products). A somewhat more computationally simple method is to use

$$Q = \frac{(k-1)\left[k\sum_{j=1}^{k} C_j^2 - \left(\sum_{j=1}^{k} C_j\right)^2\right]}{k\sum_{i=1}^{N} R_i - \sum_{i=1}^{N} R_i^2} \tag{10.7}$$

Note that these have the same denominator. The test is similar to asking whether the variance in products (columns) is larger than the variance in the individuals (rows), so it is like a χ^2 or F-ratio, which are of course related.

If an overall significant Q statistic is found, one may proceed to compare pairs of products using the McNemar approach. Of course, if one has many products, the number of pairwise comparisons will inflate your risk of type I error, so some adjustment in α may be needed. Along these lines, Lancaster and Foley (2007) suggested the SAS program Proc Multtest as an appropriate tool to do the comparisons and adjust for inflated α risk. Worked examples of Cochran's Q can be found in Gacula et al. (2009) and Siegel (1956).

An index of panelist consistency was recently suggested by Campo et al. (2010). This assumes that the panelists have evaluated at least one product as a replication. The consistency index is an average of all the replicated samples, and divides the common terms for each item (times the replicates) by the total number of terms chosen. Obviously, the larger the number of common terms across replicates, the higher the reproducibility index. The index for any given judge is

$$R = \frac{1}{N}\sum_{i=1}^{N}\left(\frac{mC_i}{\sum_{j=1}^{m} D_{ij}}\right) \tag{10.8}$$

for N samples and m replicates, and where C_i is the number of common terms for sample i and D_{ij} are the number of descriptors chosen for sample i on the jth replicate. This index has the attractive property of being bounded by zero and one. A potential difficulty is that the index is likely to shrink as the number of samples increases. Campo et al. (2010) used only one duplicate ($m=2$), so that was not an issue in their study. However, they suggested that other indices or refinements to this measure might be appropriate.

10.4 Multivariate Analyses

A number of multivariate techniques are available for analyzing CATA data. These methods can produce product and attribute maps (perceptual maps) analogous to using PCA or **generalized Procrustes analysis (GPA)** for rating scale data. After doing so, they can also be subject to external preference mapping or related techniques to find directions or positions for optimum products, and/or consumer segments. **Cluster analysis** can also be imposed on the resulting maps to get an objective idea of what products may be grouped or similar in the consumers' minds. An interesting approach would be to use the checklist for a consumer's imagined ideal product, and then submitting the ideal product checklist to cluster analysis to identify segments preferring different ideal styles of the product. The goal would be to uncover the preferred sensory profile for each group as a potential direction for product development.

One approach to analyzing frequency data is to use **correspondence analysis** (Bécue-Bertaut et al., 2008; Ares et al., 2010a; Campo et al., 2010). This is a flexible tool that operates on the total frequency counts of products (rows) by attributes (columns) and can find associations between row and column variables. At some point, the researcher needs to set a cutoff for inclusion or exclusion of items that were checked very infrequently (15% or less has been used). One might also perform a Cochran's Q analysis to see whether a given attribute was checked more or less across the set of products (that is, whether it differentiated anything). This would be analogous to using only those attributes in a PCA that were actually differentiating the products, as found in preliminary tests using the univariate analyses of variance. Correspondence analysis is available in a number of the popular statistical packages. In the R platform, it can be found in a number of libraries, including anacor (de Leeuw & Mair, 2009) and FactoMineR (see Murtagh (2005)). There is also an entire library (ca) devoted to correspondence analysis (Nenadić & Greenacre, 2007).

Another option is to use the MFA package in the R library, FactoMineR (Husson et al., 2007; Lê et al., 2008). This is an extremely versatile tool and can accommodate multiple inputs such as frequency data and liking scores. It has recently been used in a number of studies with CATA frequency counts to generate perceptual maps (Ares et al., 2010a,b; Dooley et al., 2010). As always, all libraries in the R platform collection are freeware and are thoroughly documented, often in the *Journal of Statistical Software*.

10.5 Difference from Ideal and Penalty Analysis

10.5.1 Comparisons with Ideal Products

How do we know if our product does not have enough of some important attribute? When should we work to *increase* the perception of that property? In their Pangborn Symposium presentation, Cowden et al. (2009) introduced the useful idea that consumers could be asked to check all that apply for their ideal product, and further that they could prioritize them by checking their top three. For any attribute in the CATA list, one can then compare a given product's frequency of being checked versus the frequency that the ideal product is checked. One simple analysis for this comparison would seem to be a comparison of the frequency totals using a simple binomial (or χ^2) frequency comparison, e.g. the *Z*-test on the size of two proportions. However, such a test ignores the fact that the data are usually dependent; that is, the same person has given us the CATA data for the test product and also for their ideal product. So just comparing the totals (as the only comparison) is not recommended.

In Section 10.3 we saw how the McNemar test could be used to compare the frequencies of CATA choices for two products, and how this would be appropriate for dependent or related-samples data (like a monadic sequential test). We could use a McNemar test, but we can get some further insights from casting the data into a similar 2×2 matrix. The matrix shown in Table 10.2 looks at the frequencies of people who did or did not check a given attribute for the product of interest (let us call it product X) and for the ideal.

The simple frequency count comparison is really a comparison of the marginal totals of A+B versus A+C. From this it is obvious that the comparison is not independent; they both contain category "A." This invalidates the use of a simple χ^2 or binomial test on proportions based on the totals. So, once again, one should avoid the temptation to just compare the totals for the ideal product to the totals for your test product. Perhaps more importantly, not all of these four cells are equally meaningful. So the McNemar test of B versus C is not as meaningful as it is in a real two-product comparison. We have asked for attributes of an imagined product in order to identify drivers of liking. Looking at the individual cells, they have different implications as shown in Table 10.3.

Table 10.2 A 2×2 breakdown for a comparison of a product to the ideal

	Checked for ideal?	
Checked for X?	**Yes**	**No**
Yes	A	B
No	C	D

Table 10.3 A 2×2 breakdown for a product versus ideal; interpretation of cells

	Checked for Ideal?	
Checked for Product X?	**Yes**	**No**
Yes	This cell is A-OK! (Checked for both = good news)	This cell is not very relevant
No	This cell is the troublesome one!	Also not important

Cell "B" is not relevant in the sense that some people checked this attribute as present for product X, but it is not important to them for their ideal product. In Kano terms, it is not a "must have." Or perhaps they are just not emotionally or cognitively involved with this attribute. In any event, it does appear to be so very important to the people in this cell.

The critical cell is cell "C," the troublesome one. These people felt the ideal product should have this attribute, but that product X did not. The strategic question then becomes whether this cell is a meaningfully large group. Note that if the size of cell B is about the same as the size of cell C, you could miss out on the detection of cell C by the comparing the frequencies of the marginal totals. That is because comparing A+B versus A+C might give you no difference and make you think you are matching the ideal, but there still might be a lot of potentially dissatisfied people in cell C!

There are a couple of approaches here that could be used. It depends in part upon how you phrase the question. One question is to ask whether the frequency of C is nonzero. In other words, does A+C shrink if we take out C? So the statistical test can be if the proportion of C is bigger than zero. Another related view is to ask whether the total A+C is meaningfully larger than just A. If a lot of people checked box A (both the ideal and our test product have this attribute) then maybe the size of C is not so impactful. At this time there is no agreed-upon or standard approach, so it might be prudent to look at it both ways. Discussions with management might be able to come to some reasonable criteria for what percentage of the total group in cell C is a problem, or what ratio of C to the total A+C is likely to be a real issue.

A test for nonzero C versus a test of A versus A+C will have somewhat different standard errors if you use the actual observed proportions in calculating the standard error, because the value of N is different. Of course, a χ^2 test can be based on the average frequency, but bear in mind you may be dealing with small frequencies for C, and thus you might bump into Cochran's rule about needing at least five expected observations per cell.

What about attributes that should potentially be decreased? Given some information about the ideal product, when should a checked attribute be decreased or made less perceivable? This situation is more complex than the option discussed above for increasing an attribute. Once again, we will consider the data as dependent or related-samples, meaning that the same person has "checked all that apply" for both our test product (X) and their ideal product.

Looking again at the 2×2 classification in Table 10.2, we can identify the critical cell.

In this case the troublesome category would appear to be cell B, in which the individuals had not checked this attribute for their ideal product, but indicated its presence in the test product. Cowden et al. (2009) suggested a test of the frequency A+B (total checked for test product) versus A+C (total checked for ideal), and a decrease if A+B>A+C. Note that this runs into our independence problem again. Also as stated above, cell C is the critical one for indicating *increases* in an attribute, but it is not relevant for decreases. If B is really the critical cell, then comparing A+B with A might be informative. A statistical test for nonzero B or a prior hurdle or cutpoint for B could be useful.

But further considerations apply. If the overall frequency for the ideal product is in fact fairly high (A+C high) then it might not make sense to decrease an attribute just because its frequency of CATA checking was even higher than the ideal. Looked at this way, only attributes with very low ideal-checking would seem to be potentially problematic. In other words, a more-than-ideal frequency is not necessarily a bad thing. It is also conceivable that, although the attribute was not checked much for the ideal product, its presence is not necessarily detrimental. It could be unimportant. That is, we do not have an indictment as a "negative must-have" (must *not* have) in the Kano classification (see Lawless and Heymann (2010: chapter 19)). Without further information, we really do not know if this attribute is important to people. It might not matter very much, or it could be a true defect or nuisance. Recommending a decrease or reduction in the salience of an attribute is more complex than deciding to enhance it.

This ambiguity suggests that further information is required. In an insightful modification, Cowden et al. (2009) went on to ask consumers to check their top three items that the ideal product should have, thus collecting "importance" data. A similar modification could be made to identify potential defects or nuisance factors; that is, "check your top three items that the ideal product should *not* have." Personally, I am not accustomed to putting defects, nuisance items, and product negatives on questionnaires, but your experience may be different. One wonders how the CATA items could include all the possible defects, or whether some things would not even come to mind for the average consumer ("of course it would not have rat hairs").

A graphical representation of the data can be persuasive and informative. Cowden et al. (2009) showed trendlines across a set of attributes and when checked frequencies were lower or higher than the frequency checked for the ideal product (see Figure 10.1). These

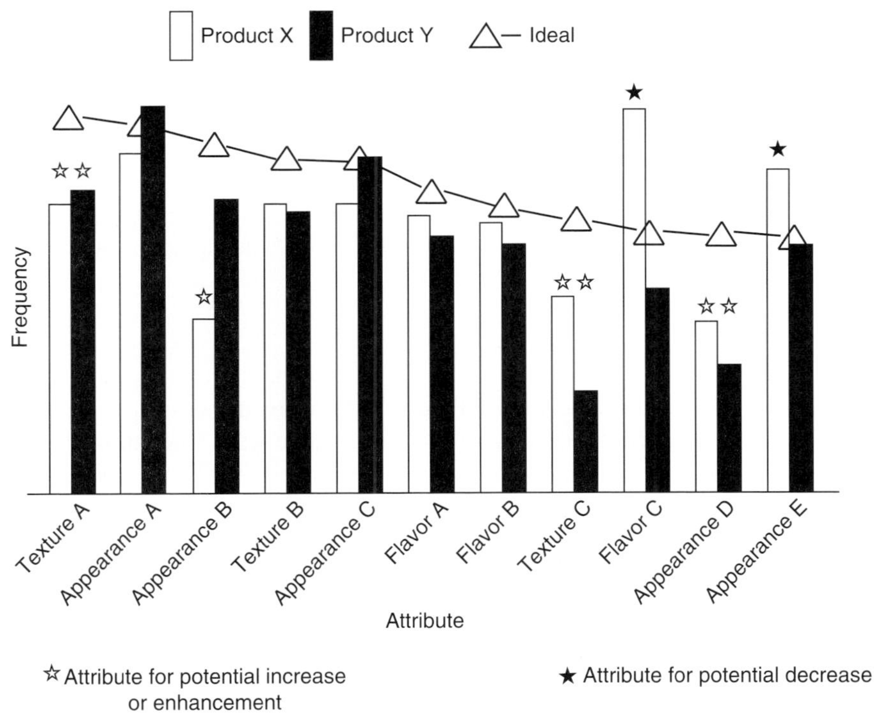

Figure 10.1 Frequency counts for two products as in Cowden et al. (2009) and a trendline for the ideal product is shown by the connected triangles. A number of attributes for product X were checked much less often than the ideal, including texture attributes A, B, and C, and appearances B and D, suggesting that these attributes could be enhanced or increased in a more appealing product. Product Y seems generally closer to the ideal with the exception of textures A and C and appearance D. Note that product X also has a flavor characteristic, C, that was mentioned more often than the ideal, suggesting a possible reduction or other modification in its overall flavor profile. This figure only shows a subset of the more frequently mentioned attributes, which are arranged in order of decreasing frequency of checked items for the ideal product. Cowden et al. also plotted upper and lower confidence bounds on the ideal trendline to aid in making recommended actions (omitted here for clarity).

trendlines suggest increases or decreases in a given product if it falls below or above the confidence interval for the ideal product. Gathering information on the worst negatives could permit the plotting of an inverse trendline for the attributes to be avoided (my cereal gets soggy too fast, for example). This might look like the mirror image of the trendline for must-haves, but is not necessarily the exact opposite since there could be segments of consumers who prefer (and avoid) different styles of products. A further valuable addition to CATA analysis, then, would seem to be analysis for sensory segments, perhaps using the ideal and must-not-have responses.

The use of ideal ratings also raises the question of whether CATA data can be used in place of **just-about-right (JAR) ratings**. JAR data also suggest when a given attribute should be increased or decreased in a product to improve or optimize the level of that characteristic. A comparison of ratings for ideal can also suggest changes as illustrated above. Furthermore, it would be possible to have checked items for "too sweet" or "not sweet enough," for example. At this time, it is not known how such an approach would be better or worse than JAR ratings. As always, the issues of reproducibility and accuracy should be considered in comparing any two methods. Logically, the two approaches should produce similar information. If so, then a secondary consideration is whether either CATA or JAR methods are easier or more user friendly for consumers. If CATA and JAR ratings produce similar information, it should be possible to impose a penalty analysis on CATA, as long as hedonic ratings are also collected. Briefly, penalty analysis computes a drop in the mean hedonic rating among those people who felt the product had too little or too much of a given attribute, compared with those people who felt it was close to JAR (Rothman & Parker, 2009). It would seem prudent to add an option such as "sweetness was just about right" to the CATA list if this was going to be attempted, rather than simply assume that if people did not check the "too much" or "too little" options they (by default) thought it was just right. Consumer responses are not always that logical.

10.5.2 Penalty-Benefit Analysis

If CATA data are combined with an overall liking (OAL) score, it is possible to do a penalty analysis, similar to the kind of mean drop calculation done with JAR ratings. With JAR ratings, the change in the mean acceptability score for consumers who are non-JAR can be calculated and compared with the scores from consumers who felt the product was just right on the critical attribute. This change is sometimes called a "mean drop." The penalty that is paid for being non-JAR can also be weighted by the percentage or proportion of consumers who felt the product was too low or too high in the attribute. A bi-plot of penalty magnitude by proportion can be very informative. Obviously, a large segment with a high penalty (large mean drop) is usually important and suggests that some modification of the product is worthy of consideration. However, even a vocal minority that is sufficiently alienated by the nonoptimal level can suggest a course of action, such as a line extension or a market segment strategy.

For CATA data, one of the responses has to be considered a baseline, from which the mean OAL score can rise or fall. It makes sense to let the baseline be the score from a group which did *not* check that particular attribute. Then, if another group checks that particular item, the mean OAL score could rise as well as fall. So perhaps it is more accurate to call this a benefit-penalty analysis rather than just penalty analysis. Plaehn (2012) developed a formal statistical model for this change, and suggested several statistical analyses for significance testing. However, it seems straightforward to do a simple independent groups

t-test to see if the mean rise or fall is significantly different than the baseline group. Essentially, you are asking whether the OAL score changed as a function of checking that CATA item. Plaehn also suggested considering a weighted effect, in which the mean score change is multiplied by the proportion of consumers checking the item.

A few considerations seem worthy of caution in this approach. First, it is possible that the mean rise or fall could be due to some other attribute that is highly correlated with the item under analysis. Thus, it is prudent to look for additional significant mean changes and see whether the groups checking both items are closely overlapping. Further information might be useful in suggesting which (or both) of the attributes might be driving the OAL score change. Second, because the t-test is essentially a test of means, it is possible that segments of opposing views might cause no difference in the mean score change. For example, if one group felt the item was beneficial but another group found it to be undesirable, then they could cancel each other's opinion in the overall mean comparison. As with JAR data, it is always a good idea to look at the distribution of responses to see whether there are differing segments.

The consideration of OAL scores can also be applied to the ideal-product checking scheme of Section 10.5.1. In that case, it seems appropriate to consider the score for an ideal product to be the baseline. The penalty then becomes the mean drop amongst those that did not check the CATA item for the actual product, versus their OAL score for the ideal product. Giving an OAL score for an imaginary product may seem a bit odd, but one could expect such a rating to reflect the maximum possible liking score for the kind of product being surveyed. Further research will help to see whether this idea is in fact feasible or whether it makes sense to consumers.

10.6 Frequency Counts in Advertising Claims

Marketers may want to make claims in advertisements based on frequency data. Examples would be "90% of regular hot dog eaters found our wiener tender and juicy" or "consumers found our product easier to prepare than the leading brand by a two to one margin." So these kinds of statements can be based on proportions or ratios. A ratio can be converted to a proportion (two to one means 2/3 to 1/3 of the sample, for example). So the critical statistical tests can be based on proportions.

A reasonable approach is to set up a null hypothesis based upon the proportion you want to claim, and then show that the proportion obtained in your data is significantly higher than that figure. This may seem a bit conservative, but it is statistically bulletproof and is likely to withstand the scrutiny of legal challenges. So we make the null state that the population proportion is equal to our proportion, and we then (if we have the data) can reject it in favor of a one-sided alternative that states the proportion is greater than our claim.

We can use the exact binomial, letting p_o be our desired proportion; the sample size is N, and we need to find the value X such that X/N is the proportion we need to get in order to be significantly higher than our claim in the actual data. We do this by summing up the values in the tail of the binomial expansion until we get to our α level of 0.05. That gives us our X value for the number of consumers who had to choose that option or check that box:

$$\text{pr}\left(\frac{X}{N} > p_o\right) = \sum_{X=k}^{N} \binom{N}{k} p_o^k (1 - p_o)^{N-k} \tag{10.9}$$

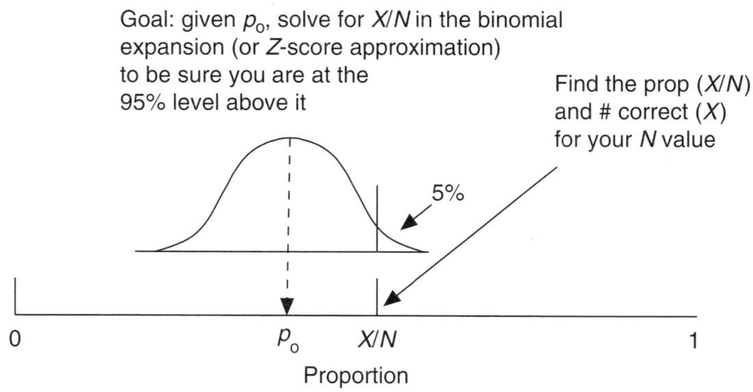

Figure 10.2 Schematic of the binomial-based test to substantiate a frequency count or proportion in advertising claims. The distribution is the expected distribution under the null for our claimed proportion p_o and showing the 0.05 cutoff for the proportion we must obtain for rejecting the null.

You can think of this as summing backwards from 100% ($k=N$), to $N-1$, $N-2$, and so on until you get to the value of k that sums to your 0.05 (total) cutoff. That would find the minimum value of X that leaves less than 5% in the tail. This is an exact binomial solution. We can also find a normal approximation to the binomial for a somewhat quicker solution. Since we are looking at the one-tailed cutoff for a Z-score of 1.645, we can solve the following inequality:

$$1.645\sqrt{\left(\frac{p_o(1-p_o)}{N}\right)} \geq \frac{X}{N} - p_o \tag{10.10}$$

And we simply find our required X for any given sample size N. Another way to think about this is that we can "prove" we are above p_o if the difference $(X/N)-p_o$ is more than 1.645 standard errors, which is basically a one-tailed 95% confidence interval. This formula has omitted any continuity correction, but for your typical consumer test with $N=100$ or 200 or more, it would probably be negligible. A schematic of this process is shown in Figure 10.2.

A quick example. Suppose we want to claim that "over 50% of women agree that (our product) ends dry skin." We test 300 consumers and 178 check the box for "ends dry skin." Is this good enough to make our claim?

Our value for X/N is then 178/300, or 0.593. So the difference from our p_o is $0.593-0.5=0.093$. Our critical difference is

$$1.645(0.45)\sqrt{\frac{1}{300}} = 0.0475$$

which we have exceeded, so we are justified in making our claim of "greater than 50%."

10.7 Conclusions

CATA data are growing in popularity as an easy way to get a consumer's qualitative impressions of a product. Statistical analyses are straightforward. However, a researcher should be careful not to use a test that assumes independent judgments when the design is a paired test or other complete block. Several multivariate techniques, notably correspondence analysis and MFA available in the R platform, can be used to construct perceptual maps, and these may resemble the product maps produced from descriptive analysis data. Such product maps could be used for preference mapping and related techniques, such as response surface or segmentation analysis. Product improvements can be uncovered in CATA data if the consumer is also asked to use the checklist for their imagined ideal product.

At this time, a number of unanswered questions remain about this technique. Can CATA data be used as a substitute for JAR ratings, and is there any advantage to doing so? How do we deal with the issue that high-verbal individuals are contributing more information to the overall CATA picture than those people who check fewer items? Is a weighting scheme required? Does it matter if we restrict the choices to a fixed number or allow people to truly check all that apply? Finally, how can we assign some importance values to the items checked so we know whether these items were not only salient, but really mattered to the consumer? Such a technique could provide more Kano-type data for identifying drivers of liking. Cowden et al. (2009) mentioned this option near the end of their presentation, and remarked that it provided additional useful information for understanding what is important to consumers, even what "delights" them.

Appendix 10.A Proof Showing Equivalence of Binomial Approximation Z-Test and χ^2 Test for Differences of Proportions

Recall that

$$\chi^2 = \sum \frac{(\text{observed} - \text{expected})^2}{\text{expected}} \tag{10.A.1}$$

and

$$Z = \frac{(x/N) - p}{\sqrt{pq/N}} \tag{10.A.2}$$

where x is the number of responses chosen for one of the two products, N is the total judgments or panelists, p is the the baseline or chance level expected under the null, and $q = 1 - p$.

Note that continuity corrections have been omitted for simplicity.

An alternative Z formula (multiply eqn 10.A.2 by N/N) is

$$Z = \frac{x - Np}{\sqrt{pqN}} \tag{10.A.3}$$

Although the χ^2 distribution changes shape with different degrees of freedom, the general relationship of the χ^2 distribution to the Z distribution is that χ^2 at df = 1 is a square of Z. Note that for df = 1 the critical $\chi^2 = 3.84 = 1.96^2 = Z^2_{0.95}$ (for one-tailed, df = 1, $\chi^2 = 2.71 = 1.645^2 = Z^2_{0.95}$).

Squaring, we get

$$Z^2 = \frac{(x - Np)^2}{pqN} \tag{10.A.4}$$

and

$$Z^2 = \frac{x^2 - 2xNp + N^2p^2}{pqN} \tag{10.A.5}$$

The proof will now proceed to show the equivalence of eqn 10.A.5 to χ^2.
Looking at any forced-choice test, the χ^2 approach requires these frequency counts:

	Correct judgments	**Incorrect**
Observed	X	N–X
Expected	Np	Nq

$$\chi^2 = \frac{(x - Np)^2}{Np} + \frac{[(N - x) - Nq]^2}{Nq} \tag{10.A.6}$$

Simplifying $(N–X)–Nq$ to $N(1–q)–X$, then, since $p = 1–q$, we have

$$(N - X) - Nq = Np - X$$

Thus, we can recast eqn 10.A.6 as

$$\chi^2 = \frac{(x - Np)^2}{Np} + \frac{(Np - X)^2}{Nq} \tag{10.A.7}$$

and then expanding the squared terms we obtain

$$\chi^2 = \frac{x^2 + 2xNp + N^2p^2}{Np} + \frac{x^2 - 2xNp + N^2p^2}{Nq} \tag{10.A.8}$$

To place them over a common denominator of Npq, we will multiple the left expression by q/q and the right expression by p/p, giving

$$\chi^2 = \frac{qx^2 + 2xNpq + qN^2p^2}{Npq} + \frac{px^2 - 2xNpp + pN^2p^2}{Npq} \tag{10.A.9}$$

Collecting common terms:

$$\chi^2 = \frac{(q+p)x^2 + (q+p)2xNp + (q+p)N^2p^2}{Npq} \tag{10.A.10}$$

Recalling that $q + p = 1$, then eqn 10.A.10 simplifies to

$$\chi^2 = \frac{(1)x^2 + (1)2xNp + (1)N^2p^2}{Npq} \tag{10.A.11}$$

and dropping the value 1 in each of the three terms in the numerator gives eqn 10.A.5, the formula for Z^2:

$$z^2 = \frac{x^2 + 2xNp + N^2p^2}{pqN} = \chi^2$$

Recall that the continuity correction was omitted for simplicity of the calculations. The equivalence holds *if and only if* the continuity correction is either *omitted from both* analyses or *included in both* analyses.

References

Ares, G., Giménez, A., Barreiro, C., and Gámbaro, A. 2010a. Use of an open-ended question to indentify drivers of liking of milk desserts. Comparison with preference mapping techniques. Food Quality and Preference, 21, 286–94.

Ares, G., Barreiro, C., Deliza, R., Giménez, A., and Gámbaro, A. 2010b. Application of check-all-that-apply questions to the development of chocolate milk desserts. Journal of Sensory Studies, 25, 67–86.

Bécue-Bertaut, M., Álvarez-Esteban, R., and Pages, J. 2008. Rating of products through scores and free-text assertions: comparing and combining both. Food Quality and Preference, 19, 122–34.

Campo, E., Do, B.V., Ferreira, V., and Valentin, D. 2008. Aroma properties of young Spanish monovarietal white wines: a study using sorting task, list of terms and frequency of citation. Australian Journal of Grape and Wine Research, 14, 104–15.

Campo, E., Ballester, J., Langlois, J., Dacremont, C., and Valentin, D. 2010. Comparison of conventional descriptive analysis and a citation frequency-based descriptive method for odor profiling: an application to Burgundy Pinot Noir wines. Food Quality and Preference, 21, 44–55.

Cowden, J., Moore, K., and Vanleur, K. 2009. Application of check-all-that-apply response to identify and optimize attributes important to consumer's ideal product. Presentation at the Eighth Pangborn Sensory Science Symposium, Florence, Italy.

De Leeuw, J. and Mair, P. 2009. Simple and canonical correspondence analysis using the R package *anacor*. Journal of Statistical Software, 31(5), 1–18.

Dooley, L., Lee, Y.-S., and Meullenet, J.-F. 2010. The application of check-all-that-apply (CATA) consumer profiling to preference mapping of vanilla ice cream and its comparison to classical external preference mapping. Food Quality and Preference, 21, 394–410.

Dravnieks, A. 1985. Atlas of Odor Character Profiles. ASTM International, Philadelphia, PA.

Gacula, M., Singh, J., Bi, J., and Altan, S. 2009. Statistical Methods in Food and Consumer Research, Second edition. Elsevier/Academic Press, Amsterdam.

Hooge, S.E. 2008. Impact of potassium chloride on saltiness, bitterness, and other sensory characteristics in model soup systems. Master of Science Thesis, Kansas State University.

Husson, F., Josse, J., Lê, S., and Mazet, J. 2007. FactoMineR: factor analysis and data mining with R. R package version 1.04. http://cran.R-project.org./package=FactoMineR (accessed 25 March 2013).

Kanji, G.K. 1993. 100 Statistical Tests. Sage Publications, London.

Lancaster, B. and Foley, M. 2007. Determining statistical significance from choose-all-that-apply questions. Presentation at the 7th Pangborn Sensory Science Symposium, Minneapolis, MN.

Lawless, H.T. and Heymann, H. 2010. Sensory Evaluation of Foods, Principles and Practices. Second edition. Springer, New York, NY.

Lê, S., Josse, J., and Husson, F. 2008. FactoMineR: an R package for multivariate analysis. Journal of Statistical Software, 25, 1–18.

Le Fur, Y., Mercurio, V., Moio, L., Blanquet, J., and Meunier, J.M. 2003. A new approach to examine the relationships between sensory and gas chromatography–olfactometry data using generalized Procrustes analysis applied to six French Chardonnay wines. Journal of Agricultural and Food Chemistry, 51, 443–52.

McCloskey, L.P., Sylvan, M., and Arrhenius, S.P. 1996. Descriptive analysis for wine quality experts determining appellations by Chardonnay wine aroma. Journal of Sensory Studies, 11, 49–67.

Murphy, C., Cardello, A.V., and Brand, J.G. 1981. Taste of fifteen halide salts following water and NaCl: anion and cation effecdts. Physiology and Behavior, 26, 1083–95.

Murtagh, F. 2005. Correspondence Analysis and Data Coding with R and Java. Chapman & Hall/CRC, Boca Raton, FL.

Nenadić, O. and Greenacre, M. 2007. Correspondence analysis in R, with two- and three-dimensional graphics: the ca package. Journal of Statistical Software, 20, 1–13.

Plaehn, D. 2012. CATA penalty/reward. Food Quality and Preference, 24, 141–52.

Rothman, L. and Parker, M.J. 2009. Just-About-Right Scales: Design, Usage, Benefits, and Risks. ASTM Manual MNL63, ASTM International, Conshohocken, PA.

Siegel, S. 1956. Nonparametric Statistics for the Behavioral Sciences. McGraw Hill, New York, NY.

Sinopoli, D.A. and Lawless, H.T. 2012. Taste properties of potassium chloride alone and in mixtures with sodium chloride using a check-all-that-apply method. Journal of Food Science, 77, S319–22.

11 Time–Intensity Modeling

Sensory systems are built to monitor varying signals. To perform that task as efficiently as possible, they adapt, which is a general term for self-regulatory processes which adjust the sensitivity characteristics of the sensory system according to the stimulus level. This means that a useful characterization of a sensory system in terms of the relation between input (stimulus level) and output (perceived intensity) should incorporate the time aspect.

Overbosch (1986:315)

11.1 Introduction: Goals and Applications

11.1.1 Overview

The sensations that arise when ingesting or tasting a food are rarely static and usually changing. The same is true of the sensory experiences we get from many personal care and other consumer products. The fragrance from a shampoo or from an air freshener may become weaker as our nose adapts to the smell. Some of these changes can be directly traced to physical changes in the product, such as texture changes in a food as it breaks down in the mouth from chewing. Others are a function of sensory physiology, such as odor or taste adaptation (Cain, 1974; Gent & McBurney, 1978; O'Mahony, 1986). Even under conditions of constant stimulation, our sensory experience varies over time. Some experiences can

Quantitative Sensory Analysis: Psychophysics, Models and Intelligent Design, First Edition. Harry T. Lawless.

build up or increase in intensity. Astringency from tannic wines is one example. Many others fade, such as most taste properties, like saltiness.

In order to form a complete picture of the sensory effects of a food or consumer product, it is often necessary to include a time factor in the evaluation. So the sensation intensity (and sometimes quality) is tracked over some time period (e.g., Neilson, 1957). The time course of sensation can be an important influence on our liking or disliking for a product. A sweetener chemical that has a lingering taste may be less appealing to consumers than one that has a time profile more like normal sugars. Wines with a "long finish" may be prized by oenophiles. A chewing gum that maintains its flavor for a longer period may be more highly acceptable to consumers. A breakfast cereal that maintains a crunchy texture even after milk is applied is probably a good thing. A hand lotion that develops a lot of "drag" as it is applied may be less liked by consumers than one that does not. So both the sensory profile and consumer acceptability are influenced by the temporal characteristics of products and the sensations they evoke.

Time as a variable, of course, is intrinsic to some sensory tests, such as packaging and shelf-life studies. However, in those cases the time span is considerably longer than the time course of the simple flavor and texture properties of a single bite of a food. This chapter will deal with the latter, the short-term flavor, odor, or texture experiences produced by a single stimulus. Of course, some flavors last longer than others. Saltiness is often characterized by a "quick hit" of taste sensation, but other tastes, such as bitterness or hot pepper burn, can last for minutes. It is also conceivable that two products could have the same sensory profile, but differ in the order of appearance or timing of sensory attributes. A sensory method that produces a static or one-time sensory profile might not capture this difference.

Measuring the time course of sensations is an important part of flavor research, especially for flavors that tend to last, such as bitterness (Pangborn et al., 1983; Leach & Noble, 1986), sweetness (Lawless & Skinner, 1979; Ott et al., 1991), astringency (Guinard et al., 1986), oral burn or heat (Lawless, 1984; Cliff & Heymann 1993a, Prescott & Stevenson, 1996), and menthol cooling (Gwartney & Heymann, 1995). The pattern of flavor release from a food as it is consumed is also a process that can be temporally tracked (Lee, 1986; Overbosch, 1987; Pionnier et al., 2004). Texture, of course, is also a temporal process because its assessment is often destructive, changing the food and changing the perception of its qualities (Moore & Shoemaker, 1981; Duizer et al., 1993; Brown et al., 1994; Butler et al, 1996; Zimoch & Gullet, 1997).

In order to measure temporal sensory changes, it is critical to assess the sensory experience at specific times, and do so as accurately as possible. Various methods are available for time–intensity tracking. These are described in various sensory textbooks such as Lawless and Heymann (2010) and Meilgaard et al. (2006), methodological papers by Dubois and Lee (1983) and Lundahl (1992), and reviews by Lee and Pangborn (1986) and Cliff and Heymann (1993b). Given the artificiality of the testing situation and the need for tight control, these methods are best done with panelists who are trained in the method, or who at least have practice in carrying out such evaluations (Peyvieux & Dijksterhuis, 2001). There are two frequently used methods for **time–intensity scaling**. One is to ask for repeated judgments at specific time intervals, using some sort of visual or auditory cue ("rate now"). This can be done using any of the common scaling techniques, such as category scales, line scales, or magnitude estimation. The second approach involves moving some analog response device, and the assessor is asked to make a more or less continuous record of the sensory experience. A typical setup requires the panelist to move a computer mouse, and they receive visual feedback by watching a pointer or cursor move on a line scale. In this case a computer can do time sampling of the response at whatever intervals the investigator has programmed, which has greatly facilitated the collection and analysis of temporal data (Guinard et al., 1985; Lee, 1985; Yoshida, 1986).

11.1.2 Issues and Distinctions

There are several important distinctions to be made in time profiling. First, is the record supposed to represent simple sensory intensity (perceived strength) or the liking/disliking of the sensation? It is possible to look at hedonic reactions or pleasantness/unpleasantness over time (Taylor & Pangborn, 1990; Veldhuizen et al., 2006), although tracking this kind of affective response as a continuous recording is somewhat rare. Of course, for some products it is common to ask consumers what they think about the aftertaste of a food. By aftertaste, we mean the sensations that are present after the food has been swallowed or expectorated. But phrased in that form, the datum is just a snapshot of one time period, rather than a continuous record. A second important question is whether the record is expected to be bidirectional or unidirectional. An example of a bidirectional record is that of a flavor experience that rises over time and then falls off and disappears. This is a common pattern for the majority of sensory phenomena in taste, smell, and flavor. Another class consists of sensory effects that are irreversible and proceed in only one direction. An example is melting behavior of a food that changes phase. Many texture characteristics are such unidirectional changes. As we masticate a food, the food does not resume its previous shape, texture, or particle sizes. However, some textural characteristics may pass through a maximum, such as the juiciness (expressed moisture) of some meat products, and thus have both a rising and falling phase. The methods and procedures for these various classes may be very similar, but sensory specialists should be fully aware of what kind of phenomenon they are dealing with. It could make a difference, for example, in what assessors are asked to do with the mouse position at the end of the record. Do panelists leave the mouse at maximum reading or return to zero? Returning the mouse to the zero reading for an attribute that is unidirectional may create problems in data handling as well as "mouse artifacts" (Lawless & Heymann, 2010). A third question is whether the method is tracking changes in a single attribute or the combination of dominant attributes at any given time period. The latter approach leads to the methods known as temporal dominance of sensations (Le Reverend et al., 2008; Labbe et al., 2009; Pineau et al. 2009), a recent methodological advance that will be discussed in Section 11.3.3.

Data handling approaches to time–intensity records have focused on two primary issues. One is the question of what parameters or operating characteristics can be extracted from a time–intensity record, such as rise time, total duration, or area under the curve. These parameters are very similar to those used in pharmacology to characterize drug delivery or bioavailability. The second issue concerns how individual time–intensity records should be combined to show an average curve. Because the data tend to be highly idiosyncratic, this is not trivial. Averaging two curves that look completely different may produce a result that resembles neither of the originals. An alternative to averaging is to find some kind of mathematical decomposition to find consistent trends in the curves, through methods such as **principal components analysis (PCA)**.

Figure 11.1 shows a typical time–intensity record for an ingested taste or flavor material or for a food product when the panelist is tracking a taste or flavor. There are three characteristic parts of this record: a rising phase, a plateau region, and a falling phase. Not all

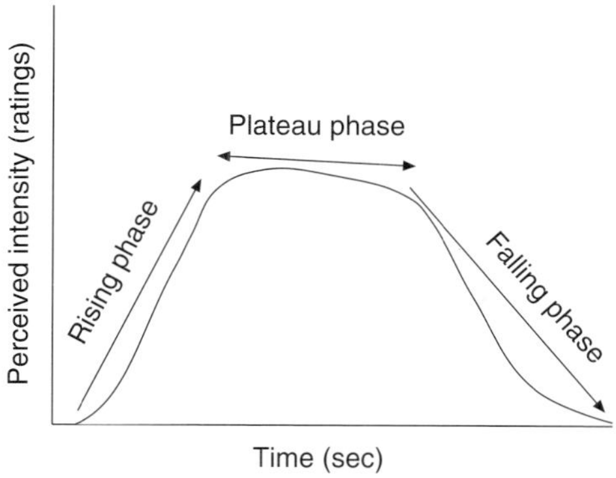

Figure 11.1 A typical time–intensity record for a taste or flavor consists of three phases: a rising phase that is often linear and of short duration, a plateau phase during which the maximum or near maximum intensity is experienced, and a falling phase where the sensation decreases, sometimes exponentially. The plateau phase may not exist in some data records, and/or can be difficult to define.

substances or sensory attributes have a plateau phase, and it may be difficult to tell when such a phase begins and ends. Typically, it is not entirely flat, but shows some drift or minor degree of decay over time. The three-part nature of a time–intensity record suggests that it could be considered as consisting of three splines with two knots or nodes (Ledauphin et al., 2005). A separate mathematical function could be fit to each segment. Note that the curve is rarely symmetric – the onset tends to be quite rapid for many substances and the decay phase much slower. It could be possible, of course, to find a single polynomial equation that fits these kinds of records, possibly with a small negative quadratic or cubic coefficient that would make the function "turn over" after reaching a peak (see Wendin et al. (2003) for examples). Such equations are typically used for the inverted U-shaped hedonic function or "Wundt curve." They take the form

$$S = k_1 t - k_2 t^2 \tag{11.1}$$

where S is the sensation intensity and t is time. Note there is no simple additive constant k_0, as intensity is thought to start from zero rather than some nonzero baseline. But such a term could be added if there was cause to believe in some background level of sensation.

11.1.3 A Quick Look at Exponential Processes

Many processes change over time based upon the current level of the dependent variable. The classical example is compound interest on a bank balance that is compounded continuously, having a derivative of the form

$$\frac{\mathrm{d}y}{\mathrm{d}t} = ky^t \quad \text{and thus} \quad \frac{\mathrm{d}y / \mathrm{d}t}{y_t} = k \tag{11.2}$$

where the instantaneous rate of change dy/dt is proportional to the value of y at time t.

Integrating the expression gives

$$y_t = y_0 e^{kt} \tag{11.3}$$

where y_o is the starting value for y and k becomes the time constant. These types of equations are ubiquitous in biological processes that grow exponentially. They are conveniently linearized to

$$\ln(y_t) = \ln(y_o) + kt \tag{11.4}$$

so that a simple least-squares fit can be made to a semi-log function of time with slope k and intercept $\ln(y_o)$. But where can these be useful in time–intensity records?

The opposite of exponential growth is exponential decay, and we merely have to make the time constant negative in order to model the falling phase of the typical flavor curve. This was suggested by Gent (1979) to model adaptation in taste; that is, the falloff in sensation under conditions of constant stimulation. A similar function was suggested by Lawless (1984) to model the decay period of sensations from various oral trigeminal irritants such as capsaicin. The resulting equation is

$$S = S_o e^{-kt} \tag{11.5}$$

where S is sensation intensity, S_o is the starting or peak intensity from which the falling phase begins, and k is the time constant. $1/k$ equals the time to reach 1/e of the initial intensity and $0.693/k$ is the half-life assuming the decay begins at time zero (Riggs, 1963). So this is not only easily linearized and fit, but provides two useful time constants to characterize the flavor disappearance. If S_o is a function of concentration, then the function can be expanded to include a power function or Beidler-type of psychophysical function (see Chapter 2).

What about the rising phase of the typical flavor function? An exponential relationship could be potentially useful here, if there is a gradual approach toward an asymptote representing the S_{max} value at the beginning of the plateau phase. For some substances and some flavor attributes, the onset is quite sudden, and a linear fit is probably acceptable for the rising slope, if it is necessary at all. For biological processes that approach an asymptote, we merely turn the exponential decay function upside down and track the decay of the difference between the current sensation level S and the maximum S_{max}. This produces

$$S_{max} - S = (S_{max} - S_o)e^{-kt} \tag{11.6}$$

where S_o is a nonzero starting intensity level. If the function starts at zero intensity (no taste) we get

$$S_{max} - S = S_{max} e^{-kt} \tag{11.7}$$

and thus the sensation intensity at any given time is determined by

$$S = S_{max} - S_{max} e^{-kt} \quad \text{or} \quad S = S_{max}\left(1 - e^{-kt}\right) \tag{11.8}$$

It is important to remember that it is not S that is appreciating exponentially, but the difference ($S_{max} - S$), that is decaying exponentially (to zero difference, and S to S_{max}).

Eilers and Dijksterhuis (2004) argued against the accuracy of an exponential decay, stating that, in their experience, the individual records were often linear in the decreasing phase. They suggested the combination of two logistic functions, one rising and one falling, and using a log transformation of intensity to get a more linear appearance in the falling phase.

That is, the growth phase would resemble a common logistic S-shaped curve – that is, $f(x)=1/(1+e^{-x})$ – and the falling phase would be approximately linear. This led to a model with five parameters as follows:

$$S = c - \ln\left[1+e^{-a_1(t-t_1)}\right] - \ln\left[1+e^{-a_2(t-t_2)}\right] \quad (11.9)$$

where c is a scaling parameter, a_x are rate constants (with $a_1>0$ and $a_2<0$, to produce the opposite or mirror image functions), and t_1 and t_2 represent the time to half maximum in each phase.

Another sensory phenomenon that shows a rising curve over time is the adaptation effects on threshold concentration (Overbosch, 1986). Instantaneous thresholds are exceedingly difficult to measure or estimate, but Overbosch published results from earlier work by Hahn (1934) on taste thresholds for NaCl and Stuiver (1958) for odor thresholds for octanol. Under conditions of constant concentration, these functions show consistent patterns. The system, being perturbed by the onset of stimulation, attempts to restore equilibrium by bringing the threshold level equal to the new stimulus concentration. When it does so, adaptation is complete. When the threshold is far from the stimulus concentration (i.e., at the beginning of the stimulation period), the threshold changes more rapidly, giving rise to the exponential curvature. When the adapting concentration is higher, the time constant is longer. It simply takes longer to adapt to a higher concentration. Overbosch reasoned that any model must accommodate these two consistent principles.

Letting C represent the adapting concentration, C^* represent the threshold concentration, C_o^* the absolute threshold (under $C=0$), and t for time, Overbosch suggested this general relationship:

$$C^* - C_o^* = C - Ce^{-kt/C} \quad (11.10)$$

or

$$C^* = C_o^* + C\left(1-e^{-kt/C}\right) \quad (11.11)$$

According to Overbosch (1986), the relationship can be related back to perceived intensity S by considering Steven's power function, $S=k'(C-C^*)^n$, where k' is now simply a proportionality constant and n is the characteristic exponent (see Chapter 2) as

$$S = k'\left(Ce^{-kt/C} - C_o^*\right)^n \quad (11.12)$$

And the time constant k can be estimated from

$$k = \frac{C}{t}\ln\left(\frac{C}{C_o^*}\right) \quad (11.13)$$

The interested reader is referred to the original paper, but should note that Overbosch's notation is different, using the letter S for (stimulus) concentration and I for perceived intensity.

11.2 Parameters Versus Average Curves

11.2.1 Extracting Parameters

Returning to our characteristic curve, there is an important choice for data handling. We can try to combine raw data curves to get an average or some measure of central tendency to

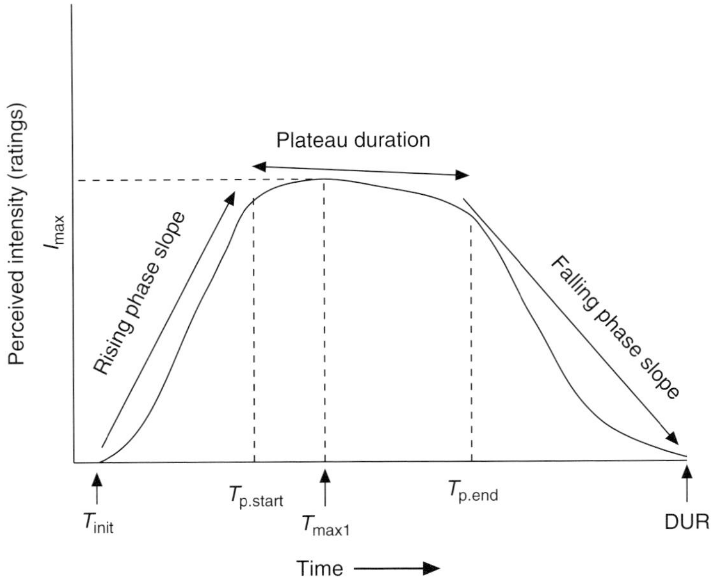

Figure 11.2 Some of the parameters that can be extracted from a time–intensity record. T_{init} is the time at the initial onset of sensation. $T_{p.start}$ is the time at the start of the plateau phase and $T_{p.end}$ is the time at the end of the plateau phase. T_{max} is the time to maximum intensity and I_{max} is the maximum intensity rating. DUR is the total duration of sensation. Another important parameter, AUC for area under the curve, is not shown.

represent the group data as a single record, or we can try to characterize each individual record at the outset. In either case, there is the opportunity to characterize a curve (either at the raw data stage or after data treatment) with some characteristic parameters. If the characterization is done at the raw data stage, these parameters from different substances or products can then be compared statistically for differences using the usual parametric statistical tests for continuous data. Of course, the assumptions for the statistical test, such as normality or normality of residuals, should always be done, especially since these are derived measures with limitations that might or might not follow the expected distributions.

Figure 11.2 shows some of the common parameters that can be extracted from a time–intensity record of the form shown in Figure 11.1. The most common characteristics include the time to maximum (often called $\boldsymbol{T}_{max}$) in the literature, which represents the rapidity of growth of the sensation, the intensity at maximum ($\boldsymbol{I}_{max}$), the total duration (**DUR**), and the area under the curve (**AUC**), a measure of the total sensory impact of the flavor. Various other measures are a bit less common, such as rising and falling slope, and the duration of the plateau, if it exists. Of course, others can be derived, such as the ratio of the areas under the rising and falling phase, a kind of look at the temporal symmetry of the experience. More can be invented, seemingly without limit. Many of these parameters may be correlated in any given study of several food products, and some workers have questioned the redundancy or need for very many of them (Lundahl, 1992). It would not be surprising, for example, to find a flavor substance with a higher peak intensity to last longer.

11.2.2 Overbosch and Lui–MacFie Analyses

Forming average curves from raw data would seem to be a simple matter. Having a common time base, one would only need to average the perceived intensities at various time intervals and plot the resulting curve. This is not without some difficulties, however. For one thing, different panelists will produce curves of varying duration. This means that, for very late time periods, most of the data will consist of zeros. Averaging those values produces very small numbers, while the median value in fact indicates that the majority opinion is that there is no taste. So the data are not normally distributed, but are positively skewed, left censored with zeros. There are statistical methods to handle such data (Owen & DeRouen, 1980), but a better solution can probably be found. Second, if two panelists produce curves with very different peak times T_{max}, averaging raw data values produces a combined curve with two peaks, but neither original data set contains such a pattern. These problems were recognized by Overbosch et al. (1986). They suggested a method for averaging curves that would avoid or attenuate the problems of producing curve artifacts, such as the production of a double-peaked average curve from two single-peaked data records.

This method is summarized in Figure 11.3. Assume there are two panelists with *TI* curves that have different I_{max} and T_{max} values. The first step is to find the geometric mean value for I_{max} for the pair and then to set both panelists I_{max} values to this geometric mean by an individual a multiplicative constant, which is applied to all values in each data set. Sometimes this process is referred to as "normalization" in the treatment of magnitude estimation data, but it should not be confused with converting data to *Z*-score values. The equation for normalization of raw data values *I* to new values *I′* is

$$I' = \frac{I_{max}}{I_{max\ i}}(I) \tag{11.14}$$

where the subscript *i* refers to the value for an individual panelist.

The more important step, and the true insight of this process, is to now average different points in the time direction. The points to be averaged are fixed proportional values of the I_{max}, starting with the time value T_{max} at the I_{max} itself. For example, the two records in Figure 11.3 would be set to the same geometric mean T_{max} value. Then the time values would be averaged at 95% of I_{max} in the rising and falling phases, 90% of I_{max}, and so on until the rising and falling phases had been sliced at *n* time intervals with *n* equal to perhaps 10 or 20. The connected points then construct an average curve. This process avoids some of the problems mentioned above.

Although this method was considered an advance, some problems remained. Different time points might have the same *I*-value, rendering it difficult to do the time averaging. The method did not truly consider the issue of multiple T_{max} values if there was an I_{max} plateau. This was addressed by placing the plateau in the falling phase, perhaps not a very good solution as it would potentially compress the plateau. Furthermore, there might be other flat sections in addition to the highest plateau. Such step functions are not uncommon in individual records. Curves with multiple peaks are difficult to analyze. Finally, missing data, particularly at the end of the record, or nonzero intensity at the end made it difficult to find values for *I* at 0% (or more) of I_{max}. For these reasons, Liu and MacFie (1990) suggested an alternative procedure.

The Lui–MacFie procedure uses a normalization for intensity, as in the individual multiplicative factors described above. Next, the procedure identifies four time markers: t_{start} for

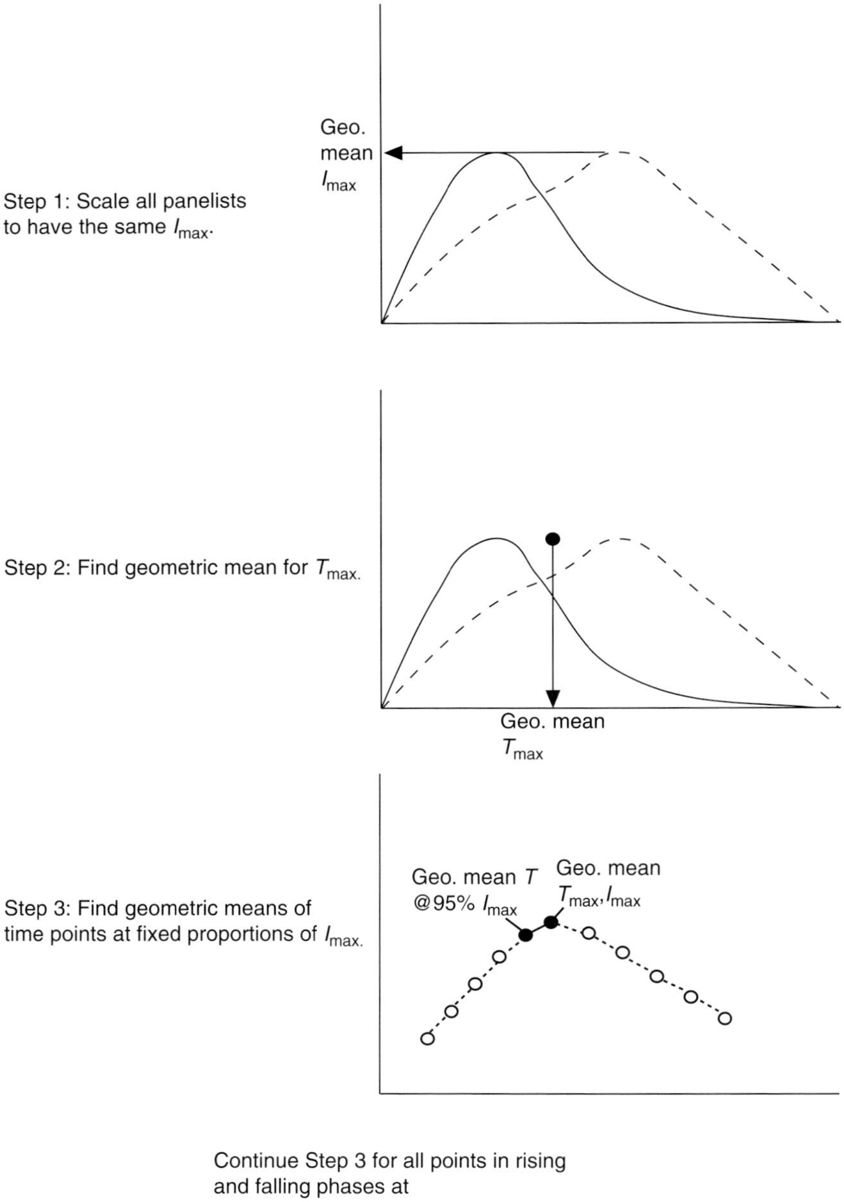

Figure 11.3 Steps in the method of Overbosch et al. for averaging time–intensity records.

the time at the first nonzero rating, t_{max} for the time at the beginning of the plateau or highest value of I, t_{dec} for the time at which the descent from the plateau begins, and t_{end} for the time at the last reading or zero reading in the declining phase. These four time markers are then averaged (geometric or arithmetic mean) to define the inflection points or nodes of the group curve. This produces a composite record of three segments. Next, each individual curve is time-normalized to set the individual curve nodes to the same value as the group nodes. The equations are similar to the intensity normalization, being a multiplicative transformation for each individual person, as shown in eqn 11.15 and eqn 11.16. The rising and falling segments are then partitioned into 20 equal time slices (or some other convenient number). For each time slice, the intensity values are averaged. Missing data are dealt with by interpolation. Letting t' indicate the transformed time value and the subscript i represent the ith panelist's individual values, and parameters without the subscript are the group means, we have

$$t' = \frac{t_{\text{max}} - t_{\text{start}}}{t_{\text{max } i} - t_{\text{start } i}}\left(t - t_{\text{start } i}\right) + t_{\text{start}} \tag{11.15}$$

for each panelist's rising phase and

$$t' = \frac{t_{\text{end}} - t_{\text{dec}}}{t_{\text{end } i} - t_{\text{dec } i}}\left(t - t_{\text{dec } i}\right) + t_{\text{dec}} \tag{11.16}$$

for the falling phase. This method takes care of some of the problems inherent in the original Overbosch procedure, but produces records remarkably similar in the published examples from Liu and MacFie (1990). For other methods of subject normalization in both the intensity and time directions, see Dijksterhuis and Eilers (1997) and Ledauphin et al. (2006). McGowan and Lee (2006) compared the Liu and MacFie method with one in which the panelists with similar curves were grouped, and then average curves were computed for each group. This eliminated the problem of what to do with panelists who never reach an extinction point, and still allowed valid comparisons among products.

11.2.3 Trapezoidal Estimation

Drawing on the idea of four critical points that could characterize the time–intensity record, Lallemand et al. (1999) proposed using a trapezoid to approximate the curve. Some ambiguities and issues remained about where to place the time markers. They proposed using four points linked to the intensity maximum. The start and end vertices of the trapezoid were the first and last points that were 5% of maximum. This tended to avoid false starts and mouse artifacts at the end of the record. The plateau region was similarly truncated at the initial point achieving 90% of maximal intensity and the last point at that level. The result is illustrated in Figure 11.4. Note that this method allows extraction of several parameters of the curve, as well as a simple method to get an estimate of AUC.

Obviously, the trapezoid is not a completely accurate representation of the *TI* record, but a sort of caricature. So it should be viewed more as a method of characterizing the critical aspects of the record, rather than an attempt to mimic it. Even if the method does not provide a completely accurate estimate of AUC, for example, it may be that the trapezoidal area is simply a reproducible translation of AUC (probably somewhat smaller). But if all *TI* records have the same percentage reduction, then they are simply a linear transformation of AUC and, thus, are still useful. As long as all products suffer the same transformation, the difference from a "true" measure is trivial. The method should be studied further to see how the trapezoidal parameters compare with those measured by other rules or algorithms. In an extensive study of ice cream texture with highly trained panelists, the authors found it to be a useful technique for comparing different products (Lallemand et al., 1999). Finding product differences is always a key criterion for any sensory method.

11.3 Other Methods and Analyses

11.3.1 Panelist Reliability

Bloom et al. (1995) discussed a method for measuring the reliability (reproducibility) of data from individual panelists serving as routine evaluators for time–intensity judgments. Of course, it is possible to look at the extracted parameters from any given subject over replicates

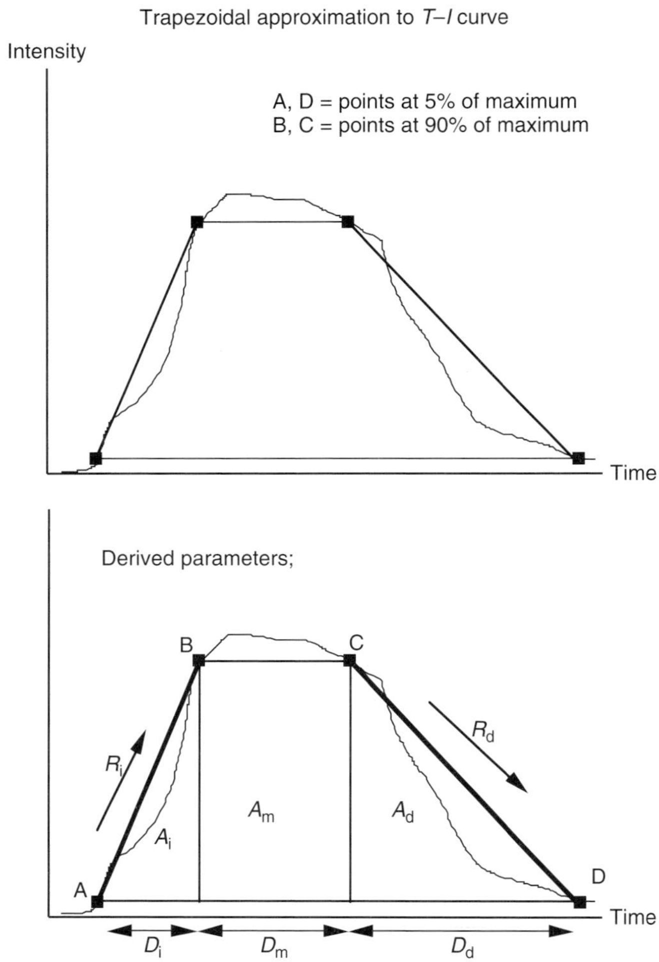

Figure 11.4 The trapezoidal method of Lallemand et al. (1999) for assessing curve parameters on *TI* records. The upper panel shows the basic scheme in which four points are found when the initial 5% of the intensity maximum I_{max} occurs, when 90% of I_{max} is first reached on the ascending segment, when the plateau is finished at 90% of I_{max} on the descending phase and the endpoint approximation at 5% of I_{max} on the descending phase. The lower panel shows the derived parameters, namely R_i, A_i, and D_i for the rate (slope), area, and duration of the initial rising phase, A_m and D_m for the area and duration of the middle plateau section, and R_d, A_d, and D_d for the rate (slope), area, and duration of the falling phase. A total duration can be found from the sum of D_i, D_m, and D_d. The total area is given by the sum of the *A* parameters or by the formula for the area of a trapezoid: Total area = $(I_{90}-I_5)(2D_m + D_i + D_d)/2$. Reprinted from Lawless and Heymann (2010) by permission of Springer Publishing.

as a measure of their consistency. However, this method uses the raw data. The method is based on slicing the time–intensity records at various fixed intervals (e.g., 10 or so) and calculating standard deviations at each time slice for each panelist. Of course, the study must involve replicated data in order to get a standard deviation.

The method is straightforward. Choose to slice the replicated time curves for a given panelist and a single product at N time intervals i (e.g., $i=1$ to 10). Calculate the standard deviation S_i at each i. Next, calculate the overall standard deviation S_{si} of the set of S_i and a mean $\overline{S}$. Now compute a Z-score for each S_i in order to normalize by the following relationship:

$$Z_i \frac{S_i - \overline{S}}{S_{si}} \tag{11.17}$$

The index of reliability TIR becomes the mean value of Z after eliminating the signs:

$$\text{TIR} = \frac{\sum_{i=1}^{N} |Z_i|}{N} \tag{11.18}$$

The normalization partly accounts for the problem of comparing different substances or concentration or ingredient levels that may have widely different $I_{\max}$ values, for example. The authors suggest that the measure is meaningful when comparing the same number of trials. The values for TIR are, of course, bounded by zero and rarely will range above a score of three.

The method does not fully address what to do with records that have different endpoints, and of course a slice taken with zero values is bound to have a low standard deviation, but may not necessarily be very trustworthy. The authors suggested a synchronization of start time for different panelists in order to compare curves, but did not mention endpoint equivalence. Perhaps some combination of adjustments, such as the Liu-MacFie method, would be helpful in that regard. Another issue is that the standard deviations are likely to increase with increasing intensity. In other words, intensity ratings are often a Poisson process, just another manifestation of Weber's law (see Chapter 2). So averaging items that are themselves level dependent would seem to suggest the need for a further adjustment, such as using the coefficient of variation (standard deviation divided by the mean) rather than the raw standard deviations themselves.

11.3.2 PCA and Noncentered PCA

A completely novel approach to the analysis of time–intensity curves was proposed by van Buuren (1992). Taking a single product, the data are cast with time blocks as rows, with as many rows as there are readings. For example, with one reading every second for 90 seconds, the data matrix has 90 rows. Columns are the individual panelists. Noting that panelists have consistent patterns or "signatures," van Buuren reasoned that it should be possible to extract common patterns as well as those that showed slightly different patterns from the group tendency. He submitted the data matrix for each individual product to PCA after column centering. By column centering we mean the normalization of the columns to a common mean, thus adjusting for different curve heights. This method is effectively a PCA on the covariance matrix, or a singular value decomposition after it is column centered (Dijksterhuis, 1993). The value of this method is that it produces factor scores for each time block, which can then be plotted as "principal curves." Van Buuren noted that the first principal curve usually collected most of the variance, with much smaller amounts for the other curves. This first curve tended to resemble the group average or most common trend. Other curves might represent different panelist signatures, such as delayed start or slow decay. However, there is no guarantee that the subsequent curves will be interpretable or represent anything meaningful. For example, Dijksterhuis (1993) performed a van Buuren type of analysis on six different bitter beverages and extracted three curves. His summary is telling: "In [principal curve 2] it is not easy to get an interpretation that makes sense, [principal curve 3] seems to contain mostly noise." (Dijksterhuis, 1993: 324).

Dijksterhuis and colleagues studied these methods further (Dijksterhuis et al., 1994; Dijksterhuis & van den Broek, 1995). In his original 1993 paper, he noted that van Buuren's method could fail to distinguish different products very effectively. He proposed a similar analysis, but without the column centering or panelist normalization. Such an analysis would retain the original curve height information in the raw data. The liability is that panelists with higher scale usage habits or greater taste sensitivity might contribute disproportionately to the principal curves (i.e., receive undue weighting). Dijksterhuis (1993) concluded that the noncentered PCA produced the best product differentiation, which was confirmed by Dijksterhuis et al. (1994). The second or subsequent principal curves using the noncentered procedure may also contain information about secondary temporal trends, such as accelerations or inflection points.

It may seem a little unusual to use panelists as if they were dependent variables in a PCA. For example, a common application of PCA analysis uses descriptive analysis means for a set of products (rows) with columns being the different attributes or rated characteristics. This produces a plot with factor loadings for the pattern of attribute correlations, and factor scores (new points in the perceptual map) for products. Similarities and differences among products are inferred from their positions in the map as well as projection of attributes into the space. Correlation arises from different products having similar scores on a pair of attributes. For example, product X is low in both sweetness and fruitiness and product Y is high in both. So in this example, sweetness and fruitiness are correlated. In the van Buuren/Dijksterhuis methods, the pattern of correlation arises from agreements of panelists about curve height at given time blocks. For example, if two panelists agree that the intensity of the product is low at time *X* and high at time *Y*, this will create a covariance pattern and allow the extraction of a principal component. Thus, it is not such a wild idea after all. A question remains about how to perform statistical tests on the resulting information, although bootstrapping or resampling methods could be applied. The user should be wary of the subjective nature of interpretation of PCA output, and give it weighting accordingly in decision making.

11.3.3 TDS Methods

The Flavor Profile Method (Cairncross & Sjöstrom, 1950) was one of the first techniques to use a trained panel to describe, quantitatively, the sensory characteristics of food products. One important aspect of the Flavor Profile was temporal information. Panelists would write down, on a blank sheet, all the sensory characteristics they perceived, in the order of appearance. Inherent in this method, then, was a sense of onset order for the different aspects of a product. With the advent of fixed ballots, as used in more recent versions of quantitative descriptive techniques, this information was lost. A recent class of methods, however, has reworked this idea to provide a profile of the dominant sensations from a product and how they rise and fall with time. These are called **temporal dominance of sensations** (TDS) methods (Le Reverend et al., 2008; Labbe et al., 2009; Pineau et al., 2009). The idea is simple. The panelists are provided with an abbreviated ballot (based upon some preliminary information) and can rate the intensity of one or two of the most dominant sensations from the provided list. When a new sensation becomes dominant, the panelist can switch to rating that characteristic.

Although it includes some intensity information, the TDS method should be viewed as primarily a qualitative, rather than quantitative, technique. The important information is that one can characterize the flow of the most important sensations in a product over time and contrast this with the temporal pattern of another product or set of products. This could be potentially valuable for complex products such as wine, which have distinctive time-related changes in sensations. For example, one might find that wine X has an initial onset of acidity, then sweetness, followed by fruity aromatics, then astringency in the finish, as opposed to wine Y that has sweetness, then acidity, then grassy and citrus notes, and finally a residual drying sensation and bitterness.

The data set thus has four variables: a fixed effect of time, a random effect of panelists, and dependent variables of attributes and intensities. Of course, one could assign zeros to attributes not chosen and then treat the complete record using attributes as a kind of dummy independent variable in the design. Various methods have been suggested for treating the data, and at least one of the major software packages for collecting and handling sensory data can now accommodate TDS experiments. Statistical treatment is not straightforward. One option is to convert the data to frequency information (similar to check-all-that-apply; see Chapter 10) where the proportion of panelists rating the product is the important information. This can then be compared with a hypothetical baseline of $1/k$, where k is the number of attributes and a simple one-tailed binomial test on proportions used for deciding statistical significance (Pineau et al., 2009). The 95% confidence limit can then be plotted as a horizontal line (as k is not changing) on a time-based plot of the dominant characteristics to show where the frequency rises above the baseline. Another option is to construct a derived measure such as the intensity multiplied by duration, then divided by the sum of the durations for that attribute (Labbe et al., 2009). This produces an intensity by persistence product, similar to that of the SMURF technique (Birch & Munton, 1981).

The idea that panelists can track quality changes over time is not new (Halpern, 1991; DeRovira, 1996). However, the TDS techniques provide additional information that is difficult to capture by traditional time–intensity ratings. *TI* ratings are usually done one attribute at a time, although some uses of dual attribute *TI* ratings have been attempted (Duizer et al., 1995, 1996; Zimoch & Findlay, 1998). TDS techniques may be able to capture a richer picture of the time properties of a product than *TI*, or at least do it with less labor-intensive methods. Pairs of products may be compared via construction of difference scores for attributes (Pineau et al., 2009). The method is still evolving, and many questions remain: How should the list of attributes be chosen? Why is $1/k$ a baseline for k attributes if more than one can be rated at time? Why is it legitimate to assign zeros to attributes not rated if their intensities may in fact be nonzero? How can individual differences be captured or accommodated? For example, it would be reasonable to think that individuals with different salivary flow rates or salivary buffering capacity would respond differently to acid stimuli over time. This could produce legitimate differences in records that could be averaged or smeared away in the composite record.

11.4 Summary and Conclusions

Spurred by interest from the late Rose Marie Pangborn, a flurry of activity in time–intensity scaling occurred in the 1980s and 1990s. Since then the tool has become a common fixture in the arsenal of sensory methods, although fewer research papers are devoted to it more recently. Its potential utility and advantages for some research questions, as well as its drawbacks, have been thoroughly discussed in sensory texts (Lawless & Heymann, 2010). Using a *TI* method involves several strategic questions: Do I need to measure changes in hedonics or just intensity? Are there one or multiple attributes that I need to track? Can I look at extracted curve features and parameters or is my goal to construct average curves? Traditional profiling or descriptive analysis may offer an alternative, particularly if there are multiple attributes to be recorded. This can simply be done by asking intensity questions at some (albeit few) fixed intervals. Continuous tracking of a single attribute offers the opportunity to see a temporal record in greater detail, but may invoke **halo effects** and **dumping effects** that artificially inflate some ratings (Clark & Lawless, 1994). So the sensory professional should weigh carefully whether the information is likely to be of sufficient value to justify the added effort needed for data collection and analysis. The critical question, of course, is whether the temporal properties, and changes in them, will affect consumer preferences.

References

Birch, G.G. and Munton, S.L. 1981. Use of the "SMURF" in taste analysis. Chemical Senses, 6, 45–52.

Bloom, K., Duizer, L.M., and Findlay, C.J. 1995. An objective numerical method of assessing the reliability of time–intensity panelists. Journal of Sensory Studies, 10, 285–94.

Brown, W.E., Landgley, K.R., Martin, A., and MacFie, H.J. 1994. Characterisation of patterns of chewing behavior in human subjects and their influence on texture perception. Journal of Texture Studies, 15, 33–48.

Butler, G., Poste, L.M., Mackie, D.A., and Jones, A. 1996. Time–intensity as a tool for the measurement of meat tenderness. Food Quality and Preference, 7, 193–204.

Cain, W.S. 1974. Perception of odor intensity and time-course of olfactory adaptation. ASHRAE Transactions, 80, 53–75.

Cairncross, S.E. and Sjöstrom, L.B. 1950. Flavor profiles: a new approach to flavor problems. Food Technology, 4, 308–11.

Clark, C.C. and Lawless, H.T. 1994. Limiting response alternatives in time–intensity scaling: an examination of the halo-dumping effect. Chemical Senses, 19, 583–94.

Cliff, M. and Heymann, H. 1993a. Time–intensity evaluation of oral burn. Journal of Sensory Studies, 8, 201–11.

Cliff, M. and Heymann, H. 1993b. Development and use of time–intensity methodology for sensory evaluation: a review. Food Research International, 26, 375–85.

DeRovira, D. 1996. The dynamic flavor profile method. Food Technology, 50, 55–60.

Dijksterhuis, G. 1993. Principal component analysis of time–intensity bitterness curves. Journal of Sensory Studies, 8, 317–28.

Dijksterhuis, G., and Eilers, P. 1997. Modelling time–intensity curves using prototype curves. Food Quality and Preference, 8(2), 131–40.

Dijksterhuis, G. and van den Broek, E. 1995. Matching the shape of time–intensity curves. Journal of Sensory Studies, 10, 149–61.

Dijksterhuis, G., Flipsen, M., and Punter, P.H. 1994. Principal component analysis of time–intensity data. Food Quality and Preference, 5, 121–7.

DuBois, G.E. and Lee, J.F. 1983. A simple technique for the evaluation of temporal taste properties. Chemical Senses, 7, 237–47.

Duizer, L.M., Gullett, E.A., and Findlay, C.J. 1993. Time–intensity methodology for beef tenderness perception. Journal of Food Science, 58, 943–7.

Duizer, L.M., Findlay, C.J., and Bloom, K. 1995. Dual-attribute time–intensity sensory evaluation: a new method for temporal measurement of sensory perceptions. Food Quality and Preference, 6, 121–6.
Duizer, L.M., Bloom, K., and Findlay, C.J. 1996. Dual attribute time–intensity measurements of sweetness and peppermint perception of chewing gum. Journal of Food Science, 61, 636–8.
Eilers, P.H.C. and Dijksterhuis, G.B. 2004. A parametric model for time–intensity curves. Food Quality and Preference, 15, 239–45.
Gent, J.F. 1979. An exponential model for adaptation in taste. Sensory Processes, 3, 303–16.
Gent, J.F. and McBurney, D.H. 1978. Time course of gustatory adaptation. Perception & Psychophysics, 23, 171–5.
Guinard, J.-X., Pangborn, R.M., and Shoemaker, C.F. 1985. Computerized procedure for time–intensity sensory measurements. Journal of Food Science, 50, 543–44, 546.
Guinard, J.-X., Pangborn, R.M., and Lewis, M.J. 1986. The time course of astringency in wine upon repeated ingestions, American Journal of Enology and Viticulture, 37, 184–9.
Gwartney, E. and Heymann, H. 1995. The temporal perception of menthol. Journal of Sensory Studies, 10, 393–400.
Hahn, H. 1934. Die adaptation des Geschmackessinnes. Zeitschrift für Sinnesphysiologie, 65, 105–45.
Halpern, B.P. 1991. More than meets the tongue: temporal characteristics of taste intensity and quality. In: Sensory Science Theory and Applications in Foods. H.T. Lawless and B.P. Klein (Eds). Marcel Dekker, New York, NY, pp. 37–105.
Labbe, D., Schlich, P., Pineau, N., Gilbert, F., and Martin, N. 2009. Temporal dominance of sensations and sensory profiling: a comparative study. Food Quality and Preference, 20, 216–21.
Lallemand, M., Giboreau, A., Rytz, A., and Colas, B. 1999. Extracting parameters from time–intensity curves using a trapezoid model: the example of some sensory attributes of ice cream. Journal of Sensory Studies, 14, 387–99.
Lawless, H.T. 1984. Oral chemical irritation: psychophysical properties. Chemical Senses, 9, 143–55.
Lawless, H.T. and Clark, C.C. 1992. Psychological biases in time–intensity scaling. Food Technology, 46(11), 81, 84–6, 90.
Lawless, H.T. and Heymann, H. 2010. Sensory Evaluation of Foods, Principles and Practices. Second edition. Springer, New York, NY.
Lawless, H.T. and Skinner, E.Z. 1979. The duration and perceived intensity of sucrose taste. Perception & Psychophysics, 25, 249–58.
Le Reverend, F.M., Hidrio, C., Fernandes, A., and Aubry, V. 2008. Comparison between temporal dominance of sensation and time intensity results. Food Quality and Preference, 19, 174–8.
Leach, E.J. and Noble, A.C. 1986. Comparison of bitterness of caffeine and quinine by a time–intensity procedure. Chemical Senses, 11, 339–45.
Ledauphin, S., Vigneau, E., and Causeur, D. 2005. Functional approach for the analysis of time intensity curves using B-splines. Journal of Sensory Studies, 20, 285–300.
Ledauphin, S., Vigneau, E., and Qannari, E.M. 2006. A procedure for analysis of time intensity curves. Food Quality and Preference, 17, 290–5.
Lee, W.E. 1985. Evaluation of time–intensity sensory responses using a personal computer. Journal of Food Science, 50, 1750–1.
Lee, W.E. 1986. A suggested instrumental technique for studying dynamic flavor release from food products. Journal of Food Science, 51, 249–50.
Lee, W.E. and Pangborn, R.M. 1986. Time–intensity: the temporal aspects of sensory perception. Food Technology, 40, 71–8, 82.
Liu, Y.H. and MacFie, H.J.H. 1990. Methods for averaging time–intensity curves. Chemical Senses, 15, 471–84.
Lundahl, D.S. 1992. Comparing time–intensity to category scales in sensory evaluation. Food Technology, 46(11), 98–103.
McGowan, B.A. and Lee, S.-Y. 2006. Comparison of methods to analyze time–intensity curves in a corn zein chewing gum study. Food Quality and Preference 17, 296–306.
Meilgaard, M., Civille, G.V., and Carr, B.T. 2006. Sensory Evaluation Techniques. Third edition. CRC Press, Boca Raton, FL.
Moore, L.J. and Shoemaker, C.F. 1981. Sensory textural properties of stabilized ice cream. Journal of Food Science, 46, 399–402, 409.
Neilson, A.J. 1957. Time–intensity studies. Drug and Cosmetic Industry, 80, 452–3, 534.
O'Mahony, M. 1986. Sensory adaptation. Journal of Sensory Studies, 1, 237–57.

Ott, D.B., Edwards, C.L., and Palmer, S.J. 1991. Perceived taste intensity and duration of nutritive and non-nutritive sweeteners in water using time–intensity (*T–I*) evaluations. Journal of Food Science, 56, 535–42.
Overbosch, P. 1986. A theoretical model for perceived intensity in human taste and smell as a function of time. Chemical Senses, 11, 315–29.
Overbosch, P. 1987. Flavour release and perception. In: Flavour Science and Technology. M. Martens, G.A. Dalen, and H. Russwurm (Eds). John Wiley & Sons, Inc., New York, NY, pp. 291–300.
Overbosch, P., Van den Enden, J.C., and Keur, B.M. 1986. An improved method for measuring perceived intensity/time relationships in human taste and smell. Chemical Senses, 11, 315–38.
Owen, W.J. and DeRouen, T.A. 1980. Estimation of the mean for lognormal data containing zeroes and left-censored values, with application to the measurement of worker exposure to air contaminants. Biometrics, 36, 707–19.
Pangborn, R.M., Lewis, M.J., and Yamashita, J.F. 1983. Comparison of time–intensity with category scaling of bitterness of iso-alpha-acids in model systems and in beer. Journal of the Institute of Brewing, 89, 349–55.
Peyvieux, C. and Dijksterhuis, G. 2001. Training a sensory panel for *T–I*: a case study. Food Quality and Preference, 12, 19–28.
Pineau, N., Schlich, P., Cordelle, S., Mathonniere, C., Issanchou, S., Imbert, A., Rogeaux, M., Eteviant, P., and Köster, E. 2009. Temporal dominance of sensations: construction of the TDS curves and comparison with time–intensity. Food Quality and Preference, 20, 450–5.
Pionnier, E., Nicklaus, S., Chabanet, C., lMioche, L., Taylor, A.J., Le Quérés, J.L., and Salles, C. 2004. Flavor perception of a model cheese: relationships with oral and physico-chemical parameters. Food Quality and Preference, 15, 843–52.
Prescott, J. and Stevenson, R.J. 1996. Psychophysical responses to single and multiple presentations of the oral irritant zingerone: relationship to frequency of chili consumption. Physiology & Behavior, 60, 617–24.
Riggs, D.S. 1963. The Mathematical Approach to Physiological Problems. MIT Press, Cambridge, MA.
Stuiver, M. 1958. Biophysics of the sense of smell. Thesis, Groningen, The Netherlands.
Taylor, D.E. and Pangborn, R.M. 1990. Temporal aspects of hedonic response. Journal of Sensory Studies, 4, 241–47.
Tuorila, H. and Vainio, L. 1993. Perceived saltiness of table spreads of varying fat compositions. Journal of Sensory Studies, 8, 115–20.
Van Buuren, S. 1992. Analyzing time–intensity responses in sensory evaluation. Food Technology, 46 (2), 101–4.
Veldhuizen, M.G., Wuister, M.J.P., and Kroeze, J.H.A. 2006. Temporal aspects of hedonic and intensity responses. Food Quality and Preference, 17, 489–96.
Wendin, K., Janestad, H., and Hall, G. 2003. Modeling and analysis of dynamic sensory data. Food Quality and Preference, 14, 663–71.
Yoshida, M. 1986. A microcomputer (PC9801/MS mouse) system to record and analyze time–intensity curves of sweetness. Chemical Senses, 11, 105–18.
Zimoch, J. and Findlay, C.J. 1998. Effective discrimination of meat tenderness using dual attribute time intensity. Journal of Food Science, 63, 940–4.
Zimoch, J. and Gullett, E.A. 1997. Temporal aspects of perception of juiciness and tenderness of beef. Food Quality and Preference, 8, 203–11.

12 Product Stability and Shelf-Life Measurement

In no way can a few skilled assessors predict whether a product will be acceptable to consumers or not, and relying on this methodology will inevitably lead to mistakes.

Hough (2010: 12)

12.1 Introduction

An important aspect of product quality is the stability of a product over time. Manufacturers must make products that meet consumer expectations. These expectations include the survival of an intact product through the distribution channels, the appearance of an intact and visually appealing product on the store shelves, and a reasonable time to expiration once the product is purchased. Of course, these expectations differ for different product categories. The shelf life of a bread product is not expected to be as long as that of a cookie (or biscuit if you are British) that would have lower water activity. Consumer information concerning shelf life has increased over the years. Depending upon the product category, the product may bear a sell-by date or a use-by date. Sometimes the language may be phrased in consumer-friendly terms that suggest that the product is still usable after a certain time, such as "best if used by______." In the USA, there is a confusing plethora of such phrases, although in other countries the shelf-life specifications are more clearly regulated and defined.

Quantitative Sensory Analysis: Psychophysics, Models and Intelligent Design, First Edition. Harry T. Lawless.

The variability in the use of such terms lends them more to consumer guidance, than hard-and-fast rules for when a product is no longer usable.

Sooner or later, all sensory professionals become involved in shelf-life or stability testing. It is one important aspect of new product development. Unfortunately, it is often overlooked in the early stages of product innovation, and sometimes there is a rush to obtain shelf-life information in the face of an impending product launch in the marketplace. A common comment from sensory practitioners is "they expect me to have a time machine." In some food companies, product stability measurement is a normal part of sensory quality control. Its repetitive nature makes it one of the less appealing and less creative tasks in the sensory evaluation world. It is easy to overlook the importance of temporal stability in consumers' expectations of product quality. Store retrievals and retail quality audits can be useful in checking for product deterioration in the distribution chain.

Shelf-life determination is an integral part of packaging technology, and many packaging textbooks will include a section on product stability measurement (Robertson, 2006). One of the important functions of packaging, after all, is to maintain the integrity of a product for an acceptable period of time; that is, to protect it from factors that would tend to accelerate its deterioration. The packaging provides a barrier to microbial changes, oxidative deterioration, light-catalyzed changes, and so forth, all of which may be important influencers on consumer rejection of a product. Furthermore, they are almost always time dependent.

Product stability measurement is discussed within the framework of general quality control in the textbook by Lawless and Heymann (2010). A general reference is the comprehensive treatise by Hough (2010), *Sensory Shelf Life Estimation of Food Products*. Hough's book deals with a variety of situations, measurements, and food products. The book also provides statistical functions in the R language that can be used for shelf-life modeling. For the sensory professional involved in stability measurement of consumer products, it is highly recommended. This chapter discusses some of the models and approaches to the determination of sensory shelf life. Shelf-life testing is in many ways similar to a routine assessment of product quality in a quality control program. A useful resource for such sensory methods is the book on sensory quality control by Muñoz *et al.* (1992). Historical papers on sensory approaches to product shelf life and design of experiments are the papers by Dethmers (1979), Gacula (1975), and Gacula and Kubala (1975). Other works concerned with shelf life are reviewed in the first chapter of Hough (2010). A general food science text that also deals with specific product categories is the text edited by Man and Jones (2000).

12.2 Strategies, Measurements, and Choices

What defines product failure? The sensory professional can use different measurements to determine shelf life. An obvious measurement is consumer rejection (Peryam, 1964). "Rejection," however, is a broad term. At face value it would seem to imply a product that would not be purchased or not consumed (i.e., discarded). However, a conservative manufacturer may choose to define rejection as some level of decrement or fall-off in consumer acceptance ratings, and not necessarily into the range of disliking. Alienation of consumers is a risky business, and one might want to insure that a point of real objectionability is never reached. An even more conservative approach is to define a rejection point in terms of discrimination. A product that has changed in any way whatsoever may be deemed unsalable or unacceptable to consumers. In that case, a simple discrimination test may be appropriate. But most products can change in some way but still be acceptable, and sometimes a combination of a discrimination test followed by a consumer acceptance test can be applied. If the product fails the discrimination test, it is then followed by a consumer evaluation, a decision-tree approach.

If the factors that drive consumer rejection are well known, a potentially more cost-effective approach for shelf-life analysis may be to use a trained descriptive panel. This approach requires a detailed understanding of the sensory changes that consumers deem important, and at what level of change a given percentage of consumers is likely to find the product unacceptable. Obviously, there are a number of important professional decisions in this kind of method. What percentage of consumer rejection do we choose as a cut-point for calibrating our descriptive measurements? Before the trained panel can be used accurately, a calibration study needs to be done to relate consumer acceptance/rejection to the key descriptive attribute levels. This can be an expensive and daunting task, depending upon the number of products and line extensions within a manufacturer's portfolio.

A sensory method related to descriptive analysis has been the historical use of trained evaluators in quality measurement. These are sometimes referred to as "expert panels," although "quality grading" may be somewhat more accurate. These evaluations are typified by the dairy science approach to quality scoring (Bodyfelt et al., 1988; Clark et al., 2009). The evaluators would be trained in recognizing all the ways in which a product could become defective. Thus, the method resembles a descriptive analysis, except that assessors are focused only on the negative aspects of a product's sensory attributes. Associated with a score for the severity of a particular defect is a point deduction system for an overall quality score, which presumably reflects a decreased appeal or increased objectionability from a consumer's perspective. For over a century, such quality grading schemes have been useful in providing a short-cut cost-effective methods for insuring product quality and for estimation of shelf life. However, these sensory grading methods have been roundly criticized by the mainstream sensory evaluation community. The quality grading systems are poorly suited to the development of new and innovative processed foods. After all, they were developed for standardized commodities. Furthermore, the scores from such expert judges do not always reflect consumer opinion (McBride & Hall, 1979; O'Mahony, 1979; Claassen & Lawless, 1992; Lawless & Claasen, 1993).

Quantitative modeling of product shelf life and time-related failure has drawn from a number of other scientific disciplines, including chemical kinetics and thermodynamics and actuarial science's measurement of death rates. This chapter will explore several of these quantitative models. The sensory professional should bear in mind that the complexity of foods does not always lend itself to modeling based on a simple chemical reaction or a single-phase monotonic statistical model for time-related change. These limitations are especially apparent when one considers the variety of simultaneous product changes that can occur, such as microbial deterioration, oxidative and other chemical changes, and water migration, to name a few. These physical changes can result in a variety of sensory phenomena, such as discolorations and other visual changes, flavor changes, and texture changes. Each of the physical and sensory changes may proceed at different rates, and thus product "failure" is rarely a simple process. This complexity becomes especially relevant if one attempts to estimate shelf life using accelerated or temperature-enhanced storage testing. It would be ludicrous, for example, to subject a food to elevated temperature testing if it was prone to a phase change at higher temperatures (ice cream comes to mind as an obvious example).

The fact that multiple processes are at work in time-related product deterioration is exemplified by the common **bathtub function**, as shown in Figure 12.1. The bathtub

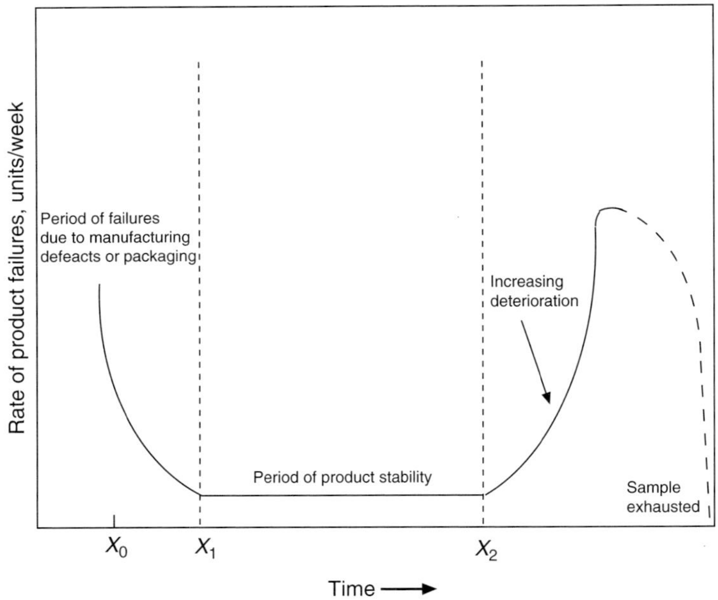

Figure 12.1 The "bathtub" function showing a common pattern of failure rates changing over the sampling time in a shelf-life study. From time X_0 to X_1, some products will fail due to improper processing or faulty packaging. This is followed by a period of fairly low failure rates when products are stable or within specification limits. At time X_2, failures start to increase markedly. Researchers fitting hazard functions or doing survival analysis for estimation of shelf life should consider using only those times after X_2 in curve-fitting hazard functions as earlier failures are due to causes other than the time-related deterioration. The bathtub function is an incomplete picture, because failures will eventually decrease as all units are expended, used up, or fail in service.

function illustrates a general kind of hazard function, in which the probability of product failure is plotted over time. In the very early stages of a product's shelf life, there may be manufacturing or production or packaging errors that lead to some number of early or immediate failures. For example, the product's inner package may fail to seal properly. Manufacturers will work hard to minimize these kinds of problems, but no process is completely foolproof. So a certain number of immediate problems are expected, and in a good quality control system the defective products are caught before they leave the factory. This period is usually followed by a period of product stability, in which failures are relatively rare. As time progresses, further chemical or physical changes progress and lead to more noticeable sensory changes. At some point the sensory changes are both noticed and objectionable to consumers; that is, a rejection threshold (Prescott et al., 2005) has been crossed, and product failures become more common. This is the latter phase of the bathtub function, and the part that is usually described by most time-related quantitative models.

The reader should keep in mind that quantitative models for product stability are primarily descriptive, and not meant to imply causal models. That is, they provide some handy simplified equations for describing time-related occurrences. But they are mathematical descriptions of empirical phenomena only. However, one is tempted to think of them as more scientific causal models, especially those that come from chemical kinetics. Consider **activation energy**. It is useful in chemistry to know about the potential energy barriers to any specific chemical reaction. But as we have noted, foods are complex systems that are changing in many ways simultaneously. We can still estimate an "activation energy" for a time-related product change using Arrhenius equations and by systematic temperature variation. Such a measurement may even provide a convenient index of the temperature stability of a product. However, it is important to keep in mind that you are probably not dealing with a single chemical reaction in any given time–temperature study.

12.3 Study Designs

Sampling over time can be done in several ways. The sensory specialist should be careful to sample enough time points to allow an accurate function to be fit to the data. A well-fitting function will allow interpolation with some degree of certainty about the estimates. If too few time points are sampled, you run the risk of having one or two outlying points give a false reading on the shape or slope parameters of the hazard or failure function. Obviously, some previous knowledge about the nature, extent, and rates of failure will facilitate a sensible sampling design. The goal is usually to sample more points around the times of increasing product failures. However, this will depend somewhat on the nature of the mathematical function. A linear failure function suggests a constant time interval for testing, while an exponential function might be better served by increasing detervals as the testing progresses.

A critical question in the design is whether you want to try to perform all of the testing on the same day, or do the sensory evaluations sequentially on each pull date from storage. The most common design is simply to pull items from storage at the specified time intervals and conduct a sensory test each time. This is not always efficient, however, as it demands that the panel (whether consumers or descriptive or discrimination panels) be assembled for each test. With an internal company panel and a regular ongoing testing program, this is not difficult. For a consumer study, however, the cost factors may influence the need to test on a single day.

Single-day testing is achieved by a **reversed-storage design**. In one form of reversed storage, all the products are made at a single time and placed under the appropriate storage conditions. For example, a bread product might be stored at 20 °C or the common temperature for distribution and store shelving. Samples are then pulled at the appropriate intervals and moved to an *optimal storage* condition. For bread, this might be a frozen storage at −18 °C. On the day of testing, all the samples are removed from storage and submitted for sensory analysis. Another option is to freeze all of the samples first, and then pull samples from storage at the appropriate intervals, to allow for different storage times at the distribution temperature. This way, all the samples have been subject to the same freeze–defrost cycle. The **optimal storage** approach assumes that any changes in the product during the optimal conditions are minimal. This may or may not be true for any given food product.

12.4 Hazard Functions and Failure Distributions

12.4.1 Curve Fitting and Statistical Distributions

Modeling the probability of product failure over time necessitates the idea that there will be a statistical distribution of product failures and the probability of failure changes over time. The two events that can be modeled are failure, and conversely, survival. The failure function can be thought of as the cumulative distribution for a product rejection event. The survival function is the probability that the rejection event will be greater than a given time T. It is assumed to

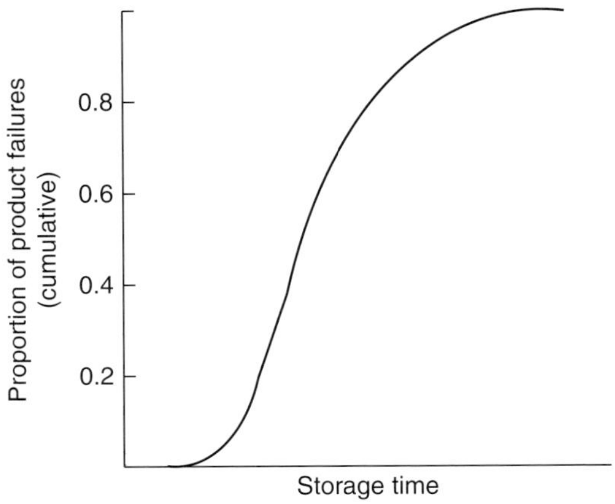

Figure 12.2 A cumulative distribution function of product failures over time.

be equal to one at time zero, and will approach zero as the shelf life is surpassed. It can also be viewed as a product acceptance function. **Hazard** is the momentary probability that the event will occur at time *T*, assuming (or conditional upon) surviving until that time.

Figure 12.2 shows a simple cumulative distribution of product failures over time. Obviously, it would be valuable to fit a function to such a curve in order to estimate the times to reach different percentages of failures that might be of interest to a manufacturer. For a food with a known and limited shelf life, a 50% consumer rejection point might be of interest. This has been defined by Prescott and others as a consumer rejection threshold (Prescott et al., 2005, Harwood et al., 2012). For a conservative and quality-oriented manufacturer with a loyal customer base, lower percentages (25% or 10%) may suggest a pull-by date that provides a better safety net against potential consumer alienation.

As with many statistical distributions, it is common to describe the events by a location and shape parameter (e.g., a mean and standard deviation). If there is a bell-shaped symmetric distribution, two options are the simple normal distribution and a logistic distribution as shown in eqns 12.1 and 12.2.

$$F(t) = \Phi\left(\frac{t-\mu}{\sigma}\right) \tag{12.1}$$

where Φ is the cumulative normal distribution, μ is the mean, and σ is the standard deviation, modeled as a function of time. Another common choice is the logistic function, which was originally used to model population growth that would approach some natural limit:

$$F(-t) = \frac{1}{1+e^{-t}} = \frac{e^{t}}{1+e^{t}} \tag{12.2}$$

This relationship is often modeled by its logistic regression form, in which a linear function of the odds ratio is used to fit the slope and intercept parameters:

$$\ln\left(\frac{p}{1-p}\right) = b_0 + b_1 t \tag{12.3}$$

where p is the probability of rejection (or failure, so $p = F(t)$) and b_0 and b_1 are the parameters to be fit. In some cases, $\log(t)$ should be substituted for t. The underlying distribution for the logistic function is similar to the Gaussian bell curve, and also produces a sigmoidal cumulative distribution, but it is somewhat "heavier in the tails."

Because many time-dependent processes are not normal or symmetric, a useful function is to model a lognormal distribution. When there are some rare products that last a very long time, compared to the bulk of the distribution, the failure or survival function should be lognormal. The failure of electric incandescent light bulbs is a common example. In this case the distribution of events are right-skewed, and become rare are time progresses. The rejection function can be given by

$$F(t) = \Phi\left[\frac{\ln(t) - \mu}{\sigma}\right] \tag{12.4}$$

Where Φ is the cumulative normal distribution, μ is the mean and σ is the standard deviation, modeled as a function of log time.

Hough (2010) uses another common function, the Weibull distribution,[1] which is frequently used in modeling product failure and other hazard functions:

$$F(t) = F_{sev}\left[\frac{\ln(t) - \mu}{\sigma}\right] \tag{12.5}$$

And $F_{sev}(x)$ is the extreme value distribution, or

$$F_{sev}(x) = 1 - \exp[-\exp(x)] \tag{12.6}$$

where exp(x) is equal to e^x and thus

$$F(t) = 1 - \exp\left\{-\exp\left[\frac{\ln(t) - \mu}{\sigma}\right]\right\} \tag{12.7}$$

This is sometimes reparameterized as

$$F(t) = 1 - \exp\left[-\left(\frac{t}{\eta}\right)^{\beta}\right] \quad \text{or} \quad F(t) = 1 - \exp\left[-(\rho t)^{k}\right] \tag{12.8}$$

where $\beta = k = 1/\sigma$, $\mu = \ln(\eta)$, and $\rho = \exp(-\mu)$.

A quick worked example. Weibull distributions are very versatile and can take on many different shapes, including bell-like symmetric distributions and various skewed-looking shapes. The keys to using them for shelf-life estimation in eqns 12.7 and 12.8 are first to remember that the location parameter μ, is not the same as the mean in a normal or Gaussian distribution, and that it is figured in ln(t) (e.g., the natural log of hours).

Suppose we have determined that the distribution for a yogurt product has the following values: $\mu = 3.39$, $\sigma = 0.79$. Using the Weibull function (eqn 12.7), what is the expected failure percentage $F(t)$ at t=25 hours? Substituting with these values we get the following expression:

$$F(t) = 1 - \exp\left\{-\exp\left[\frac{\ln(25) - 3.39}{0.79}\right]\right\}$$

[1] Weibull distribution information can be found at http://www.mathpages.com/home/kmath122/kmath122.htm and http://www.weibull.nl/weibullstatistics.htm (accessed 18 November 2012).

and since $(\ln(25)-3.39)/0.79=-0.228$, we get

$$1-e^{-e^{-0.228}}=1-e^{-0.796}=1-0.45=0.55$$

or 55% failure at 25 hours under these storage conditions. Working with the somewhat friendlier expression in terms of ρ and k, the parameters now become $k=1/\sigma=1.266$ and $\rho=e^{-\mu}=0.337$. Using the right side of eqn 12.8 then gives a value for $-(\rho t)^k=[0.337(25)]^{1.226}=0.805$ and thus

$$F(t)=1-\exp\left[-(\rho t)^k\right]=1-e^{-0.805}=1-0.45=0.55$$

Buy rearranging, the Weibull function can also be used to estimate the time necessary to reach a certain proportion of failures, once again if the shape and spread parameters are known. Solving for time t we get a somewhat unwieldy but potentially useful function for finding the time at any particular rejection proportion ($F(t)$), once k and ρ are estimated:

$$t=\frac{-\ln\left(1-F(t)\right)^{1/k}}{\rho} \qquad (12.9)$$

A quick worked example (or two). Suppose we want to know the time to reach 50% failure ($F(t)$ in the above expression). The product has the following parameters, found in previous experimentation and after fitting the Weibull function: $\mu=4.125$, $\sigma=0.539$, so $\rho=0.016$ and $k=1.855$. This value of μ corresponds to 62 hours. Plugging into eqn 12.9, the 50% failure time becomes

$$t=\frac{-\ln\left(1-0.5\right)^{0.539}}{0.016}=\frac{0.693^{0.539}}{0.016}=51.3\text{ hours}$$

Suppose we had a second product, with a shorter shelf life and a wider spread parameter as follows: $\mu=3.38$, $\sigma=0.787$, so $\rho=0.034$ and $k=1.27$. This value of μ corresponds to about 30 hours. Plugging into eqn 12.9, the 50% failure time becomes

$$t=\frac{-\ln\left(1-0.5\right)^{0.787}}{0.034}=\frac{0.693^{0.787}}{0.034}=22\text{ hours}$$

So cutting the location parameter in half and raising the spread parameter cuts the shelf life by more than half if you define the failure cutoff as 50% failure.

The parameters μ and σ can also be found by maximum likelihood methods, solving two partial differential equations (setting the derivatives to zero to find the maxima for the likelihood of any μ and σ, or $L(\mu,\sigma)$ as follows (Hough, 2010):

$$\frac{\partial \ln L(\mu,\sigma)}{\partial\mu}=0 \qquad (12.10)$$

and

$$\frac{\partial \ln L(\mu,\sigma)}{\partial\sigma}=0 \qquad (12.11)$$

which can be approached using the Newton–Raphson method or other techniques for systems of differential equations. Hough (2010) gives an example of how to do this using the R statistical package.

12.4.2 Batch Data: Using Median Ranks

The above statistical equations are useful for fitting model parameters to the proportion of failures in a single experiment. But suppose one has an extensive data set giving the mean or median failure times for a group of batches of products produced at various locations, times, or some other collection of variables such as different manufacturing lines. Another simple method for fitting to the lognormal distribution can be used when we have different batches with known failure times, which we will call the median rank method, commonly used in some failure analyses.

A simple graphical approach works as follows to find the interpolated 50% failure level. For N samples of foods sampled over time that have known failure times T_i, rank all the batches i as to the time of failure ($i=1$ to N). Calculate the median ranks, MR-values. The median rank can be found in some statistical tables or estimated as $\text{MR}=(i-0.3)/(N+0.4)$.

Plot the median rank on log probability paper versus T_i and interpolate at the 50% point. If a straight line fits the data, the 50% point can be estimated from the linear equation and standard deviations estimated from the probability paper. This is essentially a fit of log MRs to Z-scores. If one prefers a numerical parameter estimation to a graphical solution, a least-squares estimate can be found as follows:

1. Convert each T_i (time of failure) to $\ln(T_i)$, called Y_i. This will permit a fit of MR to the lognormal model.
2. Calculate the Z-score for each MR at each T_i. Call this X_i.
3. Regress Y against X using least squares to get the linear equation $Y=a+bX$. This is equivalent to finding the straight-line fit to the log probability plot.
4. Solve for $Y=0$ (Z-score for 50%) which is $X=-a/b$, to get the 50th percentile.
5. Convert back to the original units by exponentiating:

$$\text{Time at 50\% failure} = e^{X} = e^{-a/b}$$

12.4.3 Censoring in Shelf-Life Data

Let us assume you set up a shelf-life test using consumer data as input, and you test at N intervals of time t_i ($i=1$ to N). Assume there are M consumers, noted by subscript j. As the test progresses, each consumer gets to some rejection point in which all subsequent products are also rejected. That defines a rejection time t_{ij} for that particular consumer j. However, we do not know exactly when this consumer would reject the product, only that it would occur between time t_i and the previous testing interval t_{i-1}, when the consumer still found the product to be acceptable. This is an example of **interval censoring**. A consumer who found all samples to be acceptable would be right-censored and one who rejected all products would be left-censored. Hough (2010) has argued that this censored nature of the consumer data set should be taken into account and form part of the modeling for sensory shelf-life studies. Hough gives several examples of how censored data can be analyzed for shelf life using the R statistical package. At this time it is not clear how a censored-data analysis would differ from one using the first rejection point as the estimate of t_{ij} for a given consumer. Logically, one could argue that taking the first rejection point is in fact too late, as a consumer might very well reject a product between time t_i and t_{i-1}.

12.4.4 Cutoff Point Testing

Once a consumer rejection study has been performed, it may be possible to analyze the reasons for consumer rejection and then use a trained panel to estimate the degree of a particular defect (or set of defects) that are likely to produce a given amount of consumer rejection. Finding out the reasons for consumer rejection requires a careful analysis of open-ended or "check-all-that-apply" data, and relies upon consumers' abilities to verbalize the reasons for rejection. A somewhat more risky approach is simply to have the trained panel examine products stored for the same time intervals and find all the characteristics that have changed, and to what degree (e.g., browning of bananas) it occurs at the chosen time point for consumer rejection. The panel may then analyze any storage set for the critical attribute(s), but this assumes that what you are looking for is in fact the basis for consumer rejection. On the other hand, even if it is *not* the basis, it may be so highly correlated with consumers' criteria that you are fortuitously lucky and have a good predictor anyway.

The choice of a measure for the cutoff point requires careful consideration. Options include (1) a significant difference in a discrimination test, (2) some degree of difference from control product on a scaled attribute, (3) a cutoff point on an overall degree of difference scale, and (4) various forms of consumer data. Consumer data may involve a significant difference in acceptability ratings from control, a cutoff point on an acceptance score, or some percentage of consumer rejection (e.g., 10% or 25%). Giminez et al. (2007) found a statistically significant sensory difference to be too conservative a criterion because acceptance scores were still above 6 on the nine-point scale. Obviously, two products may differ but still be acceptable (Kilcast, 2000). Another option is to use any value less than 6.0 on the nine-point hedonic scale (6 = like slightly; i.e., just above neutral) (Muñoz et al., 1992). As discussed above, a simple option is consumer rejection; for example, agreeing with the statement, "I would not buy/eat this product" (Hough et al., 2003). These two measures are not necessarily equivalent, but logically they should be related. Giminez et al. (2008) found that, for certain baked products, consumers might not like the product, but they would answer "yes" when asked if they would consume it at home. Having already purchased it, they might consume it anyway so as not to go to waste. A criterion of simple consumer rejection may not be sufficiently conservative; that is, that some products may become disliked and even generate consumer complaints before the point of complete rejection is reached. Giminez et al. (2007) showed that acceptability scores can be related to percentage of rejection by logistic regression analysis, as one might expect. However, the logistic equations for two different countries (Spain and Uruguay) for a baked product were different. This serves as a warning about cultural, regional, and/or national differences.

There are advantages and disadvantages to using trained panels instead of using consumers, and these are discussed at length in many sensory texts. Briefly, the trained panel may be less costly to test on any given day than using consumers, but of course is more costly to set up in terms of screening and training time. Trained panels require monitoring and calibration, and replacement of lost panelists due to attrition. Like any analytical instrument, there are maintenance costs.

One liability in a cutoff point method is that the chosen point represents some mean percentage of consumers, but may not deal with minority opinion or segments of consumers that have different criteria for product rejection. Some people like tart green bananas, and some that are overly ripe, sweet, and brown. Many US consumers are used to oxidized flavors in cooking oils, even ones that might be rejected by a trained panel as unsuitable for sale. So your point estimate may not reflect a segment of consumers with different criteria.

12.5 Reaction Rates and Kinetic Modeling

12.5.1 Reaction Rates

One option for modeling shelf life borrows from the study of chemical kinetics: the study of reaction rates. Often, a reaction rate is temperature dependent and in most cases proceeds faster at higher temperatures. This idea is one of the foundations of accelerated shelf life testing, in which samples are stored and aged at higher than normal temperatures to approximate the results at lower temperatures and longer times. Accelerated testing and its potential limitations are discussed in Section 12.5.2.

In a first-order system, the reaction is a constant function of time, so a simple multiplicative constant provides the slope for the appearance of some reaction product (e.g., a lipid oxidation product) or disappearance of the original substrate for the reaction. Some enzymatic degradations, nonenzymatic browning reactions, and lipid oxidations follow first-order kinetics. If you are modeling a decline in some quantity, such as the disappearance of salable products due to deterioration or a decline in consumer acceptability ratings, the constant will be negative. If you are modeling an increase in some measure, such as an increasing oxidation score by a trained panel, the constant will be positive. So, for an increasing sensory score,

$$S = S_0 + kt \qquad (12.12)$$

where S is the oxidation score at time t, S_0 is the initial score at time zero, and k is the rate constant (= $\mathrm{d}S/\mathrm{d}t$).

In some other cases, the decay or growth may be exponential, which is a case of first-order kinetics. The change is proportional to the measured parameter. So $\mathrm{d}S/\mathrm{d}t = kS$. These follow the relationship

$$S = S_0 \mathrm{e}^{kt} \qquad (12.13)$$

and

$$\ln(S) = \ln(S_0) + kt \quad \text{or} \quad \ln\left(\frac{S}{S_0}\right) = kt \qquad (12.14)$$

Examples of first-order reactions include some microbial processes and their attendant sensory defects (appearance of slime, off-flavors), vitamin loss, and loss of protein quality in powdered milk. The fact that reaction rates vary with temperature provides a basis for accelerated testing. In other words, we can model changes in k values with temperature.

12.5.2 Accelerated Testing

Accelerated testing is based on the notion that temporal changes in the food can be modeled by storage at increased temperatures, to accelerate the changes that occur in the food (Mizrahi, 2000). The underpinning for this notion draws from the idea that chemical reactions must cross a potential energy barrier to proceed, and that such barriers are crossed with increasing frequency as the temperature of the reaction conditions is increased. Practically speaking, the strategic pressure to bring a new product to market as fast as possible makes the use of accelerated storage conditions very appealing. After a lengthy R&D process, the final version of the product may seem ready for the marketplace. However, a prudent manufacturer will also want to know whether the product holds up over time as a consumer might expect. Thus, there can be some frustration on the part of marketing managers when they are informed that there must be a lengthy delay to conduct a shelf-life study. Accelerated testing may allay some of this frustration.

Hough (2010) gives several reasons why accelerated testing may not be wise for every food product. The R&D team needs to consider whether their particular product is a good candidate. Situations in which accelerated testing may not work include (1) where there are multiple modes of deterioration (i.e., not a single simple chemical reaction), (2) where the acceleration factors are poorly estimated or involve high uncertainty, (3) where unforeseen variables, such as temperature differences or poor distribution channels, may affect the food, (4) where there is differential degradation of preservatives, such as antioxidants, with higher temperatures, (5) cases in which increasing temperature actually decelerates degradation, and (6) where a "masked rejection mode" may occur due to two different modes of failure that are differentially affected by time. Hough (2010) gives the example of plastic and oxidized flavors that show different temperature dependence in a mayonnaise product. Under the higher temperature conditions, the rate of plastic flavor development was higher than the rate of oxidized flavor development. This could give an inaccurate picture of flavor changes to the sensory panel. Another example is microbial processes by different strains of microorganisms in a cultured product. Thermophiles would be favored by high temperatures and psychrophiles by lower temperatures. Thus, the microbial profile at any related flavor changes would vary under accelerated conditions.

When the reaction rate k changes with temperature, the Arrhenius equation is an applicable model:

$$k = k_0 e^{-E_a/RT} \tag{12.15}$$

This may be more convenient in its derivative form:

$$\frac{\mathrm{d}(\ln(k))}{\mathrm{d}T} = \frac{E_a}{RT^2} \tag{12.16}$$

where k is the rate constant to be estimated, k_0 is a constant independent of temperature (also known as the Arrhenius, pre-exponential, collision, or frequency factor), E_a (J/mol) is the activation energy, R is the ideal gas constant (8.314 J/(mol K)), and T (kelvin) is (absolute) temperature; if E_a is expressed in cal/mol, the gas constant R is 1.985 cal/(mol K).

In order to compare the reaction rates at different temperatures, a reaction rate constant k_{ref} at the reference temperature is added to the relationship:

$$k = k_{ref} \exp\left[\frac{-E_a}{R}\left(\frac{1}{T} - \frac{1}{T_{ref}}\right)\right] \tag{12.17}$$

where $\exp(x) = e^x$. Then in its log form we have

$$\ln k = \ln k_{ref} - \frac{E_a}{R}\left(\frac{1}{T} - \frac{1}{T_{ref}}\right) \tag{12.18}$$

So, if a series of experiments is conducted at different temperatures and the reaction rates are estimated, then E_a may be conveniently found from a plot of ln(k) versus ($1/T - 1/T_{ref}$), which should form a line with slope equal to $-E_a/R$. The activation energies may be something of a fiction, in that multiple reactions are leading to the resulting deterioration of the

Table 12.1 Estimation of rate constants for a mayonnaise product stored at different temperatures

Temperature (°C)	Rate constant k oxidized score/days	ln(k)	$(1/T-1/T_{ref})$ T_{ref} = 300 K
20	0.197	−1.62	7.96×10^{-5}
35	0.577	−0.55	-8.66×10^{-5}
45	1.236	0.21	-18.87×10^{-5}

product. However, it may serve as a kind of shorthand index of the fragility of the systems being measured. If the temperature effects are known, then one can estimate how hot (and thus how fast) the accelerated conditions should proceed.

So, in practice, it is possible to conduct a storage study at three or four temperatures, find the reaction rates for each temperature, and then find the activation energy from a plot of ln(k) versus $(1/T-1/T_{ref})$ or simple regression of the two values. Hough (2010) gives the following example for development of oxidized flavor of commercial mayonnaise, with a zero-order (linear) rate constant and estimation at three storage temperatures, as shown in Table 12.1. The rate constants are found from simple plots of oxidized flavor scores against time, and $\ln(k_{ref})$ from the intercept in the equation as follows:

$$\ln k = \ln k_{ref} - \frac{E_a}{R}\left(\frac{1}{T} - \frac{1}{T_{ref}}\right) = 1.097 - 6803\left(\frac{1}{T} - \frac{1}{300}\right) \tag{12.19}$$

giving an activation energy divided by R of 6803 cal/mol, or E_a of 13.5 kcal/mol. Of course, this example is based on only three points, and it would seem prudent to test more temperatures to improve on the uncertainty involved in the several regressions that are needed (first to find k values and then to find E_a). The downside of this is that a fair amount of sensory testing may be needed to obtain the oxidation scores at all the temperatures and time intervals. Another option is to combine the estimates of k values and time and solve a nonlinear regression for both simultaneously. Hough (2010) argues that this will improve the confidence intervals around the estimates. However, some initial and reference conditions must be approximated in order to take this approach.

Hough also gives examples of where a sensory score (say, for oxidation) should also follow this relationship. The simple zero-order reaction expression is then

$$\text{Score} = S_0 + kt \tag{12.20}$$

where S_0 is an oxidation score at time zero or the initial measurement time. Be careful not to assume this is zero. The equation for the sensory score S then becomes

$$S = S_0 + k_{ref} \exp\left[\frac{-E_a}{R}\left(\frac{1}{T} - \frac{1}{T_{ref}}\right)\right]t \tag{12.21}$$

Or for first-order kinetics we have the log version:

$$\ln(S) = \ln(S_0) + k_{ref} \exp\left[\frac{-E_a}{R}\left(\frac{1}{T} - \frac{1}{T_{ref}}\right)\right]t \tag{12.22}$$

An example of oxidation scores plotted at various times and temperatures is shown in Figure 12.3. Hough (2010) provides an example of how to set up and conduct the nonlinear regression in the R language, using the mayonnaise data from his example. Many statistical

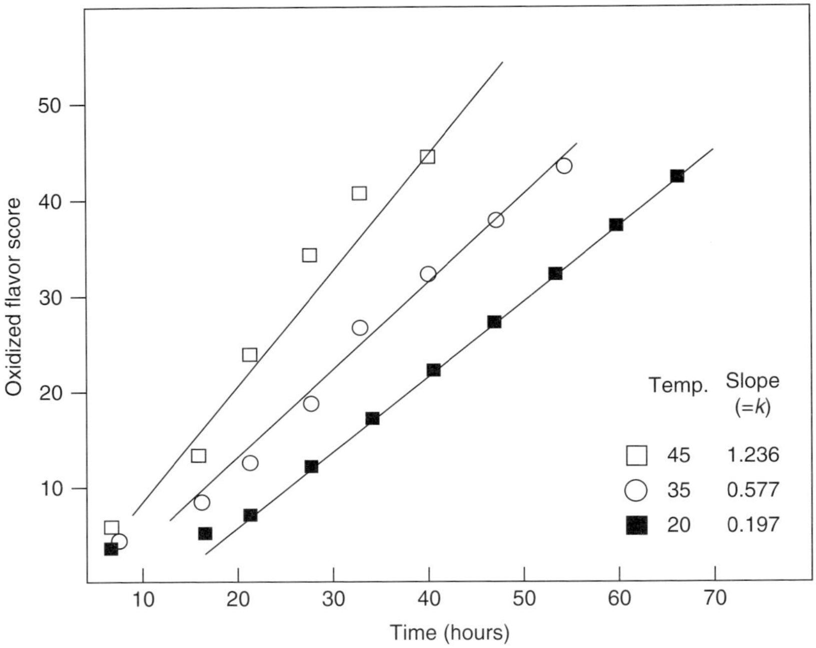

Figure 12.3 Hypothetical rate curves for a product undergoing oxidation, with oxidized scores generated by a trained panel over days, at three storage temperatures. The linear functions indicate zero-order kinetics. Rate constants can be used in the estimation of activation energy E_a, acceleration factors, and Q_{10} values.

programs are available for this type of calculation. For the mayonnaise data, the following (zero-order) expression resulted:

$$S = 1.2 + 0.216\exp\left[-8727\left(\frac{1}{T} - \frac{1}{300}\right)\right]t \tag{12.23}$$

A quick example. Using the relationship above, suppose we want to know what the oxidized score would be at 25 °C (298 K) at 90 days storage, knowing that our cutoff point was a 15 on 0 to 60-point scale. Substitution in the above equation results in

$$S = 1.2 + 0.216\exp\left[-8727\left(\frac{1}{298} - \frac{1}{300}\right)\right](90) = 14.8 \tag{12.24}$$

for the resulting score, suggesting that our shelf life has been reached at the 3 months storage at this temperature.

12.5.3 Acceleration Factors and Q_{10} Values

The Arrhenius models can also be used to define an acceleration factor used to provide a multiplier for the increase in reaction rates at temperatures other than the reference conditions. Once again, the strategic questions are "How hot?" and "How fast?" The acceleration factor is given by the following expression:

$$\mathrm{AF} = \frac{k(T)}{k(T_{\mathrm{ref}})} = \exp\left[\frac{-E_{\mathrm{a}}}{R}\left(\frac{1}{T} - \frac{1}{T_{\mathrm{ref}}}\right)\right] \tag{12.25}$$

This can be useful in calculating shelf life at different temperatures. The shelf life SL at any given temperature T can be estimated from the shelf life at the reference temperature or normal storage conditions, divided by the acceleration factor:

$$\mathrm{SL}(T_{\mathrm{ref}}) = \mathrm{AF} \times \mathrm{SL}(T) \tag{12.26}$$

Another common practice in accelerated testing is the estimation of a Q_{10} factor. This is a similar logic to the acceleration factor approach. It uses the convention of a 10-degree increase in temperature as a basis for the acceleration factor. Q_{10} is then defined as

$$Q_{10} = \frac{k_{T+10}}{k_T} = \frac{\mathrm{SL}_T}{\mathrm{SL}_{T+10}} \tag{12.27}$$

Note that the shelf-life expression is the inverse of the reaction rate expression. This makes sense. A faster reaction rate leads to a shorter shelf life.

If E_a is known (perhaps one value of going through the experiments described in Section 12.5.2), then the Q_{10} factor can be estimated from the Arrhenius expressions as follows:

$$Q_{10} = \exp\left\{\frac{E_a}{R}\left[\frac{10}{T(T+10)}\right]\right\} \tag{12.28}$$

The Q_{10} approach suggests an exponential relationship of the form

$$k = k_0 e^{bT} \quad \text{and thus} \quad \ln(k) = \ln(k_0) + bT \tag{12.29}$$

This implies that a plot of ln(k) versus temperature (as opposed to $1/T$ in the Arrhenius equation) will yield a straight line. Labuza (1982) suggested that a plot of log(shelf life) against storage temperature would also usually obtain a straight line. Your results may vary with different products.

12.6 Summary and Conclusions

Product durability has often been cited as one of the hallmarks of quality (Garvin, 1987). That is, a product is expected by consumers to be usable or remain in service for some period of time. Considering all consumer products, foods are especially fragile and prone to changes over time. Some of these changes are desirable, such as ripening of cheeses, mellowing of soft drinks, or aging of wines. Often, however, the changes are types of deterioration. At some point, foods become unsafe to consume. However, there is a difference between pathogenicity and spoilage, as well as a difference between spoilage and consumer rejection. So, long before the point that a food has obviously deteriorated, a consumer may find some changes unacceptable. Such sensory changes should be subject to measurement in the determination of a practical use-by or shelf-life dating system.

Accelerated testing presents special challenges to the research team and to the sensory specialist. Careful consideration should be given to whether or not accelerated conditions can be useful and give an accurate picture of storage changes for your product. Obviously, a product that undergoes phase changes is not a good candidate for accelerated testing. If such an approach worked for all products, you could make fine aged wine in an oven. If that were the case it would be commonplace, but it is not.

There are many sensory measurements that can be brought to bear on determining shelf life, such as discrimination test results, descriptive (trained) panel ratings of off-flavors, and consumer reaction. Ultimately, all measurements should be cross-referenced and calibrated to consumer rejection, although this is expensive to do, and thus remains somewhat rare in industrial practice. When product quality is paramount, the shelf-life program can act as a safety net. By this we mean a more stringent hurdle than what consumer testing would suggest as a cutoff. If the measured attribute provides a cutoff point that is more conservative than consumer opinion, product quality can be safeguarded. These are strategic and management decisions, and consumer alienation and franchise risk must be the ultimate consideration.

References

Bodyfelt, F.W., Tobias, J., and Trout, G.M. 1988. Sensory Evaluation of Dairy Products. Van Nostrand/AVI Publishing, New York, NY.

Clark, S., Costello, M., Drake, M., and Bodyfelt, F. 2009. The Sensory Evaluation of Dairy Products. Springer Science + Business, New York, NY.

Claassen, M. and Lawless, H.T. 1992. Comparison of descriptive terminology systems for sensory evaluation of fluid milk. Journal of Food Science, 57, 596–600, 621.

Dethmers, A.E. 1979. Utilizing sensory evaluation to determine product shelf life. Food Technology, 33(9), 40–3.

Gacula, M.C. 1975. The design of experiments for shelf life study. Journal of Food Science, 40, 399–403.

Gacula, M.C. and Kubala, J.J. 1975. Statistical models for shelf life failures. Journal of Food Science, 40, 404–9.

Garvin, D.A. 1987. Competing on the eight dimensions of quality. Harvard Business Review, 65(6), 101–9.

Giminez, A., Varela, P., Salvador, A., Ares, G., Fiszman, S., and Garitta, L. 2007. Shelf life estimation of brown pan bread: a consumer approach. Food Quality and Preference, 18, 196–204.

Giminez, A., Ares, G., and Gambaro, A. 2008. Survival analysis to estimate sensory shelf life using acceptability scores. Journal of Sensory Studies, 23, 571–82.

Harwood, M.L., Ziegler, G.R., and Hayes, J.E. 2012. Rejection thresholds in solid chocolate-flavored compound coatings. Journal of Food Science, 77(10), S390–3.

Hough, G. 2010. Sensory Shelf Life Estimation of Food Products. CRC Press/Taylor and Francis, Boca Raton, FL.

Hough, G., Langohr, K., Gómez, G., and Curia, A. 2003. Survival analysis applied to sensory shelf life of foods. Journal of Food Science, 68, 359–62.

Kilcast, D. 2000. Sensory evaluation methods for shelf-life assessment. In: The Stability and Shelf-life of Food. D. Kilcast and P. Subramaniam (Eds). CRC Press/Woodhead Publishing, Boca Raton, FL, pp. 79–105.

Labuza, T.P. 1982. Shelf-life Dating of Foods. Food and Nutrition Press, Westport, CT.

Lawless, H.T. and Claassen, M.R. 1993. Validity of descriptive and defect-oriented terminology systems for sensory analysis of fluid milk. Journal of Food Science, 58, 108–12, 119.

Lawless, H.T. and Heymann, H. 2010. Sensory Evaluation of Foods, Principles and Practices. Second edition. Springer, New York, NY.

Man, C.M.D. and Jones, A.A. 2000. Shelf-life Evaluation of Foods. Second edition. Aspen Publishing, Gaithersburg, MD.

McBride, R.L. and Hall, C. 1979. Cheese grading versus consumer acceptability: an inevitable discrepancy. Australian Journal of Dairy Technology, (June), 66–8.

Mizrahi, S. 2000. Accelerated shelf-life tests. In: The Stability and Shelf-life of Foods. D. Kilcast and P. Subramaniam (Eds). CRC Press/Woodhead Publishing,Boca Raton, FL, pp. 107–42.

Muñoz, A.M., Civille, G.V., and Carr, B.T. 1992. Sensory Evaluation in Quality Control. Van Nostrand Reinhold, New York, NY.

O'Mahony, M. 1979. Our industry today – psychophysical aspects of sensory analysis of dairy products. A critique. Journal of Dairy Science, 62, 1954–62.

Peryam, D.R. 1964. Consumer preference evaluation of the storage stability of foods. Food Technology, 18, 214–17.

Prescott, J., Norris, L., Kunst, M., and Kim, S. 2005. Estimating a "consumer rejection threshold" for cork taint in white wine. Food Quality and Preference, 18, 345–9.

Robertson, G.L. 2006. Food Packaging, Principles and Practice. Second edition. CRC Press/Taylor and Francis, Boca Raton, FL.

13 Product Optimization, Just-About-Right (JAR) Scales, and Ideal Profiling

Then the little old Woman sate down in the chair of the Great, Huge Bear, and that was too hard for her. And then she sate down in the chair of the Middle Bear, and that was too soft for her. And then she sate down in the chair of the Little, Small, Wee Bear, and that was neither too hard, nor too soft, but just right.

Robert Southey (1837)

From *The Selected Prose of Robert Southey*, Macmillan, New York. 1916, J. Zeitlin, Ed. Originally published in *The Doctor* (anon./Southey, 1837/1916)

13.1 Introduction

This chapter deals with some of the sensory testing methods used with consumers to improve or optimize products. One common goal is to find an optimal level of an ingredient, and one that is usually associated with one or just a few sensory attributes. An example would be the sourness associated with a food acidulant in a fruit beverage. Some attributes pass through a "bliss point" at which the intensity seems optimal to most consumers (Moskowitz, 1981; McBride, 1990; Conner & Booth, 1992). This situation is shown in Figure 13.1, where there is an inverse U-shaped function for the hedonic appeal plotted either as a function of the physical ingredient level (e.g., sucrose concentration), or the sensory intensity of a specific attribute (e.g., mean sweetness intensity rating). This inverted U-shaped function

Quantitative Sensory Analysis: Psychophysics, Models and Intelligent Design, First Edition. Harry T. Lawless.

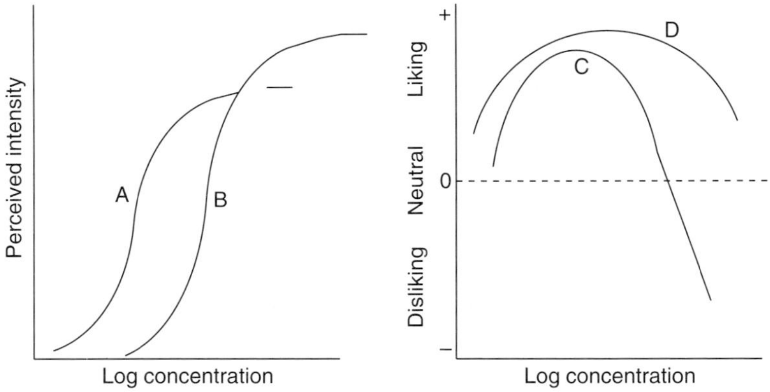

Figure 13.1 Psychophysical function for perceived sensory intensity and psycho-hedonic functions (right panel). Functions A and B are consistent with the semi-hyperbolic function of Beidler for taste receptor binding. Subtance B is less potent that substance A in terms of requiring a higher concentration to be half-maximal, but shows a higher maximum response. For the psycho-hedonic functions, substance D shows a larger region of acceptable sensory properties, while C has a sharper peak, implying less consumer tolerance of variations. Note that substance C also becomes hedonically negative (disliked) at higher levels.

is widely attributed to Wilhelm Wundt, the founder of the first experimental psychology laboratory in Leipzig in the 1880s (Beebe-Center, 1932/1965). Of course, the bliss point concept does not apply to sensory properties that are aversive from the very first time they are perceived. Still others may not offend until they reach a certain level at which point consumers will reject the product in favor of another version that lacks the offending quality. Prescott et al. (2005) outlined a method for quantification of this process, calling it the "consumer rejection threshold." So the first question in any optimization process should be, "Do I have an attribute that passes through a bliss point?" Then the question arises as to how it can be measured and identified as one that is important to consumers, or not.

Two common methods for finding the optimal level of a single attribute are just-about right (JAR) scales and ratings for an imaginary ideal product. Other methods include adjustment, in which the consumer adds or subtracts the ingredient to their preferred taste (Pangborn & Pecore, 1982; Pangborn et al., 1985; Hernandez & Lawless, 1999). Of course, it is also possible to simply apply hedonic ratings for degree of liking/disliking to an ingredient series and find the peak or optimum by ratings. Further information on acceptability scaling is found in Chapter 8 and in Lawless and Heymann (2010). The most common method is to use a nine-point liking/disliking scale or one of its variants, which are discussed in Chapter 8 along with several other techniques for assessing consumer appeal.

Using this approach with individual attributes, one can ask consumers how much they like the level of some sensory characteristic. This is not always a common practice, partly because it requires a consistent understanding of what that the word means among the members of the consumer test sample. The use of JAR scales or ideal profiling is limited to those qualities that are well understood by consumers and for which there is a general consensus as to their meaning. It is tempting to think that all consumers understand words like crispness, but remember that respondents in a test tend to be cooperative and will answer any question you put to them, whether or not they comprehend it.

The optimization process may involve multiple attributes in combination, and so it may be of interest to find the ideal points for a collection of key features of a product (e.g., Cooper et al., 1989). But which attributes are "key?" Identification of such attributes is a technical process sometimes referred to as finding the "drivers of liking" (Moskowitz et al., 2006; Ares et al., 2010). Clearly, those attributes that change your opinion of a product as their levels change are key aspects to optimize. So one can think of this as a kind of slope or rate of change issue, in the psycho-hedonic function. Similarly, there may be attributes that are highly appealing, but the level may not matter so much. That is, an attribute may have a broad optimum, where consumers do not care so much as long as it is in a certain range. So there is a question of tolerance limits for consumer liking and rejection. Booth and colleagues have developed extensive models of food acceptability based on ideal levels and deviation from ideal (Booth, 1994, 1995) and have considered the slope of the psycho-hedonic relationship as a kind of hedonic discrimination function. A function with a broad optimum may present an opportunity for reduction of an ingredient that is considered unhealthful, such as salt or fat (IOM, 2010), without any major decrease in the overall appeal of the product.

At the root of the product improvement process is the notion that there is a best version of the product that maximizes the consumer appeal. This imaginary product is called the ideal product. There are several ways to probe the characteristics of this ideal. One is to use JAR scales for individual attributes or combinations (i.e., a profile). JAR scales will indicate how an existing product should be changed in order to approach the ideal but may not give a very specific target for how much of a change is required. A second approach is to ask the consumer directly, usually during the evaluation of other actual products, what point or level on each attribute scale the ideal product would possess. Assuming consumers can do this accurately, the ideal profile is then known and appropriate changes in existing products can be made. The third option is to infer the characteristics of the ideal product indirectly, from other hedonic scores. Thus, a set of attributes can be used as regression predictors for overall liking (OAL) scores. Sometimes the profile information is used to make a perceptual map, wherein products that are similar are plotted close together and those that are different are far apart. The dimensions of this map are derived from the attribute information (i.e., their profile). Once the map or model is constructed, one can search for the areas of the map that represent the best-liked products. Using this kind of procedure, the optimal product can be inferred from the position that best corresponds to the individual consumer's liking ratings. That is, their ideal should occupy a position in the map close to products that are highly liked and be far from products that are disliked. Various versions of this mapping process will be discussed in this chapter.

It is important to remember that not all attributes are equally influential. Some may be more or less expected ("must-haves" in the Kano system; see Rivière et al. (2006)), some may increase liking as the performance or delivery or level increases ("performance attributes" in Kano), while others may not be expected, but when they do occur they delight the consumer ("delighters"). But what about product negatives? Some defects must not be present at all, a kind of "negative must-have." Still others are expected and tolerated but the product can be improved if they are eliminated or minimized (e.g., seedless fruits). Lawless and Heymann (2010) refer to these as "nuisance" attributes.

Measurement of imagined ideals is not always straightforward. Consumer reactions tend to be integrative, rather than analytical, so asking them to attend to too many individual facets of a product simply goes against human nature. Yet, people seem to be able to express what they think an ideal product would be like, and these ideals tend to be reproducible (Cooper et al., 1989; Worch et al., 2012). So this chapter will review some of the current methods used for making products that are optimized, or close to a consumer ideal.

The techniques range from simple, univariate methods (one attribute at a time) to multivariate statistical techniques used for product mapping, a kind of spatial visualization of the similarities and differences among products. Finding the optimal zones in such a map has long been a topic of interest among sensory scientists, product developers, and sensometricians (Meullenet et al., 2007).

A final concern in this area is the fact that people may disagree about where the optimum lies. Some people like salty foods, others do not (Pangborn, 1970). So segmentation and identification of market segments of consumers is key to the success of any optimization program. Segmentation was traditionally done on the basis of geographic locations, nationality, ethnicity, or other demographic variables. However, segmentation by sensory preference styles may be even more effective at providing products tailored to specific tastes (Moskowitz et al., 1985). That is, within a nation or ethnic group there may be consumers who prefer different tastes, textures, and appearance of the products, and these sensory segments may exist among multiple nations or ethnicities. If so, marketing two coffees for those people who love (versus hate) bitterness makes more sense than marketing one coffee for France and another for Germany.

13.2 Basic Equations, Designed Experiments, and Response Surfaces

Probably the simplest and least mathematical approach to optimization is simply to adjust the product until it tastes best. This is sometimes called an adjustment method and sometimes "ad libitum" mixing. Pangborn and colleagues used this procedure effectively to study preference levels for salt and for fat in beverages (Pangborn & Pecore, 1982; Pangborn et al., 1985), and Hernandez and Lawless (1999) showed how it could be used for seasoning in solid foods by a sequential weighing procedure. Clearly this is just a parallel to what any cook or chef does in the culinary arts when they prepare a food "to taste." However, the direction of adjustment is usually only increasing.[1] This can lead to some false optima. Mattes and Lawless (1985) showed that an increasing concentration series for both salt and sugar would produce a lower apparent optimum than a series based on dilution. That is, when increasing or decreasing the level of a taste like sweetness or saltiness to find the best level, consumers will tend to stop too soon. Increasing and decreasing ideal concentration levels can differ by 30% or more. This undershoot persists even in the face of increasing motivation and various attempts at eliminating adaptation effects. If the effect is commonplace, it suggests that finding the optimum would better be achieved by a randomly ordered presentation of various levels for hedonic ratings or by using JAR ratings.

As noted above, the hedonic function tends to be nonlinear, either as a function of an ingredient concentration (e.g., sucrose level) or the sensory intensity (e.g., sweetness). In order to fit the nonmonotonic relationship, a quadratic term is usually required (Moskowitz, 1981). A simple function would be

$$H = k_0 + k_1(\log C) + k_2(\log C)^2 \tag{13.1}$$

[1] However, it is possible to backtrack by some kind of neutralization; for example, to decrease the level of hot pepper burn by adding fat, a technique I have seen used by Chef Paul Prudhomme using heavy cream.

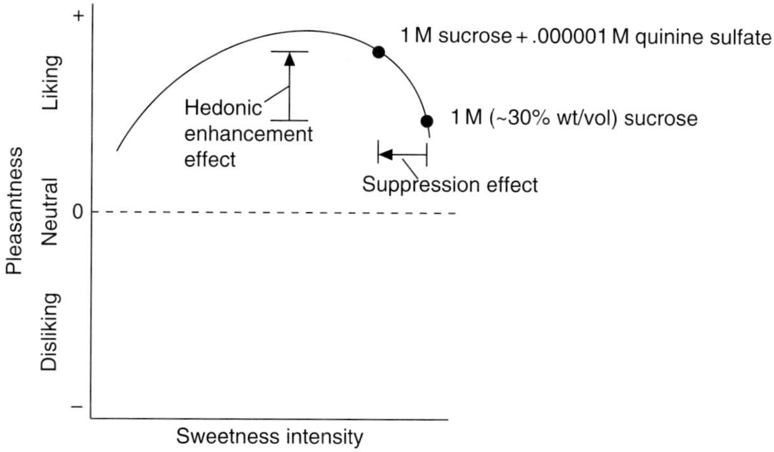

Figure 13.2 Addition of a small amount of quinine to an overly sweet stimulus will increase the overall pleasantness of the mixture, even though the quinine is bitter and unpleasant when tasted alone. This is due to a mixture suppression effect in which the sweetness is past its just right optimum, and thus a decrease in sweetness brings it closer to the optimum (see Lawless (1977) for an example).

where H is the hedonic rating (OAL), k_2 is a small but negative coefficient, and C is the concentration or weight per volume of the flavor ingredient. As the concentration increases, the negative squared term overtakes the positive linear term, and the function turns over. If the fit of the equation is good, the optimum may be found from the point at which the first derivative is zero. The psycho-hedonic function is shown in Figure 13.1.

But what about finding optimal combinations of multiple ingredients with different sensory characteristics? One can expand on the relationship in eqn 13.1 to include multiple items and thus make a response surface from the multi-attribute relationship. For two items the equation might look like

$$H = k_0 + k_1(\log C_1) + k_2(\log C_1)^2 + k_3(\log C_2) + k_4(\log C_2)^2 + k_5(\log C_1 \log C_2)^2 \quad (13.2)$$

Note that there is a multiplicative interaction term in case of mixture suppression (k_5 negative) or any kind of synergistic interactions (k_5 positive). A good example of a somewhat unexpected interaction can be found in Lawless (1977) in which addition of a small amount of bitter quinine to a sweet–bitter mixture actually increases the overall pleasantness of the mixture, even though the quinine itself is unpleasant. This is due to the mixture suppression or masking effect that decreases the sweetness of an overly sweet item, thus "backing it up" toward the hedonic optimum (see Figure 13.2). Eqn 13.2 and similar polynomials may be fit by various multiple regression or nonlinear regression modeling algorithms. For a two-variable mixture, it is easy to make a response surface plot or a contour plot to help visualize the optimum (see Figure 13.3).

Various systematic approaches to optimization with multiple ingredient or process variables can be found in Moskowitz et al. (2006: chapter 7). A case study is presented using four ingredient variables for a margarine product, two flavoring/coloring blends, the fat level, and a nutritional ingredient level. The overall model includes both linear and quadratic terms for each variable, as well as all possible two-way interactions. Several designs are illustrated, with consumers evaluating some or all of the possible combinations. One is the Placket–Burman design that works well for two levels of each variable. Another is the

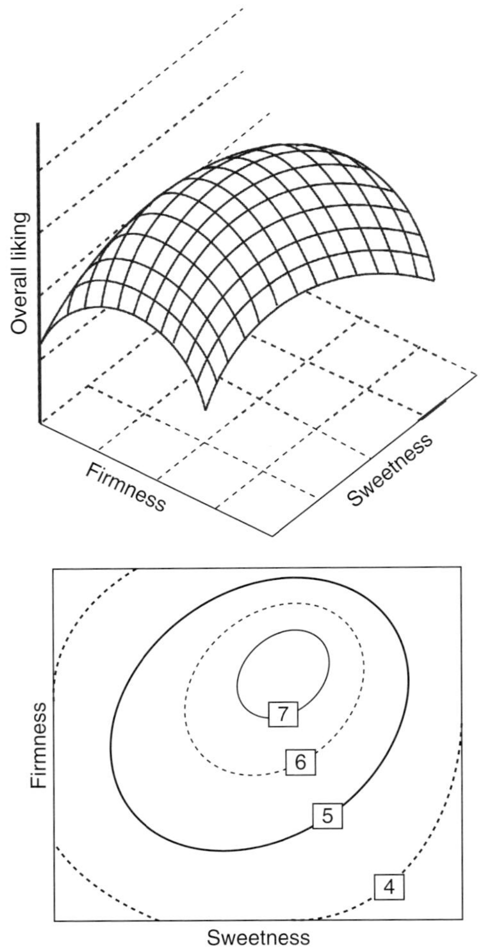

Figure 13.3 An example of a response surface plot and a contour plot for a two-ingredient mixture. The contour lines connect points of equal hedonic value.

Box–Behnken design (a variation of central composite design) which uses three levels of each variable and can allow a fit of curved response surface. For the four-variable margarine example, a full replicate design would demand evaluation of 25 products but permit estimation of all quadratic and two-way interaction terms. This is possibly too many evaluations for any single consumer, but an incomplete block variation on this design would help narrow down the number of items that have to be formulated. Another option illustrated by Moskowitz et al. (2006) is a "half-replicate" design that eliminates eight of the products, but at the expense of losing three of the six pairwise interaction terms. If it is suspected from the outset that those variables will probably not interact (in the statistical sense), then such an alternative seems a good choice. Other techniques to search for an optimum are discussed by Mao and Danzart (2008) using several design variables in a meat product varying in two processing parameters (temperature and pH) and two ingredient variables. The two methods involved an interactive search program for seeking product improvements and an exhaustive grid search technique for finding the global optimum.

Planned experimental designs with multiple factors are sometimes called mixture designs, and there is a large literature on mixtures experiments and response surfaces in food technology and quality control. The classic texts on the topic are by John Cornell (Cornell & Khuri, 1996; Cornell, 2002) and Max Gacula (1993). More recent treatments of response surface optimization from a sensory perspective can be found in Moskowitz et al. (2006), Meullenet et al. (2007) and in Gacula et al. (2009: chapter 7). An overview of common experimental designs for optimization and mixture studies is also given in Appendix B to this book.

13.3 Just-About-Right Scales

13.3.1 Basic Analysis

JAR scales (Rothman and Parker, 2009) are useful in optimizing the levels of key ingredients with a single important sensory attribute, such as the sweetness of sugars in a product. There are different versions of these scales, but the basic idea is that a consumer can express whether the attribute in question is too weak, just right, or too strong, usually in some graded manner. An alternative wording is to offer a response option such as "needs more X" or "increase X." Examples of JAR-type scales are shown in Table 13.1. JAR scales assume that there is an ideal level of a sensory attribute for each person; that is, a bliss point (McBride, 1990). With normal hedonic scaling, the bliss point shows up as a peak in the plot of acceptability ratings versus level of that attribute or versus the concentration of that ingredient (i.e., a hedonic dose–response function). Deviations from this optimum result in a decreased OAL for a product. We will return to this idea of deviation from ideal in the topic of ideal profiling later in this chapter.

From a quantitative perspective, there are several important questions that can be answered with JAR data. First, is my product not JAR? Second, considering two or more products, is one more JAR than another? Next, what does it cost me if my product is not JAR in terms of a lowered hedonic rating? The last question is the realm of penalty analysis. The first two questions are statistical in nature. Finally, do my JAR data provide any evidence of segmentation or clusters of people that prefer different styles of the product?

The first question can usually be answered by simply examining the distribution of the data along the scale points. With a continuous line scale, it is useful to divide the line into segments to see the distribution density at various positions. The company may have action standards that suggest an acceptable distribution, such as 25% below JAR, 50% at

Table 13.1 Examples of JAR scales

Simple category	Directional change
Response options:	Reponse options:
Much too salty	Decrease a lot
Too salty	Decrease
A little too salty	Decrease slightly
Just about right	Don't change
A little bit not salty enough	Increase slightly
Not salty enough	Increase
Very much not salty enough	Increase a lot

Table 13.2 Stuart–Maxwell categorization for JAR data

Product A	Product B			
	Below JAR	**JAR**	**Above JAR**	**Marginal total**
Below JAR	N_{11}	N_{12}	N_{13}	T_{1j}
JAR	N_{21}	N_{22}	N_{23}	T_{2j}
Above JAR	N_{31}	N_{32}	N_{33}	T_{3j}
Marginal total	T_{i1}	T_{i2}	T_{i3}	

JAR or within one category, and 25% above JAR. Obviously, one would like a distribution that is symmetric and light in the tails. If there is a large study with a finely graded scale, it may make sense to look at skewness and kurtosis and see whether there are deviations from normality (e.g., through a Kolmogorov–Smirnov test). If there are fewer categories of response or a small sample size, comparison with the normal distribution is less appropriate, but a χ^2 test can be used against the action standards or historical norms for that product.

The comparison of two products can be achieved by a Stuart–Maxwell test if both products have been viewed by the same consumers, as is typically done in a monadic sequential consumer test (Best & Rayner, 2001). If they have been evaluated by different groups, then a χ^2 comparison is appropriate on the two sets of frequency counts. For the correlated or complete block design, we can tabulate the frequency counts for above JAR, JAR, and below JAR for each product and the marginal totals, as shown in Table 13.2, with *i* rows and *j* columns.

It simplifies the calculation if two sets of intermediate quantities are formed: the differences of the row and column totals T_{ij} for $D_k = T_{ij} - T_{ji}$, and the averages of the off-diagonal cells (N_{ij} and N_{ji}) with opposite subscripts, giving $A_k = (N_{ij} + N_{ji})/2$. This gives three differences and three averages, or six quantities to use in the final χ^2 calculation as follows:

$$D_1 = T_{1j} - T_{i1} \tag{13.3a}$$

$$D_2 = T_{2j} - T_{i2} \tag{13.3b}$$

$$D_3 = T_{3j} - T_{i3} \tag{13.3c}$$

$$A_1 = \frac{N_{12} + N_{21}}{2} \tag{13.3d}$$

$$A_2 = \frac{N_{13} + N_{31}}{2} \tag{13.3e}$$

$$A_3 = \frac{N_{23} + N_{32}}{2} \tag{13.3f}$$

The full χ^2 calculation is then given by

$$\chi^2 = \frac{A_1 D_3^2 + A_2 D_2^2 + A_3 D_1^2}{2\left(A_1 A_2 + A_2 A_3 + A_1 A_3\right)} \tag{13.4}$$

Note that the off-diagonal averages are multiplied by the squared differences from the row and column with which they do not participate. This may seem counterintuitive but it works. Since there are two degrees of freedom, this test has a critical value χ^2 of 5.99.

For more than three categories, the Stuart test can still be used, but the formulae become more complicated and it is difficult to write out an algebraic solution for more than four categories. This was discussed further in Chapter 9. Fortunately, there are a number of statistical analysis options, including the program MHTEST in the library COIN of the R software platform. For more than two products, the Cochran–Mantel–Haenszel (CMH) analysis can be used (see Fritz (2009) for a thorough discussion). The Stuart test is a subset or special case of the CMH tests. CMH tests are available in various freeware and commercial statistical packages. Once again, this is assuming that a complete block design has been used, where all consumers see all products. If an overall significant Stuart test is found, comparisons of individual cells can be made using the McNemar test after collapsing or combining the other cells (Lawless & Heymann, 2010).

13.3.2 Proportional Odds/Hazards Models

The Stuart test is actually a test of marginal homogeneity (MH); that is, a test of the equivalence of marginal proportions. Thus, it does not take into account the ordinal nature of the JAR ratings, although there are variations of the MH tests that will consider the categories as ordered. Several alternative analyses have been presented by Meullenet et al. for significance testing of JAR data when multiple products are assessed, and these also take into account the ordered nature of the categories (see Meullenet et al. (2007: chapter 10)). These are called the **proportional odds model** (POM) and **proportional hazards model** (PHM).

The POM is a form of logistic regression, where the probability of any response score Y being in response category k or below is given by

$$\Pr(Y \le k) = \frac{1}{1+e^{-(\alpha_k+\beta x)}} \tag{13.5}$$

For j samples and i panelists, it is more convenient to write this in terms of log odds, where π_{ijk} is the probability of a score Y_{ij} in category k or below for panelist i and sample j, given a vector of predictor variables X_{ij}. In its linear form, we get

$$\ln\left(\frac{\pi_{ijk}}{1-\pi_{ijk}}\right) = \alpha + \beta X_{ij} \tag{13.6}$$

where a and b are intercept and slope parameters for a vector of predictors (such as products or consumers). This assumes equal slopes across levels of the response variable; if this assumption is not met, then the PHM may be used instead. The POM model is attractive because it can be implemented in a variety of software routines, and these provide information on overall differences and contrasts between pairs of products. The model selects one product as a baseline and then compares all other products with it. However, contrasts between pairs can be requested as well. An example is shown in Meullenet et al. (2007). Implementation in the SAS LOGISTIC procedure is also illustrated.

The PHM is commonly used in survival analysis. If we let $1-\pi_{ijk}$ be the survivor probability (analogous to one minus death probability, in this case falling into the kth category or below) and taking the natural log twice, we obtain a linear function again as

$$\ln\left[-\ln\left(1-\pi_{ijk}\right)\right]=\alpha_k+\beta X_{ij} \tag{13.7}$$

Once again, this model can be used with contrast statements in SAS, for example, to provide tests of differences among pairs of products.

13.3.3 Penalty Analysis on JAR Data

What are the consequences of having a product that is off from the JAR level? In terms of overall hedonic scores, this can be assessed if both JAR data and acceptability ratings are collected from the same individuals. The attribute in question may have little or no influence on a person's overall opinion, or it could be very impactful. This is the basis of **penalty analysis** (Schraidt, 2009). We can ask two important questions. First, what is the drop in mean hedonic scores for groups of consumers who rate the product as higher or lower than the JAR response? Second, how large are these segments? Obviously, a large drop in mean scores from a large segment suggests an urgent need for a correction in the product formulation. It is, of course, always possible that the attribute does not seem quite right, but that this matters little to a given consumer. So a connection between JAR scores and hedonic penalty is not guaranteed.

A simple penalty analysis involves three steps (Schraidt, 2009). First, classify groups as above, below, and at or near the just-right category. If there are multiple categories, a regression approach can be taken, but the simple three-group split suffices for most purposes. Next, calculate the mean hedonic scores from the acceptability scales (OAL) for each of the three groups. Third, find the differences in the OAL means of both the above-JAR group and the below-JAR group from the JAR group. Use the JAR group's actual mean and not the overall data mean for this purpose. A useful diagnostic plot can then be made of the mean drop or penalty versus the proportion of the total sample in each of the two non-JAR groups. See Lawless and Heymann (2010: chapter 14) or Rothman and Parker (2009: appendix L) for examples.

Various statistical tests and methods for modeling the penalty analysis are given in Rothman and Parker (2009). Two basic questions are whether the above- and below-JAR groups are different sizes, and whether the penalties for being above versus below are more or less severe. The first question can be answered by a simple χ^2 test or its binomial equivalent test on proportions. Another approach is to make a 2×2 classification table of the above/below-JAR versus likers/dislikers for the product (as shown by positive versus negative or neutral OAL scores) and perform a simple χ^2 test (Templeton, 2009). A significant χ^2 indicates that dislikers might be more concentrated in the above- or below-JAR groups. Another straightforward comparison is to perform an independent groups *t*-test on the OAL scores of above- versus below-JAR groups. Other important questions are: (1) Are some JAR scales showing more of a problem than others (i.e., are there different degrees of drivers of liking)? (2) Are some products significantly less JAR than others? In addition to the simple Stuart tests, there are additional multivariate models that can be used to look at these questions.

Given the fact that some JAR scales have multiple response options and, therefore, multiple groups of people with various difference-from-JAR classifications, it is also possible to do a more detailed analysis. Suppose we have seven JAR categories for response, three on either side of the optimal response. It should be possible, then, to get a graded response; for example, the farther from JAR (the center rating), the more severe the penalty might be. This possibility was embraced by Plaehn and Horne (2008) in using a regression

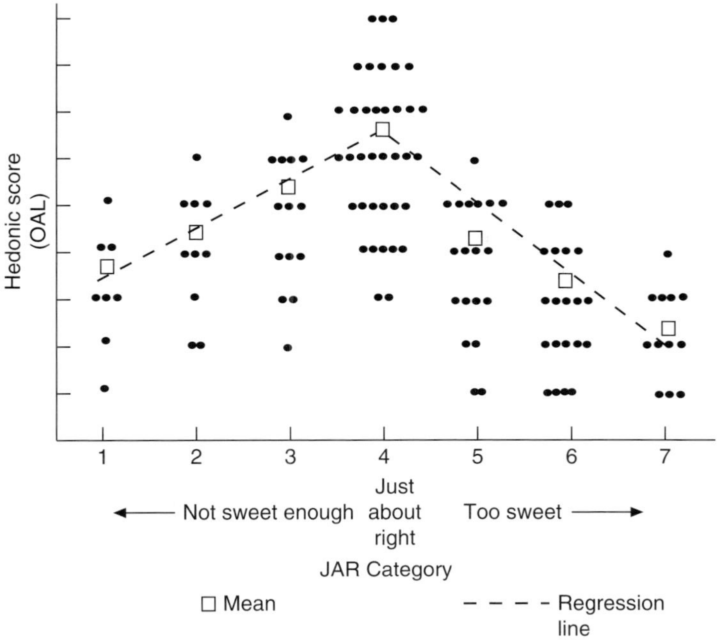

Figure 13.4 An example of penalty analysis with a two-stage regression model.

approach to testing the significance of penalty weights. One consideration is the proportion of consumers in each JAR scale category. That is, their mean drop can be weighted by the frequency count, producing a combined score based on score change and proportion. Plaehn and Horne also provided various methods for testing the significance of the regression coefficients, including resampling or bootstrap procedures.

A simple version of this model could be obtained by simply creating a scatter plot of the mean hedonic values, starting with the JAR hedonic value and then proceeding in one direction or the other along the scale. Owing to the nonlinear nature of having the peak in the middle, it would seem sensible to fit functions separately for the less than JAR section and the more than JAR section. For a seven-point scale, one could then estimate a slope value for the hedonic change from category four to category seven, and also from category one through four, by simple least-squares regression or any other suitable method, such as partial least squares (Xiong & Muellenet, 2006). Most statistical software will also provide a standard error of the regression coefficient, so it is straightforward to test for nonzero slope using a t-test from the provided standard error, or constructing a confidence interval to see if it overlaps zero. A weighted regression line could also be fit, using the proportions in each category. This is shown graphically in Figure 13.4. Note that the right-side regression line is steeper and is based on more individuals. This suggests that being too sweet is a problem for this product. Note that this model is limited to one product at a time. If multiple products were tested in a complete block design, one could expand the model to take into account the consumer effect, a product effect, and interaction terms, and thus different trends among different consumers as the ingredient level was altered. Such an approach would produce a much fuller picture of the degree of penalty and whether there were different consumer trends. For a fuller mathematical treatment of this approach, see Xiong and Muellenet (2006) and Plaehn and Horne (2008).

The example shown in Figure 13.4 illustrates that the nonlinear nature of the JAR function can be modeled by considering the two segments as separate functions. One formal mathematical approach to this uses splines, which are segments of a function separated by knots. Meullenet et al. (2007) provide examples of using "multivariate adaptive regression splines" models for JAR data. Software is available for this analysis and instructions are given in Meullenet et al. (2007: 233–4). Goodness-of-fit tests are provided for each JAR variable and for the rising and falling segments, along with slope estimates (regression coefficients). Another approach to the segmented nature of the JAR function is to use dummy variables in a covariance or partial least-squares analysis. The dummy variables effectively separate the JAR function into rising and falling phases. Meullenet et al. (2007) show examples of testing for the significance of the penalty, and also tests to compare the rising and falling phases, essentially whether having too much or too little of the attribute is more detrimental to OAL. The partial least-squares model is preferred if there is significant correlation between some of the JAR attributes, as often occurs. Meullenet et al. (2007) provide an example of the partial least-squares application using the UNSCRAMBLER software from CAMO.

13.3.4 Cautions in Using JAR Scales

JAR scales are not always appropriate. They work as intended when there is an optimum and not very well when an attribute is a case of "more is always better" or "any of this is bad." Consumers should have a clear understanding of the attribute. Of course, changing one attribute at a time for optimizatiom can be difficult. Changing the sweetness of a fruit beverage by adding sugar, for example, will also change the sourness and mouthfeel. Attitudinal factors can also play a part. For example, if a person believes that salt is unhealthy, a product could be rated as too salty even though it is actually at the sensory JAR level. One's reaction to an attribute may change over time. High sweetness may seem acceptable at first, but may be less appealing after consuming an entire portion. What seems good in a small bite may not seem so appealing in a larger portion. Unless additional intensity information is collected, JAR data can be misleading. For example, two groups of respondents might both mark, "just right" but one might think the product is very strong (but still the level they prefer) while the other group thinks the product is fairly mild (but still the level they prefer). Collecting both intensity and JAR information would help identify these differing consumer segments. Other methods are also useful for finding the optimum level of some ingredient with a key sensory attribute, notably **methods of adjustment** (Pangborn & Pecore, 1982; Hernandez & Lawless, 1999), but they are not without their pitfalls (Mattes & Lawless, 1985), as discussed above. Scaling products relative to an ideal will be discussed in Section 13.4.

The **centering bias** can also be a problem for JAR scales. The centering bias is a tendency to put the middle product in an increasing ingredient series at the just right point. This may cause an inaccurate estimation of the true optimum. Johnson and Vickers (1986) compared several methods for dealing with the centering problem (McBride, 1982; Poulton, 1989). These methods involve testing multiple ranges of products to interpolate the true JAR point. A simple way to make this adjustment is to test two overlapping series of ingredient levels, one higher than the other, and both probably containing a reasonable guess as to the JAR level. For example, one sugar series for a fruit beverage might contain 5%, 7%, 9%, 11%, and 13% sucrose (wt/vol) and the second might contain 8%, 10%, 12%, 14%, and 16%. Each series is rated by the same panel, in random order, in different sessions. The group JAR estimate can then be found for each series. A simple plot of sucrose concentration on both axes is then made to interpolate the true JAR, involving two lines. One line plots the midpoints of the two series against themselves (i.e., an identity line where $y=x$, so the points would be (9,9) and (12,12) for our example). The second line plots the experimentally obtained JAR points from the two series (usually on the y-axis) against the midpoint of the series (usually on the x-axis). Where the two lines cross indicates the JAR point which would be obtained had the series with that midpoint been fortuitously centered on the true JAR level. Of course, we did not know that going into the experiment, but the fact that it is centered eliminates the centering bias at that point; see Lawless and Heymann (2010) for a graphical example.

13.4 Ideal Profiling

13.4.1 Basic Concepts and Measurement Issues

The basic concept behind JAR scales is that the consumer has a mental image of what an ideal product would taste like, and how the current product deviates from that ideal or just-right level. However straightforward, there is a potential loss of information in using only JAR scales, because the sensory intensity itself is never specified and only the relative direction (and possibly a rough idea of the sensory distance) is known. In order to get a clearer picture of the actual profile, the alternative is to simply ask for intensity judgments of the product itself, and also where an ideal product would fall on the scale for that attribute (e.g., sweetness). The notion that consumers can accurately report on an imagined ideal product has been used for quite some time (Moskowitz, 1972; Hoggan, 1975; Szczesniak et al., 1975; Moskowitz et al., 1977; Cooper et al., 1989) and has recently revived interest (Worch et al., 2010, 2012). A good overview of the technique of **ideal profiling** can be found in the chapter by Cooper et al. (1989) in the ASTM Special Technical Publication (STP 1035) edited by Wu. A comparison of JAR modeling, regression of liking scores against attributes, and direct ideal profiling can be found in van Trijp et al. (2007).

Cooper et al. (1989) provided a case study of the orange juice category in which New Zealand consumers evaluated a set of commercial orange juice and orange drink products and also indicated the score on each scale they would give to an ideal product. The attributes were straightforward: the sweet, sour, and bitter tastes, strength of orange flavor, color, aftertaste, thickness, and pulp. Cooper et al. discussed the various ways of treating the data relative to the ideal. One can simply use the raw scores, or one can compute a difference from ideal on each scale, or a ratio. The ratio notion is appealing. One can imagine a juice that needs to be 25% sweeter to match the ideal score. If the psychophysical function for sweetness is known, it becomes straightforward to add sufficient sugar to get to the ideal level. However, ratios have odd properties. If I buy a common stock that decreases 50% in value, I have to increase the value of that stock by 100% to break even. So there is an inherent asymmetry in using ratios when one considers adjustments above or below the ideal point. The choice of whether to use raw scores or differences from ideal in the statistical analysis is tricky. Often, the standard errors of a difference from ideal may be lower, as the difference scores can remove some scale usage tendencies or number usage biases of individual consumers.

Cooper et al. (1989) also encountered the problem of having too many juices in the sample set (16) than could reasonably be evaluated by a single consumer in a single sitting. So they employed an incomplete block design, in which each consumer only saw four of the products. Owing to the incomplete design, only the mean scores were submitted to analysis for perceptual mapping, thus losing some of the individual information. Perceptual mapping will be explained in Section 13.4.2, but the basic idea is to place the products in a two- or three-dimensional space, such that similar products are mapped close together and different products are far apart. Directions through this derived space usually correspond to the original attributes. As the ideal product can be positioned, just as if it was a real product, the characteristics of this product and similarly appealing items can be visualized. Cooper et al. had a valuable insight. They examined the original data to look at the distribution of ideal scores and found that, for the variable of pulpiness, there was a tri-modal distribution. That is, some consumers liked a lot of pulp in their orange juice, some preferred none at all, and some an intermediate amount. This is a valuable lesson in looking at characteristics of your raw data, rather than assuming the mean value is a fair representation of everyone in the test! This insight permitted Cooper and colleagues to construct perceptual maps for each separate group; not surprisingly, the ideal point was positioned near very different products for the high pulp versus no-pulp groups. The interplay of perceptual mapping and identification of market segments (different consumer tastes) will be discussed further in Section 13.4.4. The finding of preferences for different pulp levels turned out to be prophetic: the US market is now populated by orange juice choices of different pulpiness.

13.4.2 A Quick Look at Perceptual Mapping

Before we get to further uses of ideal profiling, it is important to understand the basics of **perceptual mapping**. As noted above, the goal is to represent the similarities and differences of a set of products in a spatial model or map. Various types of data can be used for this purpose, notably scaled attributes by consumers or trained panels, direct estimates of product similarity, or indirect similarity measurements such as data from sorting or grouping tasks. A host of multivariate statistical techniques are available for this purpose, and each has its advantages, pitfalls, and proponents. The reader is referred to the books by Dijksterhuis (1977), Meullenet et al. (2007), and Lawless and Heymann (2010: chapter 18) for further information. If one considers each attribute to be a dimension in multidimensional space, each product then becomes a point or vector in that space based upon its rated values on the collection of attributes. The problem is that we cannot visualize more than three-dimensional models easily, so this hyperspace must be reduced somehow in order to make a map that we can see. In doing so, simplification occurs, but also loss of information. Consider looking at a person in real life (three-space) versus looking at their shadow (now in two-space). The simplification of the silhouette is accompanied by information loss.

Perhaps the first, classical, and most common method for constructing a perceptual map is to use attribute ratings, usually from a trained descriptive panel and submitting the product mean values to a principal components analysis (PCA). The PCA will look for collections of correlated attributes and, from each collection, construct a new variable called a factor or component based on the correlation pattern. For example, if products in the set that are very crisp also tend to be very dry, and those that are not dry are also not crisp, there is an across-product correlation of these two, and the PCA will find it. Once it sweeps the data set for the pattern of highest correlation and most common variance, it will sweep again to collect the next group of correlated attributes, which are uncorrelated (or "orthogonal") to the first group, and so on. The value of this procedure is that the products now each have values on these new dimensions, called their factor scores, and using these factor scores we can plot the position of each product in the new, reduced space. Thus, we have constructed a simple perceptual model with the desired characteristic that similar products are close together, and different products are far apart. Furthermore, we can project the original attributes through the space as vectors, based upon their correlation with the new factors. If all the sweet products have plotted in the upper left corner and all the unsweet products in the opposite corner of the map, the chances are that the sweetness vector points to the upper left. So the directions through the space can also be interpreted.

But this model so far is based upon the sensory characteristics of the products, not consumer appeal or differences in consumer taste preferences, so how can we find our best direction for products that have high consumer appeal? To do this, we obviously need to get consumer data on the same products, and their hedonic ratings (OAL) are the usual data to collect. Once we have this second piece of information, we can use vector projection techniques to plot directions through the space for each consumer, or for the average ratings to try to find the overall best product. The vector fitting, once again, works on the basis of correlations. In this case, the liking ratings are regressed against the coordinates of the products to see whether there is a direction through the space such that, when the product points are dropped via a perpendicular line to the vector, the positions along that vector are maximally correlated with the original (actual) liking ratings. Of course, there may be no such direction, and the spatial model we have constructed may have little or nothing to do with consumer perception or consumer preferences, but usually it does. Mathematically, this process of vector fitting is equivalent to a multiple regression as follows:

$$\mathrm{OAL} = k_0 + k_1\mathrm{DIM}_1 + k_2\mathrm{DIM}_2 + \cdots + k_n\mathrm{DIM}_n \quad (13.8)$$

where OAL is the overall liking score and DIM_x is the score for each product on the factor or dimension x. For a two-dimensional map, one can simply think of these as the x- and y-coordinates of the product in the model. If the liking scores are highly and positively correlated with the x-axis, the vector will line up near the x-axis and point to the right. If liking is negatively correlated with positions on the y-axis, the vector will point downward. If the ratings are partially, equally, and positively correlated with both x and y positions, the vector will point to the upper right quadrant at approximately a 45° angle. The simplest way to plot these by hand is to run the multiple regression (the R^2 value must be statistically significant or there is no relationship) and use the standardized regression weights for the k-values, as they correspond to the coordinates of the vector's direction in a unit space when drawn through the origin (Schiffman et al., 1981). But many multivariate analysis programs will automatically plot this for you as part of their output.

Fitting these hedonic vectors to an existing configuration is a process called preference mapping, or more specifically as "external preference mapping" since we are using data external to the consumer test to construct the original map or PCA plot. Early users of this technique discovered that individual consumers' vectors would point in different directions through the space (Nute et al., 1988, Tunaley et al., 1988). This insight provided a valuable new approach to market segmentation; that is, identification of differing consumer preferences. There are two pitfalls to this kind of vector model. First, the consumer preferences may not line up at all with the product space. Second, the vector model presumes that "more is better" and that the farther out we go along that vector's direction the higher is the product's OAL. But there can be too much of a good thing, as we saw above with the inverted U-curve for finding a sensory optimum. So an alternative model to the vector fitting is to find an ideal point in the space, rather than an ideal direction. As products move farther from this ideal point, their appeal drops off. This is the basis for a number of methods for finding consumer ideals and groups of consumer segments as discussed below.

Of course, there is no rule that a simple linear model need be used to find the best direction or the best product. If the OAL scores are fit as a response surface (the vertical direction or z-axis), a quadratic equation with interaction terms could be useful. This kind of model is attributed to Danzart (see Mao and Danzart (2008)) and discussed extensively by Meullenet et al. (2007, 2008). The model is sometimes cast as follows:

$$\mathrm{OAL}_k = \alpha + \sum_{i=1}^{n} \beta_i X_i + \sum_{i=1}^{n} \gamma_i X_i^2 + \sum_{i=1}^{n}\sum_{j=1}^{n} \delta_i X_i X_j \qquad (13.9)$$

where OAL_k are the liking scores for consumer k, and there are regression coefficients (β, γ, δ) for each principal component (i.e., each X_i) and the model has both quadratic and interaction terms. Meullenet et al. (2007) discuss two approaches to finding an optimal product based on a response surface fit to PCAs. One method (attributed to Danzart) is to triangulate three products around the ideal point, analyze their sensory properties, and construct an ideal profile based on a linear model of the three products' predictions of the intensity of each attribute. A second method is to use a generalized inverse matrix based on the PCA to calculate an optimal profile. Meullenet et al. caution that not all consumers will have an acceptable fit to the response surface, and thus it may not be possible to define an optimal point for every respondent. Furthermore, some consumers are described as "eclectic," meaning they have more than one style of product they like.

There are many other methods for constructing perceptual maps, and some of these will be discussed in Chapter 14. Hedonic scores themselves can be used to construct a map. This kind of procedure is sometimes called "internal preference mapping." The idea is that products with similar hedonic scores are plotted close together. Yet the question arises as to how the correlation is obtained. For any two products, there will be a high correlation if two conditions exist: first, the scores by any given consumer must be similar for the products; second, consumers must disagree. That is, there has to be a range of scores, and the correlation will only expand and become significant if, for example, one consumer likes both products a lot and another consumer dislikes them equally. Of course, a systematic negative correlation would also extract a factor. Thus, the map is unfortunately based in part on consumer disagreement, which may have nothing to do with common sensory properties. The two consumers may in fact have quite different sensory perceptions (we just do not know). So, in my opinion, the internal preference mapping methods should be used with extreme caution. It is possible to incorporate sensory attribute information into the hedonically derived maps by multiple regression, just as the hedonic vectors can be plotted in a sensory profile space by external mapping. If a good correspondence is found, it is evidence that the attribute is in fact an important driver of liking, which is a potentially valuable piece of information.

13.4.3 Recent Developments

The use of JAR scales for making product improvements and finding drivers of liking is only one approach to product optimization. The use of ideal profiles provides an attractive alternative in which the actual attribute intensity information is preserved. This can be done with a designed set of engineered products, or by using a commercial set of sensorially divergent items within the same product category. One approach to ideal profiling is to construct a **fishbone plot**, a graph for each product, showing the deviations from ideal for each attribute. The resemblance to a fishbone occurs because some attributes are plotted in a positive direction (they need reduction) and others negative (they need to be increased) relative to the ideal that is represented as zero. Thus, this is a deviation-from-ideal plot (Worch et al., 2010). One can also superimpose (on the fishbone) the potential gain in OAL scores (the opposite of a penalty analysis) if the product were to be altered to match its ideal profile. This is done with a separate ideal profile for each product. In a way, this makes sense because the ideal set of attributes might be different in combination for one product than another. The desired texture for an oatmeal cookie might be different from the desired texture profile of an oatmeal cookie with raisins included.

Worch et al. (2010) showed a case study of the above method and compared it with a partial least-squares analysis using dummy variables (Xiong & Meullenet, 2006) for a set of fine fragrances. In the latter approach, deviations from ideal are constructed and then the statistical analysis is used for two new "dummy" variables, one for positive deviations and one for negative deviations as shown in the OAL scores. It is, of course, possible to have both phases, as some consumers might consider the product too weak and other consumers too strong. The dummy variable can be assigned a value of zero if that particular direction is not statistically significant. Both methods showed similar "improvement" profiles for perfumes that were far from ideal, but less clear directions for those close to their ideal profile. As an alternative to the fishbone plot (basically a bipolar bar chart) one can, of course, construct a simple radar (spiderweb) plot of the product's profile and also its ideal and compare them. Worch et al. (2010) give several illustrative examples. Their complete method is mathematically complex, and will not be described fully here. It involves a PCA on the subject by attribute by product matrix, to derive a set of dimensions to represent the correlated bunches of attributes. The OAL score is then regressed against the PCA dimensions similar to external preference mapping. This information is used to develop a weighting coefficient for each attribute and a scaling factor is also introduced in the calculations. The end result is a "potential gain" value, once again something like the opposite of a mean drop or penalty. The potential gains might then be summed (assuming no attribute interactions, a big assumption) to show the overall gain possible if all the significant dimensions were optimized. Worch et al. (2010) cautioned the reader about the usual correlation or interaction of attributes, stating "the conclusions drawn here should only be taken as guidance to improvement, not as a recipe." This is remarkably similar to the concern raised by Cooper et al. (1989) many years earlier, referring to the difficulty in making specific recommendations based on a PCA plot because the PCA factors are combinations of attributes.

Several results of this study raise concerns about the method. First, the ideal product was assumed to get a value of nine on the OAL scale. This might or might not be true. Second, the correlation between the two methods in the predicted gain in liking was only 0.3, averaged across products. In summary, the agreement of the two different methods was better when the product had a heterogeneous set of attributes (some near ideal, some not, and some not influential) and were farther from their ideal.

Later, Worch et al. (2012) looked at the consistency and validity of the ideal profile method. The basic idea was to construct a product map based on the ideal profiles, once again of the perfume set. If the consumers were consistent, the ideal map should correspond to (1) their likes and dislikes for the actual products when consumers were later plotted in the map and (2) the attribute profiles when the attributes were later fit to the map. Finally, they attempted to construct the ideal product for each perfume and predict its liking rating based on a regression model for each consumer's attribute by liking data. As expected, the "constructed" ideal products had, in general, higher liking ratings than the actual products did. The attempt at constructing a meaningful PCA map of the ideals was less successful. The first dimension corresponded to an overall intensity vector, which is not uncommon for PCAs, which tend to sweep the largest amount of common variance into the first factor (a common pitfall). Thus, they were forced to look at the second and third dimensions. Although a large number of consumers fell near the origin of the space, a few outliers seemed illustrative. Worch et al. (2012) chose a few whose product preferences for the real products corresponded to the directions of their ideals, and whose attribute/profiles also matched up. Thus, if a consumer preferred honey and caramel notes in their perfume, a perfume with those notes was plotted near their ideal, and it received a high liking rating. Overall, the study raises the interesting question of whether a newly engineered ideal product would actually be liked more than the real product the ideal profile was based on. A true demonstration proof is lacking, but time will certainly tell.

13.4.4 Ideal Point Mapping Based on Liking Scores and Distance

There are many ways to construct a product map representing product similarities and dissimilarities within a set of related items. A consumer's ideal point or ideal profile should correspond to a specific section of the map, as long as those attributes that fed into the map are meaningful to the consumer and correspond to the reasons for their preferences. This is not always guaranteed. Rather than have a consumer report or specify a profile of their ideal product, an ideal position can be found in the product map as long as the consumer has provided OAL ratings for the same set that was used to construct the map. The goal is to find the position that satisfies the following condition: the distance from that point to each product provides a maximal negative correlation with their liking ratings. In other words, at that point they are nearest to the things they like and farthest from things they dislike. Finding said point can be done in many ways. Basically, we are finding the optimum in a response surface, so various hill-climbing algorithms can be brought to bear. A simple solution is to simply march through the space on an exhaustive grid search and compute the size of the correlation at each sampled point. This is a kind of "brute force" approach, not very elegant, but one that avoids local maxima and also provides a kind of slope estimate. It is possible that the top of the preference hill is very broad, for example, for a consumer who tolerates large changes around their just-right points. This approach is shown in Figure 13.5.

Meullenet and colleagues have made extensive use of this kind of **ideal point model**, which they call Euclidean distance ideal point measurement or EDIPM for short (Meullenet et al., 2008; Lovely & Meullenet, 2009). The method derives a product space from a PCA on product attributes and then finds the individual ideal points for each consumer. An important part of this analysis is the construction of a density plot showing the areas in which greater or fewer consumer ideals are positioned.

Another approach is based on difference-from-ideal. This kind of method looks at each product's hedonic score as its difference from the top of the scale (e.g., from nine in the nine-point hedonic scale). These difference scores can then be submitted to various multidimensional scaling algorithms, one of which uses a probabilistic model to construct

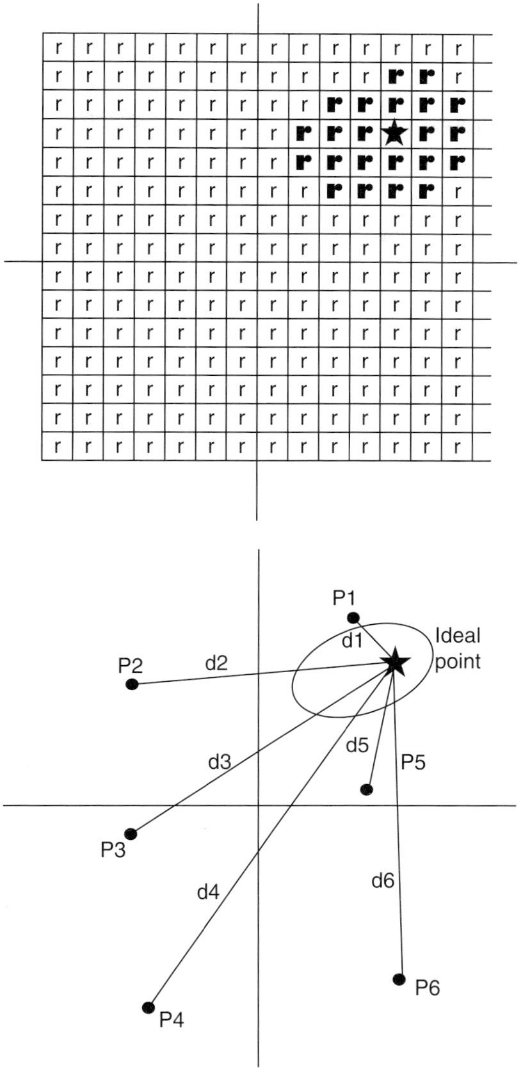

Figure 13.5 An ideal point grid search using distance and hedonic scores to find optimum zones. The upper panel shows the grid divisions, and distances to each product (e.g., d1–d6 in the lower panel) are measured and compared with the actual hedonic scores of a consumer. The consumers ideal point lies in the grid box that has the highest negative correlation *r*. The bold-face rs show an area in which there is no significant difference from the ideal point *r*-value, or is similar enough to be considered equivalent. This defines a zone or plateau in the optimization response surface as shown by the ellipse in the lower panel.

the space, called landscape segmentation analysis (LSA) (Ennis et al., 2011). Because this is a form of internal preference mapping, attribute vectors can then be projected into the space to identify potential drivers of liking. The LSA is a proprietary program but produces much useful information, such as a density plot that is useful in discovering consumer segments.

13.5 Summary and Conclusions

This chapter has discussed a number of methods used to improve or optimize products. In traditional experimental designs, products could be systematically varied in some ingredient or processing variables, and the OAL scores modeled as a linear, quadratic, or interactive combination of those design features. In a simple two-factor design, for example, it was popular to model the liking as a response surface, with attempts to find the highest point or zone in the design space (Gacula, 1993). One key aspect of a successful model is to base the choice of design variables on drivers of liking. That is, the sensory attributes that most heavily impact consumer preferences should be identified first, and then those ingredient or processing variables that affect the critical sensory attributes can (in theory) be systematically varied. In any use of hedonic scores as a measure, we bump into the issue of how discriminating a consumer may be. Some may be eclectic, as described above, and others may like all the products in the test set or hate all products, or even be indifferent. So a strategic question has to be raised as to what to do with such "nondiscriminators." Eliminating them from the analysis is one approach, although this runs counter to the notion that consumer testing should include a random sample of everyone who uses the product category. However, if one is constructing a model or a perceptual map, it may just add noise (or even give a false optimum in the center of the map) if nondiscriminators are included. This is an important strategic decision.

Using hedonic scales for optimization has two pitfalls. One is to assume that the ideal product will have a score of nine on the nine-point hedonic scale. This is a common assumption in some forms of perceptual mapping or in using scores that are calculated as a difference from ideal. It seems quite likely that no product, no matter how wonderful, could attain a mean score of nine. Even for an individual consumer, my highest liking rating for the best possible form of tomato ketchup is probably only a seven (like moderately) on the nine-point scale. Even "ideal" ketchup is not all that exciting. The second pitfall is that the common nine-point liking scale is really a truncated interval scale with eight intervals and no real zero point. There are no raw data between zero and one if one is the bottom of the scale (dislike extremely). This makes taking ratios in a model based on the nine-point scale questionable at best (but see Cooper et al. (1989)). As one example, Meullenet et al. discuss using a ratio to ideal as the model's input, where the hedonic score is divided by nine, the assumed ideal. Thus, a product scoring 9 gets a ratio of 1 and a product scoring 5 gets a ratio of 0.55 (Meullenet et al., 2007: 99). But the data should probably be recast as a zero to eight scale in order to make a meaningful ratio. The neutral fifth point now becomes a 4 and Meullenet's ratio becomes 0.50, where it should logically be for a score at the center of the scale.

A good deal of attention was paid to JAR scales, because they are very commonly used in consumer testing and the statistical treatment of JAR data has been extensively discussed in the recent literature (Rothman & Parker, 2009). JAR scores are logically related to the difference from an ideal product. So the natural extension of this approach is to have consumers profile an ideal product directly. This requires some assumptions about what consumers are capable of, and the ideal profile method has a history of criticism on fundamental grounds. Yet, there appear to be some promising results in the recent literature (Worch et al., 2010, 2012).

The second part of the chapter was devoted to methods that produce perceptual maps of a group of related products or a purposely sampled product category. Perceptual mapping for purposes of modeling consumer preferences is discussed extensively in Meullenet et al. (2007), to which the interested reader is enthusiastically referred. If these maps or models are to be used successfully in defining the attributes of preferred products, they should have a set of properties that are desirable, if not required. These define what is a useful map. These properties include the following. First, the map should reproduce sensory differences as distances. That is, similar products should be plotted close together and different products far apart. Second, areas of the map should correspond to particular profiles or combinations of properties that are actually preferred by consumers. These can be represented as directions through the space (vectors) or regions on a response surface. Third, the mapping procedure should be capable of differentiating consumer segments; that is, different sensory profiles that appeal to different consumer groups. Fourth, the map should be validated, especially if the ideal point or points are defined by imaginary profiles. That is, a map based on ideal profiles should plot the ideal point near products that are actually preferred. Fifth, the map should be capable of indicating some kind of density distribution; that is, the number or proportion of consumers whose ideal product lies in a certain region.

This chapter has dealt primarily with optimization from the standpoint of using scaled data, such as OAL scores, as the dependent variable to maximize. However, there are other alternatives, such as maximizing the acceptor set size (LaGrange & Norback, 1987) or modeling choice behavior. A large literature exists on choice behavior modeling, as this formed one of the basic research areas of early experimental psychology (would the rat choose to go left or right in the maze?). Choice, of course, is the basis for various paired comparison models, such as the Thurstonian models discussed in Chapter 4. For additional choice models, the reader is referred to Baird and Noma (1978: chapter 9), for an introduction. The basis for optimization is the idea that a sensory attribute has some optimal level of intensity that is preferred by a consumer (i.e., the inverted-U or Wundt curve). However, not all attributes are continuous and varying in sensory intensity, although this is the most common kind of attribute with foods. With durable goods, the attributes may be more categorical, discontinuous, or even binary. For example, I can order my new car with two doors or four doors, a V-6 or a four-cylinder engine, as a convertible or hard top, and with side curtain air bags or not. Various methods have been developed for dealing with combinations of attributes or features of this nature, notably conjoint measurement (Moskowitz et al., 2006).

The reader should also realize that the multivariate techniques for mapping may lead to different decisions about what is an optimum product. Comparative studies and validations are rare. Meullenet et al. (2007) compared product optimizations performed via landscape segmentation and Euclidean distance point methods and found some correspondences and some differences. Comparison of optimal chip products via Euclidean distance modeling and response surfaces showed strong similarities. Worch et al. (2010) compared conclusions from the partial least-squares dummy variable analysis of perfume preferences to the pattern seen with ideal point/fishbone analysis and found better correspondence of those products that were far from ideal, as opposed to those that were closer to optimal. Van Trijp et al. (2007) compared optimization by JAR methods, inferred ideals from regression models, and direct scaling of ideals. Similar orders of ideal points were obtained, but substantial differences existed in the directions for product improvement.

The need for caution in this area was noted by Meullenet et al. (2007: 75), who stated:

> Unfortunately, it is not yet possible to let the reader know what method is best as a comparative study, and validation of the methods is not available in the literature. These methods may yield different optimal product solutions, but unless the products were to be formulated and tested for hedonic level with consumers, there is not a good way to assess superiority of a method over another.

Several years hence, this is still true.

References

Ares, G., Giménez, A, Barreiro, C., and Gámbaro, A. 2010. Use of an open-ended question to indentify drivers of liking of milk desserts. Comparison with preference mapping techniques. Food Quality and Preference, 21, 286–94.

Baird, J.C. and Noma, E. 1978. Fundamentals of Scaling and Psychophysics. John Wiley & Sons, Inc., New York, NY.

Beebe-Center, J.G. 1932/1965. The Psychology of Pleasantness and Unpleasantness. Russell & Russell, New York, NY.

Best, D.J. and Rayner, J.C.W. 2001. Application of the Stuart test to sensory evaluation data. Food Quality and Preference, 12, 353–7.

Booth, D.A. 1994. Flavour quality as cognitive psychology: the applied science of mental mechanisms relating flavour descriptions to chemical and physical stimulation patterns. Food Quality and Preference, 5, 41–54.

Booth, D.A. 1995. The cognitive basis of quality. Food Quality and Preference 6, 201–7.

Conner, M.T. and Booth, D.A. 1992. Combined measurement of food taste and consumer preference in the individual: reliability, precision and stability data. Journal of Food Quality, 15, 1–17.

Cooper, H.R., Earle, M.D., and Triggs, C.M. 1989. Ratios of ideals – a new twist to an old idea. In: Product Testing with Consumers for Research Guidance. L.S. Wu (Ed.). ASTM STP 1035. ASTM, Philadelphia, PA, pp. 54–63.

Cornell., J.A. 2002. Experiments with Mixtures: Designs, Models and the Analysis of Mixture Data. Third edition. John Wiley & Sons, Inc., New York, NY.

Cornell., J.A. and Khuri, A.I. 1996. Response Surfaces: Design and Analyses. Second edition. Marcel Dekker, New York, NY.

Dijksterhuis, G.B. 1997. Multivariate Data Analysis in Sensory and Consumer Sciences. Food and Nutrition Press, Trumbull, CT.

Ennis, D. M., Rousseau, B., and Ennis, J. M. 2011. Short Stories in Sensory and Consumer Science. Institute for Perception, Richmond, VA.

Fritz, C. 2009. Appendix G: Methods for determining whether JAR distributions are similar among products (chi-square, Cochran–Mantel–Haenszel (CMH), Stuart–Maxwell, McNemar). In: Just-About-Right Scales: Design, Usage, Benefits, and Risks. L. Rothman and M.J. Parker (Eds). ASTM Manual MNL63. ASTM International, Conshohocken, PA, pp. 29–37.

Gacula, M.C., Jr. 1993. Design and Analysis of Sensory Optimization. Food and Nutrition Press, Trumbull, CT.

Gacula, M., Singh, J., Bi, J., and Altan, S. 2009. Statistical Methods in Food and Consumer Research, Second edition. Amsterdam: Elsevier/Academic Press.

Hernandez, S.V. and Lawless, H.T. 1999. A method of adjustment for preferred levels of capsaicin in liquid and solid food systems among panelists of two ethnic groups and comparison to hedonic scaling. Food Quality and Preference, 10, 41–9.

Hoggan, J. 1975. New product development. MBAA Technical Quarterly, 12, 81–6.

IOM (Institute of Medicine). 2010. Strategies to reduce sodium intake in the United States. National Academies Press, Washington, DC.

Johnson, J. and Vickers, Z. 1987. Avoiding the centering bias or range effect when determining an optimum level of sweetness in lemonade. Journal of Sensory Studies, 2, 283–91.

Lagrange, V. and Norback, J.P. 1987. Product optimization and the acceptor set size. Journal of Sensory Studies, 2, 119–36.

Lawless, H.T. 1977. The pleasantness of mixtures in taste and olfaction. Sensory Processes, 1, 227–37.

Lawless, H.T. and Heymann, H. 2010. Sensory Evaluation of Foods, Principles and Practices. Second edition. Springer, New York, NY.

Lovely, C. and Meullenet, J.-F. 2009. Comparison of preference mapping techniques for the optimization of strawberry yogurt. Journal of Sensory Studies, 24, 457–78.

Mao, M. and Danzart, M. 2008. Multi-response optimization strategies for targeting a profile of product attributes with an application on food data. Food Quality and Preference, 19, 62–173.

Mattes, R.D. and Lawless, H.T. 1985. An adjustment error in optimization of taste intensity. Appetite, 6, 103–14.

McBride, R. 1990. The Bliss Point Factor. Macmillan (Australia), South Melbourne, NSW.

McBride, R.L. 1982. Range bias in sensory evaluation. Journal of Food Technology, 17, 405–10.

Meullenet, J.-F., Xiong, R., and Findlay, C.J. 2007. Multivariate and Probabilistic Analyses of Sensory Science Problems. IFT Press/Blackwell Publishing, Ames, IA.

Meullenet, J.-F., Lovely, C., Threlfall, R., Morris, J.R., and Streigler, R.K. 2008. An ideal point density plot method for determining an optimal sensory profile for Muscadine grape juice. Food Quality and Preference, 19, 210–19.

Moskowitz, H.R. 1972. Subjective ideals and sensory optimization in evaluating perceptual dimensions in food. Journal of Applied Psychology, 56, 60–6.

Moskowitz, H.R. 1981. Relative importance of perceptual factors to consumer acceptance: linear vs. quadratic analysis. Journal of Food Science, 46, 244–8.

Moskowitz, H.R., Stanley, D.W., and Chandler, J.W. 1977. The eclipse method: optimizing product formulation through a consumer generated ideal sensory profile. Canadian Institute of Food Science and Technology, 10, 161–8.

Moskowitz, H.R., Jacobs, B.E., and Lazar, N. 1985. Product response segmentation and the analysis of individual differences in liking. Journal of Food Quality, 8, 168–191.

Moskowitz, H.R., Beckley, J.H., and Resurreccion, A.V.A. 2006. Sensory and Consumer Research in Food Product Design and Development. IFT Press/Blackwell Publishing, Ames, IA.

Nute, G.R., MacFie, H.J.H., and Greenhoff, K. 1988. Practical application of preference mapping. In: Food Acceptability. D.M.H. Thomson (Ed.). Elsevier Applied Science, London, pp. 377–86.

Pangborn, R.M. 1970. Individual variations in affective responses to taste stimuli. Psychonomic Science, 2, 125–8.

Pangborn, R.M. and Pecore, S.D. 1982. Taste perception of sodium chloride in relation to dietary intake of salt. American Journal of Clinical Nutrition, 35, 510–20.

Pangborn, R.M., Bos, K.E.O., and Stern, J.S. 1985. Dietary fat intake and taste responses to fat in milk by under-, normal and overweight women. Appetite, 6, 25–40.

Plaehn, D. and Horne, J. 2008. A regression-based approach for testing significance of "just-about-right" variable penalties. Food Quality and Preference, 19, 21–32.

Poulton, E.C. 1989. Bias in Quantifying Judgments. Lawrence Erlbaum Associates, Hillsdale, NJ.

Prescott, J., Norris, L., Kunst, M., and Kim, S. 2005. Estimating a consumer rejection threshold for cork taint in white wine. Food Quality and Preference, 16, 345–9.

Rivière, P., Monrozier, R., Rogeaux, M., Pagès, J., and Saporta, G. 2006. Adaptive preference target: contribution of Kano's model of satisfaction for an optimized preference analysis using a sequential consumer test. Food Quality and Preference, 17, 572–81.

Rothman, L. and Parker, M.J. 2009. Just-About-Right Scales: Design, Usage, Benefits, and Risks. ASTM Manual MNL63. ASTM International, Conshohocken, PA.

Schiffman, S.S., Reynolds, L.M., and Young, F.W. 1981. Introduction to Multidimensional Scaling. Academic Press, New York, NY.

Schraidt, M. 2009. Appendix L: Penalty analysis or mean drop analysis. In: Just-About-Right Scales: Design, Usage, Benefits, and Risks. L. Rothman and M.J. Parker (Eds). ASTM Manual MNL63, ASTM International, Conshohocken, PA, pp. 40–7.

Southey, R. 1837/1916. The selected Prose of Robert Southey, J. Zeitlin (Ed.). Macmillan, New York, NY. (Originally published in *The Doctor* (anon./Southey, 1837).)

Szczesniak, A.S., Loew, B.J., and Skinner, E.Z. 1975. Consumer texture profile technique. Journal of Food Science, 40, 1253–6.

Templeton, L. 2009. Appendix R: Chi-Square. In: Just-About-Right Scales: Design, Usage, Benefits, and Risks. L. Rothman and M.J. Parker (Eds). ASTM Manual MNL63. ASTM International, Conshohocken, PA, pp. 75–81.

Tunaley, A., Thomson, D.M.H., and McEwan, J.A. 1988. An investigation of the relationship between preference and the sensory characteristics of nine sweeteners. In: Food Acceptability. D.M.H. Thomson (Ed.). Elsevier Applied Science, London, pp. 387–400.

Van Trijp, H.C.M., Punter, P.H., Mickartz, F., and Kruithof, L. 2007. The quest for the ideal product: comparing different methods and approaches. Food Quality and Preference, 18, 729–40.

Worch, T., Dooley, L. Meullent, J.-F., and Punter, P. 2010. Comparison of PLS dummy variables and fishbone method to determine optimal product characteristics from ideal profiles. Food Quality and Preference, 21, 1077–87.

Worch, T., Lê, S., Punter, P., and Pagès, J. 2012. Assessment of the consistency of ideal profiles according to non-ideal data for IPM. Food Quality and Preference, 24, 99–110.

Xiong, R. and Meullenet, J.-F. 2006. A PLS dummy variable approach to assessing the impact of JAR attributes on liking. Food Quality and Preference, 17, 188–98.

14 Perceptual Mapping, Multivariate Tools, and Graph Theory

> *In sensory analysis, one of the main objectives is to characterize a set of products according to the way they are perceived. To do so, a common practice consists in asking subjects to rate the products on the perceived intensities of a list of attributes…* In fine, *the objective of such methodology is to obtain a product space, which is a map positioning the products that are perceived as similar close to each other, and placing apart those that are perceived as different.*
>
> Worch et al. (2013)

14.1 Introduction

Although one may question whether the ultimate goal of sensory analysis is the placement of products into a spatial model of perception, such graphs are commonplace in sensory analysis. In his book on multivariate analyses, for example, Dijksterhuis (1997) showed over 80 such plots, representing about 90% of the total number of figures in that treatise. An increasing number of multivariate statistical methods are becoming available to sensory analysts. Indeed, a major focus area for the **sensometrics** community for the last few decades has been to develop more, and possibly better, tools for taking large data sets and reducing them into a simpler picture that is understandable and hypothesis-generating. It would not be possible to cover all of them in one chapter, or perhaps even a whole book, nor would it be possible to give an up-to-date picture of the latest trendy methods. Thus, this

Quantitative Sensory Analysis: Psychophysics, Models and Intelligent Design, First Edition. Harry T. Lawless.

chapter is necessarily a snapshot in time. The goal is to look at the major types of techniques that are available, both in data collection and statistical analysis, in order to give the reader a foundation for understanding the major common techniques.

Most, if not all, **perceptual mapping** techniques result in a pictorial display, often in two or three dimensions, that show the relationships between products, and often the attributes used to evaluate them. Products that have similar characteristics are plotted close together, and products that are different are far apart. Attributes may be projected through the map as vectors. This is an important explanatory aid to the interpretation of the map, as it shows which products are high and low in certain characteristics. The picture may also help us to understand which of many attributes were important to the consumers or assessors when they first evaluated the products. An alternative to a vector model for an attribute (farther from the origin is more of that characteristic) is an ideal point model, where a point in the space shows the spot or region highest in that, and distances from that spot in any direction indicate decreasing amounts of that attribute. Some of these methods were discussed in Chapter 13.

One major distinction among these maps is those that are based on sensory information and those based on hedonic information. That is, one can build a perceptual map based on sensory similarity, or patters of similarity in liking/disliking among consumers. Often, the product positions in these two kinds of maps, if performed on the same product set, will look similar. This is an expected result if the hedonic evaluations are based on the products having certain characteristics that people like or dislike. On the other hand, the characteristics used to create a sensory similarity space may not have the same weight when it comes to determining what consumers like or not. So the correspondence may be fortuitous, but it should not be expected. Perceptual maps using hedonic information are often called **preference maps** or a prefmap for short. These were also discussed in Chapter 13 on product optimization.

Any construction of a perceptual map should be viewed as an exercise in data reduction. Consider a descriptive analysis profile of j products evaluated on k attributes. The mean scores form a $j \times k$ matrix of values, and each product can be considered a vector in k spatial dimensions. So the raw information is a k-dimensional hyperspace where each product's position is determined by its k (attribute) values. However, this is certainly difficult, if not impossible, to visualize. So, to make a perceptual map of two or three dimensions, the k-dimensional hyperspace must be reduced somehow to a smaller number of new dimensions. But information is bound to be lost. The view of the professor in the classroom is three-dimensional owing to our depth and distance perception. Consider the two-dimensional shadow she casts on the floor from the sun coming through a window. That, too, is a model of the professor, but now she is in 2-space. The shadow does not contain all of the information from the 3-space view. Such techniques are sometimes called projective, since we are projecting some information from the multidimensional data set into a pictorial representation of smaller dimensionality. So the user of any multivariate method for perceptual mapping must first understand that although these techniques can be very informative, there is no free lunch. Information is lost. Conversely, the correspondence of the new configuration to the old one is not perfect. There is always some badness of fit. Many techniques use badness of fit as a criterion to be minimized in constructing the map.

Many resources are available for those wishing to understand the basics of multivariate analyses and how they are applied to sensory data to produce perceptual maps. A good starting point is the chapter in Lawless and Heymann (2010) on multivariates, as well as the chapter on strategic research. A more complete and yet practical coverage is in the recent book by Meullenet et al. (2007). Books devoted to more specific topics include the treatise by Dijksterhuis (1997), Martens and Martens (2001) and Shiffmann et al. (1981) on multidimensional scaling, and Gower and Dijksterhuis (2004) on generalized Procrustes analysis (GPA).

An important advance in recent years has been the ability to plot confidence ellipses around the plotted product positions (e.g., Husson et al., 2004, 2005, 2006; Lê & Husson, 2008). This kind of tool can be important in understanding the degree of uncertainty around the product position or its variability. As such, it can be used to examine hypotheses about product differences. At least it may deter researchers from drawing conclusions about product differences based on positional differences in the map that may not be reliable. In these methods, the variability estimates usually come from bootstrapping or other resampling techniques. An important distinction is drawn between deterministic models, that represented products and/or consumers as points in the space, and probabilistic models that represent them as density distributions (Ennis et al., 2011b). The latter models the uncertainty around the position, a probably more realistic approach to perception.

14.2 Common Multivariate Methods

14.2.1 Principal Components Analysis

A classical method for looking at attribute data is **principal components analysis (PCA)**. PCA is a method that detects patterns of correlation among attributes. Assume, again, that we have a matrix of mean values for *j* products on *k* attributes. The PCA will extract new variables, sometimes called factors or principal components, collecting information from a set of correlated attributes. For example, there might be a negative correlation between perceived denseness of a product and its values on crispy and crunchy scales. These three characteristics might be collected into a single texture factor, with negative weight for denseness and positive weight for the other two. Sometimes, the process of factor extraction is referred to as a search for latent (i.e., hidden) variables that may be driving the set of intercorrelated measurements. For example, we might lower the pH of a product, which changes the perceived sourness, decreases sweetness, and makes it seem juicier.

The "weights" are called **factor loadings**, and are proportional to the correlation between the values of the products on the new factor (called **factor scores**) and the original values for the products on the attributes you started with. So we get two important pieces of information from a PCA: the positions of the products in the new space defined by the factors and the relationship of the factors to the original variables, which is critically important in interpreting their meanings. A third piece of information concerns the proportion of the total original variance that is captured or swept into each of the new factors. This is called the **eigenvalue**. The eigenvalue for a factor, divided by the number of original attributes, tells us the proportion of variance accounted for by that factor. Another common term for this is the inertia. Usually, factors with eigenvalues less than 1.0 are not retained in the resulting output, as they tend to have less than the average variance accounted for in the original variables. Subsequent factors are extracted after the information from the earlier factors is removed. In other words, the second factor only uses the variance left over after

Table 14.1 Sorted, rotated factor loadings* from the vegetable experiment of Stevens and Lawless (1981)

Original attribute	Factor 1 loadings	Factor 2 loadings	Factor 3 loadings
Sharp	0.947		
Flavor strength	0.886		−0.393
Sour	0.872	−0.346	
Odor strength	0.649	−0.495	−0.506
Fruity (versus vegetable)	0.614	0.572	0.452
Sweet		**0.953**	
Liking	−0.261	**0.903**	
Salty		**−0.834**	
Bitter	0.631	**−0.707**	
Smooth versus gritty			**0.914**
Eigenvalue	3.77	3.65	1.61

*Loadings with absolute value less than |0.25| have been replaced by blanks.

the first one is constructed. This creates a set of factors which are orthogonal, or uncorrelated. A quick example follows.

Stevens and Lawless (1981) gave a group of consumers 12 pureed fruits and vegetables to evaluate on 10 simple attributes, as part of a larger study on age-related changes in flavor perception. Perceived intensity of the four classical tastes, overall odor strength, flavor strength, smoothness, sharpness of flavor, liking and fruit versus vegetable character were rated on 15 cm line scales. The resulting matrix of mean values then contained 120 means. Table 14.1 shows the factor loadings for the 10 attributes on the first three factors. Given this information, the researcher will make the plot of products and attribute vectors and will begin a process of subjective interpretation of the underlying meaning of the factors. In the case of this data set, three factors were extracted from the 10 original variables, accounting for 38%, 37%, and 16% of the original variance. Note that the factors were rotated. This is an option in many PCA programs and usually is performed to maximize the amount of unique variance in the extracted factors. Now for the interpretive part that starts by examining the size of the factor loadings. The third factor is clearly related to texture, for which there was only one scale, so it is not surprising that this appears to catch some unique variance. The second factor is a little more mysterious, but note that it includes positive loadings for sweetness and liking and a high negative loading for bitter. So a good guess would be that it represents hedonic information. The first factor is more troublesome. It has fairly high loadings on six scales, many having to do with strength of sensations, so it is most likely an overall strength indicator. It is a common result that a PCA will sweep the largest amount of common variance into the first factor, which can sometimes produce an odd-seeming collection of loosely related variables. An example of the kind of perceptual map constructed from this PCA and the product scores is shown in Figure 14.1.

One further piece of information is usually available in PCA output, called **communality**. Communality shows us how much of the original variance of a given variable or attribute is accounted for in the factors you have retained. If communality is low (say, below 0.7 or even 0.5) then that particular attribute is poorly explained by the PCA model. This can be an indication that the attribute has some unique variance that is not correlated with the rest of the input variables. If it is an attribute that is important to consumers' liking or disliking of a product, your PCA model may be missing some important information.

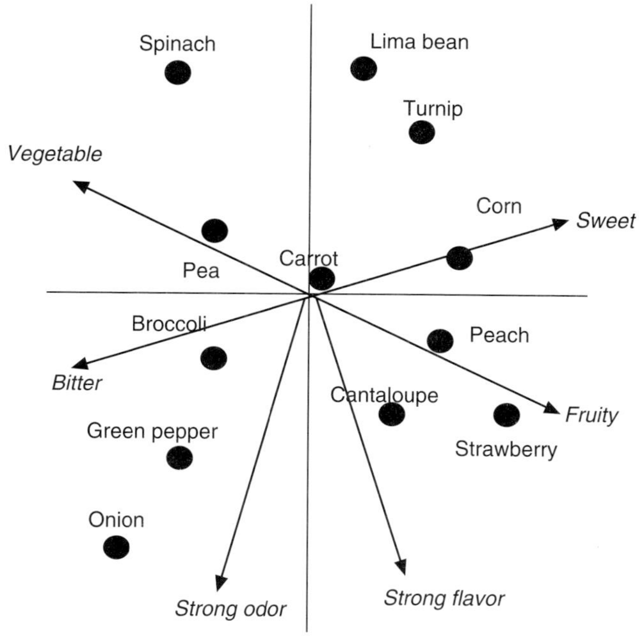

Figure 14.1 A PCA plot from some of the vegetable data in Stevens and Lawless (1981). Attribute vectors based on factor loadings are labeled in italics.

14.2.2 Multidimensional Scaling

A completely different approach to perceptual mapping is a technique called **multi-dimensional scaling** (MDS). Originating at Bell Labs in the mid-20th century, the method takes similarity information as the input and attempts to construct a perceptual space that reproduces the similarity information as accurately as possible. As noted above, there are always losses of information, and in this case distortions when the final output is achieved. In classical MDS, the input is pairwise similarity data. Usually, all possible pairs of products are judged for their similarity; for example, on a rating scale such as a line scale. The method, however, is quite flexible, in that many other derived measures of similarity, such as a correlation pattern across descriptive attributes, could also be used as input. Most programs are iterative; that is, they move the products around in the space until the badness of fit is no longer decreasing by a certain amount or criterion. When the fit essentially stops improving, the program terminates and the positions of the products are fixed. The measure of badness of fit is called stress, of which there are several different computational forms. Stress is inversely related to the R^2 value for the variance accounted for, and usually both measures are provided by an MDS program.

A common demonstration for an MDS analysis is to take the intercity mileage chart from an atlas or road map and reconstruct the positions using an MDS algorithm. The intercity distances constitute a half matrix of pairwise information, usually omitting the diagonal of a city's distance from itself, as shown in Table 14.2. Such demonstrations generally work fairly well for cities in a contiguous land mass like the lower 48 states of the USA, but less so for countries with major cities on islands and isolated land masses such as Canada (flying distance would work in that case). Figure 14.2 shows such a sample from New York State, and the resulting MDS output. Owing to the fact that there

Table 14.2 Intercity driving distances, New York State[1]

	Albany	Binghamton	Syracuse	Buffalo	NYC
Albany	X				
Binghamton	140	X			
Syracuse	146	74	X		
Buffalo	290	202	150	X	
NYC	152	180	250	377	X

[1] Cooperstown omitted for brevity.

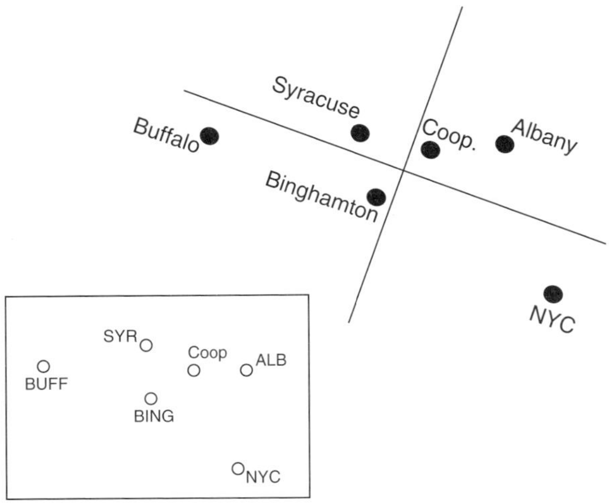

Figure 14.2 An MDS configuration based on the intercity driving distances among six cities in New York State. The output was generated by the R program isoMDS and rotated to bring the output configuration into line with the geographic positions. The inset at the lower left shows the accurate geographical positions of the cities.

are major interstate highways connecting most of these cities in roughly straight lines, the relative positions in the MDS output are quite similar to the actual geographical positions, as shown in the inset on the lower left.

One of the advantages of an MDS approach is that the critical attributes are not defined ahead of time in any way. In deciding upon the similarity or dissimilarity of any two products, the participants can use whatever criteria they deem to be important. Thus, the fundamental process is one of unbiased discovery. The method can discover what attributes might be important to consumers, say, in thinking about or tasting a group of products. This is quite a different scenario from an analysis using principal components on a set of descriptive attributes. Those attributes or scales are predetermined, and there is no knowledge ahead of time about their importance or weighting in the mind of a consumer. They are all equal in the eyes of a covariance detector like PCA.

MDS creates a configuration, and then it is up to the experimenter to interpret that configuration and see what the important characteristics might be. One common approach to this is to examine the edges, looking for products that have been plotted at opposite ends or corners, and asking how they differ. The common demonstration for this was to use Morse code symbols (Kruskal & Wish, 1978), which when scaled for similarity gave a two-dimensional configuration with dots versus dashes as one axis of differentiation and the length of the string (one to five symbols) as the other (usually orthogonal) axis. However, a more direct approach to this discovery process uses additional outside information, as discussed below.

14.2.3 Vector Projection

If additional information is available, it can be connected to the positions in an MDS configuration by a process of vector projection. This is a common method for exploring the attributes that might have been used by consumers when they were thinking about the similarities and dissimilarities of different products. The MDS experiment, then, must include a second phase or some outside data collection in order to find the values of the products on the potential attributes. One way to do this is simply to ask the subjects, following their similarity judgments, what aspects of the products they considered. The most common or most frequent attributes can then be used as rating scales by these same consumers in a subsequent session. Usually, the mean values will be used then to find the best-fitting vectors.

The mathematical process is one of multiple regression, the same as external preference mapping (see Chapter 13). This is illustrated in Table 14.3. The mean attribute ratings are regressed against the x- and y-position values for the products. If the attribute was something that was actually used by the subjects in their considerations of the product similarities, the chances are that there will be some correspondence to the product positions in the map. If not, there may be no significant R^2 or any significant regression coefficients. The regression equation for a two-dimensional configuration would look something like this:

$$A = \beta_1 X + \beta_2 Y + \beta_0 \tag{14.1}$$

where A is the attribute rating means, and X and Y are the coordinates in the spatial model (map). The β values are the standardized regression coefficients. Often, the intercept value β_0 is ignored or the vector constrained to pass through the origin. Examples can be round in Schiffman et al. (1981) and Kruskal and Wish (1978). The β weights are convenient to use since they predict the direction in a unit space when used as the terminal coordinates of the vector. In other words, simply plotting them in a map bounded by −1 to +1 (after rescaling if necessary) tells us where the vector should be plotted.

The regression coefficients tell us what direction the vector should have. For example, if an attribute is highly positively correlated with the x-axis and has zero correlation with the y-axis, then the vector will line up with the x-axis and point in the positive direction. If it is partially, equally, and positively correlated with both x and y positions of the products, it would point roughly at a 45° angle towards the upper right quadrant, and so on. Looking at Figure 14.3, the function of the multiple regression is to maximize the correlation between the original attribute values and the intersection values (I1–I6) of the perpendicular lines dropped from the product position to the vector. The vector is like a new ruler or axis, much like a principal

Table 14.3 Example for multiple regression to fit attribute vectors in a perceptual map

Product	Attribute 1 means	X-coordinate	Y-coordinate
1	(value for) P_1, A_1	P_1, X_1(value)	P_1, Y_1(value)
2	P_2, A_2(value)	P_2, X_2(value)	P_2, Y_2(value)
3	P_3, A_3(value)	P_3, X_3(value)	P_3, Y_3(value)
⁝	⁝	⁝	⁝
N	P_N, A_1(value)	P_N, X_1(value)	P_N, Y_1(value)

component. In other words, the products' "values" on this new axis bear the maximum possible correlation with the attribute means, given this position/direction for the vector.

Going back to our city map for New York State, Figure 14.4 shows some vector projections for several variables. Imagine that the cities are like food products and that the vectors arise from ratings on attribute scales (consumer or descriptive). The attributes include the actual

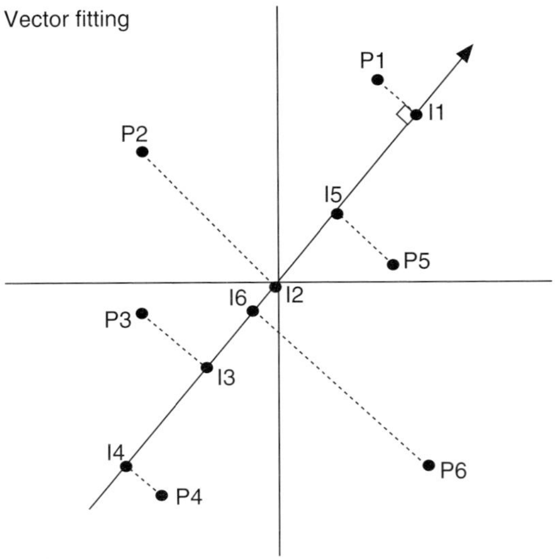

Figure 14.3 Vector projection. A multiple regression finds a direction through the space such that the values of the products (P1–P6) on that attribute (V1–V6) are maximally correlated with the values given by the intersection points (I1–I6). Much like a PCA, we are maximizing the correlation between the original values and the values the products would have on the new "ruler" of the vector.

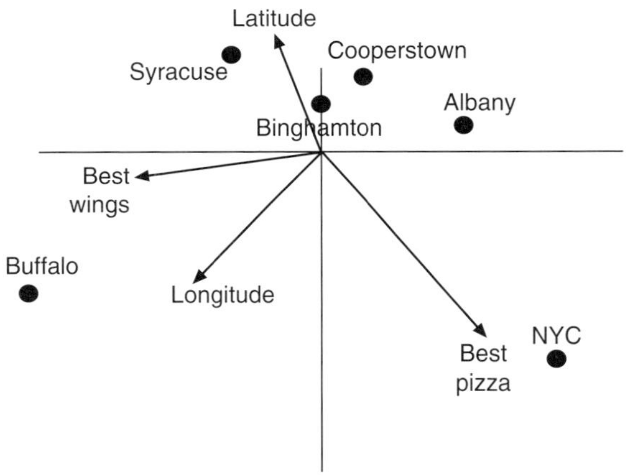

Figure 14.4 Vector projection on the configuration of Figure 14.2 (cities). Actual latitude, longitude, and consumer ratings for best chicken wings, best pizza, and major sports teams were regressed against the unrotated positions in Figure 14.2. Note that the latitude and longitude are tilted at about 30°, so the configuration in Figure 14.2 was rotated to compensate. The vector for most important sports teams did not fit the configuration at all ($R^2 = 0.17$, $p = 0.35$) since both Buffalo and New York City have major sports teams but are on opposite ends of the map (the vector actually points at the sparsely populated Allegheny mountains of central Pennsylvania).

longitude and latitude of each city that are analogous to trained panel ratings of a more objective nature. Regression weights were also calculated from consumer ratings for which cities have the best chicken wings, best pizza, and the most important major sports teams. These are analogous to consumer ratings of attributes or hedonics.

14.2.4 Generalized Procrustes Analysis

Workers in apple and cider research in the UK noticed that some of their panelists confused astringency and bitterness. This problem could be solved by panel training with reference standards, of course, but they wondered if there was a way to massage the data after the fact to translate different people's perceptions into a common pattern (Langron, 1983). Earlier work by Gower (1975) had developed a technique for taking any pair of multidimensional configurations and lining them up as well as possible by various mathematical transformations. This was the birth of Procrustes analysis, named after the Greek mythological innkeeper, who stretched his guests who were too short for his bed and amputated those who were too tall. Given two perceptual maps from two different individuals, the mathematical steps were simple. First, the configurations should be translated or moved so they have a common center. Double-centering a two-dimensional matrix is now common practice in many multivariate analyses. Second, the data could be scaled or adjusted to bring everyone's range into the same overall size. Third, the data could be rotated and/or reflected if necessary, until the maximal possible match was found. These three steps are shown in Figure 14.5. The technique became known as GPA. An important feature of this method is that the data

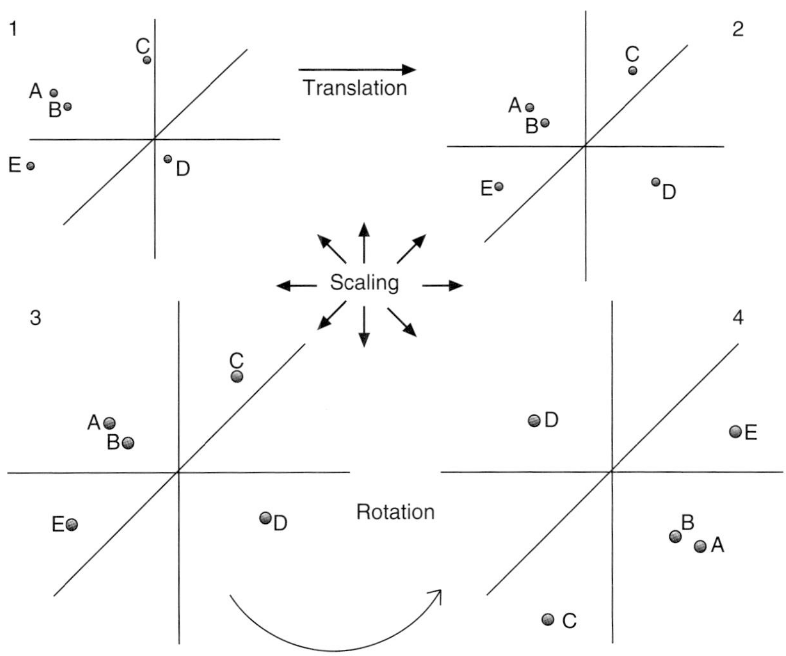

Figure 14.5 The transformation steps for a Procrustes analysis. Translation means moving the position of the origin, usually to provide a common center of gravity for each. Scaling means expanding or contracting the configuration so that the maximal distances are all in the same range. Rotation is performed to provide the maximum alignment (i.e., minimal distances between corresponding items), usually after the first two steps.

were reduced to a smaller dimensionality by techniques such as PCA. The PCA could be performed before the transformational steps, or afterwards, depending upon the software. If one person uses more attributes than another, the smaller data set would usually have to be filled in with columns of zeros or dummy variables to bring them both to the same size. The programs usually work using a least-squares criterion, which will attempt to minimize the differences between the consensus space and all the individual spaces or configurations.

This method went hand in hand with interest in a new type of descriptive analysis that could be performed by untrained consumers, namely free-choice profiling (Williams & Langron, 1984; Williams & Arnold, 1985). In this method, each consumer could choose attributes in their own words to describe the products, and use those words as attribute-intensity scales. So if one participant mistakes bitterness for astringency and another has them more or less correctly defined, it does not matter in the end because the two configurations could be rotated or reflected to match up, as long as they viewed similar apples in the same way. Some workers hailed this as a great advance, because the need for extensive (and expensive) panel training could be avoided and, furthermore, the data could be collected from consumers. Various modifications of this kind of individual language profiling continue to be developed, such as "flash profiling," a ranking method where panelists use their own terms to rank members of product set on their selected attributes (Delarue & Sieffermann, 2004).

As a view of the distilled, agreed-upon patterns among the products from a group of potentially divergent panelists, the method provides valuable information. It not only provides the view of the products in a perceptual map, but also additional information on the amount of variance extracted from the original data set, the variance that is unique to panelists, and the variance that is shared between a given product and the consensus configuration. Think of the total variance in the data set being reduced by the process of projection. This is similar to the variance accounted for in a PCA of given dimensionality. Some information is lost. What is retained can then be further partitioned in the variance present in the consensus configuration, and what is left over that is unique to the panelists or consumers. Obviously, not every consumer fits the consensus configuration well, and not every product has enough inter-judge agreement to have a stable and well-defined position in the results.

This information is present in variance tables provided in the GPA output, and clear examples are given in the classic paper by Dijksterhuis and Punter (1990). They first partition the total variance into the projection variance and the variance that is lost in this step. So the reduction in dimensionality costs you some information. The projection variance then shows up in a product table, showing how much of the variance in the consensus is contributed by that product and how much is left over as nonconsensus variance attributable to individual differences among the panelists (termed "within" in their paper). So, at a glance, you can see if a product is well-fit or not, and whether the position in the spatial model is certain or uncertain. They give an example where one cheese product had high disagreement among judges and thus was poorly fit. This cheese plotted near the center of the configuration, a case of where an outlier becomes an "in-lier" in a multidimensional configuration. This kind of result was also seen in a cheese study by Lawless et al. (1995) where one goat's milk cheese was included in a set of cow's milk cheeses, and subjects did not agree on what to do with it. Two other variance tables are informative. Dijksterhuis and Punter show a judge table, which shows the percentage of the "within" variance attributable to each judge. This is an indicator, or at least a hint, about whether a judge is in line with the rest of the panel across the product set as a whole, which could be important information to someone monitoring the agreement among a descriptive panel, or someone looking for evidence of consumer segmentation in a consumer study. Finally, there is a dimensions table that shows the consensus variance per extracted dimension. If one considers the variance explained to be an index of "importance" of that dimension, this is useful information.

GPA provided a very flexible, useful tool for looking at individual differences and the degree of group consensus. Other multivariate techniques could accommodate individual data and give some idea of the fit as well (Popper & Heymann, 1996). One such method was individual differences MDS, known for the name of its most popular program, INDSCAL. Input consisted of all the individual triangular half matrices for similarity or dissimilarity, so the data set had an added dimension. The output included a plot of the subjects or panelists on dimensions corresponding to the product plot. This showed how different subjects emphasized or de-emphasized those dimensions. Note that this approach is based on a fit of everyone to the same dimensions, and lacks the flexibility of rotation and reflection that is possible with GPA.

One potential shortcoming of GPA was the opinion that it seemed capable of making patterns out of nonsense or random data (King & Arents, 1991). In order to show the structure and trustworthy nature of a GPA output, a method needed to be developed to show what would happen with a randomized data set with similar structure. Thus, the permutation test was born. Wakeling et al. (1992) proposed a procedure that had been used previously for data sets that violated statistical assumptions like normality and used the actual data set itself to test statistical hypotheses. The method consisted of random permutations of the product rows for each assessor. Over many such randomizations, the variance accounted for by the consensus configuration can be sampled from what is apparent nonsense (a kind of null situation). The sampling distribution of the null or random data set will have some 95% confidence intervals over many resamplings, and if the consensus variance from the actual data set falls outside that interval, you can reject a null hypothesis of sorts, that the data show no consensus pattern. Recently, this approach was modified and sharpened by using specific permutations to allow more specific hypothesis tests about products, assessors, and repeatability (Xiong et al., 2008).

14.2.5 Other Methods

Another popular procedure for perceptual mapping is **partial least squares** (Martens & Martens, 2001). This technique is widely used in chemometrics. It attempts to fit the patterns of interrelationships by a set of latent variables, similar to factors. It is also possible to build models between sets of predictor variables and sets of outcome (predicted) variables or, for that matter, any two sets of variables one thinks to be related. That is, there is more than one variable at a time on the output side, in contrast to multiple regression. The technique has an advantage over PCA in that it can accommodate data sets with more observations than products. PCA generally demands a ratio of three or more products for every attribute. However, in this day and age, we can make many, many observations on every product. So data are cheap and products are expensive to make. If our data set is products in rows and attributes in columns, the partial least squares can work on short, fat data sets, where PCA needs a long and thin one.

The equivalent for PCA but using frequency count data is called correspondence analysis, as discussed in Chapter 10 (Hoffman & Franke, 1986). Another alternative to PCA is canonical variate analysis (see Heymann and Noble (1989) for a comparison). This is a robust technique that can also be used to model group differences, such as wines from different regions. It is related to discriminant analysis, a technique commonly used for authenticity testing with foods or food components (e.g., spices) that might have fraudulent claims for their origins. All of these techniques can produce perceptual maps, usually with products as points and attributes as vectors. A full discussion of multivariate statistics is beyond the scope of this book, but the student of sensory science should carefully monitor the literature on sensometrics for the newest techniques.

14.3 Shortcuts for Data Collection: Sorting and Projective Mapping

MDS never enjoyed much popularity with workers in sensory evaluation or food research. One of the problems was the intensive data collection necessary for product sets of any interesting size. The number of pairwise similarity judgments required for N products consists of $N(N-1)/2$ pairs or $N(N-1)$ total items. Thus, the number of products to be tasted increases by nearly the square of N as the set gets larger. Given that a product set of 10 or more products was in some way "interesting," the amount of tasting required for a complete block design was daunting (90 items to be tasted for a set of 10 products, and 210 items for a set of 15). However, if other methods could be used for data collection, perhaps the tasting burden could be relaxed. Of course, one approach would be to use an incomplete block design (Malhotra et al., 1988), but that allowed the possibility that different subjects with different criteria could be contributing to different product pairings, which is a "messy" situation for both the computation and later interpretation of the results.

A more rapid, cost-efficient way of data collection was **sorting** (Rosenberg et al., 1968; Rosenberg & Kim, 1975). In this method, subjects would inspect the product set (i.e., taste each item), perhaps making notes as an aid to memory and then group them based on similarities. The index of similarity then became the total number of times any two products were sorted into the same group, as summed across the subject pool. Note that this is both an index of similarity and one of inter-subject agreement. I brought the method to the chemical senses community (Lawless, 1989), and later tried it with foods with a reasonable degree of success (Lawless et al., 1995), as did other workers (Tang & Heymann, 2002; Faye et al., 2004, 2006; Cartier et al., 2006). The similarity data (usually a lower half matrix of product pairs) was simply submitted to any MDS program. Subsequent analyses could project vectors into the space, and this was a common practice once panelists' or consumers' attribute ratings were obtained in a second experimental session. Recent statistical methods have looked at both products and assessors (Abdi et al., 2007).

The technique worked well for products sets of from 10 to 20 products, of a moderate degree of dissimilarity. That is, it did not make sense to use this method with things that could not form distinct groups (i.e., had zero similarity), nor with a set that was so similar that subjects would not differentiate them. The method was user friendly and easy for most subjects to understand, since grouping or categorization is a common process in everyday human cognition. The sessions could usually be completed in about 30 minutes or less. The method was a quick way to get consumer input on the important attributes that would differentiate products. The consumer MDS space then could be compared with the PCA space of a descriptive panel, for example. Remember that there is no guarantee that the PCA space would necessarily reflect the "important" factors to consumers. PCA merely feeds off patterns of covariance and produces factors that sweep up the largest amounts of common variance. There is no insurance that the percentage of variance accounted for in a factor bears any relationship at all to what consumers might value or consider worth noting.

A second cost-efficient method appeared in the literature in the mid 1990s under the name **projective mapping** by Risvik and colleagues (Risvik et al., 1994, 1997; King et al., 1998). This idea was simple and elegant. Why not have subjects place the products on a surface and make the map themselves? Products that were similar could be positioned close together and different ones farther apart. Thus, there are *x* and *y* coordinates for each product in the individual data sets, as well as distance measures between pairs that could be submitted to MDS. This kind of data set was considerably richer in information than data provided by the sorting exercise. For sorting, the data of each individual is merely binary, consisting of zeros and ones. For any given pair, it was either placed in the same group or not.

The technique was used sporadically until interest was renewed in the past decade due to the advent of a useful statistical technique for dealing with the data, called multifactor analysis (MFA) through the work of Pagès, Lê and colleagues (Pagès, 2005; Lê et al., 2008; Perrin et al., 2008). MFA could handle the individual matrices and provided a group space. The data collection method was renamed "napping," based on the French word for tablecloth (nappe). Like sorting, the method was quite user friendly, and product sets of 10 to 20 items could easily be mapped by a consumer in an hour or less (Nestrud & Lawless, 2008). Various applications were published (Barcenas et al., 2004; Kennedy & Heymann, 2009).

Importantly, the variance accounted for by the resulting factors in MFA was related to the proportion of subjects choosing to pay attention to that stimulus dimension (Nestrud & Lawless, 2011). Thus, the resulting output was not necessarily constrained to two dimensions, even though the structure of the input data was two-dimensional. In other words, if some subjects paid attention to sweetness and mouthfeel, and others to mouthfeel and color, and still others to sweetness and color, and these three attributes were uncorrelated, the MFA could extract a three-dimensional solution. In a direct comparison of napping and sorting, the napping method was found to have more attribute vectors fit to the two-dimensional solutions than the sorting configurations (Nestrud & Lawless, 2010). Hybrid techniques, such as sorted napping, have also been used (Pagès et al., 2010).

14.4 Preference Mapping Revisited

Preference mapping was discussed in Chapter 13 (on optimization) as a way to find likely candidates for successful products with high consumer appeal. However, the methods are also a technique for perceptual mapping of both products and consumers (Nute et al., 1998; Greenhof & MacFie, 1999). There are two main approaches. One is based on objective attributes, with hedonic vectors fit later, and is called external preference mapping (see Elmore et al. (1999) for an application). The other main approach, internal preference mapping, starts with hedonic ratings and bases the map on similarities among products in terms of consumers' likes and dislikes. Other vectors representing sensory dimensions or descriptive attributes can be projected into the hedonic-based map in subsequent steps. A useful comparison of internal and external preference mapping is found in van Kleef et al. (2006), who concluded that external mapping was more useful for product development and that internal mapping was more useful for marketing and consumer insights.

External preference mapping first obtains a configuration where distance represents sensory dissimilarity (proximity equals similarity). This is typically generated by trained panel descriptive data that has been submitted to PCA. However, the map can be created by any number of methods that work with perceptual attributes or similarities, such as MDS of similarity data, MFA of napping data, and so on. The connection to consumer preferences

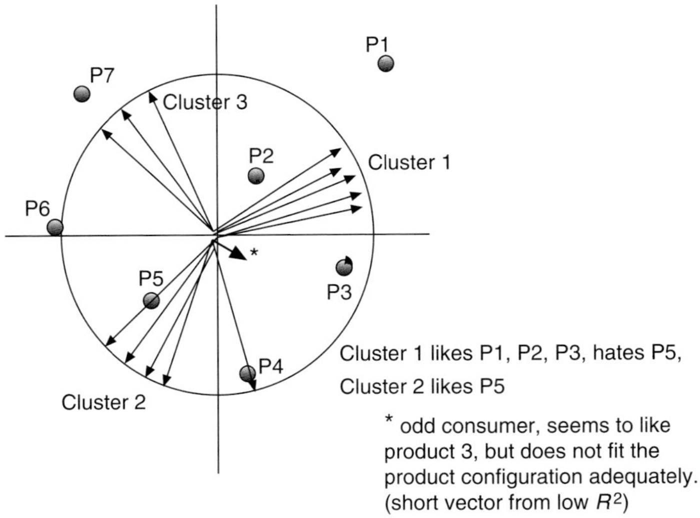

Figure 14.6 An example of preference mapping vectors showing groups of consumers with different patterns of product preference.

is achieved by simple vector projection, usually of individual hedonic ratings of the product set via the normal multiple regression approach (see Table 14.3) or any similar linear polynomial model. So each consumer is represented by a direction in the map, and clusters of consumers with similar like and dislikes will point in about the same direction. These angles, then, can provide a mechanism for clustering or grouping consumers into differing segments, as shown in Figure 14.6. Recall that adoption of vector model means that "more is better." The researcher should carefully consider whether an ideal point model may be a more accurate description of the consumer's preferences. In ideal point modeling, the distance from the ideal is proportional to decreasing liking (Meullenet et al., 2008). Also, if the major axes of the space do not correspond to the criteria used by consumers when deciding what they like, there may be little or no relationship between the map and that individual's product preferences. This would normally show up as a poor fit, or low R^2 value for that person.

Internal preference mapping starts with the hedonic data and constructs the map based on the patterns of liking/disliking. The simplest way to do this is to set up a matrix of products (rows) by consumers (columns) and perform a PCA. Since consumers now have what is analogous to factors loadings, they can be plotted in the map, as well as the products. In theory, they should lie close to the items they like and be distant from products they dislike. In the reverse order from the external map, now attribute information for each product can be regressed into the space. These can be perceptual attributes, or even physical or chemical measurements on the products. There is no limit on the amount of exploration one can do with these techniques, and the sensory specialist should not feel constrained or locked into any one type of data. On the other hand, the choice of which data to mine should be directed by sensible hypotheses about what is likely to be an underlying cause for the consumers' decisions, rather than a shotgun approach or mindless fishing expedition.

Other techniques can be used to build the internal preference maps. One option is a variation on multidimensional unfolding (Mackay, 2001), which takes attribute data in order to create a similarity map (in this case one based on hedonic similarity). If we consider each consumer–product pair as a distance indicator, it is easy to see that people should be placed in the map close to the things they like and distant from things that they dislike. Thus, the transformation is one of liking ratings in the person–product pair into distance in the model. One commercial version of this approach uses the following relationship (Nestrud et al., 2012b):

$$d_{ij} = a_i - \sum_{t=1}^{T} \left(x_{j,t} - y_{i,t} \right)^2 + \varepsilon_{i,j} \tag{14.2}$$

where d_{ij} are the hedonic ratings of respondent i for product j (the data), $x_{j,t}$ is the position of product j on dimension t, $y_{i,t}$ is the position of respondent i on dimension t, a_i is a scaling factor, ε is an error term, and where there are $1 \ldots T$ unknown (to be modeled) dimensions t.

In other words, the program is attempting to fit the person and products into a relationship where the Euclidean distance in the output map best corresponds to the original liking ratings. The fitting can be done by an iterative Bayesian process, with thousands of Monte Carlo draws on the data set and intermediate transformations, in order to converge to the optimal solution. Some of these techniques will use an ideal point model and place a predicted ideal product for that consumer into the space (Mackay, 2001).

14.5 Cautions and Concerns

14.5.1 Subjectivity and Analysts' Choice Points

Multivariate techniques entail a fair degree of subjectivity in the choices one makes and the interpretations of the results. Univariate results are generally clear with simple means, standard deviations, distributions, and tests of significant differences. The multivariate output, on the other hand, must be examined and interpreted for conclusions to be drawn. Remember, some information is always lost, so you are looking at a pattern made with reduced information. Would another scientist come to the same conclusions? At worst, it can be like looking at a Rorschach inkblot. Many possibilities come to mind, and the predisposition of the investigator may influence the interpretation. A reality check in the form of an independent interpretation from a colleague is never a bad idea.

Another problem is propagation of error. If, for example, the positions in a PCA plot are not accurate, any further analysis is going to be feeding off of poor information. For example, a subsequent cluster analysis on the product positions may give the impression of groupings that are not a good representation of consumer opinion.

There are several important choice points regarding data cleaning. Should one eliminate nondiscriminating consumers from the data set for an internal preference map? For any normal consumer test, throwing out data points would be unthinkable. After all, these nondiscriminating consumers are part of the story (yes, "no preference" is important!). But if one is building a model, it seems silly to include individuals who are not contributing any useful information about the product liking patterns, if they themselves show no pattern whatsoever. Note that this is not a question of poor fit to the model – a consumer who has a distinct but contrary pattern to the majority should never be eliminated. If you throw out such people you end up with a tautology: among people who think like this, this is how they think. Worthless. Another common situation is when consumers' liking data is fit to an external preference map. Some people's likes and dislikes may bear no relationship to the sensory map which is derived from some outside attribute information. So be it. But the lack of fit, and number of nonfitted consumers, should be noted and explained if possible. A related choice is whether to work with the covariance matrix in PCA (unscaled, raw data) or scaled data via rescaling of consumers to a common range, or use of the correlation matrix (see Meullenet et al. (2007) for examples). On the one hand, rescaling poor discriminators gives them equal weight to good discriminators, but on the other hand, all consumers now have an equal contribution.

Another important choice point is which attributes to include in the input to a PCA or similar analysis. One suggestion is to eliminate attributes that show no significant differences among products. This seems reasonable and is often done. However, there is always the chance that the significant main effect in the univariate analysis of variance is overwhelmed by a panelist by product interaction, which could be very interesting and meaningful in and of itself. It might be indicative of a consistent pattern among different persons or market segments. So the sensory scientist should be very cautious and look carefully at the results before throwing out attributes willy-nilly.

Another important choice is how many dimensions to retain in the output configuration. Of course, two or three are convenient for visualization, but convenience is a rather unscientific criterion for trying to find out the truth. As mentioned above, in PCA the "eigenvalue greater than one" is a common rule, but it tends to retain more PCs than you probably need. A second approach is the scree plot (see Lawless and Heymann (2010: chapter 17)). The scree is named for the talus or rock pile at the bottom of a cliff. The number of dimensions is plotted on the x-axis and some measure of variance accounted for (such as eigenvalue or R^2) on the y-axis. This produces a descending curve or group of line segments that shows a law of diminishing returns. If the curve has a sudden break or "elbow," that can be an indication of where the subsequent PCs are starting to give little improvement in the total information gained. The dimensionality is usually set by the last point before the break, or the point at which the break occurs. The scree plot can also be used for MDS studies with stress plotted against MDS solutions of different dimensionality (1, 2, 3, etc.). This may show when the stress fails to improve by a noteworthy amount; that is, when the curve flattens out you are not gaining much in terms of fit.

14.5.2 Sources of Covariance and Correlation

The PCA, as well as many other multivariate analyses, feeds off patterns of covariance. By "feeds off" I mean both detects and develops a model based on the strength of relationship between two or more variables. Usually, the variables in the model are attributes or scales, and the pattern is across products. However, we also have assessors in the situation, so sometimes the correlation is driven by the different likes and dislikes of different consumers, as in internal preference mapping, for example.

Perhaps the most common or traditional application is the PCA performed on product means from a descriptive analysis. In that case, we have a matrix of attributes (usually columns) by products (usually rows) as discussed above. The correlations we are looking for are correlations between pairs of attributes. So, for example, if one product is both very hard and dense and a second product is very low in hardness and denseness, there is a trend for these attributes to co-vary, as shown in Figure 14.7. If we populate the scatter diagram with even more products, the pattern of correlation emerges (Figure 14.7b). Of course, the more the envelope ellipse collapses around a central line, the higher the pattern of linear correlation and the "stronger" the relationship. The more the ellipse bows outward, tending in the extreme case toward looking like a circle, the lower the correlation.

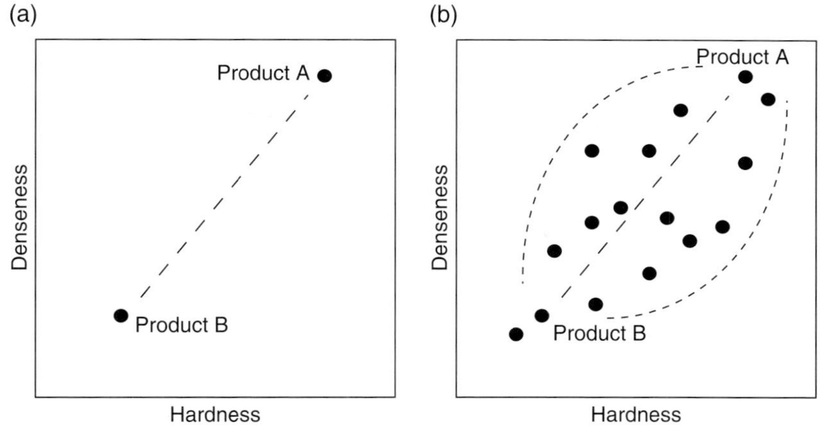

Figure 14.7 A correlation based on differences among products for two attributes. (a) Products are either high or low, but on both attributes. (b) Across many products, the two attributes in question are correlated, in this case positively, and would load on the same principal component.

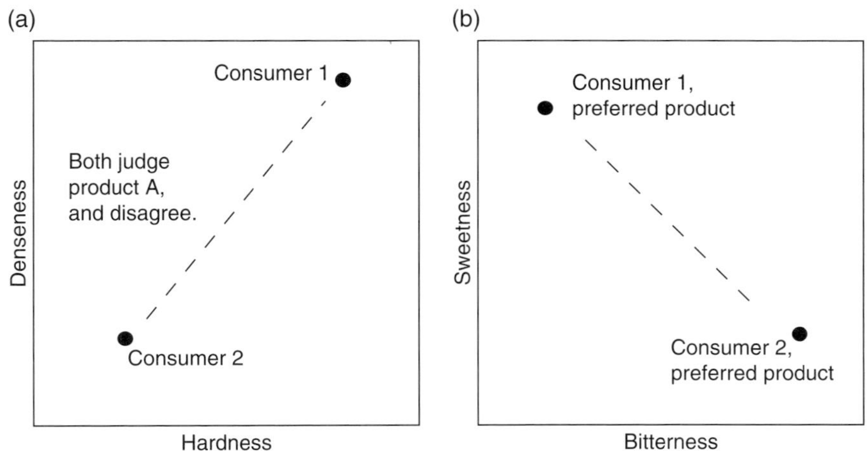

Figure 14.8 A correlation based on consumer data, where the individuals disagree (a). A perceptual map built on such correlations is not necessarily useful and must be interpreted with caution. A disagreement on hedonics (b) is useful information and can be used to build an internal preference map.

Note that we have been speaking of a single value per product on any given attribute. This is the case when we have descriptive panel data and are using the product means as input to the PCA. There is no need to consider the pattern using individual panelist scores, as any variance among panelists is simply noise or error variance. This is true if the panel is well trained and calibrated. Consumer data are another matter. Consider one single product rated on a couple of attributes by two consumers, as shown in Figure 14.8. One consumer thinks the product is high on both attributes, and the second consumer thinks they are both low. If we populate this graph with additional consumers, we might see a pattern of correlation similar to that in Figure 14.7. But now the correlation is actually arising from differences among people, or inter-person *disagreement*. In a PCA on descriptive data, this would be considered a pattern based on error! But with consumers it might reflect true individual differences. However, if your data have both different products and different consumers in the rows, you cannot tell from a simple PCA whether the correlation is coming from product correlations or consumer disagreement. That kind of analysis must be done with extreme care, and a good close look at both the product relationships and consumer patterns is warranted.

In the case of consumer likes/dislikes, consumer disagreement is now useful information. Patterns of different consumers' hedonic judgments are a completely legitimate area of investigation. So a PCA on individual consumer liking/disliking scores across a set of products makes sense. This is the basis for internal preference mapping. For example, if consumers who like milk chocolate (sweet, not bitter) are different people from consumers who prefer dark chocolate (bitter and not very sweet), then there is a pattern of negative correlation between the preferred sweetness and bitterness levels for any individual and segmentation regarding the two classes of products, as shown in Figure 14.8b. The pattern could also be seen if the liking scores were viewed as a function of bitterness and sweetness. Looking at product means, of course, would not capture this picture. So it is logical in this case to look at raw consumer data as the input, across products, with the single variable of liking as the critical attribute.

The important lesson is to know where your patterns of correlation are coming from. Do not blindly throw any data set that contains both product and consumer variance into a PCA or other multivariate program and treat it as a magical black box. You could be getting results based on product relationships, consumer disagreement, or consumer segmentation patterns. Interpretation becomes tricky and can be ambiguous.

Validation of any model is useful. One approach is called "leave one out" validation, in which the model is built, minus one product, and then that product's values (say on overall liking in a regression model) is predicted (Meullenet et al., 2007). Correspondence of the predicted value and actual value then is a form of model validation.

14.6 Introduction to Graph Theory

14.6.1 What Can I Use This For?

Not all perceptual maps are fit to Euclidean spaces with a Cartesian coordinate model. Another example of a pictorial model is found in graph theory. Graph theory is an area in which food combinations can be investigated for their appropriateness, compatibility, and/or consumer appeal. One example would be the components of a tossed salad. The methods can also be extended to ingredient combinations, especially when the items are distinguishable and remain so in the completed food matrix. Examples are pizza toppings, inclusions in ice cream (nuts, fruits, syrups), and components of a one-pot meal such as stew or stir fry. When one is developing a menu or set of meal components to be sold and/or consumed together, the prediction of the appeal or value of the combinations is obviously important.

Different approaches have been taken to look at item compatibilities,such as multiple regression and other linear model equations (reviewed in Nestrud et al. (2011)). The basic idea was to predict the overall appeal of the combination from some weighted linear combination of the values of the individual items (as evaluated separately) (Moskowitz & Krieger, 1995). Others have taken a pairwise approach and tried to look at compatibility in a multidimensional model (Klarman & Moskowitz, 1977). None of these approaches have met with exceptional success, due to factors such as the dominance of an entrée, interaction effects, contrast, and a curious à la carte effect (see Nestrud (2011) and Nestrud et al. (2011) for a discussion).

The sensory science community has recently gained interest in graph theory because of the potential savings in screening out large numbers of potential item combinations. For example, the total number of combinations of 25 candidate pizza toppings taken five at a time is 53,130. However, if the compatibility could be predicted by pairwise evaluations, the number of pairs shrinks to 300, a manageable number for paper-and-pencil evaluation by consumers. So, one of the key potential benefits of graph theory in food product development is as a screening tool to help eliminate vast numbers of unlikely combinations (Ennis et al., 2011a). This does not produce a final product for market launch, but reduces the candidate pool. The smaller number can then be taken forward, or some subset selected for further consideration.

14.6.2 Basics of Graph Theory

In graph theory, items are represented as points or nodes, and nodes that are related (or compatible in the case of food items) are connected by vertices or lines. Applications abound in many fields, such as intelligence modeling of terrorist sleeper cells and the connectivity of secret operatives. Other uses include electronic circuit-board design and models for social network connections. An interesting problem is how to represent the significant differences among means in a multiproduct study. The usual approach is to use letters to signify differences following the analysis of variance and planned comparisons via Duncan or Tukey tests or some other method. Items sharing a common letter are not significantly different. For large numbers of products, this becomes an interesting problem if one wishes to minimize the numbers of required letters. Ennis and Ennis (2012) show how graph theory can be used to address this particular application.

The historical origin of graph theory is often traced back to Euler in the 1700s. Euler contributed to many fields, but one math problem attracted the attention of numerous other theoreticians as well as amateur puzzle-solvers. Euler's problem concerned the town of Königsberg, which was traversed by branches of the Pregel River. The town was the residence of the dukes of Prussia in the 16th century and is now known as Kaliningrad in Russia. Königsberg had four distinct land masses connected by seven famous bridges. The puzzle was whether a townsperson or visitor could set foot on all four sections by crossing all of the bridges, but cross each bridge only once. Such a trail is also called an Euler trail, and Euler later provided a proof that, in the case of the Königsberg connections, it was not possible. The connections of the land masses and the bridges' representations as graph theory connections are shown in Figure 14.9. Graph theory is considered by some to be one of the founding areas or historical precedent of the mathematical field of topology and is closely related to network analysis (Kular et al., 2011).

So the connection of vertices is an all-or-nothing representation. Items are either connected or they are not. Once they pass a certain threshold or criterion, they are connected. Obviously, this tells us nothing about the strength of a connection, or its probability. This is a potential shortcoming for food research. If one was looking at the compatibility of items in a salad for instance, to any given consumer there are good matches and poor ones, and some probably in-between. Looking at group data, some items will be compatible for some individuals but not others. Once again, the degree of compatibility is graded, but connectivity itself in the simplest form of graph theory does not address this. There is an application of graph theory

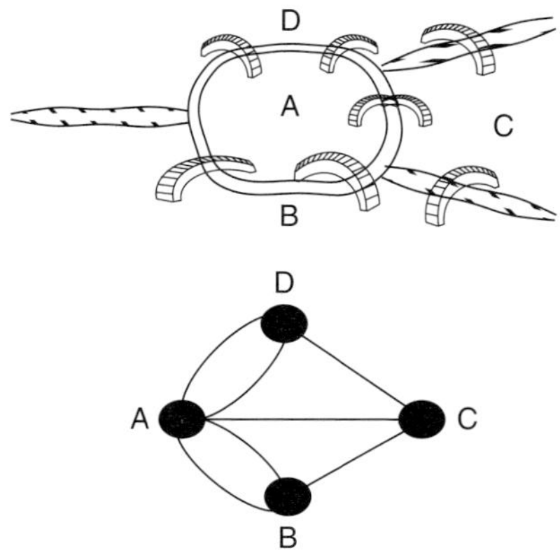

Figure 14.9 The seven bridges and four land masses of Königsberg (upper panel) and their representation as a graph (lower panel).

that adds weights to the connections, generally thought of as a kind of penalty. For example, a cable company might want to connect all of the houses in a neighborhood with no duplication or connected cycles. They might also want to do so with the connections that are least expensive in terms of distance or the need to bury cable. Such a graph is called a minimal spanning tree, connecting all vertices with the least total weight.

The method becomes useful for food combinations when one considers the collections of connected items. A group where all items are connected to one another, with no exceptions or gaps, is called a clique. A clique that is not a subset of any other clique is called a maximal clique. One might hypothesize, for example, that a collection of pizza toppings that forms a maximal clique is sufficient to warrant consideration for a well-decorated pizza, and that inclusion of any other items beyond the maximal clique would risk having some item incompatibilities. Suppose you were trying to design a salad for an airline meal and had eight components to work with. The data from a given consumer might look like the graph in Figure 14.10. The lower right-hand side shows the maximal cliques. This suggests three options: a simple apple–radish combination, a Waldorf-like salad with apple, raisin, and walnuts, or a more traditional tossed salad with tomato, pepper, cucumber, carrot, and radish.

Another useful application is in prediction of group compatibilities from the perceived harmony of smaller combinations. Again, consider salad components. If apple and carrot are compatible, and carrot and raisins, and apple and raisins, one might predict that the trio, apple–carrot–raisin would be compatible and form a clique. On the other hand, in spite of their pairwise compatibility, the trio might be offensive to some, and thus would not form a clique. The ability to predict larger combinations from pairwise information would be a great time saver in terms of consumer data collection.

14.6.3 Recent Applications and Research

Nestrud et al. (2011) looked at consumers' opinions of the pairwise compatibilities of 25 salad items and then tried to predict the compatibilities of three to eight item combinations. Consumers were asked to make yes/no decisions on item compatibilities, after elimination of items they said they would never consume on a salad. Note that this kind of item compatibility is a form of appropriateness rating (Schutz, 1988). After an individual's data were collected, a specialized clique-finding algorithm generated cliques for that person. One

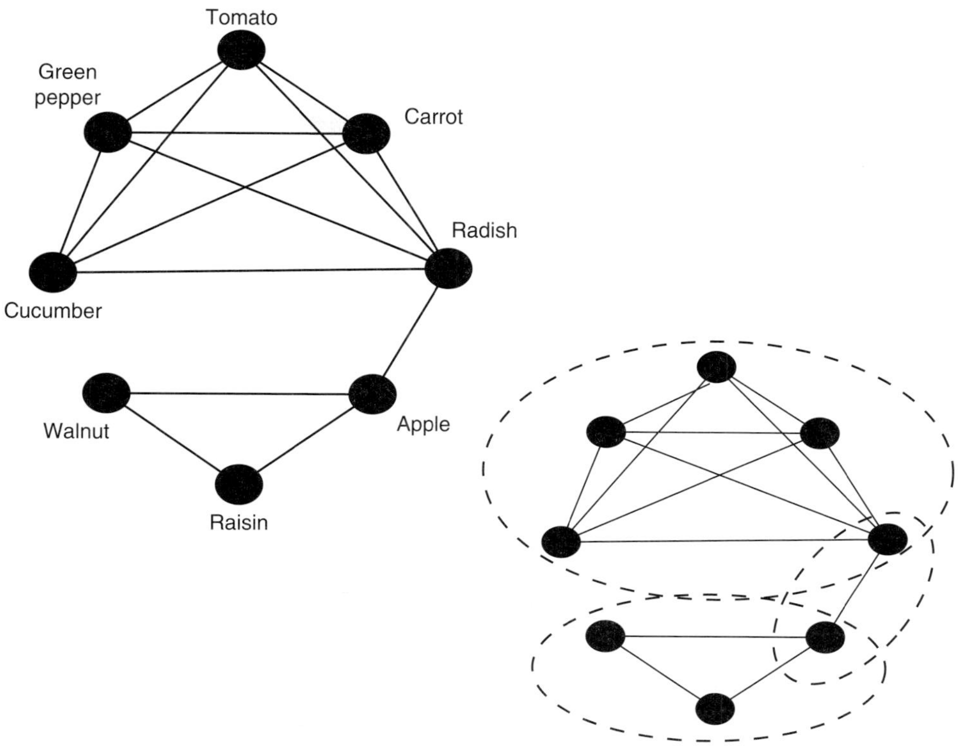

Figure 14.10 Graph representation of the salad component data of one hypothetical individual (upper left) and the same graph showing maximal cliques surrounded by the dashed ellipses (lower right).

primary test was the compatibility of cliques versus noncliques. Although some consumers deemed some noncliques to be compatible, in all cases from three to eight item combinations, the frequency of rating items as compatible was higher for cliques than for noncliques. Only 3 of 64 consumers rated more noncliques as compatible than cliques. One surprising result was that as the size of the combinations increased the probability of overall compatibility also increased. Perhaps consumers are more forgiving of an occasional odd item in a larger group, or you could just decide not to eat that item if it was present in your salad.

In a second study, Nestrud and colleagues examined the issue of whether the compatibility of larger groups of items could be predicted from their pairwise information. They dubbed this the principle of supercombinatorality (Nestrud et al., 2012a). Formally, it states that combinations that are fully pairwise compatible will be considered more compatible than combinations that are not fully pairwise compatible. In the language of graph theory, cliques are more likely to be compatible than noncliques are. They also chose to examine the value of maximal versus nonmaximal cliques (and of course noncliques) using a battery of 25 pizza toppings as the stimulus set. Subjects were asked to indicate, yes or no, whether the pairs of toppings were compatible on a tomato-sauce-based pizza with mozzarella cheese. The actual question was one of purchase intent (would or would not buy such a pizza).

The second part of the survey included maximal cliques, nonmaximal cliques, and noncliques of one to six items, except that a nonclique cannot be formed from a single item (a unique ingredient is a clique of size one, a somewhat odd idea for a clique). The overall proportions of compatible noncliques, nonmaximal cliques, and maximal cliques were 0.37,

0.65, and 0.55, respectively. This supported the principle of supercombinatorality, but with some complications. There was a slight trend across all three groups showing lower compatibility as the number of toppings increased, opposite to the effect seen with salads. Maximal cliques of size one were items deemed appropriate to consume alone, without other partners, and were polarizing. Participants either liked them (anchovy, artichoke, eggplant) or did not.

In a third research application, Nestrud (2011) examined food item combinations in a military field ration known as the MRE, for meal-ready-to-eat. This ration contains an entrée (main dish) and usually four other items, such as a side dish, snack, and beverage; however, there are 11 total categories for the other items. This leads to a situation with a menu of 12 main dishes with millions of potential combinations. Nestrud limited the choices by using only the most important pairs of categories (e.g., entrée–bread, fruit–dessert, entrée–starch, or bread–spread) in the sample set. Soldiers familiar with the items were given paper surveys to rate item pairs as compatible or not. The sample set consisted of 1370 pairs, evaluated in an incomplete block design. The data were divided into a lower half matrix for each of the main dishes (as no two main dishes would be contained in a real MRE). The cutoff or threshold was found by setting proportions smaller than a given amount to zero (thus being unconnected) when that critical amount produced cliques of size eight, the maximal number of components of the MRE. Menus were developed based upon the cliques that were formed, and the original compatibility proportions were averaged to get a measure of the overall compatibility of the predicted best combinations. This was seen as a tool for finding good combinations, without exhaustive search of millions of possibilities and also for improving some combinations that might have less than optimal scores.

14.6.4 Input, Mechanics and Software

The input format for constructing a graph is similar to that used for MDS. That is, it is usually a lower half (or upper half) matrix that specifies the pairwise connectivity of the items. For an individual, this matrix would consist of zeros and ones. For group data, the matrix consists of frequency counts of the numbers of persons who found the items compatible or an appropriate combination. Obviously, it is possible to use scaled data as well, for something like appropriateness ratings. With frequency counts or more continuous data, it is often the case that a threshold level is set, thus converting the information to a binary form. Note that the method is a lot like MDS. However, instead of a map with graded distances indicating degrees of dissimilarity, it has only binary information. Items are either connected or they are not in the examples discussed here.

A technical choice must be made regarding the threshold value for considering items compatible or not. In the case of the pizza toppings study, additional information was brought to bear. In that case, previous research indicated that most people preferred at most three items on their pizzas, and rarely more than five or six. Using this information, a threshold can be set to exclude any cliques larger than a given size of, say, six items. In the case of the MRE study, a maximal clique size of eight provided a practical benchmark. Thus, a practical criterion can be brought to bear on the threshold question, if additional information is available. Alternatively, different threshold values can be tested to see how the results and implications differ.

Various programs are available to assist in transforming the lower half matrix information into a usable graph. For example, the IGRAPH package is a free program that is implemented in various platforms, including the C language and R statistical platform. Commercially vended programs are also available, and recently the Institute for Perception group has programmed applications for use in food and consumer product research (Ennis et al., 2011a).

14.6.5 Future Promise

Graph theory provides a unique tool for investigating the potential compatibility of food items or food ingredients in more complex combinations than those that are usually considered by product developers. Its true value and additional applications may yet be undiscovered. For example, it should be possible to examine consumers with different sets of nonoverlapping maximal cliques in order to engineer a menu with wide appeal. For example, an airline might want to offer two main dishes with sides, such that the majority of consumers will either prefer one or the other, and perhaps in an approximate 50/50 ratio. This approach would be similar to TURF analysis, for "total unduplicated reach and frequency" (Miaoulis et al., 1990; Cohen, 1993). The main idea is to reach as many consumers as possible, while avoiding duplication. For example, an advertiser might want to choose TV stations or programs that appeal to different audiences in order to reach as many people as possible, but avoid showing the same advertisement over and over to the same individuals. There are obvious applications for this in market segmentation and the construction of product lines with different nonoverlapping line extensions and flanking products that appeal to different consumer groups.

Information scientists have also been applying network analysis to recipe databases. Kular et al. (2011) represented recipes as nodes and common ingredients between them as edges. Using a large-scale database the authors were able to extract features of the culinary cultures, uninhibited by more artificial (e.g., country) boundaries. Ahn et al. (2011) explored the food pairing hypothesis (Foodpairing web site: www.foodpairing.be), which states that the reason that specific culinary combinations work well is because they share common underlying flavor chemicals. They explored this relationship using known information about flavor chemicals and an online analysis of 56,498 recipes. They concluded that the foodpairing hypothesis holds true for Western cuisines, but not for Eastern cuisines, which puts the foodpairing paradigm in jeopardy. Teng et al. (2012) built a recipe recommendation engine that also predicts overall preference for recipes (e.g., "star ratings") from the ingredient relationships captured by a network structure. Other problem-solving applications are likely to be discovered.

References

Abdi, H., Valentin, D., Chollet, S., and Chrea, C. 2007. Analyzing assessors and products in sorting tasks: DISTATIS, theory and applications. Food Quality and Preference, 18, 627–40.

Ahn, Y.-Y., Ahnert, S.E., Bagrow, J.P., and Barabási, A.-L. 2011. Flavor network and the principles of food pairing. Scientific Reports, 1, article 196, doi:10.1038/srep00196.

Barcenas, P., Elortondo, F.J.P., and Albisu, M. 2004. Projective mapping in sensory analysis of ewes milk cheeses: a study on consumers and trained panel performance. Food Research International 37, 723–9.

Cartier, R., Rytz, A., Lecomte, A., Poblete, F., Krystlik, J., Belin, E., and Martin, N. 2006. Sorting procedure as an alternative to a product sensory map. Food Quality and Preference, 17, 562–71.

Cohen, E. 1993. TURF analysis. Quirk's Marketing Research. Article number 19930604, June 1993. www.quirks.com/articles/a1993/19930604.aspx?searchID=500211653 (accessed 18 September 2012).

Delarue, J. and Sieffermann, J.-M. 2004. Sensory mapping using flash profile. Comparison with a conventional descriptive method for the evaluation of the flavor of fruit dairy products. Food Quality and Preference, 15, 383–92.

Dijksterhuis, G.B. 1997. Multivariate Data Analysis in Sensory and Consumer Sciences. Wiley–Blackwell.

Dijksterhuis, G. and Punter, P. 1990. Interpreting generalized Procrustes analysis 'analysis of variance' tables. Food Quality and Preference, 2, 255–65.

Elmore, J.R., Heymann, H., Johnson, J., and Hewett, J.E. 1999. Preference mapping: relating acceptance of 'creaminess' to a descriptive sensory map of a semi-solid. Food Quality and Preference, 10, 465–75.

Ennis, J.M. and Ennis, D.M. 2012. Efficient representation of pairwise sensory information. IFPress, 15(3), 3–4.

Ennis, J.M., Fayle, C.M., and Ennis, D.M. 2011a. From many to few: a graph theoretic screening tool for product developers. IFPress, 14(2), 3–4.

Ennis, D.M., Rousseau, B., and Ennis, J.M. 2011b. Short Stories in Sensory and Consumer Science. Institute for Perception, Richmond, VA.

Faye, P., Bremaud, D., Daubin, M.D., Courcoux, P., Giboreau, A., and Nicod, H. 2004. Perceptive free sorting and verbalization tasks with naïve subjects: an alternative to descriptive mappings. Food Quality and Preference, 15, 781–91.

Faye, P., Bremaud, D., Teillet, E., Courcoux, P., Giboreau, A., and Nicod, H. 2006. An alternative to external preference mapping based on consumer perceptive mapping. Food Quality and Preference, 17, 604–14.

Gower, J.C. 1975. Generalized Procrustes analysis. Psychometrika, 40, 33–51.

Gower, J.C. and Dijksterhuis, G.B. 2004. Procrustes Problems. Oxford Statistical Science Series, vol. 30. Oxford University Press.

Greenhof, K. and Macfie, H.J.H. 1999. Preference mapping in practice. In: Measurement of Food Preferences, H.J.H. MacFie and D.M.H. Thomson (Eds). Aspen, Gaithersburg, MD, pp. 137–47.

Heymann, H. and Noble, A.C. 1989. Comparison of canonical variate and principal component analyses. Journal of Food Science, 54, 1355–8.

Hoffman, D. L. and Franke, G. R. 1986. Correspondence analysis: Graphical representation of categorical data in marketing research. Journal of Marketing Research, 23, 213–27.

Husson, F., Bocquet, V., and Pagès, J. 2004. Use of confidence ellipses in a PCA applied to sensory analysis application to the comparison of monovarietal ciders. Journal of Sensory Studies, 19, 510–18.

Husson, F., Lê, S., and Pagès, J. 2005. Confidence ellipse for the sensory profiles obtained by principal component analysis. Food Quality and Preference, 16, 245–50.

Husson, F., Lê, S., and Pagès, J. 2006. Variability of the representation of the variables resulting from PCA in the case of a conventional sensory profile. Food Quality and Preference, 18, 933–7.

Kennedy, J. and Heymann, H. 2009. Projective mapping and descriptive analysis of milk and dark chocolates. Journal of Sensory Studies, 24, 220–33.

King, B.M. and Arents, P.A. 1991. Statistical test of consensus obtained from generalized Procrustes analysis of sensory data. Journal of Sensory Studies, 6, 37–48.

King, M.J., Cliff, M.A., and Hall, J.W. 1998. Comparison of projective mapping and sorting data collection and multivariate methodologies for identification of similarity-of-use of snack bars. Journal of Sensory Studies, 13, 347–58.

Klarman, L. and Moskowitz, H. R. 1977. Food compatibilities and menu planning. Canadian Institute of Food Science and Technology, 10, 257–64.

Kruskal, J.B. and Wish, M. 1978. Multidimensional Scaling. Sage Publications, Beverly Hills, CA.

Kular, D.K., Menezes, R., and Ribeiro, E. 2011. Using network analysis to understand the relation between cuisine and culture. 2011 IEEE Network Science Workshop. IEEE Press, pp. 38–45, doi:10.1109/NSW.2011.6004656.

Langron, S.P. 1983. The application of Procrustes statistics to sensory profiling. In: Sensory Quality in Food and Beverages: Definition, Measurement and Control, A.A. Willliams and R.K. Atkin (Eds). Horwood, Chichester, UK, pp. 89–95.

Lawless, H.T. 1989. Exploration of fragrance categories and ambiguous odors using multidimensional scaling and cluster analysis. Chemical Senses, 14, 349–60.

Lawless, H.T. and Heymann, H. 2010. Sensory Evaluation of Foods, Principles and Practices. Second edition. Springer, New York, NY.

Lawless, H.T., Sheng, N., and Knoops, S.S.C.P. 1995. Multidimensional scaling of sorting data applied to specialty cheeses. Food Quality and Preference, 6, 91–8.

Lê, S. and Husson, F. 2008. SensomineR: a package for sensory data analysis. Journal of Sensory Studies, 23, 14–25.

Lê, S., Pagès, J., and Husson, F. 2008. Methodology for the comparison of sensory profiles provided by serval panels: application to a cross-cultural study. Food Quality and Preference 19, 179–84.

Mackay, D.B. 2001. Probablistic unfolding models of sensory data. Food Quality and Preference, 12, 427–36.

Malhotra, N., Jain, A., and Pinson, C. 1988. Robustness of multidimensional scaling in the case of incomplete data. Journal of Marketing Research, 24, 169–73.

Martens, H. and Martens, M. 2001. Multivariate Analysis of Quality: An Introduction. John Wiley & Sons, Ltd, Chichester, UK.

Meullenet, J.-F., Xiong, R., and Findlay, C.J. 2007. Multivariate and Probabilistic Analyses of Sensory Science Problems. IFT Press/Blackwell Publishing, Ames, IA.

Meullenet, J.-F., Lovely, C., Threlfall, R., Morris, J.R., and Streigler, R.K. 2008. An ideal point density plot method for determining an optimal sensory profile for Muscadine grape juice. Food Quality and Preference, 19, 210–19.

Miaoulis, G., Free, V., and Parsons, H. 1990. Turf: a new planning approach for product line extensions. Marketing Research, 2(1), 28–40.

Moskowitz, H.R. and Krieger, B. 1995. The contribution of sensory liking to overall liking: an analysis of six food categories. Food Quality and Preference, 6, 83–90.

Nestrud, M.A. 2011. A graph theoretic approach to food combination problems. Doctoral dissertation, Cornell University Department of Food Science. Retrieved from http://hdl.handle.net/1813/29153.

Nestrud, M.A. and Lawless, H.T. 2008. Perceptual mapping of citrus juices using projective mapping and profiling data from culinary professionals and consumers. Food Quality and Preference 19, 431–8.

Nestrud, M.A. and Lawless, H.T. 2010. Perceptual mapping of applies and cheeses using projective mapping and sorting. Journal of Sensory Studies, 25, 390–405.

Nestrud, M.A. and Lawless, H.T. 2011. Recovery of subsampled dimensions and configurations derived from napping data by MFA and MDS. Attention, Perception and Psychophysics, 73, 1266–78.

Nestrud, M. A., Ennis, J. M., Fayle, C. M., Ennis, D. M., and Lawless, H. T. 2011. Validating a graph theoretic screening approach to food item combinations. Journal of Sensory Studies, 26, 331–8.

Nestrud, M.A., Ennis, J.M., and Lawless, H.T. 2012a. A group level validation of the supercombinatorality property: finding high-quality ingredient combinations using pairwise information. Food Quality and Preference, 25, 23–8.

Nestrud, M.A., Wedel, M., Irwin, M., and Cohen, S.H. 2012b. Bayesian stochastic unfolding model for sensory dominance judgments. Oral presentation at Sensometrics, Rennes, France.

Nute, G.R., MacFie, H.J.H., and Greenhoff, K. 1988. Practical application of preference mapping. In: Food Acceptability, D.M.H. Thomson (Ed.). Elsevier Applied Science, London, pp. 377–86.

Pagès, J. 2005. Collection and analysis of perceived product interdistances using multiple factor analysis: application to the study of 10 white wines from the Loire Valley. Food Quality and Preference, 16(7), 642–9.

Pagès, J., Cadoret, M., and Lê, S. 2010. The sorted napping: a new holistic approach in sensory evaluation. Journal of Sensory Studies, 25, 637–58.

Perrin, L., Symoneaux, R., Maitre, I., Asselin, C., Jourjon, F., and Pagès, J. 2008. Comparison of three sensory methods for use with the napping procedure: case of ten wines from Loire valley. Food Quality and Preference, 19, 1–11.

Popper, R. and Heymann, H. 1996. Analyzing differences among panelists by multidimensional scaling. In: Multivariate Analysis of Data in Sensory Science. T. Naes and E. Risvik (Eds). Elsevier, Amsterdam, pp. 159–84.

Risvik, E., McEwan, J.A., Colwill, J.S., Rogers, R., and Lyon, D.H. 1994. Projective mapping: a tool for sensory analysis and consumer research. Food Quality and Preference, 5, 263–9.

Risvik, E., McEwan, J.A., and Rodbotten, M. 1997. Evaluation of sensory profiling and projective mapping data. Food Quality and Preference, 8, 63–71.

Rosenberg, S. and Kim, M.P. 1975. The method of sorting as a data-gathering procedure in multivariate research, Multivariate Behavioral Research, 10, 489–502.

Rosenberg, S., Nelson, C., and Vivekananthan, P.S. 1968. A multidimensional approach to the structure of personality impressions. Journal of Personality and Social Psychology, 9, 283–94.

Schiffman, S.S., Reynolds, L.M., and Young, F.W. 1981. Introduction to Multidimensional Scaling. Academic Press, New York, NY.

Schutz, H. 1988. Beyond preference: appropriateness as a measure of contextual acceptance of food. In: Food Acceptability, D.M.H. Thomson (Ed.). Elsevier Applied Science, London.

Stevens, D.A. and Lawless, H.T. 1981. Age-related changes in flavor perception. Appetite, 2, 127–36.

Tang, C. and Heymann, H. 2002. Multidimensional sorting, similarity scaling and free choice profiling of grape jellies. Journal of Sensory Studies, 17, 493–509.
Teng, C-Y.., Lin, Y.-R., and Adamic, L.A. 2012. Recipe recommendation using ingredient networks. arXiv. Physics and Society. Retrieved from http://arxiv.org/abs/1111.3919.
Van Kleef, E., van Trijp, H.C.M., and Luning, P. 2006. Internal versus external preference analyses: an exploratory study on end-user evaluation. Food Quality and Preference, 17, 387–99.
Wakeling, I.N., Raats, M.M., and MacFie, H.J.H. 1992. A new significance test for consensus in generalized procrustes analysis. Journal of Sensory Studies, 7, 91–6.
Williams, A.A. and Arnold, G.M. 1985. A comparison of the aromas of six coffees characterized by conventional profiing, free-choice profiling and similarity scaling methods. Journal of the Science of Food and Agriculture, 36, 201–14.
Williams, A.A. and Langron, S.P. 1984. The use of free-choice profiling for the evaluation of commercial ports. Journal of the Science of Food and Agriculture, 35, 558–68.
Worch, T., Lê, S., Punter, P., and Pagès, J., 2013. Ideal Profile Method (IPM): the ins and outs. Food Quality and Preference, 28(1), 45–59, doi: http://dx.doi.org/10.1016/j.foodqual.2012.08.001.
Xiong, R., Blot, K., Meullenet, J.-F., and Dessirier, J.M. 2008. Permutation tests for generalized Procrustes analysis. Food Quality and Preference, 19, 146–55.

15 Segmentation

Animals can be divided into: those that belong to the Emperor, embalmed ones, those that are trained, suckling pigs, mermaids, fabulous ones, stray dogs, those that are included in this classification, those that tremble as if they were mad, innumerable ones, those drawn with a very fine camel's hair brush, others, those that have just broken a flower vase and those that resemble flies from a distance.

From: *An Ancient Chinese Classification of Animals*,
cited in Aldenderfer and Blashfield (1984)

15.1 Introduction

People differ. Individual differences abound in people's sensory functions, due to age, genetic causes, life history, gender, culture, and a host of other variables. When it comes to products that people like or dislike, preference patterns are even more diverse. Human diversity can be considered a source of experimental error (i.e., uncontrolled variability) to be minimized, reduced, or eliminated, or it can be embraced and studied as interesting phenomenon in its own right (Stevens, 1991). In the case of new product development, the researchers and marketing strategists have an interesting choice: Should we construct a product that has maximum appeal to the entire pool of users of that product category? Or should we construct different products that can have even higher acceptability to groups of consumers with different tastes? The former

Quantitative Sensory Analysis: Psychophysics, Models and Intelligent Design, First Edition. Harry T. Lawless.

approach is an attempt to find products with the maximum overall acceptance pool; that is, the largest number of consumers that will like the product (Lagrange & Norback, 1987; Beausire et al., 1988). The latter approach emphasizes a potential gain in the acceptance score(s) by tailoring the product to distinct groups. Food manufacturers have largely embraced the **segmentation** approach in recent years, leading to a proliferation of different textures and flavors within a given brand and a given category. The sensory evaluation world has followed the lead of marketing research in embracing these techniques. To quote Moskowitz et al. (1985), "Marketing researchers have used clustering procedures and segmentation for close to twenty years (Plummer, 1974). They recognized decades ago that individuals differed in attitudes and usage." Classification, of course, is part and parcel of biological taxonomy, and so some of the statistical techniques such as cluster analysis have evolved for that kind of application – fitting animals, plants, organisms, and so on into logical groups.

Inherent in the idea of consumer segmentation is the notion of forming groups of people who are homogeneous and at the same time different from people in other groups. So it is important to conceptualize segmentation as an exercise both in differentiation and in finding commonalities. This is fundamental to concept research, in which different groups may like or dislike the same elements or features of a product (Moskowitz et al., 2006). One group may react positively to some brand image/concept, and another negatively. To a contrasting image, the groups may switch opinions. An important notion here is one of suppression. The overall average appeal of a concept or feature may be lackluster, but when segments are examined individually, one group may have a very strong interest in that feature (Moskowitz et al., 2006). So averaging the groups together gives a poor and inaccurate reading for that idea or characteristic.

Rose Marie Pangborn identified the potential importance of individual segments with different sensory preferences in a seminal paper in 1970. In it she showed that the common inverted U-shaped curve for liking versus sensory intensity could be masking other patterns in the data when they were averaged. Considering the attributes of sweetness and saltiness (and using prison inmate volunteers as subjects!), Pangborn found three distinct patterns of liking for increasing strength of these tastes. Some people increased their liking ratings, others decreased, and a third group showed the classic pattern of preferring intermediate levels. Another informative demonstration was seen in data collected by her student, Aileen Sontag (Trant) in her MS thesis. Sontag measured liking for different levels of sucrose in coffee over an 8-week period (Sontag, 1978; also see Pangborn (1981)). A plot of the group means showed a flattened version of the inverted U function, but individuals varied widely, some liking more sugar and some preferring less. Furthermore, the group data failed to capture the diverse patterns of changing preference among the group over the 8 weeks of usage (see Stevens (1991) for a summary and graphs).

Segmentation can be carried out using many different kinds of information. Traditional marketing research approached group differences based upon demographic information. Factors such as age, gender, or ethnic group seemed to make sense in looking at product preferences. Later, psychographic techniques became popular, and markets could be segmented by the individual's predilection to try new products, for example, or whether they could be considered a trend-setter. Lifestyle choices are another example of a kind of psychographic approach. More relevant to product development, however, are patterns of sensory segmentation such as those observed by Pangborn. Individuals in these groups could come from different demographic or psychographic segments, but they share a common hedonic pattern. Another option is to segment individuals by their own reported preferences, or purchase and consumption of distinct product types. For example, Harwood et al. (2012) segmented consumers by their reported preferences for dark over sweet (milk) chocolate. They found rejection thresholds for bittering agents in chocolate milk to be higher for those preferring dark chocolate; that is, a preference or at least a tolerance for higher bitterness levels. Others have found self-declared segmentation to be less informative than actual clustering based on data. For example, Hauswirth et al. (2010) attempted to replicate the preference splits in the famous "Pepsi challenge" of Pepsi versus Coke. On the first trial, the historical finding of a 57% preference for Pepsi (among those expressing a preference) was duplicated, but there was little or no relationship between self-stated cola preference and their choices. In the beefsteak study of Schmidt et al. (2010), discussed below, more consistent clusters were found based on the data, rather than self-reported preferences.

Sensory segments, those partitioned on the basis of sensory preferences, may cut across what are considered important geopolitical boundaries, such as nationality. In a study of coffee bitterness in different European countries, Moskowitz and Krieger (1998) showed that the five countries in the study gave rather similar inverted U-curves for bitterness preferences. However, when consumers were segmented by their actual bitterness preferences, three segments emerged, as shown in Figure 15.1. Once again, we see the kind of relationship noted by Pangborn, with three groups. One preferred intermediate levels of bitterness, another high bitterness, and a third low bitterness. Two findings were important. First, all

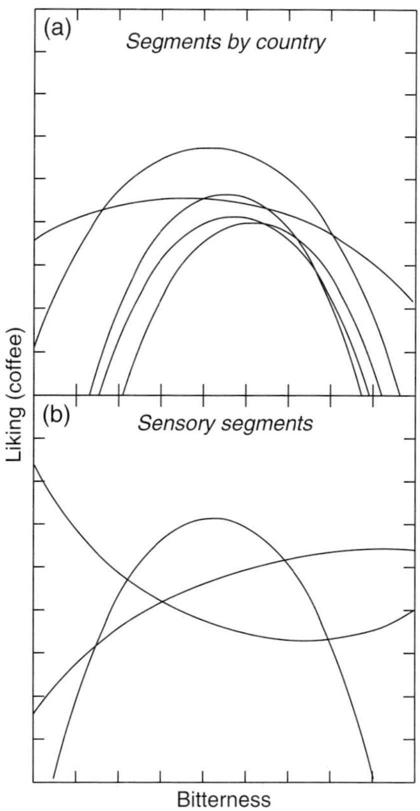

Figure 15.1 Segments for preferred coffee bitterness based on five European countries (top panel) or three sensory segments (bottom panel). Data from Moskowitz and Krieger (1998). Reprinted from Lawless and Heymann (2010) by permission from Springer Publishing.

three sensory segments could be found in all countries. Second, the hedonic scores for the most preferred levels were higher than one would obtain by creating a coffee product for each country using the overall curves, and ignoring the sensory segments. They suggested using the sensory segments as a starting point, and then fine-tuning products for individual countries. Once again, the notion that "one size fits all" does not work very well.

Sensory segments are generally studied with attributes that are called "drivers of liking." That is, a change in the intensity level of an attribute creates a change in acceptability for a consumer. Attributes that have steep or peaked functions, and/or show a lot of area under the curve, are generally considered drivers of liking (Moskowitz & Krieger, 1998; Moskowitz et al., 2006). However, the reader should keep in mind that such a relationship, even if it makes a convincing graph, is only correlational. A causal relationship may or may not exist. Another hidden or latent variable may be the ultimate cause of the apparent linkage and the true driver.

Sometimes patterns can be seen just by inspecting the data closely. After simply eyeballing the results, the primary statistical tool for categorization of segments has been cluster analysis. Like any multivariate technique, cluster analysis necessarily involves some subjective decisions on the part of the experimenter, the analyst, and the person interpreting the data. This does not mean that the methods are potentially defective, only that different choices in the distance measure or the joining rules may produce different groupings. In any complete clustering algorithm, the experimenter will have to look at the output and decide how many categories are enough. Do the groupings seem intuitively reasonable or do they seem to be force-fit? Various mathematical criteria can be brought to bear, but researchers should beware that their results may be influenced by personal hypotheses conjured up before the experimental data were ever gathered. Sometimes this is not a bad idea, if the method is viewed as hypothesis confirming (or hypothesis generating, a useful process) rather than truth standing alone. Many resources exist for the student of cluster analysis. Everitt et al.'s (2001) book is widely cited and has several worked examples on simple data sets from biology. Aldenderfer and Blashfield (1984) provide a good concise introduction from the behavioral science perspective, with suitable warnings to the student and statistician. An updated and focused treatment from a sensory evaluation perspective at this point seems lacking, and some notable sensory statistics books do not deal with segmentation and clustering at all. However, segmentation is addressed in the multivariate book by Meullenet et al. (2007: chapter 6) and landscape segmentation of consumers is illustrated in the compendium by Ennis et al. (2011). A review with examples from food science is found in the chapter by Jacobsen and Gunderson (1986) in the statistics anthology edited by Piggott. They provide a table of published papers using cluster analysis in various food applications (some sensory), plus a comparison of results using different linkage methods, and a rare example of "fuzzy clustering."

15.2 Case Studies

15.2.1 Cluster Analysis and Segmentation

A common technique for segmentation of consumer groups is through the use of cluster analysis. Like multidimensional scaling (MDS), it is a set of mathematical procedures for analyzing and simplifying the patters of similarity and dissimilarity among products and/or consumers. The various techniques and options of cluster analysis will be explored in

Section 15.3. But first we will examine a few examples of clustering and segmentation to see how the techniques can be applied to consumer acceptability data. In the last decade, a number of studies have been published, many in the area of dairy products, that have used cluster analysis or other methods of segementation in sensory evaluation (Lawlor & Delahunty, 2000; Drake et al., 2001; Ares et al., 2006; Casapia et al., 2006; Serrano-Megias & Lopez-Nicolas, 2006; Schilling & Coggins, 2007). Segmentation techniques are often used in connection with preference mapping. When connected to descriptive data or other objective measures, the potential underlying reasons or correlates of the consumers' preferences can be deciphered. Schilling and Coggins (2007) show how clustering of consumer groups can be used with simple analysis of variance (ANOVA) in order to look for perceived product differences within each group. Cluster analysis followed by ANOVA and least-significant difference tests were performed on four products: processed ham, milk at different pasteurization temperatures, shrimp with different packaging conditions, and a processed chicken product. This serves as a good illustration of how consumer preferences can be connected to processing variables. Three additional case studies will be described further here, one from the early literature that looks at preferred spice intensity levels in sauces (Moskowitz et al., 1985), a more recent study on preferences for end-point cooking temperatures in beef steaks (Schmidt et al., 2010), and a study of wine segmentation (Findlay, 2008).

15.2.2 A Case Study: Pasta Sauce

An early case study by Moskowitz et al. (1985) provides an example of how segmentation can work using a combination of liking ratings and intensity scores. For some attributes there is a firm relationship between the preferred intensity of an attribute in some product and how that preference changes as intensity changes. Thus, liking plotted as a function of sensory intensity (or the underlying ingredient level) will often produce the inverted U-shaped curve with an optimum. If the curve is steep or has a large area under the curve, it is possible that the attribute is a "driver" of liking, since its intensity appears to influence the person's preference.

In this study, a large battery of products were engineered that varied in levels of key attributes. The products were spicy pasta sauces and evaluations of all products were conducted by consumers on 22 scales for intensity of aromas, flavors, textures, and various attitudinal and liking scales. A general quadratic equation was then fit to model the U-curve, with a negative squared term to make the function decrease after the optimum as follows:

$$L = \beta_0 + \beta_1 A_i + \beta_2 A_i^2 \qquad (15.1)$$

where L is the liking ratings from one consumer, A_i is the attribute intensity rating averaged across the entire group on attribute i, and the β coefficients determine the regression equation. Once this "psycho-hedonic" function is fit, the optimum can be found for that individual on this attribute. This produces a matrix of consumers (rows) by attribute optima (columns).

The matrix of optima was then submitted to principal components analysis (PCA) to reduce the number of variables and take into account the correlations among some of the attributes. The principal components (PCs) then have factor scores for each consumer, presumably corresponding to his or her optimum levels on the constructed PC. Finally, these values were submitted to a cluster analysis to identify segments. One can then plot the average psycho-hedonic functions for each segment. The paper shows that, for the attribute of spiciness, there were three segments: one that liked increasing levels of spiciness, one that liked little or no spiciness (decreasing liking as spiciness increased), and a third group with a preferred level in the middle of the series (the classic inverted U-curve).

This paper serves as a model of one approach to sensory segmentation. Sensory segmentation again, is based on clustering of similarities of hedonic response to a varying sensory attribute. Several aspects of the study, however, deserve further thought. First, is it necessary to perform the PCA to derive factors, or would a cluster analysis based on the entire attribute set have shown the same segments? Second, how well did the quadratic functions fit various individuals? That is, were the optima clear? We do not know. Finally, note that this kind of study requires a substantial investment in engineering and producing the product set that is to be evaluated. One must insure sufficient variation to show sensory differences that are clearly perceived and potentially impactful to consumers.

15.2.3 Case Study 2: Meat Preferences

A more recent case study examined consumer preferences for beef steaks cooked to various degrees of done-ness, as measured by end-point temperature (Schmidt et al., 2010). These temperatures have official meanings in the meat science world that the consumer usually associates with color, such as 60 °C for rare steaks and 77 °C for a steak well done. Since it is known that many consumers have strong preferences for their preferred degree of done-ness in beef steaks, it seemed logical to examine these preferences and try to group consumers using cluster analysis, based on acceptability scores. Descriptive analysis was also carried out to examine objective sensory correlates of steak preferences. Steaks were cooked to five different end-point temperatures, and cooking loss and instrumental shear measurements confirmed the expected trends. Two groups were tested, one with select grade steaks and one with choice grade. The summary here will look at the common trends across both groups.

Virtually no differences were observed in overall acceptability as a function of end-point temperature, but preferences for texture showed higher texture acceptability scores for rare and medium-rare steaks. Thus, the group data, when aggregated ($N=156$), only poorly differentiated, with most scores in the range of six to seven on the standard nine-point hedonic scales. It seems that people like free steak. Next, they analyzed preferences based on self-reported preference for different degrees of done-ness. Once again, almost no trends were seen, although the small groups (too small to analyze statistically) that reported liking rare steaks showed the expected trend in decreasing scores with increasing done-ness. Also, those preferring medium-rare steaks gave significantly lower scores to well-done samples.

Rather than rely on self-reported preferences, the authors then used agglomerative clustering techniques to derive empirical groups. About 80% of the consumer groups could be separated into meaningful clusters, but 20% showed no trends and were omitted from further analysis. Among those showing preference patterns, distinct consumer groups emerged. The largest groups were (1) those showing preference for intermediate levels of done-ness, (2) liking for higher levels, and (3) liking for lower levels. Another cluster liked all steaks as long as they were not well done, and the opposite pattern (liking all but rare) was another smaller group. Another cluster liked all choice steaks, regardless of temperature, but was different from the rest of the consumers in giving higher overall scores (means of 8.5 out of 9.0 for four of the five samples!). Comparison of consumer clusters with descriptive attributes through preference mapping was also instructive. For consumers who disliked rare steaks, for example, the increased tenderness and juiciness did not overcome their dislike for bloody and metallic notes. Those preferring well-done steaks were hypothesized to appreciate the strong roasted and browned/burnt flavors based on the attribute vectors seen in the preference map.

15.2.4 Case Study 3: Wine Preferences

A clear demonstration of sensory segments was observed in an unpublished study based on several hundred drinkers of Cabernet Sauvignon, given at a recent Sensometrics meeting (Findlay, 2008). A group of Cabernet wines representing different styles were evaluated for acceptability on nine-point hedonic scales by Cabernet drinkers. Perhaps not surprisingly, the group ranges went from about 5.4 to 6.0 for the least liked wine to only 6.1 to 6.8 for the most liked. However, when the consumers were segmented by preferences, four clusters emerged with greatly expanded ranges of ratings, as shown in Figure 15.2. Bear in mind that the most preferred and least preferred wines were different for the four groups.

Two of the clusters had demographic correlates. Cluster 4, the smallest one, consisted mostly of younger women who preferred lighter styles of Cabernet Sauvignon. Cluster 3 consisted largely of older males who preferred wines with higher tannin and extract, wines that could be described as "big and chewy." Clusters 1 and 2 were true sensory segments, with one group preferring smoky and coffee notes (perhaps from barrel aging) and the other liking sweet, raisin, fruity, and eucalyptus aromas. This study serves as a very good

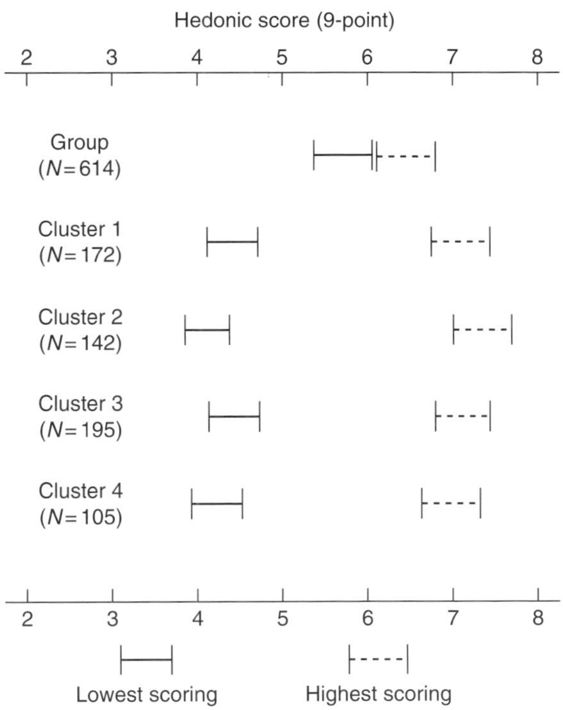

Figure 15.2 The 95% confidence intervals for the lowest and highest rated wines shown for the entire group, and for the four sub-groups of Cabernet consumers after they were grouped by cluster analysis and separated by preference styles (data from Findlay (2008)). Note that the wines are not the same for the five groups of data, as the most and least preferred changed in each segment.

illustration of how an individual segment's preferences will result in higher scores for the products that they like as opposed to the aggregated whole-group averages. Conversely, for the styles they do not like, the scores are respectively lower than the average. Note that the expansion in the negative direction for the least liked wines seems larger than the boost in the positive direction for the most preferred items. This suggests that if you were to market to the entire audience with the best (average) wine, you are likely to end up offending someone, and perhaps not delighting anyone.

15.3 Cluster Analysis

15.3.1 Types of Cluster Analysis

Cluster analyses can proceed in two directions. Some programs start with a single cluster and divide it into smaller pieces (divisive methods), and some begin with single items and proceed to join them. The joining algorithms are referred to as agglomerative (Aldenderfer & Blashfield, 1984). The most common form of cluster analysis is probably **hierarchical agglomerative clustering**, a fancy term for a joining process that proceeds in a stepwise manner. Most of these programs have some kind of objective function they are trying to minimize, such as the ratio of within-cluster variance to between cluster variance. For example, in an agglomerative technique, the program may join whatever two items or clusters have the closest neighbors at each step. Or they may seek to join clusters with the two closest centroids, given the clusters that have been formed so far.

Other programs use iterative partitioning methods. An example is *k*-means clustering, a popular technique. It starts with any arbitrary set of prespecified clusters, using part of the data. Each unassigned point or item is assigned to the cluster with the nearest centroid. New centroids are computed after a complete pass through the data. Points are then reassigned on the basis of proximity to the new centroids. The process is repeated until no points change membership. The centroids may be updated after each item, or after an entire pass through the data set. Sometimes, a poor initial partitioning or starting set can lead to suboptimal ending solutions. This is analogous to getting stuck in a local minimum when trying to minimize a badness-of-fit measure.

Since hierarchical agglomerative methods are very common, their characteristics will be examined in detail. Agglomerative methods work on the basis of similarity or dissimilarity, so the first issue concerns the choice of distance measure. Assume we have two columns or vectors of observations corresponding to the measurements on two items we would like to join (or not). Simple Euclidean distance, the square root of the sum of the squared differences, is common. Other options exist, such as Procrustean distance. This allows the two vectors to be rotated to maximum correspondence before examining the residual unshared variance (Qannari et al., 1997). If the two vectors are standardized by their standard deviations, similarity can be quantified by the correlation coefficient and distance by the square root of two times one minus the correlation coefficient, as shown below:

$$d(x,y) = \sqrt{2(1 - |r_{xy}|)} \qquad (15.2)$$

Other options are discussed in Everitt et al. (1996), as cited in Everitt et al. (2001), and may be more appropriate when the within-group variations are not equal, or when various measures feeding into the distances are highly correlated.

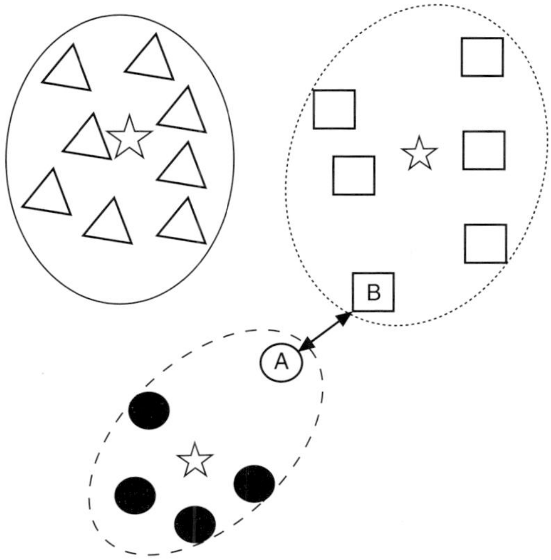

Figure 15.3 A hypothetical plot showing differences in cluster joining by a single linkage rule and a centroid linkage rule. Star symbols show the centroids of the three clusters. On the basis of single linkage, the short distance (i.e., similarity) from item A to item B would attach A to the cluster of squares (if B was already a member), whereas the centroid linkage would attach A to the cluster of filled circles.

15.3.2 Linkage Methods

Once the distance measure is determined, the next choice is the criterion or method for joining items or for joining previously formed clusters called **linkage** criteria. Many options exist, and they can produce different results. Single linkage is a simple technique. Items are joined on the basis of proximity, starting with the two that are closest. Other items may join the cluster, based on their proximity to any single item. This procedure tends to "chain" observations, and leads to long, stringy clusters. In contrast, complete linkage joins an item based on its having a certain level of similarity to all members of the cluster. Centroid linkage uses the distance from the centroid (or barycenter) of the group as a criterion for joining. Figure 15.3 shows how single linkage and centroid linkage may produce different group memberships. Complete linkage or furthest neighbor is opposite to the single linkage or nearest neighbor criterion. Now the distance between groups is defined as the distance between the two most dissimilar items. Another criterion is the average linkage, which defines the inter-group distance as the average of the distances between all pairs of individuals (with one item from each group in each pair).

Another popular criterion is **Ward's method**. It is designed to minimize the variance within clusters. It uses an error sums of squares as the objective function to minimize, also called the within-group sums of squares, and is very similar to those calculated in analysis of variance (Everitt et al., 2001), as shown in eqn 15.3. It finds the sum of squared deviations from the average value or from the mean vector (centroid). For a univariate group of values, for example, the error sums of squares would be the simple numerator of the variance estimate:

$$\mathrm{ESS} = \sum_{i=1}^{N} (x_i - \overline{x})^2 \qquad (15.3)$$

where ESS is the error sums of squares for N members of the cluster, with x_i being the ith example of member x and $\overline{x}$ is the mean. This method tends to form clusters of similar size and spherical shape when viewed as a cloud in multidimensional hyperspace (Aldenfelder & Blashfield, 1984).

Use of a correlation measure was suggested by Vigneau et al. (2001) and Vigneau and Qannari (2002). Assume a group of consumers have expressed their preferences or acceptability ratings for a group of products. Consumers are considered variables and products are observations. The merging of clusters is done so that the sum of the correlations between the centered consumer data and their respective centroid is maximized. A consumer can be moved to another cluster if their correlation with the centroid values is higher for that group, and they will be placed in the cluster for which the correlation is maximized. Of course, this creates a new centroid, so the process can be repeated until a stable solution is achieved.

The number of available linkage methods continue to proliferate in various software packages, and so this section outlines only some of the common ones, and a few of those that appear in the sensory literature. The utility and validity of the different linkage methods are debated, and empirical information on their accuracy can be found in reviews by Everitt et al. (1996) (as cited in Everitt et al. (2001)) and Aldenderfer and Blashfield (1984). Single linkage is considered the least useful in producing cluster solutions, but it is computationally simple and, like complete linkage, is invariant under any monotonic transformation. Any matrix with the same rank orders will produce the same result; scaling issues are not relevant. As noted, it tends to chain items, leading to some distant and strange-looking combinations as the joining progresses. Ward's method has proven to give good fits of solutions to data, but it may "impose" more spherical-shaped groups on the solution. If the researcher is unsure that the chosen method has produced something reasonable, more than one linkage method can be tried.

15.3.3 Output Trees and Decision Time

The output of a cluster analysis contains several important graphs and information. If the data have been input as scaled ratings on hedonic or intensity scales, a distance matrix may be available, as computation of the distance matrix is the first step in the analysis. The final view of the clusters comes from a tree structure or **dendrogram** that shows the order in which the items were joined, to what neighbors, and at what amalgamation or aggregation distance. Another important plot shows the joining distance measure as a function of the order in which things (items or previously formed clusters) were joined. Examples of a dendrogram and a joining distance plot are shown in Figure 15.4 and Figure 15.5, respectively.

There are several names for the joining distance measure, such as a fusion coefficient, amalgamation distance, or aggregation distance. Of course, early in the process, items that are quite similar (i.e., close in multidimensional space) are joined, and as the algorithm proceeds, clusters with greater overall dissimilarity are joined. This information can be useful in the most important decision made by the experimenter, namely how many clusters to accept as the final description of the groups. One common approach is similar to the examination of a scree plot. The plot of joining distances, that may be a bar graph or line graph, is examined for a sudden jump in the distance measure, usually toward the end of the process. A sudden jump in the aggregation distance indicates that items are now being joined that have high dissimilarity. In other words, they probably should not be lumped into the same group. This is very much like looking for an elbow in a scree plot, as discussed in

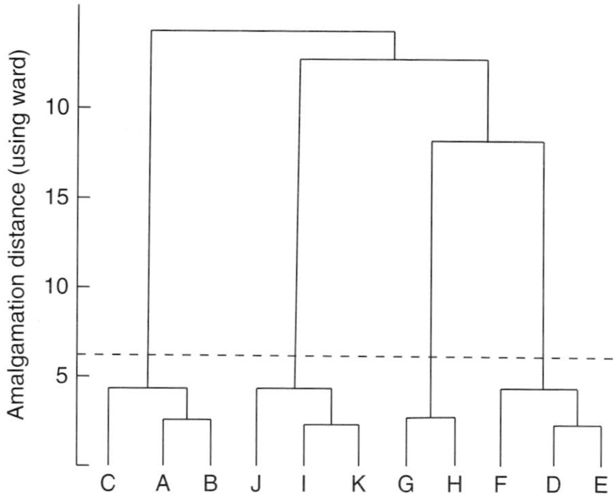

Figure 15.4 A sample dendrogram based on the snack food preference data shown in Table 15.1.

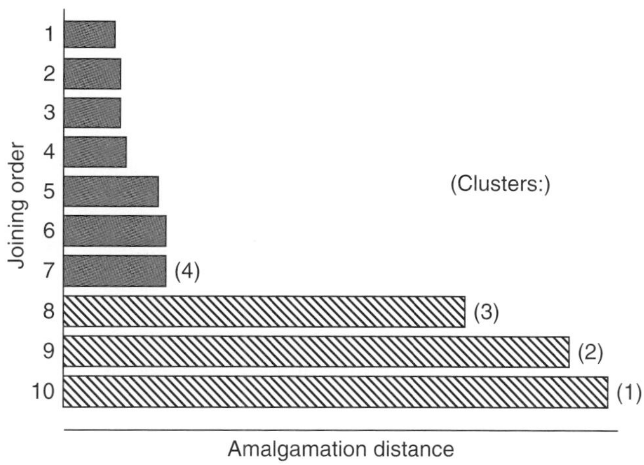

Figure 15.5 A sample plot of the joining distances for the items and clusters as the algorithm proceeds, from the snack food data (Table 15.1). At the beginning, items with high similarity are joined; towards the end, clusters with high dissimilarity are forced together.

Chapter 14. The decision to choose a cutoff can be somewhat subjective, and prior information may be brought to bear in addition to whatever the graph may show. The cutoff level is often shown as a horizontal line in the dendrogram (see Figure 15.4) as the vertical height is typically a function of the joining distance.

The overall goodness of fit should be a concern of the experimenter. A common measure of the fit of the dendrogram joining distances to the original proximity measure is called the **cophenetic correlation coefficient**. One set of items are the $N(N-1)/2$ entries of the observed or calculated lower-half similarity (proximity) matrix. These values exist for each possible pair. They are compared with the values of the cophenetic matrix, whose entries are the first level of the dendrogram at which the two items are entered into the same cluster.

Table 15.1 Data for snack food preferences submitted to cluster analysis

	Consumer										
Snack food	**A**	**B**	**C**	**D**	**E**	**F**	**G**	**H**	**I**	**J**	**K**
Peanuts	**8**	**7**	**9**	4	4	3	3	4	3	4	4
Beer nuts	**8**	**8**	**8**	4	4	4	4	3	5	3	4
Almonds	**7**	**8**	**7**	3	3	4	3	4	4	5	4
Popcorn	4	4	3	**7**	**8**	**7**	4	4	4	5	4
Cheese corn	3	3	4	**8**	**7**	**9**	3	4	5	4	5
Caramel corn	3	3	4	**6**	**7**	**8**	4	5	3	3	4
Tortilla chips	5	5	3	3	3	4	**9**	**8**	3	4	4
Corn chips	5	5	5	5	4	5	**9**	**8**	4	3	5
Potato chips	5	5	4	4	3	5	4	4	**8**	**8**	**9**
Kettle chips	6	4	4	5	4	5	4	5	**9**	**7**	**9**
Re-formed potato crisps	4	5	3	3	3	5	4	4	**7**	**8**	**8**

This produces a kind of ranking information. The correlation is usually not perfect, but values above 0.8 are considered a good fit (see Everitt et al. (2001) for an example).

15.3.4 A Simple Example of Cluster Analysis on Consumer Data

A sample data set is shown in Table 15.1 for the hedonic ratings of some hypothetical consumers as might be found on a food preference survey or some kind of category appraisal, in this case for some snack foods. Three consumers, A, B, and C, like nut products. Consumers D, E, and F prefer the popcorns. G and H like the corn-based chips and I, J, and K prefer potato chip type products. In order to get the separation for this demonstration, they have given lukewarm or slightly negative ratings to the other items outside their preferred type of snack food. The preferred items are listed in bold face in the table to show the patterns. Data were submitted to the HCLUST procedure in the R platform, after computing a distance matrix using the DIST procedure (Euclidean distance was chosen). Ward's method was chosen for the joining criterion. This is an example of hierarchical agglomerative clustering.

Given these clear preferences, the program recovers the intended groupings as shown in the dendrogram in Figure 15.4. Consumers in each of the four clusters all join their respective groups at a fairly low distance, and then the four groups are joined at much higher fusion distance. A bar plot of the joining distances in sequential order shows the large jump in distance once the program attempts to go from four to three clusters (Figure 15.5). So the dotted line in the dendrogram shows the experimenter's decision to use four clusters to represent the group structure and to ignore the later amalgamations.

Since the R code for HCLUST requires the computation of a distance matrix as one step, the data can also be input into an MDS program to visualize the cluster structure. The MDS plot is shown in Figure 15.6, as produced by the operation CMDSCAL and requesting a two-dimensional configuration. The plot has been drawn in a graphing program, from the coordinates produced by CMDSCAL. The data were originally read from a spreadsheet file as comma-delimited values (.csv). The R code sequence is shown in Table 15.2. This reanalysis by MDS is an example of how using more than one multivariate technique can

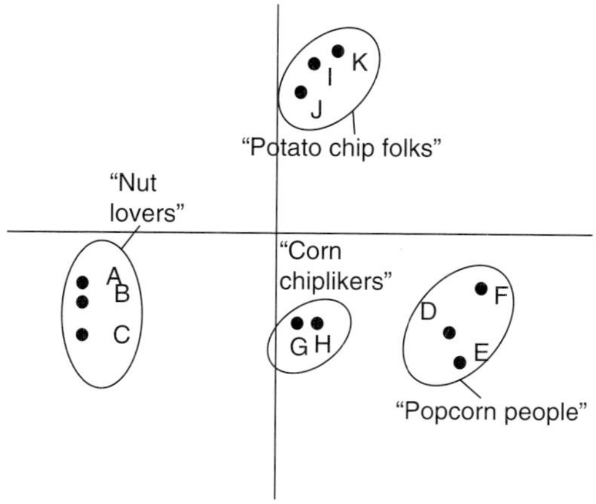

Figure 15.6 A multidimensional scaling solution for the data from Table 15.1, after transforming to a Euclidean distance matrix and using the CMDSCAL function in R.

Table 15.2 R code for the clustering procedure in Section 15.4.2

R command line	**Comment**
>snack<-read.csv("filename.csv", header=TRUE, row.names=1)	Reads data file to "snack"
>snackdist<-dist(snack, method="euclidean")	Computes distance matrix
>snackclust<-hclust(snackdist, method="wards")	Performs cluster analysis
>plot(snackclust)	Plots dendrogram

Notes.
1. Data were originally in a spreadsheet with consumers as rows, product ratings as columns, and both column headings in the first row and row headings in the first (leftmost column, hence the options header=TRUE and row.names=1.
2. The object names, snack, snackdist, and snackclust are arbitrary choices, intended as descriptive memory aids.

improve the understanding of patterns in the data. If often helps to visualize them in more than one way.

15.3.5 Programs and Options

Most software programs will offer the researcher several choices for distance measures and for the linkage criterion, with Euclidean distance and Ward's method being common in sensory segmentation studies. Most, if not all, of the larger commercial statistical packages can be used for cluster analysis as the techniques have a long history. Schilling and Coggins (2007) provide some SAS code for meshing hierarchical clustering with ANOVA. They also cite an XLSTAT web tutorial (XLSTAT, 2006) that shows how the software can be used on that platform, with various pull-down menus and dialog/choice boxes. The R platform has a variety of techniques available in the CLUSTER library, including the functions KMEANS (obviously for a *k*-means technique), HCLUST (for hierarchical clustering), and PAM, a more highly iterative version of *k*-means techniques (http://statistics.berkeley.edu/classes/s133/Cluster2a.html). A Qannari clustering method can be found in the package SENSTOOLS 3.3.1.

15.4 Other Analyses and Methods

15.4.1 A Simple Divisive Method

Everitt et al. (2001) provide the following example of a divisive method. At the beginning, all items or consumers are members of the same cluster. So a splinter group must be formed. Based on the procedure of MacNaughton-Smith et al. (1964), a singleton is first exiled based on the individual with the largest average distance from the other items. Next, the average distance from each individual in the parent group to the splinter group is found. A difference score is then generated to find the difference between the average distance to the new (reduced) parent group minus the average distance to the splinter group. The individual with the highest difference (main minus splinter) is then exiled, forming a two-item splinter group. The process continues, until only negative differences remain. At this point, everyone is closer to their respective group than to the other. The divisive procedure may then be repeated on each daughter group separately, and so on. The method is remarkably simple and not computationally intensive.

15.4.2 Taxmapping

From the realm of biological taxonomy, techniques have been developed that mimic the human perceptual tendency to group items that seem to form a clusters with open space between the clusters (Carmichael et al., 1968). Imagine you had a multivariate space with only two or three dimensions. You could examine the space for densely populated areas separated by areas of sparse population. Clusters can be formed by any method, but single linkage is a good choice here. Additions continue until some criterion is reached. A stoppage criterion is invoked when the next prospective point is considered much farther away than the last point admitted. In other words, there is some kind of discontinuity in the distance measure for the next iteration (Everitt et al., 2001). For example, up to a point, there will be small drops in the average similarity as items are added to a group. When this drop becomes large (relative to the change attributed to the last addition) the item must fail to join. Obviously, different thresholds can be set for this discontinuity calculation, probably leading to different cluster structure.

15.4.3 Others

Almost any consumer data set or statistical analysis output can be examined for patterns of clustering or grouping if consumers form part of the output or plot. Cluster analysis, for example, can be imposed on MDS output in order to provide a more objective confirmation of the groups that the experimenter sees in their spatial map (e.g., Lawless, 1989). On the other hand, MDS may not provide much of any evidence of groups at all if the items are dispersed evenly through the space. MDS programs merely try to reproduce the original similarity structure in the output. Cluster analysis, however, will always join things.

A related method to cluster analysis, and an alternative way of looking at group membership, is to use graph theoretic approaches. In graph theory, the distance matrix is often input, like the lower half matrix of an MDS data set. The programs join items based on some threshold or cutoff for their similarity or other measure of connection. Graph theory output is different from cluster analysis, in that an item may belong to more than one fully joined group (called a clique). Cluster analysis usually produces unique group membership (an item belongs to only one group), unless a probabilistic approach is taken, called fuzzy clustering (Jacobsen & Gunderson, 1986). The output of a graph theory analysis represents the items as nodes and the connections as edges, often in a circular plot of the items with the connecting edges inside. Examples can be found in Chapter 14 and the articles cited there.

Another opportunity for segmentation exists in the various methods for perceptual mapping of consumer preferences, such as internal and external preference mapping, ideal point mapping, and landscape analysis. With internal preference mapping, the consumers are placed in a spatial model such that people with similar likes will be positioned in close proximity, thus providing the opportunity to look for groups with common patterns of product preference. Since the consumers are usually positioned near the products they like best (if products are also placed in the model), their preferences may be fairly transparent, allowing some verbal description (e.g., "people that prefer only mild salsas"). In external preference mapping, consumers may be represented as vectors, and thus the angles of the vectors can be examined for common directions through the space, often that point to the preferred product or product type. In ideal point mapping, the consumer may be represented as a point, such that their dislikes are proportional to the distance in the map from each member of the product set (and thus liking is proportional to proximity). So a dense concentration of consumers in one area of the plot is indicative of a group with similar preferences. The measure of density can be important in determining the product profile that may be desirable in order to design a successful product to attract those consumers. Sometimes the density plot is represented as a contour plot, or a surface plot in three dimensions with the vertical direction representing density (see Ennis et al. (2011) for several examples). A goal of this kind of landscape analysis is to find pockets of consumers who like different sensory styles and features in their products. Multiple peaks, then, in the density plot give a picture of likely consumer segments.

15.5 Women, Fire, and Dangerous Things

Clustering methods are common statistical tools in other branches of science, including genetics and biological taxonomy. The sensory scientist should adopt methods that suit the structure of the data (interval scaling, frequency counts, etc.). Remember that clustering output is only the beginning of understanding consumer groups. Once the groups are formed, the reasons for their preferences must be studied further. This is usually achieved by reanalysis of the data using the groups as separate data sets and looking for differences. This is a process similar to **discriminant analysis**. In one common application of discriminant analysis, the group membership is known. Then the additional sensory variables are combined and reduced to provide a linear combination that gives the best possible separation and classification. Whether by univariate exploration or multivariate techniques, the resulting differences then can be used to characterize their preferences and communicate the patterns to clients and management.

The investigation and understanding of consumer segments is an important part of current product development in the foods and consumer products industries. Tailoring product sensory profiles to groups or even to individuals is a growing part of sensory science. It can offer enhanced consumer acceptance of products and thus increased profits, providing that the cost of the line extensions and background research are not prohibitive and the potential cannibalization of existing successful product lines is not likely to cause problems. George Lakoff (1987) pointed out that one group of aboriginal Australians, the Dyirbal, have a world-view category for nouns that includes women, fire, and dangerous things, and wrote a book with that provocative title (the category also includes water, violence, and exceptional animals). The work is a study of the cognitive approach to categorization, and warns us that our preconceived notions and linguistically imposed categories will affect our perception. The same risks can be encountered in any experiment that attempts to group individuals. People certainly differ. But one must proceed with caution.

References

Aldenderfer, M.S. and Blashfield, R.K. 1984. Cluster Analysis. Sage Publications, Beverly Hills, CA.

Ares, G., Gimenez, A., and Gambaro, A. 2006. Preference mapping of dulce de leche. Journal of Sensory Studies, 21, 553–71.

Beausire, R.L.W., Norback, J.P., and Maurer, A.J. 1988. Development of an acceptability constraint for a linear programming model in food formulation. Journal of Sensory Studies, 3, 137–49.

Carmichael, J.W., George, J.A., and Julius, R.S. 1968. Finding natural clusters. Systematic Zoology, 17, 144–50.

Casapia, E.C., Coggins, P.C., Schilling, M.W., Yoon, Y., and White, C.H. 2006. The relationship between consumer acceptability and descriptive sensory attributes in Cheddar cheese. Journal of Sensory Studies, 21, 112–27.

Drake, M.A., McIngvale, S.C., Gerard, P.D., Cadwaller, K.R., and Civille, G.V. 2001. Development of a descriptive language for Cheddar cheese. Journal of Food Science, 66, 1422–7.

Ennis, D.M., Rousseau, B., and Ennis, J.M. 2011. Short Stories in Sensory and Consumer Science. The Institute for Perception, Richmond, VA, pp. 46–7, 50–1. (Originally published as "Drivers of liking for multiple segments. IFPress 4(1), 2–3, 2001" and "Identifying latent segments, IFPress 6(1), 2–3, 2003.")

Everitt, B.S., Landau, S., and Leese, M. 2001. Cluster Analysis, Fourth edition. John Wiley & Sons.

Findlay, C.J. 2008. Consumer segmentation of BIB liking data of 12 Cabernet Sauvignon wines: a case study. Presentation at 9th Sensometrics Meeting, 20–23 July, Brock University, St. Catharines, Ontario, Canada.

Harwood, M.L., Ziegler, G.R., and Hayes, J.E. 2012. Rejection thresholds in chocolate milk: evidence for segmentation. Food Quality and Preference, 26, 128–33.

Hauswirth, J., Sinopoli, D., and Lawless, H.T. 2010. Does loyalty dictate blind preference? Poster presented at the Society of Sensory Professionals, Napa, CA, 28 October.

Jacobsen, T. and Gunderson, R.W. 1986. Applied cluster analysis. In: Statistial Procedures in Food Research. J.R. Piggott (Ed.). Elsevier Applied Science, London, pp. 361–408.

Lagrange, V. and Norback, J.P. 1987. Product optimization and the acceptor set size. Journal of Sensory Studies, 2, 119–36.

Lakoff, G. 1987. Women, Fire and Dangerous Things. University of Chicago Press, Chicago, IL.

Lawless, H.T. 1989. Exploration of fragrance categories and ambiguous odors using multidimensional scaling and cluster analysis. Chemical Senses, 14, 349–60.

Lawlor, J.B. and Delahunty, C.M. 2000. The sensory profile and consumer preference for ten specialty cheeses. International Journal of Dairy Technology, 53, 28–36.

MacNaughton-Smith, P., Williams, W.T., Dale, M.B., and Mockett, L.G. 1964. Dissimilarity analysis. Nature, 202, 1034–5.

Meullenet, J.-F., Xiong, R., and Findlay, C.J. 2007. Multivariate and Probabilistic Analyses of Sensory Science Problems. IFT Press/Blackwell Publishing, Ames, IA.

Moskowitz, H. and Krieger, B. 1998. International product optimization: a case history. Food Quality and Preference, 9, 443–54.

Moskowitz, H.R., Jacobs, B.E., and Lazar, N. 1985. Product response segmentation and the analysis of individual differences in liking. Journal of Food Quality, 8, 169–81.

Moskowitz, H.R., Beckley, J.H.J., and Resurreccion, A.V.A. 2006. A Sensory and Consumer Research in Food Product Design and Development. Blackwell Publishing/IFT Press, Ames, IA.

Pangborn, R.M. 1970. Individual variations in affective responses to taste stimuli. Psychonomic Science, 21, 125–8.

Pangborn, R.M. 1981. Individuality in responses to sensory stimuli. In: Criteria of Food Acceptance: How Man Chooses What He Eats. J. Solms and R.L. Hall (Eds). Forster Verlag, Zurich, pp. 177–219.

Plummer, J.T. 1974. Concept and application of life style segmentation. Journal of Marketing, 38, 33–7.
Qannari, E.M., Vigneau, E., Luscan, P., Lefebvre, A.C., and Vey, F. 1997. Clustering of variables, application in consumer and sensory studies. Food Quality and Preference, 8, 423–8.
Schilling, M.W. and Coggins, P.C. 2007. Utilization of agglomerative hierarchical clustering in the analysis of hedonic scaled consumer acceptability data. Journal of Sensory Studies, 22, 477–91.
Schmidt, T.B., Schilling, M.W., Behrends, J.M., Battula, V., Jackson, V., Sekhon, R.K., and Lawrence, T.E. 2010. Use of cluster analysis and preference mapping to evaluate consumer acceptability of choice and select bovine M. longissimus lumborum steaks cooked to various end-point temperatures. Meat Science, 84, 46–53.
Serrano-Megias, M. and Lopez-Nicolas, J.M. 2006. Application of agglomerative hierarchical clustering to identify consumer tomato preferences: Influence of physiochemical and sensory characteristics on consumer response. Journal of the Science of Food and Agriculture, 86, 493–9.
Sontag, A.M. 1978. Comparison of sensory methods: discrimination, intensity and hedonic responses in four modalities. MS thesis, University of California, Davis.
Stevens, D.A. 1991. Individual differences in taste and smell. In: Sensory Science Theory and Applications in Foods. H.T. Lawless and B.P. Klein (Eds). Marcel Dekker, New York, NY, pp. 295–316.
Vigneau, E. and Qannari, E.M. 2002. Segmentation of consumer taking into account external data: a clustering of variables approach. Food Quality and Preference, 13, 515–52.
Vigneau, E., Qannari, E.M., Puneter, P., and Knoops, S. 2001. Segmentation of a panel of consumers using a clustering of variables around latent directions of preference. Food Quality and Preference, 12, 359–63.
XLSTAT. 2006. Tutorial XLSTAT 7.5. http://www.xlstat.com (accessed 4 Octonber 2012).

16 An Introduction to Bayesian Analysis

Although statistical tests provide the right tools for basic psychophysical research, they are not ideally suited for some of the tasks encountered in sensory analysis.

M. O'Mahony (1986: 401)

Statistical decision rules can be generated by any philosophy under any collection of assumptions. They can then be evaluated by any criteria, even those arising from an utterly different philosophy ... The Bayesian approach is an excellent "procedure generator," even if one's evaluation criteria are frequentist...

Carlin and Louis (1996: 12)

16.1 Some Binomial-Based Examples

16.1.1 Notation and Statement of the Problem

In the simple binomial version of **Bayes' rule**, we have two states of the world; let us call them A and not-A, with probabilities $p(\text{A})$ and $p(\text{not-A})$. We also have two decisions we can make based on our test outcomes. Let the critical one that we are interested in be called "B." Given that a decision has been made, we would like to know whether

Quantitative Sensory Analysis: Psychophysics, Models and Intelligent Design, First Edition. Harry T. Lawless.

it corresponds to the accurate state of the world (A). Most of our statistical practice is aimed at figuring the chances of making decision B (or not-B) when A is true. This is the notation $p(B|A)$ or "the probability of B given A." When A is the null hypothesis, and we reject it, we would like to keep the mistake down to a minimum level, which we call α, and estimate by our probability level. But management may be interested in the reverse. Having made the decision, what are the chances I am wrong? So they would like to know $p(A|B)$, the chances of A being true when I have already declared "B" to be the outcome of the test.[1]

Here is an example. Suppose that event A is the actual existence of a defective product. Event B is a test outcome in our quality control (QC) lab that indicates a product is defective, according to the test criteria. The probability of (A | B) can be rephrased as "Given that my test has detected a defective product, what are the chances it is truly defective (in fact)?" In other words, what is the chance that my test result is correct? Eqn 16.1 shows how knowledge about the incidence of event A (called a prior probability) and the accuracy of our test (detecting a defect when it does exist) can be used to answer the question by Bayes' rule:

$$p(\mathrm{A} \mid \mathrm{B}) = \frac{p(\mathrm{B} \mid \mathrm{A})\,p(\mathrm{A})}{p(\mathrm{B} \mid \mathrm{A})\,p(\mathrm{A}) + p(B \mid \text{not A})\,p(\text{not A})} \tag{16.1}$$

In order to see how this can be useful in a sensory evaluation setting, three short examples will be given: one from a QC perspective, one from an ingredient supplier situation, and one from consumer testing. Two of these examples were taken from actual consulting situations in the food industry, with the particular details modified to protect the innocent.

16.1.2 Case Study 1. The Incidence Issue, or Why I Don't Like QC Tests

Considering the binomial version of Bayes' theorem, we can see where problems can arise under two conditions: where the incidence of an event is quite low, and where the probability of a false alarm (type I error) is significant. The chance of a type I error is universally nonzero in sensory work, but because we do not often know the frequency of a true null, its frequency is difficult to estimate. Let us consider the "event" to be the detection of a substandard product in QC. The job of the sensory evaluator is to ring an alarm bell when a sample comes through the system that fails some QC test. For the purposes of this example, we can think of it as detecting that a product is different from a standard one.

Let us assume that, in the track record of this testing program, 1000 tests are conducted over some period of time. We know from over the years of manufacturing the product in question that there is significant variability and that approximately 5% of products will actually be different than the standard in the long run. Over 1000 tests, we would expect that 950 products are really acceptable, and about 50 of them are true failures. Let us further assume that our long-term α level is 5% and our long-term β level is 10% (probably an optimistic estimate in real life). Looking at our α and β levels, the long-run mistakes will consist of 5% times 950 or about 48 false alarms, and 10% times 50 or 5 mistakes where a defective product slips through. Perhaps the latter part is not bad.

[1] Many people confuse the probability of A | B with B | A. An example is the notion that given that I have rejected the null hypothesis at $p<0.05$ (95% "confidence") the probability of the alternative hypothesis being true is $(1-p)$ or 95%. Utterly incorrect. Consider the following scenario. Given the observation of a dead man on my front lawn, there is about a 5% chance he was shot in the head. Given the observation of a man shot in the head on my front lawn, there is a 95% chance he is dead.

Table 16.1 A Hypothetical set of outcomes from a QC test

True state of the world	Test outcome		Incidence
	Difference detected	**No difference**	
Difference	45	5	50
No difference	48	902	950
Total	93	907	1000

Table 16.2 The QC situation with Bayesian notation inserted

True state of the world	Test outcome		Incidence
	Difference detected	**No difference**	
Difference	$p(B \mid A) = 0.90$ $(1-\beta)$	β risk = 0.10	$p(A) = 0.05$
No difference	$p(B \mid \text{not-A}) = 0.05$		$p(\text{not-A}) = 0.95$
Total	$p(B)$		

Table 16.1 shows a summary of the situation so far. The interesting question arises, from a Bayesian point of view, is what is the chance of making a mistake, given that we have rejected the null, declared a difference, and rung the alarm bell that says the product we just tested is defective?

Note that over 90% of our tests yield valid results. However, the problem crops up when we declare that the product is different or out-of-specification. In over half the cases (48 out of 93, or about 52% of the time), we have rung the alarm bell when no problem really exists. We are correct only 48% of the time. Note that this will happen even in a perfect world, where we have conducted a state-of-the-art sensory test using all of the good principles and practices dictated by the textbooks, including a correct statistical analysis. How could things go so wrong? The credibility of the sensory department could be at stake with upper management. How can they trust this decision when it is wrong over half the time?

To connect the table to Bayes' theorem, we need to insert our notation (Table 16.2). Let "B" be the decision to reject the null and "A" be the event of there being a true difference. We would like to know the probability that there is a true difference, given that we have detected one (rejected the null, rang the alarm bell). In other words, the probability of A given B or $p(A \mid B)$, as in the opening example. Eqn 16.2 shows the formal calculation:

$$\begin{aligned} p(A \mid B) &= \frac{p(B \mid A)\,p(A)}{p(B \mid A)\,p(A) + p(B \mid \text{not } A)\,p(\text{not } A)} \\ &= \frac{0.90(0.05)}{0.90(0.05) + 0.05(0.95)} \\ &= \frac{0.045}{0.045 + 0.0475} \\ &= 0.486 \end{aligned} \tag{16.2}$$

So once again we are right less than half the time. This situation gets worse if α floats up, or if the incidence goes down. Conversely, as the incidence of our event approaches 50%, all of the outcomes fall in line with our expectations based on the α level; for example, only about 5% errors when $p(\mathrm{A})$ is 0.5.

Note that this process turns the usual scientific decision process on its head. Rather than finding the probability of an event given a true null, we are asking the probability of the null, or its alternative, given a certain event (and knowing something about the a priori probability). But this is what management wants to know. Have I made the right decision? What are the chances I am wrong, given this course of action? This example shows how the prior probability of A, which we have called incidence, can be a big problem. This is even more serious with medical diagnostic testing for rare diseases, when the test has a false-alarm rate of say 1%. Given 10,000 tests for HIV, with an incidence of 1 in 10,000, you would end up with 100 false positives for about one true detection. Imagine 100 people getting bad news that is incorrect and perhaps devastating. This is why it is always a good idea to repeat testing with that kind of false positive rate and low incidence.

16.1.3 Case Study 2. The α Blues, or Let's Play "What If?"

The scenario. An ingredient manufacturer has a major contract with a snack chip manufacturer in order to supply a flavor delivery system consisting of a spice blend powder. Both companies have small trained panels for QC that evaluate key attributes. They share the same training and test protocol and panel agreement is good. However, as a final pass before using any submitted lot, the manufacturer conducts a triangle test in the finished product with 36 judges. The cutoff for acceptance is 14 correct out of 36 or less, or if 15 are correct, the test is repeated and the total correct must be less than 28 out of 72.

The manufacturer has been rejecting a number of lots based on these triangle tests. The unfortunate supplier has conducted a series of diagnostic programs to identify the problem with no success. Almost none of the measured instrumental or sensory properties of the spice blend flavor system are correlated with the rejected lots so far. Furthermore, there is no consistent pattern. If you were hired as a consultant to advise on this situation, what would you do? Can you see the problem? All the information you need is in the paragraph above.

The key. If you think about it, the chance expectation for a triangle test with 36 judges is only 12 correct. So the cutoff that the manufacturer has set is actually quite close to chance performance. Consult any table for the minimum number required for a significant difference given 36 judges. The critical number is 17 for a significant difference at $p<0.05$. The actual statistical risk of a false positive, for example in a test using identical samples, is 30% for 14/36, 19% for 15/36, and 20% for 28/72. Depending upon the incidence, then, there is a significant chance of any rejected lot being from a perfectly acceptable batch. Using Bayes' theorem, the following calculations show how bad the situation can get.

Example 1. Assume 10% incidence rate of lots that are actually different and 90% not perceivably different. Assume α equals 30% (14/36 criterion) and β (chance of missing a difference) is 10%. The example assumes 100 lots submitted for testing. Table 16.3 shows the resulting false-alarm rate.

So there is a 75% false-alarm rate, due to what you would expect with a test on 90 (even identical) samples and 27 false positives. Another way to think of this is that the probability of being correct is only 25% once we have declared a difference. In the notation from the example

Table 16.3 False-alarm rate calculations for 10% incidence and α of 0.30

	Test outcome (conclusion)		
Actual situation	**Products found different**	**Products found not different**	**Total samples (lots tested)**
Products different	9	1 (β = 10% × 10)	10 (10% incidence rate)
Products not different	27 (30% × 90)	63	90
Total	36		100
False-alarm rate	**27/36 (=75%)**		

Table 16.4 False-alarm rate calculations for 1% incidence and α of 0.30

	Test outcome (conclusion)		
Actual situation	**Products found different**	**Products found not different**	**Total samples (lots tested)**
Products different	1	0	1 (1% incidence rate)
Products not different	33 (30% × 99)	66	99
Total	34		100
False-alarm rate	**33/34 (=97%)**		

shown in the first case study, let B represent the declaration of a difference as a test result and let A represent a true difference. The Bayes' theorem equation works out as follows:

$$
\begin{aligned}
p(\mathrm{A}\mid\mathrm{B}) &= \frac{p(\mathrm{B}\mid\mathrm{A})\,p(\mathrm{A})}{p(\mathrm{B}\mid\mathrm{A})\,p(\mathrm{A}) + p(\mathrm{B}\mid\text{not A})\,p(\text{not A})} \\
&= \frac{0.90(0.10)}{0.90(0.10)+0.30(0.90)} \\
&= \frac{0.09}{0.09+0.27} \\
&= 0.25
\end{aligned}
\tag{16.3}
$$

Example 2. Assume 1% incidence rate of lots that are actually different and 99% not perceivably different. Assume α equals 30% (14/36 criterion) and β (chance of missing a difference) is 10%. The example assumes 100 lots submitted for testing. Table 16.4 shows the resulting false-alarm rate. Note that, as the incidence goes down, the false-alarm rate gets worse.

Example 3. Assume 10% incidence rate of lots that are actually different and 90% not perceivably different. Assume α equals 20% (28/72 criterion) and β (chance of missing a difference) is 10%. The example assumes 100 lots submitted for testing. Table 16.5 shows the resulting false-alarm rate.

Some people describe this as a game of playing "what if." Even if we do not have prior knowledge of the exact incidence, different plausible values can be plugged into this simple system to

Table 16.5 False-alarm rate calculations for 10% incidence and α of 0.20

Actual situation	Test outcome (conclusion): Products found different	Products found not different	Total samples (lots tested)
Products different	9	1 ($\beta = 10\% \times 10$)	10 (10% incidence rate)
Products not different	16 (30% × 90)	63	90
Total	27		100
False-alarm rate	**16/27 (= 66%)**		

see what occurs. In all three examples, the inflated α levels, combined with low incidence, create a lot of potential problems where none really exist. You might argue that such low incidence is not a good assumption, but remember that there are internal QC panels in both companies that have already "passed" these lots. So it is quite likely that most of the time they are dealing with acceptable submissions.

16.1.4 Case Study 3. Protect the Franchise For the Name Brand, at All Costs

This example will show an even more egregious inflation of the α level in consumer testing, and the consequences for product improvement programs. In this situation, the management of a company was interested in preserving the market share and profitability of several well-established brands in a single product category. In other words, they had a goodly share of the overall corporate profit from these brands, each of which had a loyal consumer following. Let us refer to the situation as a "franchise." The company also had a large R&D arm, and this group was engaged in product improvements and cost reductions. The scenario, however, created highly conservative decision criteria, because the products in question were so valuable. Any product change that offended the loyal consumer base could be devastating. Therefore, a stringent criterion was set in making any product changes. This criterion involved a statistical test of the proportion of consumers who would change their minds from liking the product to disliking it. There are a number of ways to measure such a change of heart, as in a simple hedonic score decrement, or changing from a top-two box score on a five-point scale for purchase intent to one of the bottom two boxes, for example. Whatever, the metric, it is possible to measure consumer dissatisfaction. In order to protect the researchers from missing an effect of consumer dissatisfaction, the α level was allowed to float up to 0.50.

Such a high α level will help keep β risk down, as α and β are inversely related in the power calculations, all other things being equal. But an α level of 0.5 will almost always insure a false-alarm rate of 50% under a true null. That is, at least 50% of the "detections" of consumer dissatisfaction could be spurious, and an acceptable change in the product could be missed. So franchise risk was protected, at the expense of opportunity risk. R&D could spend a large amount of their budget on a product improvement, for example, a nutritional improvement such as sodium, sugar, or fat reduction, or increased fiber or protein. Or they might have reformulated for cost reduction purposes. Unfortunately, their new product version would have to pass through the gauntlet of the very strict consumer test, with a high chance of being rejected due to a false alarm.

A few "what if" experiments can identify the potential problems in this system. Let us assume that the incidence of a poor product that will alienate consumers occurs in 50% of the products sent over from R&D to the consumer test. Let us also assume that, through some miracle, β risk is limited to 5% (not likely, but possible in a very large test). In 1000 tests over time we have 500 bad products and 500 good ones. Given the 5% β risk, 25 bad products will slip through the net and 475 will be legitimately rejected. Unfortunately, 250, or half the good ones, will also be rejected due to the 50% false-alarm rate. You can see that we have a ratio of 250 mistaken decisions to 475 good ones, or false rejection of about one out of three product improvements. Letting A be incidence of a bad product and B be the decision to reject, the Bayes calculation looks like this:

$$\begin{aligned} p(\mathrm{A}\,|\,\mathrm{B}) &= \frac{p(\mathrm{B}\,|\,\mathrm{A})\,p(\mathrm{A})}{p(\mathrm{B}\,|\,\mathrm{A})\,p(\mathrm{A}) + p(\mathrm{B}\,|\,\text{not A})\,p(\text{not A})} \\ &= \frac{0.95(0.50)}{0.95(0.50) + 0.50(0.50)} \\ &= \frac{0.475}{0.475 + 0.25} \\ &= 0.655 \end{aligned} \tag{16.4}$$

So over one-third of the rejected products failed for no good reason. Suppose R&D does a really good job and has some sensory testing before they send the product out to consumers, so they are fairly sure it is OK. This might correspond to a p(A) of 10% (90% good product changes now). Our equation becomes

$$\begin{aligned} p(\mathrm{A}\,|\,\mathrm{B}) &= \frac{p(\mathrm{B}\,|\,\mathrm{A})\,p(\mathrm{A})}{p(\mathrm{B}\,|\,\mathrm{A})\,p(\mathrm{A}) + p(\mathrm{B}\,|\,\text{not A})\,p(\text{not A})} \\ &= \frac{0.95(0.10)}{0.95(0.10) + 0.50(0.90)} \\ &= \frac{0.095}{0.095 + 0.45} \\ &= 0.174 \end{aligned} \tag{16.5}$$

This means that about one out of six products (17.4%) was rejected for good cause, and five out of six were lost opportunities. R&D was sent back to the benchtop and their work rejected (and research costs lost) for no good reason. This could potentially be quite discouraging to researchers, unless they understood the kind of system of conservative brand maintenance they were working under.

16.1.5 Summary

These three case studies show applications of the simple Bayes equation for binomial outcomes. They can be derived intuitively from the 2×2 contingency tables, if α and β are known, and an assumption can be made about the marginal incidence rates. If an assumption cannot be made, the researcher can still insert various values (i.e., play "what if?") to see the effect. In the following sections, a somewhat more formal statement of the Bayesian approach will be used, for preference and then discrimination, first using beta distributions.

16.2 General Bayesian Models

16.2.1 When do you have Prior Information?

In Section 16.1, incidence was used as a marginal parameter to generate or estimate some prior information before the outcomes were examined. In this section, Bayesian analysis will be formally introduced. For the reader who has used only frequentist statistics with a null and alternative hypothesis, some of this terminology is new. However, if you think back to the simple examples shown in Section 16.1, the notion of using prior information makes sense.

The Bayesian approach combines the prior information (which may be more or less influential) with the current data to estimate a posterior distribution of the parameter of interest. In some respects, this is similar to a maximum likelihood approach. In maximum likelihood, we look at the data and infer what kind of distribution this result may have arisen from. That is, what is the value that would produce the highest probability from an underlying distribution that could reasonably be presumed to have generated such data? Bayesian analysis goes a step further, by using prior information to sharpen the posterior estimates, when the data and prior are considered together.

There are many cases in sensory testing where prior information exists. For example, you may have an ongoing program evaluating apple cultivars for various sensory quality factors. At the typical agricultural research station, there is a standing panel of assessors. The scale usage of such assessors may be known, perhaps in the form of a product-by-panelist interaction. Such an interaction could describe the slope of a panelist's values across a set of products, and how it may be steeper (more discriminative) or flatter (less discriminative) than the panel average. Nonyane and Theobald (2008) discussed just such a situation, and showed how Bayesian analysis can reduce the variability estimates of posterior distributions by including information from prior experiments. The reduction in error is nontrivial, with a reduction from 10% to 5% of full scale range in their examples.

Another example is in preference testing. Suppose the introduction of a new entrée into a military field ration was known to produce a consistent 55% win in a paired preference test, compared with the previous version. The win might be attributed to a novelty effect, which could dissipate or "wear off." Anything above this level could be considered practically (not statistically) significant, and Bayesian inference can be used to test the current result combined with the 55% prior expectation. An example is shown below from Carlin and Louis (1996).

16.2.2 A General Equation

Following the notation of Carlin and Louis (1996), a general equation for the posterior distribution is given as

$$p(\theta \mid y) = \frac{f(y \mid \theta)\pi(\theta)}{m(y)} \tag{16.6}$$

where θ is our theoretical proportion or test cutoff in the binomial situation, but it could be a vector in some other experimental measure, and $\pi(\theta)$ is the prior, which is assumed known or guessable, and y is our observation (the data). The denominator $m(y)$ is like summation of

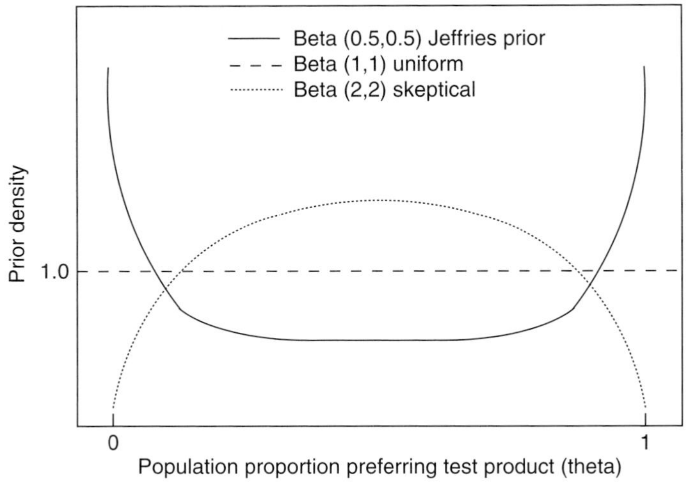

Figure 16.1 Prior distribution choices, using beta distributions, from the preference testing example of Carlin and Louis (1996).

all the possibilities, which were only two in the simple examples in Section 16.1. With multiple possibilities this becomes an integral of the marginal density of the data y as follows:

$$m(y) = \int f(y \mid \theta)\pi(\theta)\mathrm{d}\theta \tag{16.7}$$

This can be difficult to solve, although Carlin and Louis state that accurate estimation is readily available through Monte Carlo methods and they provide examples in chapter 5 of their text.

16.2.3 Choice of Prior Distribution

Now that we have a general statement of the Bayesian relationship, the next hurdle is to decide on a reasonable prior distribution. Carlin and Louis give three examples, taken from a simple taste preference test, one being the **uniform prior** (or noninformative prior) of a flat distribution (see Figure 16.1). Another is a distribution with heavy tails that could occur if you have segments of the population that either love or hate your test product, sometimes called a **Jeffreys' prior**. Finally, they give an example of a **skeptical prior**, which is a wide distribution centered on 0.5, somewhat like a null hypothesis in traditional statistics. These three distributions make very different assumptions about the situation. They are all in the family of beta distributions, with two parameters, a and b. The use of beta distributions will be discussed further in Section 16.4.

16.2.4 A Simple Example

Carlin and Louis illustrate a hypothetical meat patty taste test with 16 judges. Presumably, this would be a small internal panel used for screening purposes. In spite of the unrealistically small N, the outcome is informative. Thirteen of 16 judges prefer the test item, and the binomial probability of this under a null of $p=0.5$ is 0.022. What does the Bayesian analysis tell us? The posterior distributions can be calculated using the binomial expansion, and the

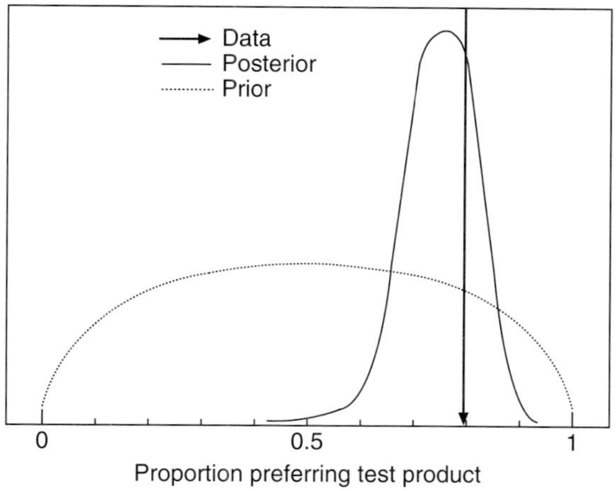

Figure 16.2 Posterior distribution based on the preference testing outcome of 13/16 choices for the new product, and the "skeptical" prior. Note that the posterior mean is drawn towards the prior, but not too far from the data value of 0.81. A larger *N* would move closer to the data (i.e., less influence of the prior) and produce a sharper distribution (lower variance).

a and *b* values from the priors. For various values of θ, our theoretical values for the proportion preferring the new product, the value for $p(x|\theta)\pi(\theta)$ is approximately equal to

$$\begin{aligned} p(\theta \mid x) &\propto \theta^{x+a-1}(1-\theta)^{N-x+b-1} \\ &\propto \text{Beta}(x+a, N-x+b) \end{aligned} \tag{16.8}$$

where Beta is shorthand for a beta distribution calculation based on *a* and *b*. In this case $N=16$ and $x=13$. Figure 16.2 shows the posterior distributions based on the skeptical prior. The probability density functions and cumulative density functions can be found in many statistical software packages.

The three distributions barely overlapped the normal null hypothesis cutoff of 0.5 at all in their lower tails, the skeptical prior being perhaps the leftmost. If we assume a more conservative cutoff of 0.6, and using a uniform prior, the posterior will give a 95% probability that θ is greater than 0.6. Carlin and Louis frame the result of $\theta>0.6$ as an example of a "substantial" win that might be required for a positive executive decision about the meat product. The modal values for all three posteriors hover in the range of 0.75 to 0.82. The modal values are the maximum likelihood estimates given these models. Perhaps a 13 to 3 win is quite obviously significant and important to most sensory professionals, but the Bayesian analysis gives us several informative ways of looking at the evidence beyond simply rejecting a null, which is rarely true anyway.

16.3 Bayesian Inference Using Beta Distributions for Preference Tests

16.3.1 A Look at Beta Distributions

Bi (2011) also suggested using a beta distribution for both prior and posterior distributions, perhaps because of their versatility, and the fact that they have only two hyperparameters that are easily related to the mean and variance of the posterior distribution. He illustrated this application using simple preference (Bi, 2003). Beta distributions can also be applied to forced-choice difference testing. Once again, the beta distribution parameters are usually noted as lowercase a and b. The beta distribution can take on many shapes, but a case of $a>b$ will generally produce what would be called negative skew in a normal distribution and positive skew when $b>a$. Various examples are given in Bi (2011: figure 1).

If we are dealing with a two-tailed situation and trying to estimate a population proportion p, the Bayesian approach will first search for a prior distribution and, once that has been specified, look at the data to obtain a posterior distribution. The prior distribution will have a mean μ given by $a/(a+b)$. Thus, if $a=b=1$, we get our familiar null hypothesis expectation of $\mu=0.5$. The variance is given as

$$\sigma^2 = \gamma\mu(1-\mu) \tag{16.9}$$

which looks similar to the binomial $p(1-p)$ except with the new parameter γ. If you are familiar with the beta binomial model, you may recall that γ describes a consistency parameter in that model. In this case, $\gamma=1/(a+b+1)$. Using the mean and variance of p, the hyperparameters can be found from

$$a = \mu\left[\frac{\mu(1-\mu)}{\sigma^2} - 1\right] \tag{16.10}$$

and

$$b = (1-\mu)\left[\frac{\mu(1-\mu)}{\sigma^2} - 1\right] \tag{16.11}$$

If one has no information about a reasonable prior distibution, Bi recommends a "non-informative, uniform prior" with $a=b=1$ (or $\mu=0.5$), and $\gamma=1/3$, and thus $\sigma^2=1/12$. Noninformative and uniform seems a little redundant to me. If there are good estimates of the mean and variance from prior knowledge, you can calculate the parameters. Bi (2011) provided R and S-plus codes for estimation of a, b, μ, and γ. However, a numerical estimation from the equations above is not difficult.

16.3.2 Getting Down to Brass Tacks: The Posterior Distribution

The more valuable information, however, comes from the posterior distribution. The posterior distribution for our estimate of the population proportion p is also a beta distribution, but with new hyperparameters. Given a sample size n and an observed number of preference wins x, the new hyperparameters are $a^*=a+x$ and $b^*=n-x+b$. The more familiar mean and variance of the distribution of p are then given as follows:

$$\mu^* = \frac{x+a}{n+a+b} \tag{16.12}$$

and

$$\sigma^{2*} = \gamma^*\mu^*(1-\mu^*) \tag{16.13}$$

and

$$\gamma^* = \frac{1}{n+a+b+1} \tag{16.14}$$

One can see that as n becomes large, the variance will shrink as a function of γ^*, as it should.

At this point the posterior distribution is known, and some decisions can be made. First, we know the mean, which is an estimate of p. The mode can also be used; that is, the value of p with the highest density m, where $m=(a^*-1)/(a^*+b^*-2)$. In the case of a preference win, one might have a cutoff value or some criterion, and a positive result (and subsequent product development action) would be obtained if 95% of the posterior distribution lies above the cutoff, for example.

16.3.3 Making Decisions: Credible Intervals and Bayes' Factors

Two other guidelines for decision-making are commonly used in Bayesian inference. These are (1) the use of **credible intervals** and (2) the use of a **Bayes factor** to estimate the relative value of two competing models. The credible interval is a function of the posterior distribution. Remember that this could be a beta distribution, so it may not be symmetric. One approach to determining a credible interval is to make the tails equal in area (probability). Another is to use the highest posterior density interval, so that every point included has a higher probability density than any point excluded. The areas of the two tails need not be the same for the latter approach, which basically finds some height of the curve so that the sum of the two tails equals one minus the credible interval area. This can allow a more formal statement for decision-making. In the case of a preference test with an assumption of 0.5 as a population proportion suggesting no significant preference, a positive decision could be reached if the credible interval does not contain 0.5.

For equivalence testing, the interpretation is straightforward. Given an equivalence zone of $0.5 \pm \Delta$, equivalence can be concluded if the posterior probability that p lies in the interval equals $1-\alpha$. This is appealing as an equivalence test, as it does not depend upon failure to reject a null or any power calculations. Another approach is to perform two one-sided tests (the "TOST" approach). In this method, if both the posterior probabilities that p lies in the interval between zero and $0.5+\Delta$, and in the interval between $0.5-\Delta$ and one are larger than $1-\alpha$, then the hypothesis of nonequivalence can be rejected in favor of the hypothesis of equivalence or similarity (Welleck, 2003; Bi, 2011). Note that the latter approach seems to mix the frequentist notion of a rejectable null with the use of a posterior distribution.

Another way to make decisions using the posterior distribution is to calculate different prior and posterior probabilities for two competing models. Such models, of course, must be tied to some managerial action, such as the adoption of a new improved product based on a prototype test, versus rejection of the prototype. The Bayes factor is a ratio of two odds ratios, which are sometimes complementary (e.g., p versus $1-p$). If we have two competing models, M_1 and M_2, the denominator is the odds ratio of the posterior probability for model M_1, given the data, to the posterior probability of model M_2, given the data. The numerator is the ratio of the prior probabilities for the two models. You can think of this as a test of the information gain, given the data and the relative fit of the two models. In other words, the Bayes factor B is given by

$$B = \frac{P(\mathrm{M}_1 \mid \mathrm{data}) / P(\mathrm{M}_2 \mid \mathrm{data})}{P(\mathrm{M}_1) / P(\mathrm{M}_2)} \tag{16.15}$$

Bayesian theorists have constructed rules of thumb for the weight of evidence based upon values for twice the natural log of B. If $2(\ln B)$ is between 2.2 and 6 (i.e., B is in the range of

3 to 20), that is considered positive evidence in favor of M_1. Values larger than 6 are considered even stronger evidence, and so on (Kass and Raftery, 1995). Likelihood ratio tests are common in statistical decision-making.

16.4 Proportions of Discriminators

One of the first uses of Bayesian reasoning in sensory science was in a paper published by Carbonell et al. (2005). Until that time, most researchers had used Abbott's formula or a transformation of it to figure the proportion of true discriminators given a certain outcome in a forced choice test (see Chapter 3). Simply stated, the proportion is given by

$$P_d = \frac{P_{observed} - P_{chance}}{1 - P_{chance}} \tag{16.16}$$

Note that this model separates assessors into two piles, one group with zero discrimination ability in this particular test and one group with perfect discrimination. It does not consider the probability of discrimination as a continuous variable, as might be done with replicated experiments.

Given x discriminators in a test with N assessors, the probability of y correct responses is given by a simple binomial expression as follows:

$$p(y \mid x) = \binom{N-x}{y-x} \left(P_{chance}\right)^{y-x} \left(1 - P_{chance}\right)^{N-y} \tag{16.17}$$

What we would like to know is the most likely number (or proportion, x/N) of discriminators if we have seen a certain outcome of y correct choices. So this is an ideal situation to apply Bayes' rule to find the probability of $x \mid y$. Taking Bayes' rule in this notation as

$$p(x \mid y) = \frac{p(x)\,p(y \mid x)}{p(y)} \tag{16.18}$$

and

$$p(y) = \sum_{x=0}^{y} p(x)\,p(y \mid x) \tag{16.19}$$

Now using eqn 16.17, and assuming we have a triangle test as in Carbonell et al. (2007), the above three equations combined give us a solution:

$$p(x \mid y) = \frac{p(x) \cdot \binom{n-x}{y-x} \left(\frac{1}{3}\right)^{y-x} \left(\frac{2}{3}\right)^{n-y}}{\sum_{u=0}^{y} p(x) \cdot \binom{n-u}{y-u} \left(\frac{1}{3}\right)^{y-u} \left(\frac{2}{3}\right)^{n-y}} \tag{16.20}$$

Of course, we need some assumption about $p(x)$ (and $p(u)$) in order to carry out this calculation. Assuming equal probabilities for all values of x is like assuming a uniform prior. This simplifies eqn 16.20 to

$$p(x \mid y) = \frac{\binom{n-x}{y-x}\left(\frac{1}{3}\right)^{y-x}\left(\frac{2}{3}\right)^{n-y}}{\sum_{u=0}^{y}\binom{n-u}{y-u}\left(\frac{1}{3}\right)^{y-u}\left(\frac{2}{3}\right)^{n-y}} \tag{16.21}$$

For a hypothetical triangle test with two assessors and one correct answer, the probability then of one discriminator is

$$p(x=1 \mid y=1) = \frac{\frac{2}{3}}{\frac{4}{9}+\frac{2}{3}} = \frac{6}{10}$$

which seems reasonable. Note that x, the number of discriminators, cannot exceed y, the number of correct answers.

Carbonell et al. go on to show the densities and cumulative distributions in a table, for a triangle test with 100 panelists and 42 correct answers. The distribution of x, of course, ranges from 0 to 42. The highest values (most likely) for $p(x \mid y)$ occur at $x=13$ and $x=14$, and they are equal to the fourth decimal place. This is a bit different than the Abbott's formula solution, which states that $P_d = (42/100 - 1/3)/(2/3)$ or 0.13. So with $N=100$ and $y=42$ we expect 13 discriminators, not 13 or 14. Given that expression 16.21 is easily evaluated in many statistical packages or easily programmable in a spreadsheet, this approach would seem to have wide application. The advantage over using a simple Abbott's formula solution is that the posterior probabilities can be used to get credible intervals for the range within P_d should like with given tail probability cutoffs. This is more informative and better justifies decisions than having a fixed criterion for P_d such as "if we get 15% discriminators we will not make this product change." Of course, all the normal concerns about discriminator theory apply here, as discussed in Chapter 6 on equivalence testing.

16.5 Modeling Forced-Choice Discrimination Tests

Bayesian methods can also be applied to provide posterior distributions for simple forced-choice discrimination tests. However, a complication arises in that there is a floor restriction due to the chance probability level. This is the same issue faced by beta binomial models, and thus brought the consideration of chance-corrected beta binomials for replicated discrimination. The assumption is that the population proportion correct is limited by the chance probability value as a lower limit. In statistical parlance this is sometimes referred to as an "independent background effect" (Bi, 2007). That is an appealing terminology, as it applies generally to many experimental situations. For example, in toxicology, a certain proportion of the control group will die. In pharmacology, a certain proportion of the placebo group will improve or be cured. These are part of the historical basis for Abbott's formula. Many fields must take background effects into account in determining the true size or significance of any experimental outcome.

Note that, for any single experimental trial, it is quite possible to have performance below chance. The same can be said for any set of replications from a single observer, if perhaps that judge is looking for the wrong or irrelevant attribute. It can even occur in an entire data set if, for example, a mistake was made in the test kitchen or prep area. Like it or not, such events do occur, and any statistical model should allow for below-chance performance when the actual data are considered. This is not so for the theoretical population proportion. A problem is that the beta distributions considered so far in this chapter do not have any such restriction, and can vary between proportions of zero and one. That is certainly acceptable for a two-tailed preference test, but not the one-tailed situation in discrimination methods such as the triangle test. So a way must be found to limit or truncate the posterior beta distributions at or near the chance probability level.

Several publications by Bi (2007, 2011) have addressed this issue. The observed proportion correct p_c is a function of the chance probability level, here denoted as C to follow Bi's notation, and the proportion of discriminators p. So $p_c = p + C(1-p)$, as in a rearrangement of Abbott's formula. Assuming that the prior distribution of p is a beta distribution with parameters $a, b > 0$, the density function for p_c as a function of x correct choices out of N total is similar to a binomial probability, with a few tweaks. According to Bi (2007), the function follows this relationship:

$$P(p_c \mid x, n, a, b, C) = \frac{1}{(1-C)^{n+b+a+1} W} p_c^x (1-p_c)^{n-x+b-1} (p_c - C)^{a-1} \tag{16.22}$$

So we have our familiar expansion of $p^x(1-p)^{n-x}$ with a few adjustments, including the additional term $(p_c - C)^{a-1}$ to help correct for chance and a weighting term W given by

$$W = \sum_{i=0}^{x} \binom{x}{i} \left(\frac{C}{1-C}\right)^{x-i} B(a+i, n+b-x) \tag{16.23}$$

And $B(a',b')$ is a beta distribution, in this case evaluated at each parameter $a' = a + i$ and $b' = n + b - x$. Bi's papers also give the cumulative probability density function, which is more useful for testing various cutoffs and credible intervals. He also provides S-plus and R routines to evaluate the beta distributions.

The result of this adjustment for chance is a new beta-type distribution, some of which are truncated at the chance probability level, and thus initially higher than the conventional beta distribution at levels slightly above chance. The combined truncation and elevation effect is more pronounced when N is small (e.g., only 10 judges), C is high (e.g., 1/2 rather than 1/3), and x is smaller (x/N is closer to chance). So the advantage of this adjustment is most valuable with a small panel, and has little or no effect when the panel size approaches 100, and/or as the observed proportion moves up away from the chance proportion. It is also a more important adjustment when the forced-choice test has a higher chance probability value (1/2 instead of 1/3).

Model testing can of course be done using Bayes' factors as measures of likelihood (and log likelihood). For example, if we wished to model a certain proportion of discriminators, one need only compute the value of the observed proportion correct that would correspond p_c, and plug p_c and $1-p_c$ into the posterior distribution calculations to get the posterior odds ratio (and of course divide by the prior odds ratio). The prior odds for the model should be the maximum tolerable proportion of discriminators, according to Bi's example; for example, 0.2 for 20% discriminators (prior odds then are 0.2/0.8). Given a certain observed value, then, for x given N subjects, the cumulative distribution value for the posterior can be constructed. Bi also provided a calculation for experimental sample size N in the 2007 paper.

Once again, if concerns are raised about the validity of proportions of discriminators as a yardstick, then the Thurstonian δ values can be used instead, by simply converting those values into the appropriate level for the various models' p_c values and thus obtain the prior probabilities. An example is shown in Bi (2011). This model testing approach can also be used to determine equivalence, once an upper limit for the value of empirical equivalence is defined.

16.6 Replicated Discrimination Tests

One of the shortcomings of the discriminator/nondiscriminatory dichotomy is that it assigns a success rate of either zero or one to assessors. This may be a reasonable starting point to describe a momentary experience in a single test. However, sensory specialists recognize that their panelists are neither infallible nor totally worthless, but lie somewhere in between. If we have replicated tests, the presence of the middle ground becomes apparent, and estimable. Such reasoning led to a flurry of papers devoted to replicated discrimination tests originating from the sensometrics community.

Shortly following Carbonell et al.'s seminal paper, other theorists attempted to extend the Bayesian modeling to replicated discrimination situations. Meyners and Duineveld (2008) employed an approach similar to Carbonell et al.'s, but extended it to predict the discrimination rates for an individual, followed by a combination of the individual distributions to get a population-wide empirical distribution. Insofar as the model includes two separate and sequential steps, it can be considered a hierarchical model. For each panelist i there are k_i replications (sometimes an equal number, but sometimes not) and x_i correct answers from which we must derive a probability distribution for y_i true discriminations, and y_i must not exceed x_i. This leads to an equation very similar to 16.21:

$$p(y_i \mid x_i, k_i) = \frac{\binom{k_i - y_i}{x_i - y_i} p_{\text{chance}}^{x_i - y_i} \left(1 - p_{\text{chance}}\right)^{k_i - x_i}}{\sum_{u=0}^{k_i - u} \binom{k_i - u}{x_i - u} p_{\text{chance}}^{x_i - u} \left(1 - p_{\text{chance}}\right)^{k_i - x_i}} \tag{16.24}$$

where p_{chance} is the chance probability, such as 1/3 in the triangle test. Dividing each y_i by k_i for each judge then gives the distribution of discrimination probabilities. In the second step, these distributions are averaged to obtain the population distribution. For an equal number of replications, the simple mean will do, and a weighted mean for different numbers of replications.

Meyners and Duineveld then gave several examples. In the first one, they took some data from Hunter et al. (2000) based on a triangle test with 23 judges and 12 replications each. So the probability distribution for the estimated discrimination densities will range from 0 to 12. The posterior distribution obtained provided several important pieces of information. There was a large chance (68%) for 1–6 discriminations out of 12. Some of the judges were quite discriminating, with the 90th percentile occurring at about 2/3, meaning 10% of the population might very well discriminate on eight or more trials. They then compared their posterior distribution with a beta binomial model. The beta binomial model overestimated low numbers of discriminations and underestimated the potential for highly discriminating judges. They commented that this was due to the fact that no beta distribution exists that can fit their posterior. This is one of the few published examples of where a beta distribution, which is often considered a very versatile choice, might fall short.

The hierarchical model was further explored in a later paper by Duineveld and Meyners (2008) which included a review of the various approaches taken up to that point. Owing to the computational complexity and difficulty in finding a mathematical solution, they employed a Monte Carlo Markov chain sampler. The interested reader is referred to that paper for further details.

A further addition to Bayesian modeling occurred with a paper by Bayarri et al. (2008). Their advance was to consider the probability of success or failure (to choose the correct sample) as informative to the next trial. Given their model, the marginal probability of a success following a success is higher than a success following a failure, and vice versa. In this model, all posterior distributions for any given trial, except the first, are also prior distributions for the following trial. A positive aspect of this sequential modeling is that it allows prediction of a panelist's success probability on the next (even hypothetical) trial.

What sort of criteria should be applied to a good Bayesian model or analysis? A number of criteria are suggested by the papers reviewed here, at least for replicated discrimination:

- The model should accommodate different rates of discrimination among assessors (i.e., panelist heterogeneity).
- The model should use information gathered from previous replicates to form new informative priors.
- The model should not consider replicated judgments from a single assessor as independent events, but if possible should use the sequential information to guide the model development.

16.7 Bayesian Networks

16.7.1 Combining Variables in Causal Chains

The utility of Bayesian analysis is further enhanced by the construction of networks of interacting variables in probabilistic causal chains, called **Bayesian networks**. The ability to do both forward reasoning (from cause to effect) and backward reasoning (from effect to cause) is a potentially powerful tool. One obvious application is in medical diagnosis. Given a set of symptoms and lab results, what disease is the patient most likely to have that caused this set of outcomes? In food research, we can apply these techniques to questions of food choice, liking, purchase intent, or any number of behavioral variables. A causally linked set of independent and dependent variables can be used to construct an informative network. Phan and colleagues (from Wageningen and Unilever) are strong proponents of this approach and have produced various publications illustrating their functions and applications. A good starting point is the chapter by Phan et al. (2010).

16.7.2 An Example of a Bayesian Network Model

The example illustrated here is taken from Phan et al. (2010). The article itself should be consulted for definitions and further clarification. The hypothetical system they set up was with two independent variables – snack type (sweet or salty) and eating situation (alone or with friends) for 200 imaginary teenagers – and three dependent variables – liking, intake and purchase intent. A sample of 20 observations was made by the Bayesian network program HUGIN, and the independent variables were initially set at 0.5, 0.5 for

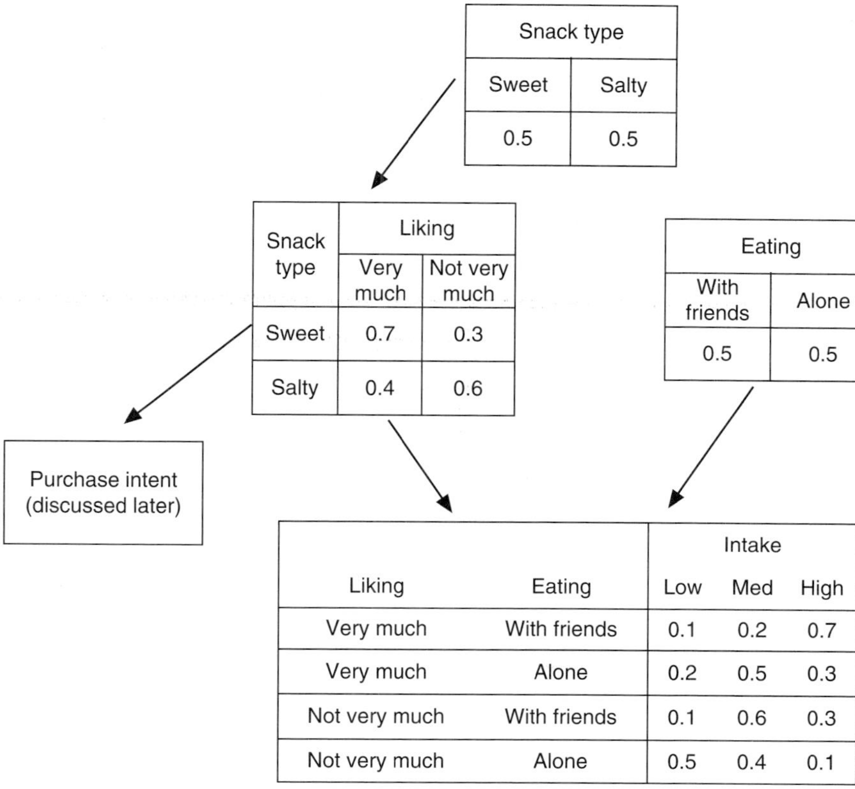

Figure 16.3 An example of a Bayesian network, from Phan et al. (2011).

each option (sweet versus salty and alone versus with friends). Sampling the data set gave the probabilities shown in Figure 16.3 with the causal relationships indicated by arrows. The cause (start of the arrow) will be referred to as a parent node, and the effect as a child node.

Working down the causal chain is straightforward. However, Bayes' theorem allows us also to reason backwards. For example, knowing something about the consumption amounts, we can ask whether the teens were more likely to be eating a salty versus a sweet snack, or whether they were more or less likely to be eating alone or with friends. The network connections and probability calculations allow us to play the game of "what if" by changing values and seeing the effects (or lack thereof) on other variables in the system.

The data from the dependent variables was originally nearly continuous, being a 100-point scale for liking and intake measured in grams. However, for the purposes of this example, the data have been discretized or dichotomized into mutually exclusive categories. This is a common practice with many of the Bayesian software programs, and also a potential shortcoming, as information is lost when the data are so bluntly categorized.

16.7.3 Using Marginal and Joint Probabilities

Give the input probabilities (based on the hypothetical data) one can proceed to calculate both marginal and joint probabilities. For the variable of liking as a function of snack type,

Table 16.6 Marginal and joint probabilities for the variables of liking and snack type

	Liking		
Snack type	**Very much**	**Not very much**	**Marginal**
Sweet	0.35	0.15	0.5
Salty	0.20	0.30	0.5
Marginal	0.55	0.45	1.0

these are shown in Table 16.6. Of course, the variable of snack type was set at 0.5, 0.5 by the experimental design, but the liking probabilities are still free to be influenced.

The marginal probabilities are the probabilities for the states of a given variable, when the states of the other variables are unknown. These become useful, however, in calculations of how the other probabilities could change, if one of the variables is fixed or known. Note that if we fix one of the upstream variables, the child nodes will change value. For example, if we fix the variable of snack type as sweet, the probability for liking very much now goes to 0.7. These kinds of changes can be useful in the game of "what if" and the Bayesian network programs will calculate them (although in a system this simple it is easy to do by hand).

These become even more useful when we fix a downstream variable and ask about the probable state of the parent node; in other words, reasoning from effect to cause. This is a primary goal of a Bayesian network system. The laws of joint probability state the following:

$$\begin{aligned} p(\mathrm{A,B}) &= p(\mathrm{A \mid B})p(\mathrm{B}) \\ &= p(\mathrm{B \mid A})\, p(\mathrm{A}) \end{aligned} \tag{16.25}$$

where $p(\mathrm{A, B})$ is the joint probability of both A and B occurring as indicated in the interior cells of the conditional probability tables. Rearranging, we can find the conditional probability of some state A given B by dividing the joint probability by the marginal probability:

$$p(\mathrm{A \mid B}) = \frac{p(\mathrm{A,B})}{p(\mathrm{B})} \tag{16.26}$$

So in the example, if we knew that the snack was liked very much, we could ask what the probability would be that the teen is eating a sweet snack. So the conditional probability for this occurrence is

$$\begin{aligned} & p(\text{snack type} = \text{sweet} \mid \text{liking} = \text{very much}) \\ &= \frac{p(\text{snack type} = \text{sweet} \mid \text{liking} = \text{very much})}{p(\text{liking} = \text{very much})} \\ &= \frac{0.35}{0.55} = 0.6364 \end{aligned} \tag{16.27}$$

In other words, if we have fixed the outcome as "like very much," the chance that our teen is eating a sweet snack is about 64%, or an odds of about two to one.

16.7.4 Bayesian Structure and Connections

Three types of connections can be found in the Bayesian networks. Given three nodes (X, Y, and Z) they can be serial ($X \rightarrow Y \rightarrow Z$) meaning "$X$ influences Y which influences Z," or diverging ($X \leftarrow Y \rightarrow Z$) or converging ($X \rightarrow Y \leftarrow Z$). If we add another node to the network for purchase intent, it would be affected by liking, so there is a serial connection from snack type to liking to purchase intent, a serial connection from snack type to liking to intake, and a converging connection on intake from liking and eating situation.

These connections have interesting properties (Phan et al., 2010). Changes in eating situation do not affect purchase intent, liking, and snack type, if only the marginal probabilities are known for the converging node of intake. If, however, the value of the convergent node is known (let us say intake = 100% medium), then changing the eating situation will now affect the other three. In Bayesian network terms, eating situation is conditionally dependent to the other three when values of intake are provided.

In the serial chain from snack type to liking to purchase intent, a known value of purchase intent can determine new values or changed values for the other two. This is another example of reasoning from effect to cause. If, however, the central node is fixed (liking = 100% very much, for example), then changes in purchase intent have no effect on the probabilities of snack type. This might be called a kind of blocking effect. In probability theory it is called conditional independence. Intake will also be affected by changes in a known value of purchase intent, so this effect-to-cause influence works through diverging connections as well. The blocking effect of a known or fixed value of liking will also occur for intake; that is, it remains unchanged if purchase intent is changed but liking is known (see Phan et al. (2010: figure 17.11) for examples). So diverging connections show the blocking effect as well as serial connections. Obviously, knowing the dependence, independence, or conditional dependence of the nodes in a network has important consequences on predictions.

16.8 Conclusions

The advantages and applicability of Bayesian models to sensory data and to product decisions seems readily apparent. Gradual improvements in modeling have occurred and no doubt will continue. The sensometrics community is interested in these types of models, and they find applicability in discrimination and preference testing, replicated experiments, shelf-life estimation (Calle et al., 2006), and estimation of panelist effects in ongoing work with standing laboratory panels (Nonyane & Theobald, 2008). Sensory practitioners will be quick to point out that many of the discrimination testing examples include large numbers of replicates, which are quite rare in industrial food testing situations, where zero, one, or two replications would be most common.

Applications to preference testing and the assessment of consumer consistency are obvious areas for using Bayesian modeling. Of note is the recent research by Dubnicka (2013), who addressed simple preference testing, replicated forced-choice preference, and non-forced preference. A noteworthy aspect of this approach is that it allows testing a variety of hypotheses of interest. Examples included (1) whether the probability of a consumer switching from product A to product B is greater than the reverse, (2) whether the probability of consistent choice is greater than the probability of switching, and (3) what the result is of an overall comparison of consistent choice of product A over B. Some of these had previously been addressed by McNemar tests. However, the frequentist McNemar does not provide the depth of information present in a Bayesian posterior distribution.

References

Bayarri, I., Carbonell, L., and Tarrega, A. 2008. Replicated triangle and duo-trio tests: discrimination capacity of assessors evaluated by Bayes' rule. Food Quality and Preference, 19, 519–532.

Bi, J. 2003. Difficulties and a way out: a Bayesian approach for sensory difference and preference tests. Journal of Sensory Studies, 21, 584–600.

Bi, J. 2007. Bayesian analysis for proportion with an independent background effect. British Journal of Mathematical and Statistical Psychology, 60, 71–83.

Bi, J. 2011. Bayesian approach to sensory preference, difference and equivalence tests. Journal of Sensory Studies, 26, 383–399.

Calle, M.L., Hough, G., Curia, A., and Gomez, G. 2006. Bayesian survival analysis modeling applied to sensory shelf life of foods. Food Quality and Preference, 17, 307–312.

Carbonell, L., Carbonell, I., and Izquierdo, L. 2007. Triangle tests: number of discriminators estimated by Bayes' rule. Food Quality and Preference, 16, 117–120.

Carlin, B.P. and Louis, T.A. 1996. Bayes and Empirical Bayes Methods for Data Analysis. Chapman and Hall/CRC Press, Boca Raton, FL.

Dubnicka, S.R. 2013. A Bayesian approach to analyzing replicated preference tests. Journal of Sensory Studies, in press, doi:10.1111/joss.12033.

Duineveld, K. and Meyners, M. 2008. Heirarchical Bayesian analysis of true discrimination rates in replicated triangle tests. Food Quality and Preference, 19, 292–305.

Hunter, E.A., Piggott, J.R., and Lee, K.Y.M. 2000. Analysis of discrimination tests. In Société Française de Statistique (ed): Actes des 6émes. Journées Européennes Agro-Industrie et Méthodes Statistiques, Pau, January 19–21.

Kass, R.E. and Raftery, A.E. 1995. Bayes factors. Journal of the American Statistical Association, 90, 773–795.

Meyners, M. and Duineveld, K. 2008. Approximating the distribution of discriminating rates in replicated difference test using Bayes' rule. Food Quality and Preference, 19, 135–138.

Nonyane, B.A.S. and Theobald, C.M. 2008. Multiplicative models for combining information from several sensory experiments: a Bayesian analysis. Food Quality and Preference, 19, 260–266.

Phan, V.A., van Boeckel, M.A.J.S., Dekker, M., and Garczarek, U. 2010. Bayesian networks for food science, theoretical background and potential applications. In: Consumer-driven Innovation in Food and Personal Care Products. S.R. Jaeger and H. MacFie (Eds). Woodhead Publishing, Oxford, pp. 488–513.

Welleck, S. 2003. Testing Statistical Hypothesis of Equivalence. Chapman and Hall/CRC Press, Boca Raton, FL.

Appendix A: Overview of Sensory Evaluation

Successful sensory testing is driven by setting clear objectives, developing robust experimental strategy, applying appropriate statistical techniques, adhering to good ethical practice and successfully delivering actionable insights that are used to inform decision-making.

Kemp et al. (2009)

A.1 Introduction

Sensory evaluation is a child of the industries that manufacture beverages, foods, and consumer products. The techniques have evolved since the mid 1990s, and many of them still are practiced in the forms in which they were first published. The goal of these kinds of tests was to get an insight into human perception of the products that could be used to guide management decisions. Often, these decisions concerned the development and introduction of a new processed food to the consumer market. The tests were also designed for purposes of quality control and stability (shelf-life) testing. The main idea was to provide information that would lower the risk in making decisions about a product, such as whether to sell it. The specific questions involved were: (1) Was it was the same as or different than an existing product? (2) What were the perceived characteristics of this product? (3) Would consumers like it? To answer these questions, the methods of discrimination testing, descriptive

Quantitative Sensory Analysis: Psychophysics, Models and Intelligent Design, First Edition. Harry T. Lawless.

analysis, and affective testing were developed. They will be discussed briefly in the sections that follow. More detailed information can be found in textbooks on the subject, including Lawless and Heymann (2010), Stone et al. (2004), Meilgaard et al. (2006), and the shorter works by Chambers and Wolf (1996) and Kemp et al. (2009).

I have assumed that most readers of this work will be familiar with the basic types of sensory tests, how they are conducted, and how the data are analyzed. However, not everyone who picks up this book will have previous experience or formal studies of the subject matter, and so a short introductory summary is given in this appendix in order to provide a background for such individuals. It is also assumed that the reader is familiar with basic statistical terminology, such as the null hypothesis, mean, standard deviation, and correlation. If not, the appendices to Lawless and Heymann (2010) give a useful introduction focused on statistical applications for sensory evaluation.

A.1.1 The Central Dogma

The central dogma of sensory evaluation is that the test method must be designed to match the objectives of the test. If one wishes to know whether or not there is a perceivable difference between two products, then a discrimination or simple difference test is needed. If consumer acceptability or preference is unknown, then a consumer test is required. The metaphor is often used that the methods are part of a "sensory toolbox." You would not use a hammer when a screwdriver is needed. So the tool is fit to the problem at hand.

As a corollary to this central concept, certain types of assessors must be used in each type of test. For a descriptive analysis panel, persons must have been screened to make sure they have normal sensory acuity, proper motivation, and no health issues that would contraindicate their consumption of the products. However, they do not need to be regular purchasers of this product in question. They are merely functioning as analytical instruments; sensory meters of sorts. The opposite is true for consumer tests. They must be regular and perhaps frequent users of the product or the product category being tested. However, they are more or less randomly sampled from the population of such users, and hence their sensory acuity is not a criterion for participation. A third principle is that we do not ask the wrong questions of the wrong panel. If we have a trained analytical descriptive panel, we do not ask them whether they like the product. That is not their job. Conversely, we are very careful about asking consumers for analytical diagnostic evaluations of specific sensory properties. They have not been trained to use a common vocabulary, and so asking questions about an attribute like astringency (which often must be taught to a panel with examples) would not make sense.

A.1.2 Blind Testing and Independent Judgments

Two guiding principles are as important as the central dogma. The first is blind testing. That is, the tester must not be aware of the identity of the products in the test, nor which one is a new test product versus a control sample. They are given only enough information to make a judgment in the proper frame of reference. We do not tell them the product concept. An example would be this: "We have a new, whole wheat, high fiber, low fat microwavable frozen pizza that your kids can safely prepare for themselves after school." That kind of concept testing is the province of marketing research. Instead, the product comes to the taste testing booth with the designation "Pizza # 357." Note that a random three-digit blinding code is used to label the sample. The second guiding principle is that of independent judgments. We do not want the assessors to discuss the properties and come to a group decision.

It is their individual opinions and data that matter. The statistical assumption is that the observations are independent. So their judgments must not be influenced by other persons in the test.

A.1.3 Facilities and Controls

In order to conform to the principles of blind testing and independent judgments, sensory testing in industry has followed some general practices. Independent judgment is facilitated by having assessors separated. This is usually achieved by seating them in a private booth, with barriers between booths to prevent interaction with other persons. Sensory evaluation booths are often connected via a pass-through window or other opening to a test kitchen or preparation area. The opening usually has a door that remains closed most of the time, so that the tester cannot see the prep area or get an unwanted hint about the identity of the products. The identity of products is further obscured by labeling the samples with random three-digit codes.

Samples are prepared in a uniform manner, and should have the same volume (size) and serving temperature. They should only differ in the variable under investigation. In other words, there should be no unwanted systematic or random variation that would add error variability to the situation or cause perception of a spurious or irrelevant difference. Of course, many products such as a snack chips have normal variation and this is to be expected. But it would be very bad practice to have the test product of a different size than the control product, unless size was in fact the variable under study. A laboratory manual with standard procedures should be part of any ongoing testing program, so that the sample preparation is standardized and repeatable.

A.2 Discrimination and Simple Difference Tests

A.2.1 Objectives

The primary goal of a discrimination test is to provide scientific evidence that two products are perceptually different. This can be rephrased as "there are at least some people in the population that can tell the difference." An important part of the testing is that it must be objective. Differentiation is not a matter of opinion, but a behavioral demonstration of the ability to discriminate is required. For this reason, most of these tests take the form of a multiple choice test, with a known chance probability level of getting the correct answer by guessing. Performance above these levels is considered evidence for a difference. When the level of performance is high enough, and the number of judges is sufficient, then a statistical test can be performed to show that the observed proportion of correct answers would only occur 5% of the time or less, under a true null hypothesis. In the case of a multiple choice format, the null states that the proportion correct in the general population equals the chance probability level. Note that this is a mathematical equality, and not a verbal statement like "there is no difference."

Discrimination tests can also be used to amass evidence that two products are equal or have some acceptable degree of sensory similarity. This is a trickier decision, and is more fully discussed in Chapter 6 on equivalence testing. A simple failure to find any statistically significant difference is weak evidence and often ambiguous. A failure to reject the null can happen for lots of reasons, including (1) you did a sloppy test that introduced sources of unwanted variability, (2) you did not test enough judges to provide a powerful and sensitive test, and (3) there really is no difference.

A.2.2 Participants

The most common scenario for a discrimination testing panel in a major food company is a reservoir of employees, often in a research setting. A subset of the employee panel can be called for any given test, until the necessary quota is reached. The panel size ranges from about 40 to 80, with the larger panels being used for equivalency testing. The panelists need not be users of the product, but they must have no objections to consuming the kind of food in that day's test. Often, they are screened for basic sensory acuity, and for any health reasons or dietary restrictions (religious, medical, or otherwise) that would prohibit them from eating that product.

In another scenario, consumers of that product type are recruited for the discrimination test. They are generally not screened for basic sensory function. Thus, this kind of panel is generally less sensitive than a screened employee panel, who may become quite discriminating due to the years of practice they may get taking taste tests. A more discriminating group of testers can provide better insurance against missing a difference (type II error) but can result in some differences being detected that most consumers might not see. A consumer panel, on the other hand, is more predictive of the discrimination abilities of the general population. The results have a greater chance of type II error, but a lower risk of finding a spurious difference; that is, a false alarm or type I error. The risks associated with each of these errors needs to be considered in setting up the type of panel the client needs. The employee panel is often thought of as a "safety net." That is, if a screened and discriminating panel cannot tell the difference between the products under controlled conditions, a consumer panel in the real world is unlikely to see the difference either. This logic seems good, but it is not airtight, because every test has some probability of error.

A.2.3 Methodologies

There are four main categories of methods used for simple difference tests. They are sorting, matching, forced-choice or attribute-specific, and response choice tests. The first three require the panelist to choose a specific product from a group of blind-coded items. The response choice tests involve choice of a response option, rather than pointing to a product. These have different statistical analyses and require a control product or control pair so as to provide a baseline of response to control for response bias. Each of these categories will be described next, and details of methods and analysis can be found in Lawless and Heymann (2010: chapter 4).

The most common sorting test is the triangle procedure, which has a long history in sensory testing. Three products are presented with random blinding codes, and two of them are duplicate items. The task is usually phrased as "choose the one item that is most different from the other two." A sample ballot for a triangle test is shown in Figure A.1. This task is also called an oddity task, for obvious reasons (choose the odd sample). The chance probability is one-third. Assuming there are two products in the test (e.g., a control product and a test item), the identity of the duplicate item is counterbalanced so that half the panelists get the control as a duplicate and half get the test product as the duplicate. Products will also be presented in different random orders to each person, or in

Welcome
Today's test is

FUNISTRADA

Taste the samples from left to right as indicated on this sheet.
You (may) (may not) go back and re-taste the samples.

Please eat a bite of cracker and rinse your mouth with water before tasting each sample.

CIRCLE THE NUMBER OF THE SAMPLE THAT IS MOST DIFFERENT FROM THE OTHER TWO.

837 456 925

Figure A.1 A typical ballot used in a triangle test.

different positions on a tray. A second kind of sorting test is the tetrad. The tetrad test has four items, with two pairs of duplicates. The job of the panelist is to sort them into two piles on the basis of similarity, so the duplicates are sorted together correctly. This also has a chance probability of one-third. These are good tests for overall difference, when the nature of the difference is unknown before the test, or is likely to be a complex set of changes across several attributes. Recent modeling suggests the tetrad test is generally more sensitive than the triangle (see Chapter 5), but it does not have the track record of decades of use, and obviously involves tasting one extra sample. In both of these tests a response is forced. That is, the panelist must respond by guessing if unsure. They are not allowed to say, "I can't tell" or "I have no response."

Matching tests are also quite popular for testing for overall difference. The duo–trio test is the oldest. In this procedure, a reference sample is presented to the tester, sometimes following a warm-up sample. After the person has a chance to inspect the reference item, two test items are presented, one of which matches the reference, and the panelist is to choose the item that is most similar to the reference. A second kind of test, the ABX test, is common in psychophysics and is the reverse of the duo–trio. Both test and control products are presented as references or whatever two versions of the product are sent to the sensory lab for testing. After the inspection period, one of the two items is presented with a blind code to each panelist, and he or she must match the third item to the correct reference. The identity of the third item, of course, is varied across panelists so the test product and control product appear an equal number of times. In the dual standard test, once again both items are presented as reference samples, and then both items are presented with blind codes, to be matched to the correct reference. In all three of these tests, choosing the correct match has a chance probability of one-half.

The third major category of difference tests is called forced choice. One of the test items is considered to be a priori higher in some attribute and the others are duplicated versions of the less intense or baseline products. The panelist is instructed to "choose the product that is highest in attribute X." If there is one test item and two control or baseline products, the task is called a three-alternative forced choice test or 3-AFC. If there is one test item and one baseline, it is called a 2-AFC or paired comparison test for differences. More baseline products can be added, resulting in 4-AFC and so on. This changes the chance probability level and increases the difficulty of the test. Because the sorting and matching tests also involve a forced choice, this terminology is somewhat unfortunate. For this reason, some sensory scientists prefer the term "attribute-specific" tests. The AFC tests are generally considered to be more sensitive than the nonspecific tests (see Chapter 4). However, the attribute that is changing must be known, and it must be a sensory property that the panelists understand and can easily recognize (such as sweet taste).

The fourth category involves a choice of response, rather than pointing to a product in a test set. These are the A–not-A test and the same–different test. Assume once again that there is a control product and a test product. If we designate the control product as "A" and the test product as "B," the requirements of these tests become clear. In the A–not-A test, the panelist is given a chance to inspect the control product, in order to learn what "A" is like. This inspection may be more or less stringent and time consuming, but the familiarity of the panelist with "A" must not be assumed. The panelist is then given blind-coded test products, one at a time, and must decide after each one whether it is an example of "A" or something else (hence not-A). Products A and B are given an equal number of times. The presence of both A and B is important to get a baseline of response frequencies for each item (you cannot just give B and see what people call it). If the two products are presented once to each panelist, the response matrix forms a 2×2 chart, and the McNemar test is an appropriate statistical test. If they are only presented once (half the panelists see A and half see B) a χ^2 test is appropriate, or a Z-test on proportions (see Bi (2006) for other models).

The same–different test requires a similar baseline condition. Now products are presented in pairs, and the response options are "same" or "different." The question, of course, is the frequency with which the AB pair is called "different," but this must be compared with the frequency with which the AA pair (control paired with itself) is called "different."

A similar requirement is present in any test with a rated degree of difference (also called a DOD test). Instead of two response options, the panelist may now signify some graded degree of perceived difference for each pair. This can be done on a line scale, or a category scale with verbal anchors ranging from "exact match" to "extremely different" or some similar wording. Now the question becomes whether the rated difference is higher for an AB pair than for an AA pair (or BB if that is a control condition). If each panelist sees each kind of pair once, a paired t-test is appropriate for scaled data. Similar designs and analyses are done for scales with sureness or certainty ratings, as discussed in Chapter 9.

A.2.4 Statistical Analysis

Because the chance probabilities and null hypotheses are known, the choice tests are straightforward to analyze. Under a true null, the expected data set would follow a binomial distribution. However, since the panel size is usually around 50 persons or above, the *Z*-score approximation to the binomial distribution can be used. The critical *Z*-value for a one-tailed test at $p=0.05$ is 1.645. So the difference between the observed proportion correct and the chance proportion must satisfy the following inequality:

$$\frac{\left(P_{\text{observed}} - P_{\text{chance}}\right) - \dfrac{1}{2N}}{\sqrt{p(1-p)/N}} \geq 1.645 \tag{A.1}$$

where N is the number of panelists and p is the chance probability level. The term $1/2N$ is a correction for continuity. If P_{observed} is equal to the count of the number correct X divided by N, we get the following formula, which is preferred by some authors and students:

$$\frac{(X - Np) - 0.5}{\sqrt{Np(1-p)}} \geq 1.645 \tag{A.2}$$

Because the chance probability level is known for each test, simple significance tables are found in most textbooks, showing the minimum number of correct choices that are required (X) as a function of N in order to satisfy the inequality. The tests are one tailed, because only performance above chance is predicted, and the alternative hypothesis states that the population proportion correct would be greater than the chance probability level (rather than not-equal-to). Analyses for replicated tests can use the beta binomial models, as discussed in Chapter 5 and in Bi (2006).

A.3 Descriptive Analysis

A.3.1 Objectives

Simple difference tests provide evidence that at least some persons can tell two products apart. But they do not tell you in what ways the products differ. This is the goal of a descriptive analysis. Descriptive analysis provides a complete specification of the sensory properties of a food or consumer product. This is achieved by two important processes: first, understanding the terms that will be used to describe the product and, second, specifying the strength or perceived intensity of each attribute. Note that this is a psychophysical model in most cases. The attribute varies from weak to strong sensations in that class of products, or perhaps from no sensation of that type to a strong sensation. Descriptive analysis is not concerned with liking or preference for the product, nor is it concerned with feelings or emotions. It answers the "what" in what is different about these products.

The tool is a versatile one. It can be used for any situation in which an ingredient, processing, or packaging variable has changed, as well as for shelf life and stability testing. Multiple products can be submitted for testing. Replicated measurements are common. Generally, a panel will be trained only to evaluate one class or category of products. So in a large food manufacturer, there may be several different panels operating for different product lines. Of course, the participants must be trained to understand the terms that are used to describe the products. This is often done through the use of examples or reference standards. In some versions of these techniques, they are also calibrated to use the intensity scale by means of reference examples of items that are weak or strong in that specific attribute. Statistical analysis is necessary and is discussed further below. Further information on descriptive analysis procedures can be found in Lawless and Heymann (2010: chapter 10).

A.3.2 Participants

Setting up a descriptive panel takes a lot of time for screening and training. Once the panel is in place, it is typically used several times a week. So, two important parts of qualification for panel work are (1) the motivation to participate and (2) supervisory approval for the time commitment if the panelists are employees (a common situation). For screening, the panelist must be known to have some normal functions of sensory acuity for the senses involved. Tests may involve discrimination of different products that have been spiked with different flavors, have different textures, or any combination as a function of different processing conditions. It is best to use products from the category they will actually evaluate after training. The persons do not have to be sensory superstars, but must have a roughly normal sense of taste and smell, for example. If a large panel is part of screening, the top scoring individuals may be invited for further service. They should also have good verbal ability, ability to describe their perceptions, be cooperative in a group situation, and not be afraid to voice their opinions. They need not be consumers of the product category, but they must have no objections to eating or drinking whatever the products are in the future tests. Typically, panel size is 8 to 12 individuals, but it is a good idea to have a reservoir of new potential trainees to replace persons who may drop out, leave the company, go on long-term leave of absence, and so on.

A.3.3 Methods

Most of the current descriptive methods are versions of the quantitative descriptive method outlined by Stone et al. (1974), although the historical antecedent was a method of flavor profiling (Caul, 1957). Once the recruits have been qualified, the training phase begins. There are two versions of training, which have been called "consensus training" and "ballot training." In consensus training, the panelists taste a wide range of products from the product category and volunteer terms that describe the product's appearance, aroma, flavor, texture, mouthfeel, and residual characteristics as applicable. The list is refined by eliminating redundant terms, vague terms, and those that refer to likes or dislikes. Reference standards may be found to illustrate good examples of products having those qualities, and these may be from other product categories or even be single chemical compounds in the case of flavors. Examples of reference standards can be found in the classic paper on the wine aroma wheel by Noble et al. (1987), and their use is discussed by Rainey (1986). The terms should only refer to one specific sensory attribute, and not be a combination such as "creaminess." The early stages of panel training proceed with a lot of group discussion, rather than individual testing. Hence, the notion of a consensus is applied to the results of the training. Note that the training is simultaneously calibrating the attributes and concepts of the panelists, while building the eventual ballot for formal evaluations and data collection.

Once the terms are found, anchor words must be assigned to the low and high ends of each scale; for example, no sweetness to extremely strong sweetness. The anchor words need to be logical and chosen sensibly. In some forms of descriptive analysis, where there is a universal intensity scale, the intensity points are trained by example, and all flavor, taste, and aroma scales have the same references (usually on a 15-point or 150-point scale). For example, a "2" in intensity is exemplified by a 2% sucrose solution, among many other examples. See Meilgaard et al. (2006) for extensive lists of references for a universal intensity scale.

The second form of training is called ballot training, in which the terms are prespecified. Once again, panelists may taste a large variety of samples from the product category during training. Examples are shown to illustrate the different characteristics. In both methods, panelist consistency is checked during training, as well as panelist agreement. That is, panelists should be able to reproduce their own judgments on blind samples and panelists should agree within a certain range that the sample should be at a certain point on the scale. Panelists who are too low or too high may be offered additional training. The ins and outs of panel calibration are discussed in Chapter 7.

Once the panel is sufficiently "in tune," the actual evaluations may begin. Samples need not be presented to panelists in a group any longer, and they may fill out their ballots on an individual basis, as long as the product handling, preparation, and conditions of serving are identical. Samples are presented in random or counterbalanced orders, usually with three-digit random codes as identifiers. Scales are typically presented by computer screen and responses recorded by mouse clicking, but paper ballots may be used in some circumstances. An example of a ballot for peanut butter is shown in Figure A.2.

A.3.4 Statistical Analysis

The classic "bread and butter" technique for statistical processing of descriptive data is the analysis of variance (ANOVA; Lea et al., 1998; Naes et al., 2010). Generally, products and panelists are factors in the ANOVA model, and often replications. The specification of panelists as a factor allows the removal of inter-person variation from the error term. If all panelists see all products (usually an equal number of replicates) the analysis is what is referred to as a repeated-measures analysis, and/or a complete block design. These designs are highly desirable for reasons discussed in Chapter 9. After estimation of the panelist main effect, the remaining source of error variance is the product by panelist interaction, so this becomes the error term, or denominator of the F-ratio for products. This is because panelists are considered what is known as a random effect, or a random sample of all such possible panelists. This is in contrast to a fixed effect, where the treatment levels consist of specific values of some variable, like 2%, 4%, and 6% sucrose. The ANOVA models are different in that the construction of the error term is different, and so the practitioner must be careful to specify the correct model. For a case where products are a fixed effect and panelists are random (a common occurrence), this is sometimes referred to as a type III or "mixed" model.

Once a significant difference among products is found in the ANOVA, a second task begins. Now the mean values must be compared using variations of the t-test. Different techniques exist for this, most of which adjust for the fact that you are performing multiple paired comparisons, and thus the chance of a type I error is being inflated. Some are more conservative than others and require a larger difference between the means in order to be statistically significant. Common choices are the least significant difference test (LSD, very

PEANUT BUTTER DESCRIPTIVE BALLOT

1. Appearance

a) Coloration

❑ ❑ ❑ ❑ ❑ ❑ ❑ ❑ ❑ ❑ ❑ ❑

very yellow — very brown

b) speckles

❑ ❑ ❑ ❑ ❑ ❑ ❑ ❑ ❑ ❑ ❑ ❑

none — many

c) visible oil

❑ ❑ ❑ ❑ ❑ ❑ ❑ ❑ ❑ ❑ ❑ ❑

none — a lot

d) particles

❑ ❑ ❑ ❑ ❑ ❑ ❑ ❑ ❑ ❑ ❑ ❑

smooth — grainy

2. Aroma

e) fresh peanut

❑ ❑ ❑ ❑ ❑ ❑ ❑ ❑ ❑ ❑ ❑ ❑

none — very strong

f) roasted

❑ ❑ ❑ ❑ ❑ ❑ ❑ ❑ ❑ ❑ ❑ ❑

none — very strong

g) oxidized

❑ ❑ ❑ ❑ ❑ ❑ ❑ ❑ ❑ ❑ ❑ ❑

none — very strong

3. Flavor

h) sweet

❑ ❑ ❑ ❑ ❑ ❑ ❑ ❑ ❑ ❑ ❑ ❑

very weak — very strong

i) salty

❑ ❑ ❑ ❑ ❑ ❑ ❑ ❑ ❑ ❑ ❑ ❑

very weak — very strong

j) fresh peanut

❑ ❑ ❑ ❑ ❑ ❑ ❑ ❑ ❑ ❑ ❑ ❑

very weak — very strong

k) roasted

❑ ❑ ❑ ❑ ❑ ❑ ❑ ❑ ❑ ❑ ❑ ❑

very weak — very strong

Figure A.2 Sample descriptive analysis ballot for peanut butter.

4. Texture/mouthfeel

l) thickness

❑ ❑ ❑ ❑ ❑ ❑ ❑ ❑ ❑ ❑ ❑ ❑

very thin very thick

m) oily

❑ ❑ ❑ ❑ ❑ ❑ ❑ ❑ ❑ ❑ ❑ ❑

not oily very oily

n) adhesiveness

❑ ❑ ❑ ❑ ❑ ❑ ❑ ❑ ❑ ❑ ❑ ❑

not sticky very sticky

o) melting rate

❑ ❑ ❑ ❑ ❑ ❑ ❑ ❑ ❑ ❑ ❑ ❑

very slow very fast

p) drying

❑ ❑ ❑ ❑ ❑ ❑ ❑ ❑ ❑ ❑ ❑ ❑

not drying very drying

5. Residual Characteristics

q) bitterness

❑ ❑ ❑ ❑ ❑ ❑ ❑ ❑ ❑ ❑ ❑ ❑

none very strong

r) mouthcoating

❑ ❑ ❑ ❑ ❑ ❑ ❑ ❑ ❑ ❑ ❑ ❑

not coating very coating

s) residual flavor

❑ ❑ ❑ ❑ ❑ ❑ ❑ ❑ ❑ ❑ ❑ ❑

very weak very strong

Figure A.2 *(Cont'd)*

liberal), Duncan's multiple range test (or "studentized range," a good choice after a significant F-ratio), and the Tukey "honestly significant difference" (HSD). Most of these can be prespecified and chosen in statistical programs.

There is potentially a lot of information coming out of a descriptive analysis. So presentation of the results presents a challenge. A simple technique is to present the means in a table and use letter superscripts (typically a, b, c, etc.) to indicate significant differences. Products that do not differ share common superscripts. Those without an overlapping letter designation are significantly different. Other common displays include bar charts with error bars or letter superscripts, and spiderweb or radar plots, which produce a polygon for each product. In general, it is not necessary to include attributes that showed no significant difference among the products. Communicate the headlines. Also bear in mind that this kind of panel can become highly discriminating. Thus, a statistically significant difference is not necessarily a difference that is practically meaningful or important to consumers.

A.4 Affective Tests

A.4.1 Objectives

The general objective of an affective test is to find out the degree of consumer appeal of a product. This is the third major category of sensory testing. Difference tests can tell you if there is any change, descriptive tests can tell you how the product changed, and affective tests can tell you if it matters. Like other sensory tests, they are usually done on a blind basis; that is, with minimal concept information. This is in contrast to market research tests, which may be done with the full-blown concept being presented to the participants, in addition to a taste test. These tests can be divided into two general methods, one measuring acceptability of the products on a scale for liking/disliking and the second using a choice paradigm, where the best-liked product from a pair or group is indicated. The latter is a preference test. Affective tests are also called hedonic tests and are what most laypersons associate with a taste test. Because they generally use consumers of the product category, they are also called consumer tests.

There are several common scenarios for an affective test. One is the test of a new product prototype in a product development scenario. There may be a necessity to get a "read" on the new product and diagnose any potential problems before more expensive market research testing. The prototype may then go through further modifications before another consumer test. Minor product changes in an existing brand may need to be checked to make sure there are no objectionable sensory developments. These scenarios include ingredient substitutions and cost reductions. Shelf-life testing (stability testing) may involve some consumer evaluations, sometimes after a discrimination test or descriptive panel has found differences at a certain time point. Large-scale consumer tests may also be done with a finished product that is about to be introduced as a new product or a line extension to an existing product. Consumer testing is also required for advertising claim substantiation, and may be done against specific competitors in order to justify the claim. An example would be a statement like, "Taste America's best beef frank!"

These tests occur in different venues. Sometimes, users of the product are recruited from an employee pool to conduct a rapid in-house test. These people are less representative than a true consumer sample, but the turnaround time is quicker, and security is of course tighter. The second type of setting is in some central location, where consumers are recruited or invited to participate in a taste test in a sensory testing or market research facility. These are known as central location tests (CLTs). Many market research test agencies are available to do this kind of testing, and some have lists of consumers who have previously filled out product usage questionnaires. Thus, recruiting can be targeted and cost effective, although it is more of a convenience sample than anything you would call random. The CLT setting affords the tester good control over product preparation, serving, and eating conditions. The third scenario occurs when the product is sent home to the panelist. These are home-use tests (HUTs; also called IHUTs for in-home). These are, of course, more expensive than CLTs, but more realistic. The product may be used by all family members, and for an extended period. The questionnaire is usually administered at the end of the usage period.

A.4.2 Participants

The participants are screened for product usage. That is, they must be users of the product category. A category is a grouping of products that serve the same purpose, and often appear on the same part of the retail shelves. So cold breakfast cereals are a product category. If needed, the category may be further subdivided to narrow down the sample to members of the actual target market. So they may be screened to be users of pre-sweetened breakfast cereals, or even sweetened flaked cereals. Frequency of usage is an important consideration. If I only eat cold cereal once a year, I am probably not qualified to be in the test. The screening questionnaire may seek to find people with moderate to high usage of the product type. The questionnaire may also seek to find users of specific brands, either the company's own products or those of key competitors. In a CLT or HUT, the screening and recruitment process often seeks to have certain quotas of different age groups, genders, or other demographic variables, such as income groups. The typical group size is 100 or more consumers per product. In advertising claim substantiation, it is common to have 300 or more consumers per cell if there are different paired preference tests.

A.4.3 Methods

Consumer acceptability tests generally use the nine-point hedonic scale developed by the US Army. The scale points are shown in Table A.1. The phrases were chosen to represent approximately equal psychological distances, although the interior three are actually somewhat more closely spaced. They are generally assigned the numbers one (for dislike extremely) through nine (for like extremely) for purposes of data analysis. Consumers are asked to use this scale to give their overall opinion of the product. Consumers may also be asked to rate liking for appearance, flavor, texture, and so on, although their ability to understand and make distinctions about more specific attributes is questionable. But the consumer has an immediate and integrative reaction to a product, that could be called the "yum or yuck" response. Other related scales include satisfaction scales and appropriateness for a given situation.

Liking may also be probed concerning specific attributes using just-about-right (JAR) scales. These scales range from "not enough X" to "just about right" to "too much X." They give immediate direction about any needed product changes in order to optimize it. Of course, the consumer must understand the attribute in question, so this is limited to simple common terms that consumers can recognize clearly in the product. The goal is to make a product that has its data distribution centered on the JAR point or category, and is symmetric and peaked.

Preference tests are straightforward. Both products are tried by the consumer. Then they are asked to choose the product they liked the best. Sometimes a "no preference option" is allowed, but this complicates the analysis and is not favored by most sensory practitioners.

Table A.1 The nine-point hedonic scale

Assigned value	Phrase
9	Like extremely
8	Like very much
7	Like moderately
6	Like slightly
5	Neither like nor dislike
4	Dislike slightly
3	Dislike moderately
2	Dislike very much
1	Dislike extremely

However, the option may be required for some advertising claim substantiations. The orders of presentation, of course, are balanced across the participants, because there may be a preference bias for the first item. If more than two products are presented, the consumer may be asked to rank them from most liked to least liked. In another form, so called best–worst scaling, they are asked to choose the best liked and least liked. This test is still rare in food research.

Two major options are available for consumer test design. In monadic testing, each consumer only evaluates a single product. Usually more than one product is tested; thus, the design requires a different group of consumers for each product. In monadic sequential testing, the consumer receives one product at a time for evaluation, but is asked to evaluate more than one item. It is common with two or three products to have all consumers evaluate all products, but with more items, an incomplete block design can be used. A preference test is not possible with a monadic design, as there is nothing to compare with. If products are presented at the same time in a CLT, that is referred to as a side-by-side test. However, because we cannot taste more than one product at a time, it is simply another form of a sequential test, albeit with a short time interval. It does permit a direct visual comparison of the two items.

A.4.4 Statistical Analysis

For acceptability data gathered with the nine-point scale, parametric statistics are generally used, with *t*-tests for two items and ANOVA for more than two. After a significant ANOVA result, means may be compared by various post hoc tests, such as Fisher's LSD, Duncan, or Tukey tests. It is important, however, to look at the distribution of scores in the data set, for there may be pockets of consumers who like the item strongly and some who dislike the item strongly. Thus, the mean value can be misleading, and the researchers must be aware of the possibility of some consumer segmentation.

For the JAR data, a simple χ^2 test can be performed if the design is monadic; that is, there are different groups of people trying each product. Sometimes the data are collapsed into three categories for simplicity: below JAR, at or near JAR, above JAR. If the same persons try both products, as in a monadic sequential test, the Stuart–Maxwell statistic is appropriate, as discussed in Chapter 9. If hedonic scores are also gathered in the same study, it is possible to calculate the mean drop or penalty difference among persons who scored the product below or above JAR (as compared with the mean hedonic score for people at or near JAR for that product). This can give two important pieces of information. First, how many people are off-JAR for this product, and how potentially detrimental is that result in terms of the overall product appeal. Further information on JAR analysis can be found in the ASTM document edited by Rothman and Parker (2009).

The analysis of paired preference data is straightforward, as long as a choice is forced. Under a null hypothesis of equal preference for each product, a *Z*-score can be constructed from the normal approximation to the binomial, or an exact binomial probability can be computed using appropriate software. The *Z*-formula does not deviate much from the exact binomial probability, because the sample size is generally very large. The *Z*-formula is similar to eqn A.1, except that the test is now two-tailed and the critical *Z*-value is now 1.96. So in order to be statistically significant, the following inequality must be satisfied:

$$\frac{P_{\mathrm{w}} - 0.5}{0.5 / \sqrt{N}} \geq 1.96 \qquad \text{(A.3)}$$

where P_w is the proportion for the winning product, the one gathering the most preference votes, and N is the number of consumers in the test. Sometimes the continuity correction ($1/2N$) is subtracted from the numerator, but this becomes negligible as the sample size gets large.

For tests with a no-preference option, various strategies may be applied to return the analysis to the binomial situation in order to apply eqn A.3. The votes may be ignored (thrown away), decreasing N accordingly. A conservative approach is to allot them equally to the two products, which of course dilutes the signal-to-noise ratio, as it follows the null hypothesis. Various other options are discussed in the body of this book, including Thurstonian analysis of the three-choice frequencies (see Chapters 4 and 8).

For ranking tests, various rank sum statistics can be applied, such as the Friedman analysis of variance on ranks and the Kramer rank sum test. Simple lookup tables for rank sum differences can be found in Lawless and Heymann (2010) and in the useful paper by Newell and MacFarlane (1987). Best–worst scaling is discussed in the paper by Jaeger et al. (2008).

A.5 Summary and Conclusions

Many years ago, Pangborn lamented three continuing problems in sensory evaluation. They were (1) lack of clear objectives, (2) a tendency to use one test method repeatedly regardless of the problem at hand, and (3) lack of proper selection of respondents in the test (Pangborn, 1979). Although these problems persist, training in sensory evaluation and adherence to good principles and practices will help avoid serious mistakes and useless data. The goal of this appendix was to give an overview of the basic methods used for different test objectives. The reader without formal training in applied sensory testing is encouraged to consult the texts mentioned in the opening and listed in the reference list below.

Many other types of testing are sometimes done under the umbrella of sensory evaluation. One example is threshold testing, which strives to determine the minimum amount of a substance that may be detectable by taste or smell, for example. However, many threshold tests use forced-choice procedures, and thus are just special cases of repeated discrimination testing. Another procedure is time–intensity scaling, in which the assessor indicates the intensity of a taste or flavor as it rises and falls once the food is tasted. The goal is to track the time course of the sensation, and identify temporal characteristics such as the total duration of the experience. But these are simply extensions of scaling, much like descriptive analysis. A third example is quality testing. But this involves some degree of difference scale, along with specification of the intensity of any key components and/or defects (Lawless & Heymann, 2010: chapter 17). Thus, it is a combination of a difference testing method and a descriptive method. So the fundamental three categories of testing (along with scaling, see Chapter 7) are still the best starting points for one's conceptual orientation to the field.

References

Bi, J. 2006. Sensory Discrimination Tests and Measurements. Blackwell Publishing, Ames, IA.
Caul, J.F. 1957. The profile method of flavor analysis. Advances in Food Research, 7, 1–40.
Chambers, E.C., IV, and Wolf, M.B. 1996. Sensory Testing Methods. Second edition. ASTM Manual Series MNL 26. ASTM, West Conshohocken, PA.

Jaeger, S.R.; Jørgensen, A.S., Aaslying, M.D., and Bredie, W.L.P. 2008. Best–worst scaling: an introduction and initial comparison with monadic rating for preference elicitation with food products. Food Quality and Preference, 19, 579–588.
Kemp, S.E., Hollowood, T., and Hort, J. 2009. Sensory Evaluation. A Practical Handbook. John Wiley & Sons, Ltd, Chichester, UK.
Lawless, H.T. and Heymann, H. 2010. Sensory Evaluation of Foods. Principles and Practices. Second edition. Springer Science + Business, New York, NY.
Lea, P., Naes, T., and Rødbotton, M. 1998. Analysis of Variance for Sensory Data. John Wiley & Sons, Ltd, Chichester, UK.
Meilgaard, M., Civille, G.V., and Carr, B.T. 2006. Sensory Evaluation Techniques. Third edition. CRC Press, Boca Raton, FL.
Naes, T., Brockoff, P.B., and Tomic, O. 2010. Statistics for Sensory and Consumer Science. John Wiley & Sons, Ltd, Chichester, UK.
Newell, G.J. and MacFarlane, J.D. 1987. Expanded tables for multiple comparison procedures in the analysis of ranked data. Journal of Food Science, 52, 1721–1725.
Noble, A.C., Arnold, R.A., Buechsenstein, J., Leach, E.J., Schmidt, J.O., and Stern, P.M. 1987. Modification of a standardized system of wine aroma terminology. American Journal of Enology and Viticulture, 38(2), 143–146.
Pangborn, R.M. 1979. Physiological and psychological misadventures in sensory measurement or the crocodiles are coming. In: Sensory Evaluation Methods for the Practicing Food Technologists. M.R. Johnson (Ed.). Institute of Food Technologists, Chicago, IL.
Rainey, B.A. 1986. Importance of reference standards in training panelists. Journal of Sensory Studies, 1, 149–154.
Rothman, L. and Parker, M.J. 2009. Just-About-Right Scales: Design, Usage, Benefits, and Risks. ASTM Manual MNL63. ASTM International, Conshohocken, PA.
Stone, H., Sidel, J., Oliver, S., Woolsey, A., and Singleton, R.C. 1974. Sensory evaluation by quantitative descriptive analysis. Food Technology 28(1), 24, 26, 28, 29, 32, 34.
Stone, H., Bleibaum, R., Sidel, J., and Thomas, H. 2004. Sensory Evaluation Practices. Third edition. Elsevier/Academic Press, Amsterdam.

Appendix B: Overview of Experimental Design

It is up to the researcher to identify what wins, not the developer, although the developer and the marketer can automatically dictate, by fiat, what they will test. They aren't necessarily right, even when they have the power to be so.

Moskowitz et al. (2006: 175)

B.1 General Considerations

The choice of products to test during a new product development cycle can be done in different ways. In the past, a common approach was simply to make what seemed to be the best candidate items at the benchtop using trial and error. The potential new products might be evaluated in more or less formal circumstances, sometimes by the head of the company, or even his/her spouse. In contrast to this kind of haphazard product selection, others favored a more systematic approach to (1) screen variables for their impact on sensory properties and/or consumer acceptance and (2) study those variables deemed to be important in some systematic experimental design.

There are two topics in decision-making for a sensory specialist in choosing an experimental design. One important consideration is the way the samples will be presented. For

Quantitative Sensory Analysis: Psychophysics, Models and Intelligent Design, First Edition. Harry T. Lawless.

example, we have to decide in what orders they will be evaluated by each person, how many samples they can evaluate in one sitting, and how many sessions will be required. The order of presentation usually involves a choice of counterbalancing schemes, such as a Latin square, or some method for randomization. The number of samples and how many are evaluated by each assessor involves a choice of blocking, such as a complete block design (all panelists see all samples) or an incomplete block design (panelists only see a subset of the total group of items). Chapter 9 discussed the value of complete block designs as a way to eliminate or factor out the individual peculiarities in scale usage or overall assessor sensitivity when looking for differences among products.

Carr (2010) refers to these considerations as "experimental designs of panels," although they really have to do with samples and how they are distributed to assessors. He contrasts these with the second important area of design: that is, the selection of the experimental variables to be studied and how they will be chosen at specific levels of each variable. The choices of levels of each independent variable are the kinds of experimental designs discussed in that chapter. Carr refers to these concerns as experimental designs of samples. The formal name for this field is *design of experiments* (DOE). A systematic approach to these designs can enhance the information output of any study, and can help the sensory specialist and client make intelligent and informed choices when taking a product further in the development or optimization process.

An important philosophical background for designed experiments is a psychophysical orientation. That is, we are trying to study variables, not products. A two-product test is a rather uninformative way of doing research. It is much more powerful and useful in the long run if product developers send products for testing that vary systematically in some key variable. Such variables include different levels of some ingredients, different mixtures of multiple ingredients, and/or process variables that change the product characteristics. Of course, variables can be changed one at a time and studied in simple two-product tests, but this will miss some important information. The interaction effects when two ingredients are varied cannot be determined by a one-variable-at-a-time approach. Interaction effects occur when the result of changing one variable depends upon the specific levels of a second variable. The increase of sweetness caused by adding more sugar to my product will be less noticeable at a high level of acid (as opposed to a low level). Food products are replete with examples of interaction effects. You cannot reduce the sodium in a product without changing a host of other sensory effects in addition to just lowering saltiness. Furthermore, the effects of sodium reduction will vary depending upon what else has been modified in the product.

The decision process and choice of variables and levels is the main concern of DOE. Various efficient designs have been developed that are commonly used. Usually, this process is not entirely the responsibility of the sensory specialist, but occurs in a negotiation with the client who has requested the testing. Often, this client is a product development specialist or a group of such individuals. Other participants in a matrixed team may become involved as well, such as marketing or brand managers and consumer insights researchers. And let us not forget statisticians. The overall goals of the study will influence the design strategy. Are we screening variables as to their importance or impact on sensory characteristics or consumer acceptance? Are we trying to specify the interactions or two or more variables? Are we trying to fit a quadratic surface to a two-variable experiment in order to find an optimal product? Each of these suggests a different design.

Further resources on DOE can be found in the chapter by Carr (2010), and in the statistics text by Naes et al. (2010: chapter 12). The statistics book by Gacula et al. (2009) includes a chapter on factorial experiments and a chapter on response surface designs. The earlier work by Gacula (1993) discusses optimization designs. The goal is to take an efficient and systematic approach to understanding the variables of interest. A key idea is that more than one variable is changed in these designs, rather than studying their effects individually. A concise description of the terms and designs discussed here can be found in the National Institute of Standards and Technology, *e-Handbook for Statistical Methods* (NIST/Sematech, 2012). Chapter 5 in the e-Handbook is devoted to DOE, with an orientation toward process optimization. Printer-friendly versions of the web site's section are available.

B.2 Factorial Designs

The starting point for any consideration of experimental design is the simple *factorial design*. Assume we have two variables of interest that are known to change some sensory response. In a simple beverage system, this could be sugar and acid. Note that the independent variables are not sensory; that is, not sweetness and sourness. They are the physical variables you are manipulating in the product. Obviously, changes in the levels of these variables have sensory consequences. So far, so good. The simplest design is to choose two levels of each of these independent variables, so that four products will be constructed. These can be designated as low and high levels for simplicity, although the actual levels you choose may not be low or high in any absolute sense. But the products should be differentiated by the assessors. It they are too confusable, the data will not show any important trends.

Figure B.1 shows a graphic way of looking at this experimental design, as four points that form the vertices of a square or rectangle. The simple factorial has one important property: it can be used to detect interactions between the two variables. If we only studied one variable at a time, and held the second variable constant, no interaction effect could be detected.

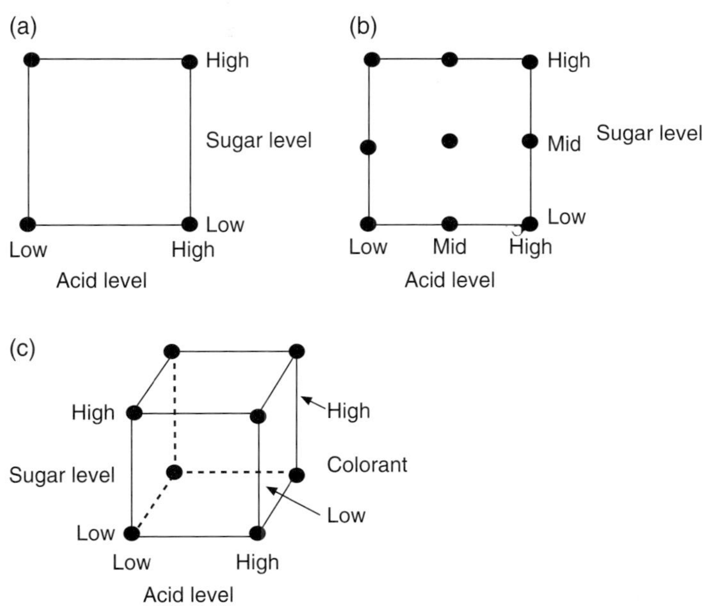

Figure B.1 (a) A simple 2×2 factorial design, also called a 2^2 design. (b) A 3×3 design. (c) A factorial design with three factors at two levels, or 2^3 design. Note that low levels of colorant consist of the plane in the front of the cube, and high levels the plane in the rear of the cube.

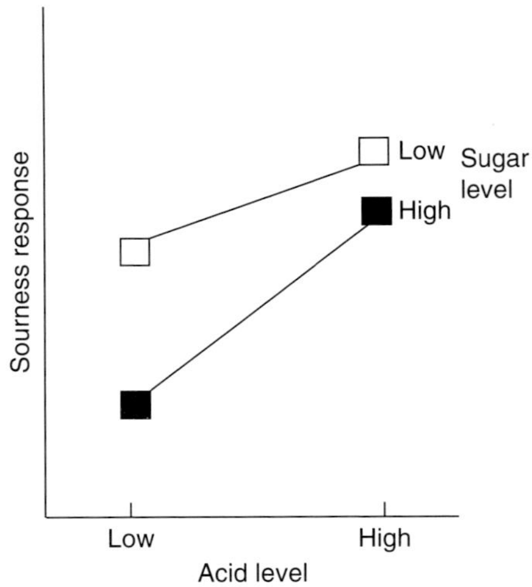

Figure B.2 A sample of an interaction effect. The added sugar level decreases the sourness, and this is more pronounced (a bigger change) at the higher level of sugar content. The effect of increased acid is greater at lower levels of sugar.

Remember that an interaction is a change in the response to one variable that depends upon the level of the second variable. Figure B.2 shows an interaction effect when both acid level and sugar level are varied, and sourness is measured. Interactions are ubiquitous in food research. The effect of adding sugar is to suppress the sourness. The suppression is stronger when sucrose is higher and acid is lower. The rate of growth of sourness as a function of acid level is also different at the higher level of sugar. So the impact of each variable depends upon the specific level of the other variable.

The two-level design also has one deficiency: you can only assess a linear trend at each level, because there are only two points. In order to assess a curvilinear trend or quadratic effect, a third level of each variable is required. This is shown in Figure B.1b. For experiments designed to screen variables or do some preliminary investigation of interactions, two levels may be sufficient information. For an optimization study, it is usually desirable to see curvature in the response surface, so more than two levels typically become part of the design. Efficient designs for fitting surfaces, such as the central composite design, are discussed below.

B.3 Fractional Factorials and Screening

If there are more than two variables to study, the factorial design can get quite large, quite fast. For this reason, investigators sometimes employ a kind of incomplete design known as a fractional factorial. Let us assume there are three variables to study at two levels of each, designated L for low and H for high. This is a $2 \times 2 \times 2$ factorial design, or 2^3. Thus, there are eight products to consider and all would be evaluated in a full factorial. The eight products are shown in Table B.1. This complete design is capable of estimating all the linear main effects of ingredients A, B, and C, all the two-way interactions, and the one three-way interaction. Sometimes in screening experiments, people are only interested in the main effects of the single variables and choose to ignore the interactions. It is possible that the interactions

Table B.1 Design of a full 2×2×2 or 2^3 factorial experiment

Product	Level of ingredient A	Level of B	Level of C
1	H	H	H
2	H	H	L
3	H	L	H
4	H	L	L
5	L	H	H
6	L	H	L
7	L	L	H
8	L	L	L

Table B.2 Levels in a one-half fraction of the design in Table B.1

Product	Level of A	Level of B	Level of C
1	H	H	H
(2 omitted)	—	—	—
(3 omitted)	—	—	—
4	H	L	L
(5 omitted)	—	—	—
6	L	H	L
7	L	L	H
(8 omitted)	—	—	—

are not very strong or not very important. So a smaller design can be used, without having to test all eight possible combinations. This kind of partial design is common in techniques of conjoint measurement in marketing research, especially on durable goods with various features like automobiles and washing machines (Moskowitz et al., 2006). Software exists to provide estimates of the various factors.

Table B.2 (and see Figure B.3) shows four products from the full design that could be used in a fractional factorial, in this case the one-half fraction of the full 2^3 design (the example is from Carr (2010)). We've chosen four corners of the design and each variable has two products at its two levels, so there is a good chance we can get a stable estimate of the main effects of variables A, B, and C. For example, we can estimate the effect of going from a low to a high level of factor A by contrasting the mean response to products 1 and 4 (the high levels of A) with the mean response to products 6 and 7 (the low levels of A). Similar contrasts can be constructed with factors B and C.

But there is no free lunch. The contrast of 1 and 4 with 6 and 7 may tell us about the effect of A, but there is also a B×C interaction potentially contributing. Thus, in the fractional factorial, some of the interaction effects are confounded or mixed in with the estimates of the main effects. The term for this is "aliasing." What happens when we try to estimate the effect of A by comparing the average of products 1 and 4 with the average of 2 and 6? Table B.2 shows that there is a contrast of the two products in which B and C are both either both low or both high, with the other two products in which one variable is low and the other is high. That is, when we try to estimate the main effect of A, we are also using products that tell us about the B×C interaction.

In order to understand how this occurs, a further look at interactions is necessary. To do this, I will use a graphic interaction detector, illustrated in Figure B.4. If there is no

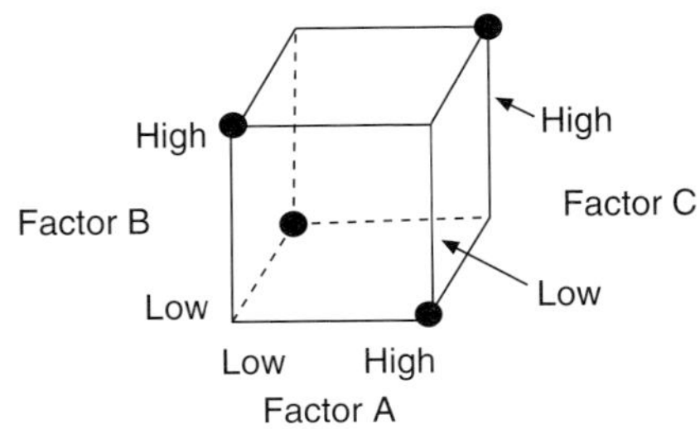

Figure B.3 The products in the fractional factorial design discussed in Section B.3 and shown in Table B.2. Note that each variable has two products at each level; therefore, there is redundant information about the main effects. However, these are each confounded with the two-way interaction with the other remaining variables.

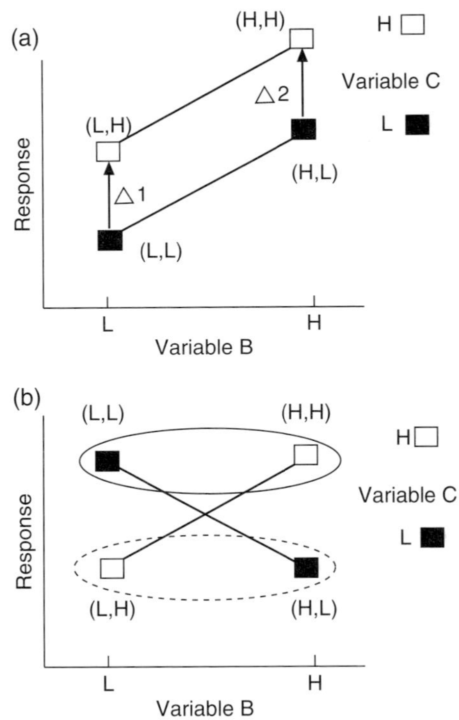

Figure B.4 The interaction detector. (a) No interaction. The trend lines are parallel. The effect of changing one variable is the same for both levels of the other variable. Another way to show this is that the average of changing both levels (average of product (L, L) and product (H, H)) is the same as the average of changing one at a time (average of product (H, L) and product (L, H)). (b) Crossover interaction. The trend lines are not parallel, and the response to one variable depends entirely upon the level of the other variable. Now the average of changing both levels (average of product (1, 1) and product (2, 2) as shown by the solid ellipse) is quite different from the average of changing one at a time (average of product (2, 1) and product (1, 2) as shown by the dashed ellipse).

interaction between two variables, the effect of going from (L, L) to (L, H) is the same as going from (H, L) to (H, H), as shown by the vertical arrows in Figure B.4a. Another way to look at this is that the average of (L, L) and (H, H) should be the same as the average of (L, H) and (H, L).[1] The upper panel (a) shows the comparison when there is no interaction. Note that the trend lines are parallel. The change in one variable does not depend on the level of the other variable. Everything looks additive. The lower panel (b) shows the situation with a pronounced crossover interaction. Now the contrast of the (L, L)(H, H) combination or average with the average of the other two shows a very big difference. This contrast is shown by the two ellipses in Figure B.4b.

The key fact to note is that this comparison made by our automatic interaction detector is the same comparison we made to estimate the effect of factor A. That is, the comparison of the average of products 1 and 4 with the average of products 6 and 7. So both the test of the main effect and the interaction detector are making the same comparison. In the fractional design, we cannot tell whether the change in the dependent variable is due to A or due to the B×C interaction. If the products have the strong crossover interaction shown in Figure B.4b, then the difference between these two pairs of products could be totally due to the B×C interaction, and variable A could have no effect whatsoever! So the A effect and the B×C effect are confounded or aliased. This is not such a bad situation if the interaction is believed to be inconsequential, but that is a risky assumption with most food products. However, if additional information is available that supports the assumption, it may be worth making in order to save on time and costs in the screening study.

A special case of fractional designs used for screening is the Plackett–Burman design, developed by two statisticians working for the British Ministry of Supply after World War II (NIST/Semantech, 2012). These designs have the property that, for any two variables, the four combinations of values (e.g. (L, L), (L, H), (H, L), and (H, H)) will occur an equal number of times. Thus, the designs work well for multiples of four variables, although partial sections of the design matrix can be used for other numbers of variables to be screened. As in the case of fractional factorials, some of the interactions are confounded with main effects. So these designs are used when there are a large number of variables to screen (say, 12) and the interactions are not likely to be practically significant. Another design used for screening that is suitable for situations with more than two levels of some variables is Taguchi's orthogonal arrays (NIST/Sematech, 2012). Once again, the interaction effects are not considered important.

B.4 Central Composite and Box–Behnken Designs

Another common variation on the factorial design is one that adds a center point. The addition of more levels of each factor allows for estimation of quadratic or curvilinear relationships. One design used for fitting response surfaces in product optimization is the central composite design (Carr, 2010; Naes et al., 2010). Figure B.5 shows a central composite design for a two-factor study (upper panel). A repeated center point is added, and four axial or "star" points which are products with higher or lower levels of one of the variables than the vertex points, but an intermediate level of the other factor. Replication of the center point is

[1] If the change from (L, L) to (L, H) is the same as the change from (H, L) to (H, H) then the following relationships must hold for the dependent variable: (L, H) − (L,L) = (H, H) − (H, L), and thus (L, H) + (H, L) = (H, H) + (L, L). Dividing each side by two give us the averages, which must also be equal.

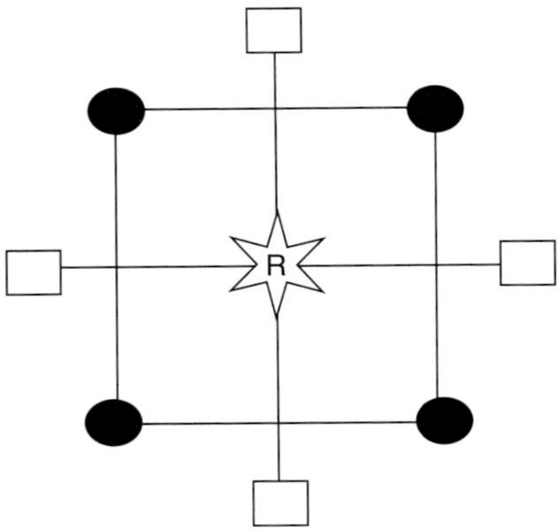

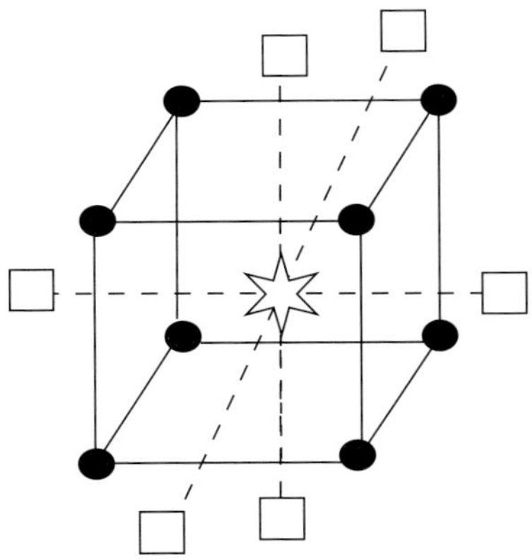

Figure B.5 A central composite design (upper figure) for a two-factor experiment consists of products on the four vertices, four "star" products with higher/lower values of one variable and intermediate values of the other, and a center point that is repeated. The lower figure shows a central composite design for a three-factor study.

customary to get a better estimate of the random error. The lower panel of Figure B.5 shows a three factor central composite design. Now the linear effects, quadratic effects, and interactions can all be estimated.

The central composite design offers some efficiency over factorial designs, especially as the number of factors increase. Consider a three-factor design. To estimate quadratic effects, low, medium, and high levels are needed for each factor, resulting in 3^3 or 27 products to construct and evaluate. In the central composite design with three factors, there are only 15 products to make, and 16 to evaluate if the center point is replicated.

Another efficient design with three levels of each factor is the Box–Behnken design. Instead of products at the vertices and axial points, the products are designed with high or low versions of two of the three factors and an intermediate level of the third factor. So, in the cubic visual model, they occupy positions in the middle of each edge of the cube. Carr also includes a center point in the Box–Behnken design. As with the central composite, the linear, quadratic, and interaction effects can all be estimated. This may be a good choice in cases where the extreme products of the central composite are difficult or impossible to manufacture. The central composite has five levels of each variable, while the Box–Benhken has three. So, for a three-factor study, there are only 12 products to be made, rather than 16 (plus repeated center points). So there is a small gain in efficiency along with a savings of not having to construct extreme low- or high-valued products.

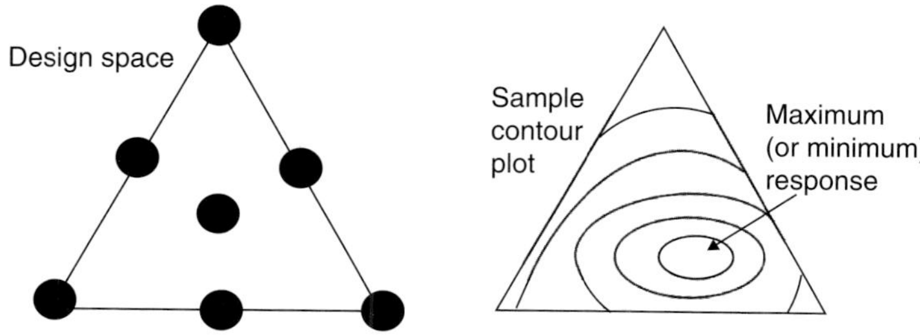

Figure B.6 A three-variable mixture design (left panel) where the filled circles represent products. The vertices are one-component products, the center of the lines show two-component mixtures, and the three-component mixture lies at the center point. The right panel shows a response contour plot where the lines connect points of equal value on the dependent (measured) variable. Typically, the values of the variable are also shown on each contour line.

B.5 Mixture Designs

If there are three factors in an experiment, and the design can be constrained to produce a constant sum of the three variables, a mixture design can be used. This is usually visualized as a triangle. The three factors must sum to 100%, so including more than one component requires lowering at least one of the others. For example, a sweetener blend may be produced with a total allowable amount of each compound. The maximum amount of each is such that when it is present on its own it is equivalent to 10% sucrose by weight. Typically, each variable will include a mixture in which it is present at 0% and 50%, and the three-way mixture will have one-third of the levels for each component. Another example could be a set of three antioxidants or preservatives, which must not exceed 0.1% of the total product weight. So the three components are varied in their percentages, and a fourth, dependent (outcome) variable is measured. Figure B.6 shows a mixture design triangle and a contour plot for the response surface. Products are represented as points in the design, so that two-component mixtures occur on the edges, one-component products at the vertices, and three-component mixtures inside the triangle. Sometimes it is impossible to manufacture some of the design points, so that only a part of the triangle is used in the experiment (see Carr (2010) for an example). This kind of experiment produces a response surface that can be used to find a maximum response at some optimal combination. In other cases, the dependent variable may be some undesirable sensory consequence that you are trying to minimize.

B.6 Summary and Conclusions

In setting up a systematic design, the researchers should ask several important questions. First, do we understand what variables are likely to be important in driving the attribute we are trying to measure? If not, a screening experiment may be called for. Second, do we need to estimate curvilinear effects or just linear trends? Third, can I use an optimal or efficient design, rather than a full factorial? Fourth, are interaction effects likely to be important, or can I ignore them in a design that has some confounding?

The success of any experimental design depends upon several factors and statistical assumptions. First, the independent variables of interest must be chosen wisely. They must of course be influential on the dependent variable(s) whether we are measuring sensory properties or consumer acceptance. Second, the levels chosen must be distinguishable, and must change those properties. They should also be at some reasonable levels; that is, a real product could conceivably be made at those settings. However, this concern should not constrain the experimenter too much because products that are insufficiently separated will not show trends. Carr (2010) cites the general rule as "be bold, but not foolhardy." It is also important to remember that the products you construct for the experiment are not likely to be optimal. But if everything is "pretty good," the response will be flat and no learning will take place.

Other assumptions are important (NIST/Sematech, 2012). The dependent variable must be measurable. In the case of a trained panel, that means that they understand the attribute, show reasonable agreement, and produce reliable data. In the case of a consumer study, this means the participants understand the task at hand and can use the scale or response instrument correctly. The products or processes you study must be stable and reproducible. The response should be one that is described by a smooth curve, without too many reversals or discontinuities. Finally, the model residuals (predicted minus observed values) should be well behaved. This usually means that they are normally distributed with known and/or constant variance. This consideration is an important statistical assumption in many models and analyses, including ANOVA.

Fractional factorials are often used in screening tests when there are many variables to sort through. They have the liability that some effects are confounded or aliased, usually interaction effects confounded with main effects. This can be acceptable if there is good reason to believe that the interaction effects are small or nonexistent. However, in foods, this is not often the case, as ingredient or processing changes often have multiple consequences that depend upon the levels of other treatments or factors in the system. The degree of confounding or aliasing is sometimes called the resolution of the design, and a full factorial will have infinite resolution (no confounding). Nonetheless, the fractional designs are popular in areas such as conjoint measurement and in the development of optimal product concepts with different combinations of features (Moskowitz et al., 2006).

This chapter serves as an introduction to the major kinds of designs that are common in product development using DOE. A variety of other designs are possible, such as rotatable designs that provide equal predictive power at all constant distances from the center point. These can be used for response surface optimization and are discussed more thoroughly in Gacula et al. (2009) in their chapter on response surface experiments.

References

Carr, B.T. 2010. Statistical design of experiments in the 21st century and implications for consumer product testing. In: Consumer-driven Innovation in Food and Personal Care Products. S.R. Jaeger and H. MacFie (Eds). Woodhead Publishing, Cambridge, UK, pp. 427–469.

Gacula, M.C., Jr. 1993. Design and Analysis of Sensory Optimization. Food and Nutrition Press, Trumbull, CT.

Gacula, M., Singh, J., Bi, J., and Altan, S. 2009. Statistical Methods in Food and Consumer Research. Second edition. Elsevier/Academic Press, Amsterdam.

Moskowitz, H.R., Beckley, J.H., and Resurreccion, A.V.A. 2006. Sensory and Consumer Research in Food Product Design and Development. IFT Press/Blackwell Publishing, Ames, IA.

Naes, T., Brockoff, P.B., and Tomic, O. 2010. Statistics for Sensory and Consumer Science. John Wiley & Sons, Ltd, Chichester, UK.

NIST/SEMATECH. 2012. e-Handbook of Statistical Methods, http://www.itl.nist.gov/div898/handbook/ (accessed 6 November 2012).

Appendix C: Glossary

A–not-A test A test in which one product at a time is shown to assessors, and they must respond indicating whether it is an example of a product they have previously inspected (hence "A") or something different (hence "not A").

Abbott's formula A method for correcting for the effect of chance or guessing in a forced-choice procedure.

ABX test A forced-choice discrimination test in which two reference items are presented, and then the assessor must match a third test item to one of the references.

Acceleration factor A factor showing how reaction rates change under storage conditions at different temperatures. It is used to estimate the storage conditions for accelerating shelf-life measurements.

Activation energy The energy required to start a chemical reaction, and/or the potential energy barrier to the reaction. It can also indicate the sensitivity of a chemical reaction to changes in temperature, and is a parameter in accelerated shelf-life modeling.

Adaptation A decrease in the intensity of a stimulus under conditions of constant stimulation.

Alienation The process of causing negative perception of a product due to a changed formula, process, or package that influences the sensory profile, usually among brand-loyal users or heavy users of that product, so that purchase intent, return purchase, and/or sales are hurt by the change that has been made.

Anchoring Providing a subject or assessor with verbal or physical examples of scale points or positions on a line scale representing different perceived intensity levels.

Ascending forced-choice method of limits A method for threshold measurement in which the stimulus intensity is raised until detection occurs, or until all the levels have been tested. The panelist must indicate detection by choosing the correct (target) sample from among a series of blank or baseline samples (typically one out of three). Choice is forced on all trials; the respondent must guess if unsure.

Quantitative Sensory Analysis: Psychophysics, Models and Intelligent Design, First Edition. Harry T. Lawless.

AUC An abbreviation for area under the curve in a time–intensity record.

Bathtub function A function describing the probability of product failure over time. It is characterized by a steep initial decline due to early product failures, then a period of stability, followed by an increasing phase of product deterioration.

Bayes' rule A method for calculating the probability of an outcome, based on prior knowledge or information. More specifically, if an event A is observed and is conditional on a situation B, the probability of A given B can be calculated if the overall probability of A, B, and the probability of B given A are known. Also known as Bayes' theorem.

Bayes' factor A ratio of two odds ratios, giving a value indicating the likelihood of one model over another competing model, given some set of observed data. It is a kind of likelihood ratio test.

Bayesian network A model of events or observations in a causal chain, used to model the probabilities of conditional outcomes.

Best–worst scaling A method for partial ranking in which a participant or consumer chooses the item liked best and least from a set of items. When using the method to rank intensities or other properties, it is sometimes call max-diff for maximum difference.

Beta binomial model A statistical model for replicated choice tests, such as discrimination tests, that incorporates a beta distribution to account for assessor variability and a binomial distribution to describe possible outcomes under a true null hypothesis.

Beta criterion A criterion or cutoff set in a signal detection study, where there are two responses allowed, and values (sensations or experiences) less that the criterion level will evoke one response, and values above the criterion will evoke the other. In other words, a criterion based upon sensation level.

Blind-labeled, blind testing Sensory tests with samples labeled only with random three-digit codes or other meaningless identifiers. Participants are only given enough information, such as the general category of the food, in order to judge it in a reasonable context.

Borda count A system for combining the ratings of expert wine judges after converting them to ranks, as an alternative to averaging.

Case V A situation in Thurstonian modeling in which the observations of two items have equal variance (equal variance of their discriminal dispersions) and the observations are uncorrelated (i.e., do not influence one another). Case V is thus a set of simplifying assumptions.

CATA An abbreviation for check-all-that-apply.

Category ratio scale A line scale developed by G. Borg for measuring perceived exertion, in which the verbal labels were spaced according to their ratio or proportional values. A forerunner of the labeled magnitude scale (LMS).

Centering bias When using just-about-right (JAR) scales, the tendency to rate the middle product in a series of products with varying ingredient levels as just about right.

Check-all-that-apply (CATA) A checklist form of response choice, in which more than a single choice is allowed.

Cluster analysis A large set of statistical techniques for grouping items on the basis of the similarity in their respective patterns of data, such as a correlation between two items in their attribute scores.

Cochran's Q test A statistical test for data consisting of binary information on more than two products, when evaluated by all assessors or consumers. An example would be checklist data (CATA).

Communality The amount of variance (sum) contained in the factors in a principal components analysis (PCA) from a given attribute or input variable; in other words, the variance retained.

Complete block design An experimental design in which all assessors evaluate all of the test products, allowing a within-subjects or repeated measures analysis of variance, and allowing the efficient partitioning of assessor variance from error variance.

Constant error In the method of constant stimuli, the difference between the point of subjective equality and the value of the standard stimulus (in physical units).

Cophentic correlation coefficient In cluster analysis, a measure of how well the dendrogram and/or solution fits the original data.

Correspondence analysis A technique similar to principal components analysis, but operating on frequency count data, rather than scaled or continuous data.

Credible interval In Bayesian analysis, an interval of the posterior distribution in which the test parameter will occur a certain proportion of the time. It is analogous to the confidence interval in (frequentist) statistics.

Criterion In signal detection theory, a level of sensation intensity, set by the subject, that separates a positive from a negative response.

Cross-adaptation A decrement in the perceived intensity of one stimulus, when it follows another stimulus that causes a state of adaptation. The decrement is found by comparison with some baseline condition in which there is no first stimulus, or a first stimulus such as water in a taste experiment that produces no adaptation.

Cross-modality matching A psychophysical procedure in which the subject adjusts a stimulus in one modality to match the perceived intensity of a stimulus from another modality that was presented by the experimenter.

d' (d-prime) In signal detection theory, the measure of observer sensitivity (or of stimulus detectability) measured independently of response bias or the subject's criterion.

Degree of difference (DOD) Commonly, this kind of test is done on a rating scale and comparisons are made between two products. The assessor will indicate on the scale whether the two items have no difference or some increasing degree of difference.

Dendrogram A graph or tree structure showing the distance at which items are joined to other items or to previously formed clusters.

Dependent *t*-test A *t*-test performed on two means based on data collected from the same individuals (i.e., paired observations). Also called a paired *t*-test.

Detection threshold The point at which 50% of the test population can discriminate the stimulus from a control stimulus, often called a blank or background stimulus. Also, the point at which a single observer can discriminate on 50% of trials.

Difference threshold The amount of change in stimulus intensity necessary to be correctly discriminated from a baseline stimulus on 50% of trials. Synonymous with the just-noticeable difference (JND).

Differencing strategy A hypothetical strategy in the triangle and other tests, in which the assessor considers the overall difference between two products as a basis for their response choice.

Dirichlet multinomial An analysis used for replicated testing with more than two responses allowed, for example, in preference testing with a no-preference option. Like the beta binomial, it adjusts for the degree of panelist consistency/inconsistency.

Discriminal dispersions In Thurstonian theory, the distribution of sensations or events caused by repeated observations of the same item.

Discriminant analysis A set of multivariate techniques for modeling group membership on the basis of a reduced set of predictor variables. More generally, the methods can be used to construct perceptual maps similar to the output of a principal components analysis.

DOD Abbreviation for degree of difference (test).

Dual pair test A test involving two pairs of products or stimuli. One pair contains physically identical items and one contains physically different items. The task of the assessor is to indicate which of the two pairs is the same or different. Traditionally, they are presented in different intervals, so that in psychology this procedure is known as 4IAX for "four interval AX test."

Dual standard test A forced-choice discrimination test in which two reference items are presented, and then the assessor must match each of two test items to one of the references. The test items are intended to be identical to the references. Therefore, only one can be matched to each reference.

Dumping effect The artificial inflation of an attribute rating due to the presence of a second attribute for which there is no response scale.

Duo–trio test A forced-choice discrimination test in which one reference item is presented, and then the assessor must match one of two test items to the reference item.

DUR An abbreviation for the total duration of a time–intensity curve; time to extinction.

Eggshell plot A graphical method for examining the rank order consistency of panelists across a group of products. The plot of multiple assessors resembles the lower half of a cracked eggshell.

Eigenvalues A specification of the amount of variance accounted for by a factor in a principal components analysis. The eigenvalue divided by the number of original variables is the percentage of variance accounted for by that factor.

Factor loadings The relationship between PCA factors and the original variables, roughly equal to a correlation coefficient, and thus varying from -1 to $+1$.

Factor scores The values of products, items, or observations on the PCA factors. These can be used to plot the products in a perceptual map using the PCA factors as axes.

Factorial design An experimental design in which all levels of each factor or treatment are combined with all levels of the other factors or treatments.

False alarm In signal detection theory, a positive response to a noise trial, or the act of calling a noise stimulus a signal; that is, an incorrect positive response, analogous to a type I error in statistics.

Fishbone plot A plot used in ideal product profiling, showing deviations from ideal in positive and negative directions for any product as an extended vertical bar chart.

Forced choice Any method involving a selection of response alternatives, usually representing several samples or products, when a selection is obligatory. Nonresponse is not permitted.

Franchise risk The risk of offending consumers with an inferior product or one that does not live up to their expectations, thus decreasing the chance of repeat purchase and causing lower sales.

Functional measurement A psychophysical theory derived from using integrated judgments of pairs of items or more complex stimuli, where the overall or average magnitude of the items is judged. The data provide evidence for a combination rule as well as the linearity of a response output function.

γ(gamma) A parameter in a beta binomial model describing the randomness versus consistency of the performance of assessors over replications. γ varies from zero (completely random behavior) to one (completely consistent behavior).

Generalized labeled magnitude scale (gLMS) The generalized labeled magnitude scale, anchored to the greatest imaginable sensation *of any kind.*

Generalized Procrustes analysis (GPA) A multivariate technique for data reduction, combining a principal components analysis with operations of scaling, translation, and rotation/reflection in order to achieve a maximum correspondence of the configurations of different assessors.

Graph theory A mathematical theory used to model the pairwise relationships between items or products using points or nodes to represent items and vertices or edges to represent connectedness or some index of relationship and/or similarity.

Halo effect The artificial inflation of ratings of one response attribute due to ratings given on another unrelated attribute; a spurious correlation. Also, the inflation of hedonic responses due to one positive attribute that the respondent deems to be highly important.

Hazard The momentary probability that a product will fail at time T assuming it has not failed previously. A hazard function can describe this probability or the cumulative probability of failure.

Hedonic scale A numerical scale used to indicate degree of liking and/or disliking. In other words, a scale for product acceptability.

Hierarchical agglomerative clustering A class of techniques for joining items into groups in a stepwise manner, based on the similarities of the items; a version of cluster analysis.

Hit In signal detection theory, a correct positive response to a signal trial.

Ideal point model A perceptual map indicating the position of ideal products for each consumer. Other actual products are plotted so that the distance to that ideal point shows a maximum negative correlation with the consumer's acceptability ratings, so that nonpreferred items are far from that person's ideal and preferred items are close.

Ideal profiling Ratings of an imagined ideal product on attribute scales to determine the sensory qualities of a product that the consumer would consider optimal.

I_{max} An abbreviation for the intensity at the maximum of a time intensity curve.

Independent *t*-test A statistical test used to compare two means, when the data for each item are collected from a different group of assessors. Also called a between-groups *t*-test.

Interval censoring A situation arising when an observer responds at some time t or concentration c, but may have perceived the event or sensation between time, t and $t-1$ or between concentration c and $c-1$ in an experiment. That is, the true value for detection is unknown, but occurs within a known interval.

Interval of uncertainty A range of stimulus intensities over which the response is "equal to" rather than "greater than" or "less than." Used in some classical psychophysical methods for difference threshold measurement.

Interval scale An assignment of numbers to stimuli, usually representing perceived intensity, such that differences in the numbers represent constant differences between the perceived intensity of the items. The scales may have an arbitrary zero point, analogous to centigrade and Fahrenheit scales for temperature.

Jeffreys' prior A prior distribution (often a beta distribution) in which extreme low and high values are much more probable than any observations falling in the middle.

Just-about-right (JAR) ratings A scale used to indicate whether a single sensory attribute is too weak, too strong, or at or near the consumer's optimum level, hence "just (about) right."

JND See difference threshold.

***k*-visit model** A model developed by George Ferris to find estimated true proportions of people preferring a product consistently over multiple tests, and allowing a no-preference response.

Labeled affective magnitude scale Also called the LAM scale. A hybrid line scale with verbal anchors for likes and dislikes, spaced according to their scaled values via magnitude estimation.

Labeled magnitude scale (LMS) A hybrid line scale for perceived intensity with verbal anchors spaced according to their scaled values via magnitude estimation. The upper anchor is "greatest imaginable."

Linkage Methods or criteria for joining items or joining clusters in agglomerative cluster analysis.

Magnitude estimation A system for judging the intensity of stimuli in which the assigned numbers are in proportion to ratios between the intensities. For example, if one stimulus seems twice as strong as the previous item, it will be given a rating twice the value assigned to the first.

Magnitude production A psychophysical method in which the experimenter tells the subject a number representing the magnitude of the stimulus, and the subject adjusts the stimulus to conform. If the next number stands in a given ratio to the last, then the perceived intensity would be adjusted to reflect that ratio.

Max-diff scaling See best–worst scaling.

McNemar test A test for complete block data based on frequency counts, when there are only two responses, and usually two products. A special case of the Stuart–Maxwell test.

Metathetic continua A class of sensory qualities that vary in kind, place, or type of sensation, such as pitch or color (hue), rather than perceived intensity. Just-noticeable differences are generally subjectively equal on metathetic continua.

Method of constant stimuli A psychophysical method in which a randomly ordered sequence of stimulus pairs is presented to a subject for a decision about which one is greater, or if a standard is specified within the pair, whether the comparison item is greater than or less than the standard. The comparison items will be varied; the standard item is held constant. A plot of responses is used to determine the difference threshold.

Method of limits A classical psychophysical method, in which the stimulus intensity is raised or lowered until a change in response is noted. Often used in various forms to study the detection threshold.

Method of triads A forced choice test in which three (usually physically different) items are presented and the assessor must indicate which is most different from the other two. The triangle test is a special case of the method of triads.

Methods of adjustment A class of methods in which the assessor has control over the stimulus or product and can change the intensity of an attribute, either to match a second item or to adjust the product to its optimal taste.

Modulus A standard stimulus used in magnitude estimation that is usually assigned a number value. All subsequent stimuli are rated relative to the modulus.

Multidimensional scaling (MDS) A set of techniques for constructing spatial models of stimuli or products, using similarity information as input. The result is a model (in *n* dimensions) in which similar items are positioned close together and different items are far apart in other words, where dissimilarity is proportional to distance.

Multiple factor analysis (MFA) A multivariate technique for analyzing data from projective mapping and other related techniques, producing output similar to PCA and GPA.

***n*-AFC test** A forced-choice test with *n* alternatives, from which one correct test item must be selected.

***n*-AFC-R test** A forced-choice test in which one correct test item must be matched to a reference or "reminder" sample. An example is the duo–trio test.

Nine-point hedonic scale A scale with nine alternatives, consisting of four adverbs, slightly, moderately, very much, and extremely, modifying the words like and dislike. The ninth phrase is intended to be neutral: Neither like nor dislike. The scale was invented at the Army Quartermaster Food and Container Institute and became an industry standard for measuring consumer acceptability of food products.

Noise A concept in signal detection theory referring to the distribution of sensory experiences in the absence of a stimulus or signal. A noise trial refers to the presentation of the baseline or background stimulus, without the signal present.

Nominal scale A method of assigning numbers to objects, products, or stimuli, based upon identity (versus nonidentity). Numbers function merely as labels.

Observer A general term used to describe the person used as an experimental subject in a psychophysical study.

Opportunity risk The risk of missing a difference that could be a product improvement (a form of type II error). Opportunity risk can also occur when a false positive result occurs, such as declaring a difference when none exists. For example, in a cost reduction effort, the change would have gone unnoticed by consumers, but the sensory test showed a false positive result so the change was not made, and no cost savings were obtained.

Optimal storage Conditions such as very low temperature storage in which few, if any, product changes are expected to occur over time. Such conditions facilitate the use of different sampling strategies for shelf-life testing, such as a reversed storage design.

Ordinal scale An assignment of numbers to stimuli, usually representing perceived intensity, such that the numbers express rank order properties (greater than or less than).

Overdispersion A statistical concept describing a situation in which the data have additional influences on variability, such as assessor variation in a discrimination test.

Panelist A member of a panel of individuals used to obtain data in a sensory evaluation study or session. Most often a panelist is considered a member of a trained or specially selected or screened group with specific sensory abilities. However, some authors also speak of consumer panelists, who are untrained and more or less randomly sampled from a population. The latter usage should probably be avoided and the term consumer used instead.

Partial least squares A set of data analysis techniques, often used in chemometrics and sensometrics, that can be used to make a spatial model similar to the output of PCA. More generally, the methods can be used to relate two multivariate sets of data such as a predictor block and an outcome or observation block.

Partitioning The process of accounting for variance in a study by attributing it to a cause (such as panelist variance) and mathematically removing it from another source of variance, such as the error variance estimate.

Penalty analysis Analysis of the loss or mean drop in hedonic scores, as a function of being nonoptimal in some attribute as indicated by JAR ratings. The analysis also usually includes the frequency or proportion of the nonoptimal ratings on that attribute.

Perceptual mapping A class of techniques for visualizing the similarities and differences among products by placing them in a map or spatial model, often using multivariate statistical methods. Products are plotted as points and attributes as vectors. Similar products lie close together and different products far apart.

Principal components analysis (PCA) A method for data reduction using multiple dependent variables as input, providing factors as output that represent collections of correlated input variables. The relationship of the input variables to the new factors is specified in factor loadings, and the positions of products on the new variables are specified in factor scores. Each factor has a proportion of original variance it has captured, specified as an eigenvalue.

Proportional hazards model A logistic regression model used to analyze JAR data, adapted from survival analysis.

Proportional odds model A logistic regression model used to analyze JAR data and used to compare products.

Proportion of discriminators A proportion derived from the proportion correct in a discrimination test, which has been adjusted for chance performance or the probability of guessing correctly. In other words, An estimation of the most likely proportion of assessors discriminating two products, given a certain observed proportion and after correction for chance or guessing.

Point of subjective equality In the method of constant stimuli, the level of the stimulus, as interpolated in the data, at which response equals 0.50 or 50%.

Power function In psychophysics, a mathematical function describing perceived intensity as a function of physical intensity raised to some power, a characteristic exponent of that sensory quality or modality. The data from magnitude estimation methods tend to conform to a power function and are thus linear in a log–log plot.

Preference mapping The development of a spatial model to represent the preferences of consumers for a set of products. When superimposed upon an existing configuration, such as those developed from a PCA of descriptive data, it is called external preference mapping. When the map is solely derived from the consumer hedonic data itself, it is called internal preference mapping.

Projective mapping A task in which assessors or consumers are asked to place products on a two-dimensional surface, where the distance between items is a representation of their sensory dissimilarity. The technique is also known more recently as napping, from the French word for tablecloth.

Prothetic continua A class of sensory qualities that vary in intensity. JNDs might or might not be subjectively equal on prothetic continua.

Psychometric function A plot or curve fit describing the probability of response as a function of some other variable, such as stimulus intensity.

Psychophysical function A function describing the relationship between energy and the perceived intensity of a stimulus. Energy may be specified in the form of a correlated or proxy variable, such as concentration or molarity in the case of a chemical stimulus.

***R*-index** A measure of stimulus detectability or discriminability used in signal detection studies, and derived from rating scale data such as sureness or certainty ratings. The *R*-index varies from 0.5 to 1.0 and is roughly analogous to the performance expected in a two-alternative forced-choice task.

Rank-rating A scaling method in which product or stimuli are placed on a scale visible in front of the subject or assessor, usually on a piece of paper, and repositioning is allowed as further samples are evaluated.

Ratio scale An assignment of numbers to stimuli, usually representing perceived intensity, such that the numbers express ratios or proportions between the perceived intensity of the items. The term is also used to refer to data generated by giving ratio instructions to subjects in a psychophysical study. Such data may or may not have ratio properties.

Recognition threshold The minimal amount of stimulus or energy that can be accurately named by a test subject or group of subjects.

Regression effect In magnitude estimation and magnitude production, the tendency of subjects to use a smaller range for the continuum that they have control over, leading to different power function exponents for estimation and production, on the same continuum.

Rejection threshold The concentration or energy level of a substance added to a product or to a stimulus that is first (statistically) significantly not preferred to a control item, product, or stimulus that does not contain the substance. This has also been estimated as the point of 75% preference for the control in a paired preference test, rather than the point of first statistically significant preference.

Response output function A function describing the process of number assignment by a subject to a perception of intensity. A judgment function.

Reversed-storage design An experimental design for shelf-life studies in which different products are placed into storage at different times, to allow testing at one final time period or on the same day. A staggered start in a foot race is a parallel idea.

ROC (or receiver operating characteristic) curve In signal detection theory, a plot of the proportion of hits versus the proportion of false alarms, as the subject changes criteria. The area below the curve is monotonically related to d', the measure of observer sensitivity (or stimulus detectability).

Same–different test A difference test in which pairs of products are presented to assessors, and they must respond as to whether they seem the same or different. In other words, this is a binary (yes/no) version of the degree of difference test.

Segmentation The grouping of consumers that is usually based upon similarities in their patterns of liking and preferences across a set of products.

Sensometrics The development and application of specialized data analysis and statistical techniques for sensory evaluation and consumer studies of products.

Signal detection theory A psychological theory of sensory behavior, originally based upon observers' responses to weak stimuli, that attempts to separate observer sensitivity from response bias.

Skeptical prior An assumed prior distribution (often a beta distribution), usually centered around an expected mean and having wide variance, used in Bayesian analysis.

Skimming strategy A hypothetical strategy in a discrimination test, in which the observer uses the absolute intensity of the sensation as a basis for making a later response choice. For example, in the triangle test, a person could decide to chose the strongest or weakest stimulus, regardless of whether it was the most different (odd sample).

Sorting A task in which assessors or consumers are asked to group products on the basis of sensory similarity. Sorting data can be used as input to MDS in the construction of perceptual maps.

Staircase procedure A method for measuring thresholds in which a stimulus is lowered in intensity when detection occurs, and is raised in intensity on trials where the stimulus is not detected. The graphical display of the sequence, when points are connected, resembles a staircase that alternately rises and falls.

Stimulus A physical object or source of energy used to create a sensation in an observer or experimental subject.

Stuart test A test used for complete block data based on frequency counts, when there are more than two responses, and usually two products. Also called the Stuart–Maxwell test.

Subject A person participating in an experimental session or psychophysical study.

Supertaster A person characterized by (1) high responsiveness to *n*-propyl thiouracil (PROP), (2) a higher than normal density of fungiform papillae, and (3) higher responsivness to other sensory stimuli, especially tastes.

τ (tau) criterion A hypothetical measure of sensory difference, describing an interval or range of perceived difference, in which two pairs of products will be responded to as if they were the same.

Temporal dominance of sensation A method for rating the most dominant perceived qualities of a product as they change over time, usually choosing from a pre-set list.

Terminal threshold The range of concentrations or energy at which a stimulus ceases to increase in perceived intensity. A type of saturation. This is rarely studied, and usually the sensation changes in quality (e.g., it may become painful), making this phenomenon difficult to study.

Tetrad method A discrimination test procedure in which two pairs of duplicate items are presented to the assessor, and they must be correctly sorted. This is a recent alternative to the triangle procedure and has the same chance probability of 1/3.

Thurstonian models A class of psychophysical models based upon using variability as a metric. The theory is closely related to signal detection and can be used to derive a method-free index of sensory discriminability or difference.

Time–intensity scaling A class of methods for measuring the response, based on perceived intensity of a taste, flavor, or texture attribute over the duration of the sensation.

T_{max} An abbreviation for the time to reach maximum intensity in time–intensity scaling.

Triangle procedure A sorting task involving two duplicate items and one that is physically different. The task of the assessor is usually phrased, "choose the item that is most different from the other two."

***t*-test** A statistical test examining the difference between two means. Under a true null hypothesis, the results follow the distribution of the statistic *t*.

Type I error A statistical concept in hypothesis testing. It is the result of rejection of a null hypothesis that is true. Also called a false alarm.

Type II error A statistical concept in hypothesis testing. It is the result of accepting the null hypothesis when it is false. In simple difference testing, it declares no difference when one actually exists. Also called a miss.

Type zero error Asking the wrong question to begin with.

Uniform prior A distribution in which all outcomes are equally probable; that is, a flat distribution, also considered uninformative in Bayesian analyses.

Universal intensity scale A scaling method for perceived intensity that has been anchored with examples during panel training, and may be applied to different sensory qualities and different sensory modalities. In other words, a single numerical scale for all sensations.

Utility function In economics, a function describing the subjective value of items which vary in monetary value.

Ward's method In cluster analysis, a linkage criterion based on minimization of the variance within clusters.

Warm-up samples Item or items given before the test samples in order to allow the assessor to adjust to the nature of the samples, and/or to acclimatize the palate.

Weber's law A psychophysical rule that states that the increase in stimulus energy needed to create a JND is a constant percentage or proportion of the starting level. In other words, in absolute terms, the physical size of the JND increases as the stimulus energy goes up. The proportion, if determined experimentally, is called a Weber fraction.

Within-subjects ANOVA An analysis of variance from a design in which all assessors evaluate all of the products; that is, a complete block. Also called a repeated measures analysis of variance. The main effect of assessor differences can be partitioned from the overall error term.

Index

Quantitative Sensory Analysis: Psychophysics, Models and Intelligent Design, First Edition. Harry T. Lawless.